Operator Theory: Advances and Applications
Vol. 134

Interpolation Theory, Systems Theory and Related Topics

The Harry Dym Anniversary Volume

Daniel Alpay
Israel Gohberg
Victor Vinnikov
Editors

Springer Basel AG

Editors:

Daniel Alpay
Department of Mathematics
Ben-Gurion University of the Negev
P.O. Box 653
Beer Sheva 84105
Israel
e-mail: dany@cs.bgu.ac.il

Victor Vinnikov
Department of Mathematics
Ben-Gurion University of the Negev
P.O. Box 653
Beer Sheva 84105
Israel
e-mail: vinnikov@cs.bgu.ac.il

Israel Gohberg
School of Mathematical Sciences
Raymond and Beverly Sackler
Faculty of Exact Sciences
Tel Aviv University
Ramat Aviv 69978
Israel
e-mail: gohberg@math.tau.ac.il

2000 Mathematics Subject Classification 47-06; 47A57, 47N20, 47N70, 65D05, 65E05

A CIP catalogue record for this book is available from the
Library of Congress, Washington D.C., USA

Deutsche Bibliothek Cataloging-in-Publication Data

Interpolation theory, systems theory and related topics : the Harry Dym
anniversary volume / Daniel Alpay ... ed. – Basel ; Boston ; Berlin :
Birkhäuser, 2002
(Operator theory ; Vol. 134)
ISBN 978-3-0348-9477-7 ISBN 978-3-0348-8215-6 (eBook)
DOI 10.1007/978-3-0348-8215-6

ISBN 978-3-0348-9477-7

Originally published by Birkhäuser Verlag in 2002
Softcover reprint of the hardcover 1st edition 2002

Printed on acid-free paper produced from chlorine-free pulp. TCF ∞
Cover design: Heinz Hiltbrunner, Basel

ISBN 978-3-0348-9477-7

9 8 7 6 5 4 3 2 1 www.birkhauser-science.com

Contents

Editorial Introduction

This volume is based on the proceedings of the Toeplitz Lectures 1999 and of the Workshop in Operator Theory held in March 1999 at Tel-Aviv University and at the Weizmann Institute of Science. The workshop was held on the occasion of the 60th birthday of Harry Dym, and the Toeplitz lecturers were Harry Dym and Jim Rovnyak. The papers in the volume reflect Harry's influence on the field of operator theory and its applications through his insights, his writings, and his personality. The volume begins with an autobiographical sketch, followed by the list of publications of Harry Dym and the paper of Israel Gohberg: *On Joint Work with Harry Dym.*

The following paper by Jim Rovnyak: *Methods of Kreĭn Space Operator Theory,* is based on his Toeplitz lectures. It gives a survey of old and recents methods of Kreĭn space operator theory along with examples from function theory, especially substitution operators on indefinite Dirichlet spaces and their relation to coefficient problems for univalent functions, an idea pioneered by L. de Branges and underlying his proof of the Bieberbach conjecture (see [9]).

The remaining papers (arranged in the alphabetical order) can be divided into the following categories.

Schur analysis and interpolation

In *Notes on Interpolation in the Generalized Schur Class. I*, D. Alpay, T. Constantinescu, A. Dijksma, and J. Rovnyak use realization theory for operator colligations in Pontryagin spaces to study interpolation and factorization problems in generalized Schur classes.

In his paper *A Generic Schur Function Is an Inner One*, V. Katsnelson uses the Schur parameters to put a probability measure on the set of all Schur functions, and studies the genericity of inner functions by the methods of multiplicative ergodic theory.

A. Kheifets, *Abstract Interpolation Scheme for Non Analytic Problems*, develops a generalization of the abstract interpolation problem of Katsnelson–Kheifetz–Yuditskii (see [14, 15]) to handle non analytic interpolation problems such as the Nehari interpolation problem. One of the key ideas is a systematic replacement of unitary colligations, or equivalently conservative input/state/output systems, by generally non-orthogonal (non-causal) scattering systems as introduced by Adamyan–Arov [1].

Several complex variables and Riemann surfaces

In *Concrete Interpolation of Meromorphic Matrix Functions on Riemann Surfaces*, J.A. Ball, K.F. Clancey, and V. Vinnikov investigate the problems of interpolating matrix pole-zero data with multiple-valued meromorphic matrix functions on compact Riemann surfaces. This is related on the one hand to homogeneous interpolation problems for rational matrix functions as studied in [5], and on the other hand to the study of vector bundles on compact Riemann surfaces initiated by André Weil in [18] and actively pursued in the last two decades in algebraic geometry (see, e.g., [17]).

The paper by M.F. Bessmertnyĭ, *On Realizations of Rational Matrix Functions of Several Complex Variables*, is an English translation, prepared by Daniel Alpay and Victor Katsnelson, of a part of a Ph. D. thesis that was written in Russian in 1982 and has never been published. It deals with realization theory for rational matrix functions of several complex variables, especially for functions satisfying positivity conditions; its publication now is especially suitable because of a recent surge of activity in the area — the works of Agler–McCarthy [4], Alpay–Kaptanoğlu [2], Ball–Sadosky–Vinnikov [6], Ball–Trent [7], Kalyuzhniy [12, 13] — inspired by the work of Agler [3].

Matrix theory

The paper by D. Hershkowitz, *On the Spectral Radius of Multi-Matrix Functions*, deals with the behaviour of the spectral radius of a matrix with positive entries under multivariable matrix functions, and some other related questions. Mostly a survey, it contains also original results.

A Class of Robustness Problems in Matrix Analysis, by A. Ran and L. Rodman, is a survey of a class of perturbation problems that has been extensively studied by the authors and their collaborators over a period of several years. The stage is set by posing an abstract "metaproblem" followed by a careful review of results concerning the pervasive question of the stability of invariant subspaces.

System theory

The main part of the paper *Stable Dissipative Linear Stationary Dynamical Scattering Systems* by D.Z. Arov is an English translation, prepared by D.Z. Arov and J. Rovnyak, of a highly influential article originally published in Russian in 1979; it deals with (linear time-invariant) dissipative input/state/output systems, and their role in electrical networks (Darlington synthesis), operator theory, and function theory. There are two new appendices, the first one by D.Z. Arov providing a commentary and an update of the results, and the second one by D.Z. Arov and J. Rovnyak showing some directions for generalizations and further development.

In *Chains of Space-Time Open Systems and DNA*, M.S. Livšic discusses a striking resemblance between chains of overdetermined multidimensional (space-time) systems, that appear in the spectral analysis of tuples of nonselfadjoint and nonunitary operators [16], and chains of nucleotides in molecular biology. He shows that some important properties of the DNA can be given a natural explanation using the methods of system theory.

Differential equations and mathematical physics

The paper *Dual Discrete Canonical Systems* by L. Sakhnovich discusses the notion of dual canonical systems in the discrete case. The notion was introduced in the continuous case in a recent paper of Dym and Sakhnovich [10], generalizing (in that case) the notion of dual string equations which was introduced by Kac and Kreĭn for scalar strings [11].

In *Finite Section Method for Linear Ordinary Differential Equations on the Full Line*, I. Gohberg, M.A. Kaashoek, and F. van Schagen study solutions of linear ordinary differential equations on the full line as limits of solutions of corresponding equations on smaller intervals (with appropriate boundary or initial conditions). Both the time-invariant and the time-varying cases are considered.

C. Calude and B. Pavlov, *The Poincaré–Hardy Inequality on the Complement of a Cantor Set*, derive the Poincaré–Hardy inequality (an important tool in classical analysis, as well as in quantum mechanics, mathematical hydrodynamics, and quantum scattering) in $\mathbb{R}^3$ on the complement of a Cantor set. The approach to the problem is via a certain relevant dynamical system, inspired by Carleson [8].

Non-Selfadjoint Sturm-Liouville Operators with Multiple Spectra by V. Tkachenko is related to the spectral theory of non-selfadjoint Sturm-Liouville operators. While it was generally believed that an operator with a complex potential can have spectral points of an arbitrary multiplicity, not a single explicit example of, say, operator on a finite interval with multiple Dirichlet or Neumann spectra was previously known. Among other results, this paper constructs a Sturm-Liouville operator with an arbitrary given (symmetric) Dirichlet spectrum λ_n subject only to a restriction dealing with a suitable asymptotic behavior of λ_n.

References

[1] V.M. Adamyan and D.Z. Arov, *On unitary coupling of semiunitary operators*, Matem. Issled. **1** (1966), 3–64 (Russian); translated in Amer. Math. Soc. Transl. Ser. 2 **95** (1970), 75–129.

[2] D. Alpay and H.T. Kaptanoğlu, *Sous espaces de codimension finie dans la boule unité et un problème de factorisation*, C. R. Acad. Sci. Paris Sér. I Math. **331** (2000), 947–952.

[3] J. Agler, *On the representation of certain holomorphic functions defined on a polydisc*, Topics in Operator Theory: Ernst D. Hellinger Memorial Volume (L. de Branges, I. Gohberg, and J. Rovnyak, eds.), Operator Theory: Adv. Appl., vol. 48, Birkhäuser Verlag, Basel, 1990, pp. 47–66.

[4] J. Agler and J.E. McCarthy, *Nevanlinna-Pick interpolation on the bidisk*, J. Reine Angew. Math. **506** (1999), 191–204.

[5] J.A. Ball, I. Gohberg, and L. Rodman, *Interpolation of rational matrix functions*, Operator Theory: Adv. Appl., vol. 45, Birkhäuser Verlag, Basel, 1990.

[6] J.A. Ball, C. Sadosky, and V. Vinnikov, *Scattering systems with several evolutions and multidimensional input/state/output systems*, preprint.

[7] J.A. Ball and T.T. Trent, *Unitary colligations, reproducing kernel Hilbert spaces, and Nevanlinna-Pick interpolation in several variables*, J. Funct. Anal. **157** (1998), 1–61.

[8] L. Carleson, *Selected problems on exceptional sets*, Van Nostrand, Princeton, 1967.

[9] L. de Branges, *Underlying concepts in the proof of the Bieberbach conjecture*, Proceedings of the International Congress of Mathematicians (Berkeley, California, 1986), Amer. Math. Soc., Providence, 1988, pp. 25–42.

[10] H. Dym and L.A. Sakhnovich, *On dual canonical systems and dual matrix string equations*, Operator Theory, System Theory and Related Topics (The Moshe Livšic Anniversary Volume) (D. Alpay and V. Vinnikov, eds.), Operator Theory: Adv. Appl., vol. 123, Birkhäuser Verlag, Basel, 2001, pp. 207–228.

[11] I.S. Kac and M.G. Kreĭn, *On the spectral functions of the string*, Russian translation of F.V. Atkinson, Discrete and Continuous Boundary Problems, Mir, Moscow, 1968, Supplement II, pp. 648–737 (Russian); translated in *Amer. Math. Soc. Transl.* (2) **103** (1974), 19–102.

[12] D.S. Kalyuzhniy, *Multiparametric dissipative linear stationary dynamical scattering systems: discrete case*, J. Operator Theory **43** (2000), no. 2, 427–460.

[13] ———, *Multiparametric dissipative linear stationary dynamical scattering systems: discrete case. II. Existence of conservative dilations*, Integral Equations Operator Theory **36** (2000), no. 1, 107–120.

[14] V.E. Katsnelson, A.Y. Kheifets, and P.M. Yuditskii, *An abstract interpolation problem and the theory of extensions of isometric operators*, Operators in Function Spaces and Problems in Function Theory (V. A. Marchenko, ed.), Naukova Dumka, Kiev, 1987, pp. 83–96, 146 (Russian); translated in *Topics in Interpolation Theory* (Harry Dym et al., editors), *Operator Theory: Adv. Appl.* **95**, Birkhäuser Verlag, Basel, 1997, pp. 283–298.

[15] A. Kheifets, *The abstract interpolation problem and applications*, Holomorphic Spaces and Their Operators (S. Axler, J. McCarthy, and D. Sarason, eds.), Math. Sci. Res. Inst. Publ., vol. 33, Cambridge University Press, Cambridge, 1988, pp. 351–379.

[16] M.S. Livšic, N. Kravitsky, A.S. Markus, and V. Vinnikov, *Theory of commuting nonselfadjoint operators*, Mathematics and Its Applications, vol. 332, Kluwer, Dordrecht, 1995.

[17] C.S. Seshadri, *Fibres vectoriels sur les courbes algébriques*, Astérisque **96** (1982).

[18] A. Weil, *Généralization des fonctions abéliennes*, J. Math. Pures Appl. **17** (1938), 47–87.

Daniel Alpay, Israel Gohberg, Victor Vinnikov

Acknowledgements

It is a pleasure to thank all those who have provided the funding for the conference in March 1999 and were instrumental in helping with the organization:

- President of Tel-Aviv University,
- Rector of Tel-Aviv University,
- Raymond and Beverly Sackler Faculty of Exact Sciences, Tel-Aviv University,
- School of Mathematical Sciences, Tel-Aviv University,
- Arthur and Rochelle Belfer Institute of Mathematics and Computer Science, Weizmann Institute,
- The Maurice and Gabriela Goldschleger Conference Foundation at the Weizmann Institute of Science,
- Faculty of Mathematical Sciences, Weizmann Institute,
- Department of Theoretical Mathematics, Weizmann Institute,
- Department of Mathematics and Computer Science, Ben-Gurion University.

We would also like to thank Ms. Miriam Abraham of the Faculty of Mathematical Sciences, Weizmann Institute, for putting her excellent typing skills (and great patience) at our disposal for several papers in this volume.

Harry Dym

Operator Theory:
Advances and Applications, Vol. 134, 1–17

Looking Back

Harry Dym

I have been asked by the editors to write a few biographical remarks. I have spent most of my professional life in the Department of Mathematics (née Department of Pure Mathematics) of The Weizmann Institute. Thirty odd years seem to have sped by, although the days, weeks, and months often went by very slowly. Moreover, the events that led to my being at the Institute seemed to fall into place by chance, not by design, at least not by my design. I never particularly wanted to be a mathematician, nor did I plan on an academic career. But that is getting ahead of the story.

Initial conditions

To begin before the beginning: My parents were both born in Poland: my father, Isaac Dym, in Lisko, my mother, née Anne Hochman, in Kalusz. Their immediate families moved to Vienna during the First World War, probably to escape from the front lines and/or the invading Russian army. I know very little about the extended families of my parents. A book that my cousin Miriam came across recently lists more than thirty Dyms from the Lisko region who perished in the Second World War.

My parents were two quite different kinds of people. My mother was a do-er, an activist and a supreme organizer. In today's world, she probably would have been the CEO of some large company. My father was more of a scholar. He was an avid reader and, in his spare time, was almost always found with a book in his hands. In his youth, Jewish orthodox families did not encourage their children to study secular subjects. Nevertheless, my father completed a doctoral dissertation in Economics at the University of Vienna, presumably as an external student. I assume that my parents met in the office that my father managed, since my mother worked as a secretary in that office. One story has it that she organized a strike of all the other secretaries to improve their conditions. I do not know if the strike was successful or not. Perhaps the only way to bring this strike to an end or to avoid future strikes was to marry her. The fact that she was also a rather attractive young woman must have made this an agreeable solution. (To be honest, I don't know if the story is true, but knowing my mother, it is certainly plausible.) I arrived on the scene a few years later, on January 26, 1938.

I wish to thank Renee and Jay Weiss for endowing the chair that supports my research.

In the spring of 1939, when the local situation became "uncomfortable", my parents and I flew to Hungary for a two-week "vacation". They had been waiting for entry permits to England, but decided that the situation in Vienna was too dangerous and asked the British Consulate to forward the permits to the Consulate in Budapest. The two-week stay in Hungary was extended illegally to many weeks and then, asking the local British Consulate to forward the permits, my parents proceeded to Trieste, where the borders had opened. Sometime later the permits arrived in Trieste. However, the consular official did not want to issue them because they were supposed to be issued in Vienna. A loud vocal argument ensued and the Chief Consul, hearing the commotion, came out to see what all the fuss was about. Fortunately, he directed his subordinate to authorize the permits, otherwise this tale might never have been written.

My parents entered England as domestic servants: housemaid and butler.

England

My father's career as a butler was short lived. His employer dismissed him when he discovered that he had a university education. He spent the war years working as a baker in Leeds, where we lived. In 1944 the family was enriched by the birth of my brother, Lionel Clive (who later changed his name to Clive Lionel). My father, who was a Zionist, wanted to immigrate to Israel in 1948. However, he was discouraged by relatives who were living there and also by a second cousin from the US, Anna Rogoff, who visited us in England after visiting Israel. Aunt Anna, as we called her, encouraged my parents to move to New York.

New York

In 1949 the family immigrated to New York. I attended Manhattan Day School (a Jewish Parochial School) for two years and then went on to The Bronx High School of Science from 1951 to 1955 during the day and to Herzliah Hebrew Teachers Institute for two evenings per week and Sunday mornings. In High School I did rather well in the standard Math and Physics courses, but was never invited to take any of the honors courses in math (not that I had any ambitions in this direction) because of mediocre grades in French. At that period I developed a fascination for electronic devices. I can't remember how or why it began (it had nothing to do with school), but I do rememember often spending many hours scouring Cortland Street, a downtown New York electronic parts center at that time, for inexpensive resistors, capacitors, inductors and vacuum tubes. These were assembled with mixed degrees of success, following plans in popular electronics magazines. Thus, electrical engineering was a rather natural choice of vocation. In any event, at that time, youngsters with an aptitude for math and physics were steered into engineering, which was viewed as being economically secure. In my circle of aquaintances, no one had ever heard of mathematics per se as a career.

Accordingly, I went on to study electrical engineering at the Cooper Union School of Engineering from 1955 to 1959.

Cooper Union

Cooper Union was my mother's discovery. In 1955, it was one of two private schools in the US that did not charge tuition. Consequently it tended to have fairly able students. In my year, the Engineering School admitted 97 Freshmen (Freshpersons?). The entering class, which was distributed more or less equally in chemical, civil, mechanical and electrical engineering, included six young women. We had to specify our choice on the application forms. Unfortunately, large chunks of the curriculum in those days left much to be desired.

The first two years of study were devoted mostly to math, physics and chemistry with frontal lectures, recitation sections and exams. This part was fairly normal. The physics was probably very good. The math was mostly calculus plus a semester of differential equations. It was, at least as far as I remember, largely a cookbook approach, learning techniques to solve problems. Basically the strategy was to look for model problems and imitate. I suppose that the instructors tried to make the material plausible, but we probably had little patience for long-(or even short)-winded explanations. I cannot speak for the others, but I certainly had no understanding of (or interest in) limits. I am pretty sure that I did not really understand derivatives or integrals either, though I could compute them. Integrals always existed. After all you could look them up in a table. We did not study linear algebra, complex variables, probability or numerical methods, let alone the more exotic subjects such as topology, geometry or modern algebra.

The last two years were devoted mostly to engineering subjects. We worked our way through fat books on magnetism, electric circuits, electric machines, electronics and transmission lines, among others. There were hardly any lectures. Classes were typically three-hour affairs. We would come in and, sitting in groups around tables, work on problems, calculating away with our thirty dollar Keuffel & Esser slide rules. (In those days a subway ride was a dime and a typical text book ran less than ten dollars.) The worst was the weekly (third year) Electric Machines Lab Report, which through years of "consulting" with the work of previous generations had become immensely long. Much was written; little was really understood. In retrospect, it probably was an exercise in obfuscation, at least for most of us. Not having a clear idea of what we were doing (or why) we tended to bury it in long rambling discourses adapted from sources presumed to be reliable that hopefully covered the issues. It probably insured that no one would read the reports, at least not too carefully.

The program we followed was a little old fashioned. It included surveying (two weeks in a sleep away camp — great fun), drafting (less fun), descriptive geometry (even less fun) and was presumably designed to enable you to solve a wide range of problems with the aid of handbooks. In my time there was no flexibility in the program. Everything was completely regimented except that in the fourth year, we were allowed to select the humanities course of our choice from an offering of four.

Getting through Cooper Union was an exercise in survival. It was something akin to Basic Training in the army, except that you could go home at night. There was even the same sense of camaraderie that comes from being with the same group in a hazardous environment for an extended period of time.

To be fair, the system was not without merit. It was a difficult program and those of us who made it through to the end, were very pleased with ourselves for having completed it. But, as our youngest son Michael once remarked while doing physiotherapy after a foot operation: "Just because it hurts, does not mean that its good for you." What was lacking was an attempt by the Faculty of Engineering to transmit the benefits of their accumulated experience and personal vision. I am sure that I would have gotten more out of my stay at Cooper Union had I put more into it. However, I was not a particularly diligent student, being more interested in other things that included the athletic program and a certain young lady.

Towards the latter half of my sojourn at Cooper I became more serious and even applied to a couple of Graduate Schools. Someone must have written a good letter of recommendation on my behalf, because, in spite of an abundant collection of "Gentlemen's C's" in my first two years, I was awarded an assistantship for graduate study in electrical engineering at Caltech. In May 1959, a week before graduation, I married the young lady (née Irene Lillian Rosner). Some dozen days later, we set out to drive across country to Pasadena, California, where I had a summer job at the Jet Propulsion Laboratory. My ambition in life at that time was to design "pulse circuits." However, the summer job at JPL, attempting to do just that, cooled my enthusiasm. It was too much like "black magic." I never could get anything that I designed to work for two days in a row.

Caltech

At Caltech, I took my first "real" Math course, Math 108, which was an introductory course on Analysis based mostly on the book *Mathematical Analysis*, by T. Apostol. The course was beautifully taught that year by James Knowles. Although I did not know it at the time, this was probably the first step in a career transition. Another significant course for me that year was Statistical Communications Theory. It was largely based on the well known book by Davenport and Root. The material was, as I recall, fascinating but rather murky, at least for someone with my limited math background.

Caltech also opened up new horizons for Irene, who was a biology major. She obtained a position as an assistant in the laboratory of Matthew Meselson. This was just a few years after Watson and Crick elucidated the double helix structure of DNA. Matthew Meselson and Frank Stahl had just proven that the two strands of the DNA molecule separate when it replicates. The experiments in the lab that were related to this work were very exciting.

The offer of a fellowship for her to pursue doctoral studies at Harvard, and the news that our first child was expected, arrived at about the same time. Irene opted to take time out to raise our progeny. Later, when we moved to the Boston area and Matt moved to Harvard, she carried out a number of special projects for him on a part-time basis, but gave up the idea of graduate study. Some years after our third son was born, she switched directions to pursue a career in art and claims to have no regrets.

In June 1960, I graduated with an MSc in Electrical Engineering and joined the Technical Staff of the MITRE Corporation in Bedford, Massachusetts. One of the attractions of MITRE was a program that allowed staff members to take one course per semester during working hours at local universities, including MIT.

MITRE

At MITRE I had the good fortune to work intensively with Ed Arthurs, then an Assistant Professor in the Electrical Engineering Department at MIT. Ed had been hired as a consultant to help prepare a theoretical analysis of digital data equipment that was being tested by my department. This was roughly a two-year project and was a wonderful experience for me. I was the liason between him and the department. It was like having my own personal tutor and gradually I started to fill in some of the many gaps in my education and began to move forward. Eventually these efforts led to a long paper that was awarded a prize as the best paper of the year 1962 in the IRE Transactions on Communications Systems. One of the Faculty members in the EE Department at MIT even took 60 reprints off my hands to distribute to his class. It hurt to part with them at the time, but today I realize that he did me a great favor. Otherwise I would have yet another pile of clutter in my office.

In the Fall of 1960 I started taking math courses at MIT as a special student; one per semester. The first year I took Real Variables. The second year I took Complex Variables the first semester and a course on Information Theory with Claude Shannon in the second semester. I remember that I did not think that I had done particularly well on the final exam in Complex Variables. I did not do all the problems and wasted far too much time trying to get a proof of the Cauchy–Schwarz inequality to come out. To my surprise I got an A. It seems that I had solved a complicated conformal mapping problem that no one else had.

Going to school while working full-time was not easy. It involved driving into Cambridge from Bedford, racing to class and then driving back to Bedford. The main problem, aside from finding a place to park, was the lack of interaction with other students that is so useful in the learning process and helps one focus on the essence. Somewhere in this time period the idea of returning to school on a full-time basis began to germinate. The Math department was more attractive than the EE department, because the latter had an extensive set of qualifying exams in subjects that held little interest for me. In the early Spring of 1962, armed with my A in Complex Variables, I went to speak to G.B. Thomas, the Admissions Officer for the Math Department, to inquire about the possibility of admission sometime in the distant future. He was vaguely encouraging, but indicated that the sooner I applied the better my chances. With encouragement from my "significant other" I went for it. I suspect that my parents and in-laws thought I was crazy. I didn't ask.

MIT

In the Fall of 1962 I took a leave of absence from MITRE and returned to full-time graduate study in the Department of Mathematics at MIT. They were willing to let me try my luck, but were not willing to support me. Fortunately (thanks to the recommendation of E. Arthurs), I was awarded a Research Assistantship in the Information Theory Group of the Research Laboratory for Electronics. As I remember, the assistantship carried a stipend of $360 per month from which MIT took back on the order of $110 per month for tuition and $120 per month for housing. That didn't leave much to support the family, which included a young Jonathan (born November 1960) and a young David enroute (to be born December 1962). I was able to supplement this with summer work at MITRE. Nevertheless, I was eager to get through and go back to work full-time as quickly as possible. In view of the MSc from Caltech (which covered the minor requirements) and the courses that I had taken as a special student, I was able to complete the course requirements in the first year by taking a year-long course in Probability with Dan Ray, and semester-long courses in Algebra, the Theory of Distributions, Topological Groups with K. Iwasawa and Fourier Analysis with Norbert Wiener. I really was not ready to absorb most of this stuff and none of it, except for the Probability, had any long-term effects. There is, as one learns with experience, a vast difference between being able to follow the formal logic of a proof and developing a feel for the material. Luckily, most of the grades were based on problem sets rather than exams, so I was able to get by reasonably well. My objective at that time was to learn more mathematics to apply to problems in Statistical Communications Theory. The main gap to be filled was Stochastic Processes. The Algebra was supposed to help in coding theory, but I lost interest in that early on. I might have cottoned to it better if the lecturer had not tried to cram a year's worth of material into one semester.

In the Spring of 1963, Henry McKean agreed to take me on as a PhD student. The first step was to pass oral exams. My committee was K. Iwasawa, I.E. Segal and Henry himself. The former two assigned reading material based on one semester courses that they had given. Henry wanted more: The Carus monograph by M. Kac on Statistical Independence, Paul Lévy's book "Processus Stochastiques et Mouvement Brownien", Dynkin's papers on Markov processes (as reproduced in Loève) and large chunks of both the first volume of Courant-Hilbert and of Hoffman's book on Hardy spaces. At that time, Henry was putting the finishing touches on his book with Ito on Markov processes. This is a marvelous book, chock-full of information and ideas, but not always easy to read. It has been somewhat unkindly said that Henry wrote it in Japanese and Ito translated it into English. I can testify that this is not true. The Oral Exams were taken in pieces, since Henry was at Rockefeller University during the academic year 1963–1964. The final hurdle was overcome in December 1963 when Henry came through Cambridge on his way to New Hampshire for the Christmas holidays. Henry suggested a number of possible directions for a thesis. I settled on the study of the trajectory in phase space of an ordinary differential equation with constant coefficients that was driven by white noise. This topic was not deemed appropriate for continued support by the heads of the Information Theory Group at that time and the Research Assistantship from RLE was not renewed for a third year. However, by this time I was far enough along to get Math Department support as a Teaching Assistant.

As I remember, work on the PhD thesis went reasonably well for a while and then got bogged down. Nevertheless, sometime towards the middle of 1964–1965, Henry suggested that I report on what had been achieved to that point to the Probability Seminar. I was not particularly eager to do this. However, to paraphrase The Godfather, it was an an offer that was difficult to refuse. At the end of the talk one of the other graduate students began to ask a number of questions about one of the conjectures that I had raised. It seemed to me that he was exhibiting excessive interest in what I considered to be my turf. This really annoyed me (though I don't think I showed it). I went home and, late that evening, resolved the conjecture. That essentially completed the thesis.

The thesis seemed to be well thought of and was published in the Transactions of the American Math. Society. It generated a number of attractive job offers, including an option to stay on at MIT for two more years as an Instructor, which I accepted. Thus, began the drift from Industry to Academia.

At MIT, in the academic year 1965–1966, I began to read the Acta paper on trigonometric approximation by Levinson-McKean and tried my hand at some of the early de Branges papers on Hilbert spaces of entire functions. I also wrote an expanded set of lecture notes on a course that Henry was giving on Fourier Analysis. This was to evolve over a number of years and a number of different courses that each of us gave into the book *Fourier Series and Integrals.* Shortly after the beginning of the first year of that appointment, just after we had moved into a house that we really liked in Brookline, Henry invited me to join him at Rockefeller University for the academic year 1966–1967.

Rockefeller University

I spent the year at "RockTech", as it was affectionately called in some circles, working with Henry on applications of de Branges spaces of entire functions to prediction theory. We had actually begun to work on this the year before at MIT, but now there was real progress. That was the good news. The bad news was that none of the fine job offers/inquiries that I had turned down the year before were repeated. I also messed up a job interview at Bell Labs. I think, in the arrogance of youth, I was much too forthright about what I would and would not like to work on.

It seemed like a good time to travel. I applied for a visiting position at the Hebrew University in Jerusalem. As fate would have it, there happened to be a visiting postdoc at Rockefeller from Israel, by the name of Moshe Kugler. When I mentioned to him that I had applied to the Hebrew University, he encouraged me to apply to the Weizmann Institute. I had never heard of the Institute at that time. He supplied a name (Joseph Gillis) and an address and very shortly after some interchanges of correspondence (shortly in those days meant a few weeks; no E-mail back then) I was offered a postdoctoral fellowship in the Department of Applied Mathematics. The offer was sent by telegram and I was given a week to answer. Since the object was to travel and the prospects at Jerusalem were still uncertain, I accepted.

The Weizmann Institute

The Weizmann Institute was really an odd choice for me at the time. I had never heard of the Institute or of anyone who worked there. I spent most of that year working on applications of Hilbert spaces of entire functions to the spectral theory of second order differential operators. No one else at the Institute was really interested in such things. Many years later Doron Zeilberger happened to see the paper that emerged from this work and was prompted to apply for graduate study at the Institute because of it (though he ended up working on something else). Although there was no mathematical interaction for me, the atmosphere was pleasant and there was a great outdoor swimming pool. Moreover, in those days, Rehovot had a certain rustic charm. (There was only one traffic light.) After a while (it took a few months) Irene and I began to look favorably on the possibility of returning some time in the future, if the opportunity were to arise.

New York again

In late August 1968 we returned to New York, where I had accepted a position as an Assistant Professor of Mathematics at the City College of the City University of New York. I also had an offer from NYU, which was more attractive mathematically, but at a much lower salary. Since it is expensive to live in New York with two small children I opted for CCNY. It was a good choice. The department was friendly and the heads were very generous in assigning me a low teaching load and also allowing me to teach special topics courses. The collaboration with Henry McKean on prediction theory resumed.

In the meantime there were new developments in Rehovot. The Institute had decided to open a second department of Mathematics. It was to be headed by Samuel Karlin. Shlomo Sternberg and Yitz Herstein had also agreed to participate in this "noble venture", as Sam called it, at least on a part-time basis. All these "Chiefs" needed some "Indians" and Sam invited me to be one (of the Indians). The target date was the Fall Semester 1970. Having by this time reentered the Riverdale community and feeling very much a part of it, it was not an easy decision, but finally we said yes. After all, if it was a complete disaster, we could always return. Moreover, to paraphrase one of our young men, "we had to pick up Michael", who was to be born in Rehovot on August 1972.

Sam was very excited about what he was going to accomplish at the Institute. I remember receiving a phone call from him one stormy Sunday evening (Purim, 1970). He had just returned from Israel and was staying with friends in the Village. He asked me to come down to discuss plans for the new department. This meant taking a bus to 236th Street and Broadway, a train to West 4th street and then a walk of a few blocks; at least an hour and a half in each direction. I made the trip. When I got there Sam said he was too tired to discuss plans. He suggested that I ride with him to the airport on the following day. I passed that one up.

Back to Rehovot

I was the first member of the newly formed Department of Pure Mathematics to arrive in Rehovot. Sam came a few weeks later, Yitz came in the second semester and Shlomo came in the summer. Sam and I never collaborated together on mathematical problems. However, he did get me interested in tennis, which became a major obsession for me for more than twenty years. I spent most of the first few years at the Institute finishing off projects that had been started with Henry McKean. The Fourier Series book was sent to the publisher in the Fall of 1971, and the book on Prediction was shaped when Henry visited Israel for 5 months in the Spring of 1973. In those days, manuscripts were typed on electric typewriters which had special inserts for math symbols, and drafts of manuscripts and/or lists of comments were sent back and forth through the mail. A slow business compared to today. A close to final draft was submitted to Edwin Beschler, then of Academic Press, for review in late September 1973, a few days before the Yom-Kippur war broke out. Although there were still several months of revisions ahead, plus galleyproofs and page proofs, the end was in sight. It was time to look for something new. By this time, my interests had shifted from probability to analysis. Because of intensive work on the books, I only published one short paper in the period between 1970 and 1976. Fortunately, I was awarded tenure a year after arrival. Otherwise, I might be driving a semi-trailer today, which was one of my childhood ambitions.

New vistas

Martin Kruskal spent the academic year 1973–1974 on sabbatical at the Weizmann Institute. In the course of the year he gave a number of interesting lectures on isospectral problems for the Schrödinger equation and the connections with the KdV equation. After so many years of living with the string equation, it seemed natural to explore analogous questions in this setting also. A few hours of calculations (and miscalculations) led to the conclusion that if the density of the string $r(x)$ is parametrized by t and allowed to evolve according to the partial differential equation $r_t - r^3 r_{xxx} = 0$, then the eigenvalues of the string equation $-r^{-1}y'' = \lambda y$, with appropriate side conditions, would stay fixed in time. In 1974, Martin reported on these calculations in a series of lectures at the Batelle Institute and called the PDE the Harry Dym equation. The name stuck, even though I never wrote any papers on the subject. (Actually a draft of a paper which explored a number of questions related to the theory of such equations was prepared in collaboration with Martin. But Martin took it back with him to Princeton, where it is presumably still collecting dust in his office.)

The other new projects that I got involved in were an outgrowth of the interest in reproducing kernel Hilbert spaces and inverse spectral problems that had been kindled by the work on prediction.

One of the first of these was carried out with Naftali Kravitsky ($Z''L$), my first PhD student. We studied the effects of small perturbations on the principal spectral function $\Delta(\lambda)$ of the vibrating string equation upon the mass distribution $m(x)$ of the string. The main result was a prescription for computing at least the initial segment of the perturbed string. The Gelfand-Levitan procedure for reconstructing the potential of the Schrödinger equation emerged as a pleasing byproduct of the main theorem. A basic tool was the abstract method of triangular factorization of Gohberg and Kreĭn that was developed in their monograph on Volterra Operators. This was my first encounter with the name Israel Gohberg.

The relevance of factorization to inverse problems is a theme which was returned to in a paper with Andrei Iacob. Indeed, this paper was written largely to emphasize the connections between factorization and three basic techniques for solving inverse problems: the method of Gelfand-Levitan, the method of Kreĭn and the method of Marchenko. For pedagogical reasons the paper focused mostly on discrete problems on the line and on the circle. A second rather long paper with Andrei focused on inverse problems for canonical systems of differential equations of the form

$$J\frac{dy}{dt} = V(t)y(t,\lambda) + \lambda y(t,\lambda) \quad (t \geq 0)$$

with spectral densities of the form $I_n + K(\lambda)$, where $K(\lambda)$ is of Wiener class. A byproduct of the analysis was a linear fractional representation for the set of all solutions of a continuous version of the Carathéodory interpolation problem. The basic strategy was to identify solutions of the interpolation problem with solutions of the inverse spectral problem. (The maximum entropy solution of the covariance extension problem had a particularly pleasing interpretation.) This identification depended in part on a theorem of M.G. Kreĭn which was not quite correct in the asserted generality. Nevertheless the representation formula was valid and was later justified by other methods.

Another project that was initiated in that period started from the observation that a continuous version of the strong Szegő formula (due to Marc Kac) could be recast in the language of operators acting on Paley-Wiener spaces. From there it was a short jump to try to generalize to the setting of de Branges spaces of entire functions. This involved a healthy dose of reproducing kernel space theory and estimates of traces and determinants for assorted classes of operators. The bible for the estimates was the wonderful monograph by Gohberg and Kreĭn on nonselfadjoint operators. This marked my second encounter with the name Israel Gohberg. Little did I suspect at the time that there was more to come, much more.

The Odessa connection

I never met M.G. Kreĭn, but he was to be a major influence on my mathematical life. It began when Loren Pitt discovered the Doklady note of M.G. Kreĭn, "On a fundamental approximation problem in the theory of extrapolation and filtration of stationary random processes," which had appeared in translation. This discovery caused a major reorientation of the work with Henry on prediction and much effort was required to fill in the missing details in that note and a number of other Doklady notes, none of which contained proofs. We didn't know who Kreĭn was at the time and Henry even called up Peter Lax to find out if Kreĭn was reliable. The answer was, of course, yes. The next influence was through the two marvelous Gohberg–Kreĭn monographs referred to earlier. And then, lo and behold, fate smiled, and Israel Gohberg showed up in person at the Weizmann Institute. The story was that in 1974, Israel Gohberg immigrated to Israel and accepted a full-time position at Tel Aviv University and also, in the Spring of 1975, a half-time position at The Weizmann Institute. Israel and I started to work together on assorted problems of extension in the Fall of 1976. It was a bit sporadic at first. He had a lot of invitations, I had some military obligations, but the die was cast. The more intensive phase of this collaboration probably began in the Fall of 1977. Israel used to come to the Institute twice a week and we would sit together several hours each time working together. This was a marvelous way to enter more deeply into the mathematical world of M.G. Kreĭn, guided by one of his foremost disciples. Israel was (and still is) a wonderful teacher, both as a lecturer before a large audience and as a collaborator with an audience of one.

Our work together focused on assorted classes of extension problems, mostly in the context of matrix valued functions in a Wiener algebra.

The first problem we considered was to establish conditions to guarantee the existence of an invertible $n \times n$ matrix valued function $f(\zeta)$ in the Wiener algebra on the circle such that the Fourier coefficients f_j of $f(\zeta)$ are specified for $|j| \leq n$

and the Fourier coefficients g_j of $g(\zeta) = f(\zeta)^{-1}$ of the inverse are equal to zero for $|j| > n$. It was also required that $f(\zeta)$ admit a factorization. Necessary and sufficient conditions for the existence of solutions to this problem were obtained. A number of analogues were developed in other settings in a subsequent sequence of papers that were devoted to the Wiener algebra on the line, banded matrices and Fredholm integral operators, respectively. The first three of these papers contain maximal entropy principles: The band extension maximizes the entropy. The fourth paper contained the first abstract formulation of what came to be called Band Extension Problems. Interestingly enough, the third paper in this series, which was a relatively easy by product of the first two, is the one that seems to have attracted the most attention. We later went on to consider a number of other problems of extension, including triangular extensions, contractive, isometric and unitary extensions, with and without factorization indices, for assorted classes of operators. The results were pleasing, but the collaboration slowed down when the Institute went through a financial crisis and disbanded all part-time positions, and came to a halt a couple of years later.

The intensive collaboration with Israel lasted for close to ten years. We got together a few years later on another problem that began with a question that Israel raised at a conference in Winnipeg and was supplemented by a number of meetings that were scheduled around Israel's visits to his dentist in Rehovot. The friendship continues.

In 1990, Michael Shmoish, a student of Ju.M. Berezanskii, and hence a mathematical grandson of M.G. Kreĭn, immigrated to Israel. He ended up at the Institute, and wrote a nice thesis on inverse problems for block Jacobi matrices and related issues.

In 1991, there was a rumor afoot at the MTNS meeting in Okeba, Japan, that both Vadim M. Adamyan and Damir Z. Arov were to attend. That was two out of the three authors of the famous AAK cycle of papers on matrix (and operator) versions of the Nehari problem that was one of the cornerstones of H^∞ control. This was shortly after Peristroika and possibly the first time that they had ever been allowed to leave Odessa to travel to the West. The rumor turned out to be correct. A. Nudelman, Lev Sakhnovich and E. Tsekanovskii were also part of the group, but they were less well known to that community. Dima Arov, using Israel Gohberg as a translator, expressed an interest in visiting Rehovot. Between his knowledge of English and my knowledge of Russian, it wasn't exactly clear to me how we would communicate, but I figured we could overcome that difficulty somehow. There were plenty of Russian students at the Institute who spoke English and could help out if need be. A visit was arranged and that marked the beginning of an intensive collaboration that is still running hot to this day. At later periods both Lev and Vadim also visited Rehovot, as did Israel Kac, yet another member of the Kreĭn circle.

The Delft connection

The Delft connection was another act of fate. In some sense the wheels began to turn when I wrote to Paul Fuhrmann for a reprint. He invited me to BeerSheva to give a seminar. At BeerSheva I met Abie Feintuch and we got to talking. Some months later, in a follow up conversation by phone, Abie told me that he had met someone called Patrick Dewilde at the second MTNS conference in Lubbock Texas and that Patrick was interested in Prediction Theory. This led to an attempt to invite Patrick (who I thought to be at Louvaine) to visit Israel for ten days under the auspices of the Belgian-Israel Cultural Exchange program. The correspondence was initiated in Autumn 1977.

However, it turned out that Patrick had already moved to Delft. Undaunted by the facts, we submitted an application to the Belgian authorities anyway. There was no response. At this point fate and Tom Kailath took a hand. Both Patrick and I were guests of Tom in the summer of 1978 at Stanford. This is where the romance with J-inner functions began. It wasn't love at first sight, but there was an attraction.

I spent a good part of that summer working through the manuscript of a paper by Dewilde, Vieira and Kailath that dealt with recursive extraction of elementary J-inner sections with a single pole at infinity. Nevertheless, I suspect that this effort might well have ended with the summer had not the Belgian authorities agreed (several months later) to support Patrick's visit to Israel. The visit took place in the Spring of 1979 and during that time a couple of open questions were resolved. This in turn led to an invitation to Delft and marked the beginning of a long and fruitful collaboration that lasted many years.

Patrick got me interested in interpolation theory. Our first paper together was a generalization of the Dewilde, Vieira and Kailath paper referred to earlier. The objective was to approximate a given function (first scalar and later matrix valued) S of Schur class or Z of Carathéodory class by a rational function of the same class that agreed with the given function at a prescribed set of points (and in the matrix case in a prescribed set of directions at each of these points) and to estimate the error. The methods were recursive. The approximant was constructed by a variant of the Schur–Nevanlinna algorithm. The recursive procedure produced a rational J-inner matrix valued function, the entries of which were used to display solutions via a linear fractional transformation of the Redheffer form. A characterization of the maximum entropy solution was given. The first two papers focused on matching points in the interior of the open unit disc. The third dealt with matching at one or more boundary points. In the latter case, the J-inner matrix valued functions alluded to above were "Brune Sections", i.e., Blaschke–Potapov factors of the third kind. Oddly enough, the Pick matrix did not figure in any of these problems.

The problems considered in these papers were all special cases of the LIS (lossless inverse scattering) problem of network theory, one formulation of which is: Given S in the Schur class, find a J-inner W such that $S = T_W[S_L]$ is a linear fractional transformation of a passive "load" S_L. In these papers, the LIS problem was solved by setting up a (tangential) interpolation problem with a finite number of constraints that were obtained from the given S. We worked on a number of other problems together, the last of which focused on interpolation in the setting of upper triangular operators, part of this was in collaboration was with Daniel Alpay. But more about him later. Our active collaboration died down when Patrick became head of DIMES at Delft and had very little time to call his own.

The French connection

The second dominant influence in my mathematical life, not counting people that I worked with, is undoubtedly Louis de Branges. His work on reproducing kernel Hilbert spaces of entire functions played a significant role in the study of prediction with Henry McKean. Applications of his abstract characterization of reproducing kernel Hilbert spaces of J-inner matrix valued functions to assorted problems of interpolation and extension has been one of the major themes in a good part of my own work. (Reproducing kernel Kreĭn spaces were useful in assorted studies of the zero distribution of various classes of matrix valued functions, some of which were carried out with Nicholas Young. The all important R_α operator even entered in another project that was carried out with Malcolm Smith and Tryphon Georgiou.)

The initial exploration of de Branges' work on reproducing kernel Hilbert spaces of J-inner matrix valued functions began, appropriately enough, with another Frenchman, Daniel Alpay. Daniel first came to the Institute from Paris to study for his MSc degree. Daniel had also studied electrical engineering as an undergraduate and and seemed to like the same kind of mathematics that I did. He stayed on to do a PhD with me and at a later stage was a postdoc at the Institute.

Our first project together focused on an abstract version of an inverse scattering problem that Patrick and I had worked on earlier. But this was more operator theoretic and was heavily based on de Branges' work. In the writing of this paper I think we both learned the power and the beauty of de Branges' abstract characterization of the reproducing kernel Hilbert spaces alluded to earlier. It was to be a major theme in our future work, both together and individually. There were also interesting applications to the theory of models of operators that were close to unitary and and operators that were close to self adjoint, which in turn intersected with work of Moshe Livsic and M.G. Kreĭn. We later went on to put the Schur algorithm into an abstract reproducing kernel Hilbert space setting and to further generalize some of these ideas to the setting of Pontryagin spaces.

There were a number of other projects that we worked on over the years including general classes of realization formulas, generalizations of the Iohvidov laws and interpolation theory for upper triangular operators. The latter was in collaboration with Patrick Dewilde. It was in fact based on work that Patrick and Daniel had initiated. I came on board a little later.

The first paper with Daniel was one of the first in our department to be typed in TeX,[1] using the computer as a word processor. In those days there was one printer in the Institute. It was housed in the Computer Center, about a five minute walk away.

The Kharkov connection

The Kharkov connection began with a letter from Victor Katsnelson that expressed an interest in visiting the Weizmann Institute. I knew the name from a private translation of one of his papers by Tsuyoshi Ando. The visit led in due course to a full-time appointment some years later and then to a progression of visitors from Kharkov, including V. A. Marchenko, Sasha Kheifets, Peter Yuditskii and Vladimir Dubovoj. I was particularly impressed by the work of Katsnelson, Kheifets and Yuditskii on what they called "The Abstract Interpolation Problem". To my mind, it is one of the most elegant and far reaching approaches to interpolation problems that is currently available. In work with Boris Freydin (my most recent PhD student, who was also from Kharkov) we managed to adapt these methods to bitangential problems in the setting of upper triangular operators. I also worked intensively on interpolation problems for degenerate Pick matrices and on boundary interpolation with another former resident of Kharkov, Vladimir Bolotnikov, during the two years he spent as a postdoctoral fellow at the Institute. Vladimir was actually a mathematical grandson, having completed his degree with Daniel Alpay at Ben Gurion University of the Negev.

The UCSD connection

I first met Bill Helton in 1976 at a meeting in Oberwolfach that Israel Gohberg helped to organize. Both of us were working on generalizations of the Szegő formula at the time and thought that it would be fruitful to get together. It took almost twenty years to work this out. A major incentive to finally do something about it arose when our oldest son Jonathan moved with his family to Los Angeles for an extended stay. Since 1996, I have spent several weeks of each year at UCSD working with Bill, in directions that are far different from what we originally envisioned. So far reproducing kernels have not yet intervened. But its been good fun and, for me at least, a nice way to enter a new area of mathematics in which I had not been active before.

[1] It was typed by Mrs. Ruby Musrie, our department secretary, who deserves a special note of thanks for having converted thousands of pages of scrawl into elegant manuscripts. In fact, in thirty odd years I got a lot of help from all our secretaries, as well as our support staff, and I am grateful to them all. But Ruby bore the brunt.

Odd recollections and thoughts with no connection

- The student who sat on my left in my home room in my first year of High School was also born in Vienna and his parents turned out to be friends of my parents. They had lost contact some dozen years before.
- The student who sat behind me in the same class before he moved to New Jersey was David Trutt, who went on to do a PhD with de Branges in Purdue. (We figured this out years later by backtracking when he visited the Institute.)
- One of the real hazards of Cooper Union was the Electric Machines Lab. I still remember the day when there was a big flash, followed by a bang as the circuit breakers went; seeing my lab partner, Steve Hofstein, standing with a quizzical look on his face and singed eyebrows after he had inadvertently brought the tips of two cables together that should have been kept apart.
- In the days of the German Democratic Republic, contact with Israel was strictly forbidden for citizens of that "Democracy." Nevertheless, Bernd Kirstein and I used to exchange reprints through mutual friends in Holland, some through Patrick and some through Rien Kaashoek. Even these shipments were intercepted and carefully inspected. Several years later, Bernd told me that one shipment was approved because Dym was regarded as a good Dutch name. Little did they know that the name is an acronym in Hebrew coming from the letters Daled, Yud, Mem.
- In a paper that I am working on, I just spent the best part of three days trying to straighten out an erroneous minus sign. It seems like an odd way for a supposedly grown person to spend one's time. On the other hand, planes have been known to crash because of a mistake in sign, or something equally foolish, in a critical computer program.
- A cartoon in the New Yorker in 1957 during the height of the competition between different branches of the US Armed Forces to put an American satellite into orbit shows an American Army General looking at the Russian Sputnik. The caption: "Thank God, for a moment I thought it was the Navy's."
- Irene's dictum: "Let's throw everything away. Then there will be room for what's left."
- It has been my observation that those who think they know all the answers, don't know all the questions. A Chinese proverb puts it nicely: "Trust only those who doubt."
- As one gets older one realizes that there are questions that one will never be able to answer. I, for one, have never understood why I have so many single socks.

The last word

Objects seen through a rear view mirror are distorted. Our memories are selective and play tricks on us. Consequently, everything written above should be taken with a grain (perhaps several grains) of salt. I have focused mainly on collaborations and how they came into being and not so much on projects that I carried out on my own steam. My intentions were not to cast judgement, but rather to indicate how I muddled into mathematics as a career and some of my experiences enroute, including the mishaps. There were downs as well as ups, but on the whole it wasn't a bad run. What I can say with certainty is that I owe a great deal to the colleagues that I collaborated with and to my former students, teachers all. Thank you. [2]

Harry Dym
Department of Theoretical Mathematics
The Weizmann Institute of Science
Rehovot 76100
Israel

[2]I owe even more to that young lady that I met while running a waterfront in the Catskills long long ago, but this is not the place to go into all that.

Operator Theory:
Advances and Applications, Vol. 134, 19–24

List of Publications of Harry Dym

Articles

1. (with E. Arthurs): On the optimum detection of digital signals in the presence of white Gaussian noise, *IRE Trans. Comm. Systems* **10** (1962), 336–372.
2. Distance properties of tree codes, *MIT Res. Lab. for Electronics, QPR*, **70** (1963), 247–261.
3. (with E. Luks): On the mean duration of a ball and cell game; a first passage problem, *Ann. Math. Stat.*, **37** (1966), 517–521.
4. A note on limit theorems for the entropy of Markov chains, *Ann. Math. Stat.*, **37** (1966), 522–524.
5. Stationary measures for the flow of a linear differential equation driven by white noise, *Trans. Amer. Math. Soc.*, **123** (1966), 130–164.
6. On a class of monotone functions induced by ergodic sequences, *Amer. Math. Monthly*, **75** (1968), 595–601.
7. (with H.P. McKean): Applications of de Branges spaces of integral functions to the prediction of stationary Gaussian processes, *Illinois J. of Math.*, **14** (1970), 299–343.
8. An introduction to de Branges spaces of entire functions with applications to differential equations of the Sturm-Liouville type, *Advances in Math.*, **5** (1970), 395–471.
9. (with H.P. McKean): Extrapolation and interpolation of stationary Gaussian processes, *Ann. Math. Stat.*, **41** (1970), 1817–1844.
10. An extremal problem in the theory of Hardy functions, *Israel J. Math.*, **18** (1974), 391–399.
11. (with D. Zeilberger): Further properties of discrete analytic functions, *J. Math. Anal. Appl.*, **58** (1977), 405–418.
12. Trace formulas for a class of Toeplitz-like operators, *Israel J. Math.*, **27** (1977), 21–48.
13. (with I. Shapiro): Conditions for restricted translation operators to belong to S_p, *Proc. Amer. Math. Soc.*, **63** (1977), 251–258.
14. Trace formulas for a class of Toeplitz-like operators II, *J. Funct. Anal.*, **28** (1978), 33–57.
15. (with N. Kravitsky): On the inverse spectral problem for the string equation, *Integral Equations Operator Theory*, **1** (1978), 270–277.

16. Conditions for a class of stationary Gaussian processes to be Kolmogorov mixing, *Ann. Prob.*, **6** (1978), 159–161.

17. (with T. Dreyfus): Product formulas for the eigenvalues of a class of boundary value problems, *Duke J.*, **45** (1978), 15–37.

18. A problem in trigonometric approximation theory, *Illinois J. of Math.*, **22** (1978), 402–403.

19. (with N. Kravitsky): On recovering the mass distribution of a string from its spectral function, in: *Topics in Functional Analysis, Adv. Math. Supp.*, **3** (1978), 45–90.

20. Trace formulas for blocks of Toeplitz-like operators, *J. Funct. Anal.*, **31** (1979), 69–100.

21. Trace formulas for pair operators, *Integral Equations Operator Theory*, **1** (1978), 152–175.

22. Applications for factorization theory to the inverse spectral problem, *International Symposium on Mathematical Theory of Networks and Systems*, **3** (1979), 188–193.

23. (with I. Gohberg): Extensions of matrix valued functions with rational polynomial inverses, *Integral Equations Operator Theory*, **2** (1979), 503–528.

24. On the absolute convergence of a multiple integral of $sin\, x/x$, *Amer. Math. Monthly*, **87** (1980), 53–54.

25. (with I. Gohberg): On an extension problem, generalized Fourier analysis, and an entropy formula, *Integral Equations Operator Theory*, **3** (1980), 143–215.

26. (with I. Gohberg): Extensions of band matrices with band inverses, *Linear Alg. Appl.*, **36** (1981), 1–24.

27. (with P. Dewilde): Schur recursions, error formulas and convergence of rational estimators for stochastic stationary sequences, *IEEE Trans. Inform. Theory*, **27** (1981), 446–461.

28. (with P. Dewilde): Lossless chain scattering matrices and optimum linear prediction: The vector case, *Int. J. Circuit Theory Appl.*, **9** (1981), 135–175.

29. (with S. Ta'assan): An abstract version of a limit theorem of Szegő, *J. Funct. Anal.*, **43** (1981), 294–312.

30. (with A. Iacob): Applications of factorization and Toeplitz operators to inverse problems, in: *Toeplitz Centennial*, **OT4**, Birkhäuser (1982), 233–260.

31. (with I. Gohberg): Extensions of triangular operators and matrix functions, *Indiana J.*, **31** (1982), 579–606.

32. (with I. Gohberg): Extensions of matrix valued functions and block matrices, *Indiana J.*, **31** (1982), 733–765.

33. (with I. Gohberg): Extensions of kernels of Fredholm operators, *J. d'Analyse Math.*, **42** (1982/83), 51–97.

34. (with I. Gohberg): Unitary interpolants, factorization indices and infinite Hankel block matrices, *J. Funct. Anal.*, **54** (1983), 229–289.
35. (with I. Gohberg): Hankel integral operators and isometric interpolants on the line, *J. Funct. Anal.*, **54** (1983), 290–307.
36. (with I. Gohberg): On unitary interpolants and Fredholm infinite block Toeplitz matrices, *Integral Equations Operator Theory*, **6** (1983), 863–878.
37. (with A. Iacob): Positive definite extensions, canonical equations and inverse problems. in: *Topics in Operator Theory Systems and Networks*, **OT12**, Birkhäuser (1984), 141–240.
38. (with P. Dewilde): Lossless inverse scattering theory, digital filters and estimation theory, *IEEE Trans. Inform. Theory*, **30** (1984), 644–662.
39. (with D. Alpay): Hilbert spaces of analytic functions, inverse scattering and operator models, I, *Integral Equations Operator Theory*, **7** (1984), 589–641.
40. (with D. Alpay): Hilbert spaces of analytic functions, inverse scattering and operator models, II, *Integral Equations Operator Theory*, **8** (1985), 145–180.
41. On a Szegő formula for a class of generalized Toeplitz kernels, *Integral Equations Operator Theory*, **8** (1985), 427–431.
42. (with I. Gohberg): A maximum entropy principle for contractive interpolants, *J. Funct. Anal.*, **65** (1986), 83–125.
43. (with D. Alpay): On applications of reproducing kernel spaces to the Schur algorithm and rational J unitary factorization, in: *I. Schur Methods in Operator Theory and Signal Processing*, **OT18**, Birkhauser (1986), 89–159.
44. (with I. Gohberg): A new class of contractive interpolants and maximum entropy principles, in: *Topics in Operator Theory and Interpolation*, **OT29**, Birkhäuser (1988), 117–150.
45. Hermitian block Toeplitz matrices, orthogonal polynomials, reproducing kernel Pontryagin spaces, interpolation and extension, in: *Orthogonal Matrix-valued Polynomials and Applications*, **OT34**, Birkhäuser (1988), 79–135.
46. On reproducing kernel spaces, J unitary matrix functions, interpolation and displacement rank, in: *The Gohberg Anniversary Collection*, **OT41**, Birkhäuser (1989), 173–239.
47. On Hermitian block Hankel matrices, matrix polynomials, the Hamburger moment problem, interpolation and maximum entropy, *Integral Equations Operator Theory*, **12** (1989), 757–812.
48. (with D. Alpay and P. Dewilde): On the existence and construction of solutions to the partial lossless inverse scattering problem with applications to estimation theory, *IEEE Trans. Information Theory*, **35** (1989), 1184–1205.
49. (with N. Young): A Schur-Cohn theorem for matrix polynomials, *Proc. Edinburgh Math. Soc.*, **33** (1990), 337–366.

50. RKHS, LIS and interpolation, in: *Signal Processing, Scattering and Operator Theory, and Numerical Methods*, Proceedings of the International Symposium MTNS-89, Vol. III, Birkhäuser, Boston, 1990, pp. 65–78.

51. On reproducing kernels and the continuous covariance extension problem, in: *Analysis and Partial Differential Equations: A Collection of Papers Dedicated to Mischa Cotlar*, Marcel Dekker, New York, 1990, pp. 427–482.

52. (with D. Alpay): Structured invariant spaces of vector valued rational functions, Hermitian matrices and a generalization of the Iohvidov laws, *Linear Alg. Appl*, **137/138** (1990), 137–181.

53. (with D. Alpay): Structured invariant spaces of vector valued rational functions, sesquilinear forms and a generalization of the Iohvidov laws, *Linear Alg. Appl.*, **137/138** (1990), 413–451.

54. (with D. Alpay and P. Dewilde): Lossless inverse scattering and reproducing kernels for upper triangular operators, in: *Extension and Interpolation of Linear Operators and Matrix Functions*, **OT47**, Birkhäuser (1990), 61–135.

55. A Hermite theorem for matrix polynomials, in: *Topics in Matrix and Operator Theory*, **OT50**, Birkhäuser (1991), pp. 191–214.

56. (with N. Young): Factorization and the Schur-Cohn matrix of a matrix polynomial, *Integral Equations Operator Th.*, **15** (1992), 1–15.

57. (with D. Alpay): On reproducing kernel spaces, the Schur algorithm, and interpolation in a general class of domains, *Oper. Theory: Adv. Appl.*, **OT 59**, Birkhäuser, 1992, pp. 30–77.

58. (with P. Dewilde): Interpolation for upper triangular operators, in: *Time-Variant Systems and Interpolation*, **OT56**, Birkhäuser, 1992, pp. 153–260.

59. (with D. Alpay): On a new class of reproducing kernel spaces and a new generalization of the Iohvidov laws, *Linear Alg. Appl.*, **178** (1993), 109–183.

60. (with D. Alpay): On a new class of structured reproducing kernel spaces, *J. Funct. Anal.*, **111** (1993), 1–28.

61. Remarks on interpolation for upper triangular operators, in: *Challenges of a Generalized System Theory* (P. Dewilde, M.A. Kaashoek and M. Verhaegen, eds.), North Holland, Amsterdam, 1993, pp. 9–24.

62. On the zeros of some continuous analogues of matrix orthogonal polynomials and a related extension problem with negative squares, *Comm. Pure Appl. Math.*, **47** (1994), 207–256.

63. Shifts, realizations and interpolation, redux, in: *Oper. Theory: Adv. Appl.*, **OT 73** (1994), 182–243.

64. Discursive Review of *The Commutant Lifting Approach to Interpolation Problems* by Ciprian Foias and Arthur E. Frazho, *Bull. Amer. Math. Soc.*, **31** (1994), 125–140.

65. Shifts, reproducing kernels, and interpolation, a tutorial, in: *Systems and Networks: Mathematical Theory and Applications*, Volume I, Key Invited

Lectures (U. Helmke, R. Mennicken and J. Saurer, eds.), Akadamie Verlag GmbH, Berlin, 1994, pp. 85–99.

66. (with T.T. Georgiou and M.C. Smith): Explicit formulas for optimally robust controllers for delay systems, *IEEE Trans. Aut. Control*, **40** (1995), 656–669.
67. (with I. Gohberg): On maximum entropy interpolants and maximum determinant completions of associated Pick matrices, *Integral Equations Operator Theory*, **23** (1995), 61–88.
68. More on maximum entropy interpolants and maximum determinant completions of associated Pick matrices, *Integral Equations Operator Theory*, **24** (1996), 188–229.
69. (with D. Alpay): On a new class of realization formulas and their application, *Linear Algebra Appl.*, **241–243** (1996), 3–84.
70. (with B. Freydin): Bitangential interpolation for upper triangular operators, in: *Topics in Interpolation Theory, Oper. Theory: Adv. Appl.* **OT95** (1997), 105–142.
71. (with B. Freydin): Bitangential interpolation for triangular operators when the Pick operator is strictly positive, in: *Topics in Interpolation Theory, Oper. Theory: Adv. Appl.* **OT95** (1997), 143–164.
72. A Basic Interpolation Problem, in: *Holomorphic Spaces* (S. Axler, J. McCarthy and D. Sarason, eds.), Cambridge University Press, Cambridge, England, 1998, pp. 381–423.
73. (with D. Z. Arov): *J*-inner matrix functions interpolation and inverse problems for canonical systems, I: Foundations, *Integral Equations Operator Theory* **29** (1997), 373–454.
74. (with D. Z. Arov): On three Kreĭn extension problems and some generalizations, *Integral Equations Operator Theory* **29** (1998), 1–91.
75. (with V. Bolotnikov): On degenerate interpolation, entropy and extremal problems for matrix Schur functions, *Integral Equations Operator Theory* **32** (1998), 367–435.
75. (with J. W. Helton and O. Merino): Algorithms for solving multidisk
76. On M. G. Kreĭn's contributions to prediction theory, in: *Operator Theory and Related Topics, Vol. II Oper. Theory: Adv. Appl.*, **OT118**, Birkhäuser, Basel, 2000, pp. 1–15.
77. (with D. Z. Arov): *J*-inner matrix functions, interpolation and inverse problems for canonical systems, II: The inverse monodromy problem, *Integral Equations Operator Theory* **36** (2000), 11–70.
78. (with D. Z. Arov): *J*-inner matrix functions, interpolation and inverse problems for canonical systems, III: More on the inverse monodromy problem, *Integral Equations Operator Theory* **36** (2000), 127–181.
79. (with D. Z. Arov): Matricial Nehari problems, *J*-inner matrix functions and the Muckenhoupt condition, *J. Funct. Anal.* **181** (2001), 227–299.

80. (with D. Z. Arov): Some remarks on the inverse monodromy problem for 2×2 canonical differential systems in: *Oper. Theory and Analysis, Oper. Theory: Adv. Appl.*, **OT122**, Birkhäuser, Basel, 2001, pp. 53–87.
81. Structured matrices, reproducing kernels and interpolation, in: *Structured Matrices in Mathematics, Computer Science and Engineering I*, Contemporary Mathematics, **Vol 280** (V. Olshevsky, ed.), Amer. Math. Soc., Providence, R.I., 2001, pp. 3–29.
82. On Riccati equations and reproducing kernel spaces, in: *Recent Advances in Operator Theory, Oper. Theory: Adv. Appl.*, **OT124**, Birkhäuser, Basel, 2001, pp. 189–215.
83. Reproducing kernels and Riccati equations, *Int. J. Appl. Math. Comp. Science*, **11** (2001), 35–53.
84. (with D. Z. Arov): J-inner matrix functions, interpolation and inverse problems for canonical systems, IV: Direct and inverse bitangential input scattering problems, *Integral Equations Operator Theory*, in press.
85. (with D. Z. Arov): J-inner matrix functions, interpolation and inverse problems for canonical systems, V: The inverse input scattering problem for Wiener class and rational $p \times q$ input scattering matrices, *Integral Equations Operator Theory*, in press.
86. (with J. W. Helton and O. Merino): Multidisk problems in H^∞ optimization: A method for analysing numerical algorithms, *Indiana J.*, in press. item87.(with V. Bolotnikov): On boundary interpolation for

Books

1. (with H.P. McKean): *Fourier Series and Integrals*, Academic Press, New York, 1972.
2. (with H.P. McKean): *Gaussian Processes, Function Theory and the Inverse Spectral Problem*, Academic Press, New York, 1976.
3. *J Contractive Matrices, Reproducing Kernel Hilbert Spaces and Interpolation*, CBMS Regional Conference Series, American Mathematical Society, Providence, R.I., 1989.

Operator Theory:
Advances and Applications, Vol. 134, 25–30

On Joint Work with Harry Dym

Israel Gohberg

1. How it started

I immigrated with my family to Israel at the end of July 1974. In the beginning we studied Hebrew very intensively. I also started to look for work in Israeli institutions of higher education; very soon I received an offer from the Tel-Aviv University which I decided to accept. In March 1975 I was invited by the Dean of the Faculty of Mathematics, Professor S. Karlin, to take a part time position at the Weizmann Institute of Science. I accepted this position and started working there two days a week. The Pure Mathematics Department in the Weizmann Institute of Science was very small; apart from S. Karlin the Dean, Harry Dym and Yakar Kannai were the only senior researchers. There was also a group of doctoral students. I started to lecture different courses in advanced operator theory and applications.

I met Harry and from our conversations I understood that he was very well informed in operator theory in general and in the work of the school of M.G. Krein especially. Already then he was the author of two books with H.P. McKean and was active in research.

I accepted a Ph.D. student. This was Sofia Levin and I started to work with her. During one of my visits to the Institute Harry expressed interest in joint work with me. I was also interested in this offer and we started to look for an appropriate problem.

I soon found such a problem during my visit to Amsterdam. The problem was proposed by a colleague from the Free University, Professor G.Y. Nieuwland. He in his turn obtained the problem from a colleague who was working in theoretical chemistry.

2. Band extension problems

The first problem consisted of the following: a function $k(t)$ $(-T < t < T)$ has to be extended to the full line to $f(t)$ in such a way that the function $1 - \hat{f}(\lambda)$, where $\hat{f}$ is the Fourier transform of $f(t)$, is positive (or more generally different from zero on the line) and the function $1/(1 - \hat{f}(\lambda))$ has the form $1 - \hat{g}(\lambda)$, where $\hat{g}$ is the Fourier transform of a function $g(t)$ that vanishes outside the interval $(-T, T)$. Both functions f and g belong to $L_1(-\infty, \infty)$.

In a short time we had a solution to this problem and we started to write it down. We did not succeed in finishing this work before the summer. One of the reasons was that Harry decided to take a four month sabbatical. The last half was spent in Stanford with Tom Kailath.

By a coincidence, in Stanford Harry discovered that the thesis of Tom Kailath's doctoral student, A.C.G.Vieira, was relevant to our problem. In fact he was dealing with a matrix discrete analogue of the above mentioned problem for the positive definite case. More than that, in applications this problem is important and the solution is called the maximum entropy solution, or the autoregressive extension of statistical estimation theory. The discrete case of the scalar solution was solved and analyzed before by J.P. Burg in 1975. He came to it within the framework of spectral analysis in geophysics problems. After Harry returned to Israel we wrote our first joint paper [1] where we solved the generalized problem of extension of matrix valued functions, including the positive definite case with the maximum entropy solution. Explicit formulas for the solution based on Szegő orthogonal polynomials was also presented.

The following year, 1980, we published the paper [2] which contained the complete solution of the continuous analogue in the matrix valued case. This is a large paper (more than 70 pages) and it contains probably the first solution of the maximum entropy extension problem in this setting, together with a new definition of entropy under some natural technical conditions.

As a byproduct of the two papers described we obtained new results in the theory of completion of finite matrices. The results were published in 1981 in [3]. The problem of extension in this case is the following: Let a symmetric band of width $2m+1$ in an $n \times n$ matrix with complex entries be given and let the rest of the entries of this matrix be unspecified. The problem is to complete the matrix in such a way that the inverse of the completed matrix is a symmetric band matrix of $\leq 2m+1$. Of special interest is the case where the completion is additionally required to be positive definite. In this case under natural conditions the solution exists it is unique and can be characterized to have the maximum determinant between the determinants of all other positive definite completions. An explicit algorithm for this solution is also presented. This result contains Burg's maximal entropy inequality in theory of covariance extensions. This is a result that follows from the case that the band is Toeplitz and in this case the solution is also Toeplitz. The described results are also generalized for block matrices. This paper became much more popular than the first two. A number of interesting results for the more general non-band case were obtained by other colleagues. Till today the non-band case in general has not been solved to the end.

Our next paper [4] can be considered as the solution of a continuous analog of the previous problem. It is about extensions of kernels of Fredholm integral operators given in a band. The positive case generalized Burg's maximal entropy inequality. This result can be considered for the time dependent noncovariance case. In this paper is developed the beginning of the general theory of extensions and completions in an abstract algebra with multiplication subject to some special features that generalize the features in the concrete examples. This abstract approach served to clarify the band extension and completion problem and to unify the results of the latter paper with the previous ones. The abstract approach became popular. It was used as a basis for a far-reaching development. This led to the band method presented in a number of papers of I. Gohberg, M.A. Kaashoek and H. Woerdemann, and of J. Ball, I. Gohberg and M.A. Kaashoek, in which new extensions and interpolation problems were solved. The results of the beginning of this section intersect with some results of D.Z. Arov and M.G. Krein.

3. Working together

In the first years of cooperation both of us made serious efforts to progress in the extension and completion problems mentioned above. The problems were new in an area which we had not considered before and we worked with interest and enthusiasm. We presented these results at different conferences and our results were nicely received by our colleagues.

I came to the Weizmann Institute twice a week and most of this time was used for joint work with Harry. A small part of the time I spent with my graduate students. Soon they were three, Sofia Levin, Israel Koltracht and Nir Cohen. The joint work with Harry was very pleasant. The work was continuing also during the lunch break and during the tea break in the afternoon. Sometimes we worked in unexpected places. I remember a few hours work in the foyer of the Van Gogh museum in Amsterdam (while Harry's wife Irene was enjoying the exhibition). Harry is a very fine coauthor; he is talented, has good taste and a wide knowledge in theoretical mathematics as well as in applications. He is hardworking and has a wonderful command of English and he very easily puts mathematics on paper.

I learned many things from Harry in mathematics and also in everyday life. I was used to the Soviet mentality and rules of behavior. Harry helped me to understand the new situation and to become used to it. In view of our friendly relations I could ask his advice on any question without hesitation. For instance, he was the first to notice and explain to me the difference between the practice regarding very good Ph.D. students in USSR and in the West after graduation. In the USSR the best Ph.D. students were kept for permanent work in the university (chair) where they studied. In the West on the other hand they would have to leave and spend at least a short time in other universities. There is a big difference between the USSR and the West in the evaluation of various areas of mathematics and mathematicians. Harry explained these things which looked like contradictions to me. On my part I told Harry a lot about M.G. Krein, his work and his school, about the difficulties of Jewish life in the USSR. All of this interested him. He especially enjoyed hearing jokes from the USSR.

4. Triangular extensions

Let $f(z)$ ($|z| = 1$) be a function with specified Fourier coefficients f_j ($j = 0, 1, 2, \ldots$) and $|f_0| + |f_1| + |f_2| + \cdots < \infty$ and let $\psi_{-1}, \psi_{-2}, \ldots$ also given complex numbers with $|\psi_{-1}| + |\psi_{-2}| + \cdots < \infty$. The problem consists of specifying the Fourier coefficients of $f(z)$ with negative indices in such a way that $f(z) \neq 0$ ($|z| = 1$) and the Fourier coefficients of $1/f(z)$ with negative indices to be equal to $\psi_{-1}, \psi_{-2}, \ldots$. Of special interest is this problem with the additional condition that $|f(z)| = 1$ ($|z| = 1$).

Our next two papers [5, 6] were dedicated to different generalizations of this problem. We solved it in the block discrete case as well as for the matrix continuous analog. In the latter case with the additional condition this result was stated by M.G. Krein and F.E. Melik-Adamyan without proof in their study of scattering theory. This is probably the first published proof of this theorem. We also solved the finite matrix block analog of the triangular extension problem. As far as we know this was a new result for matrices. The triangular completion problem for scalar matrices is stated in the following way: Let the entries of the upper triangular part (including the diagonal) of an $n \times n$ matrix be specified. Complete the matrix in such a way that it is invertible and the inverse has apriori given entries in the lower triangular part (without the diagonal). We also solved the problem of completing a matrix to be unitary if the entries of the upper triangular part is given. In the triangular extension problems some technical conditions were required. In particular the canonical factorization or the partial indices equal to zero were required for the solution.

5. Unitary interpolants and factorization indices

Three papers [7, 8, 9] deal with the problem of extending a matrix function $f(z)$ ($|z| = 1$) with specified Fourier coefficients $f_0, f_1, \ldots$; $|f_0| + |f_1| + \cdots < \infty$ to a unitary matrix function without assumptions of canonical factorization as in the previous section. The solution if it exists certainly admits a factorization in general with nonzero partial indices. In paper [7] are described all unitary interpolants. One of the central results is the expression of the number of nonnegative factorization indices of the interpolants and their individual size via the given data $f_0, f_1, f_2, \ldots$. The set of the negative indices when not empty can be chosen arbitrarily and hence in this case there exist an infinite number of unitary interpolants. Paper [8] contains the matrix continuous analogs of the previous results. In paper [9] is considered a more general problem when the condition $|f_0| + |f_1| + \cdots < \infty$ is eliminated and the factorization is replaced by generalized factorization. The results of these papers intersect with results of F.E. Melik-Adamyan and M.G. Krein and is related to a paper of J. Ball.

In 1983 Harry and I organized a workshop on applications of linear operator theory to systems and networks in the Weizmann Institute; as we now call it, an IWOTA workshop. It was the second in this series and it was a satellite workshop just before the MTNS conference in Beersheva. The workshop attracted mathematicians and engineers. A volume of the proceedings was published in the OT series — OT12 [13].

6. Contractive interpolants and a maximum entropy principle

This section is based on two papers [10, 11]. In paper [10] are studied all $n \times n$ matrix contractive interpolants on the unit circle when the Fourier coefficients with positive indices are given. It turns out that for this problem a maximum entropy solution can be found with an appropriate entropy formula and inequality. The solution is obtained by a reduction to a generalized band problem. Paper [11] contains further generalizations of these results.

7. Nevanlinna-Pick problem and maximum entropy

Our last paper was written after a long break. Starting with 1984 I did not work regularly in the Weizmann Institute. The Institute was going through a financial crisis and all part time positions were disbanded. For a while, by inertia, I continued to visit the Institute and by the way continued to work with Harry. During these visits we wrote papers [10, 11]. Then the breaks became longer, but we again started to work systematically for a period in 1995. My dentist's office was located in Rehovot and for some part of 1995 I had to visit him at least once a week. Sometimes I would visit Harry in his office before the dental appointment, sometimes after. This time we worked on the Nevanlinna-Pick problem for matrix valued functions in the disc and we wrote paper [12]. In this paper we studied maximum entropy solutions and an extremal problem for the Pick matrix. A generalization for the half plane was also obtained.

8. This is not the end

Harry Dym is a very good friend and an excellent coauthor. We worked together for almost twenty years. Some of the periods were more intensive, some less. In parallel with this research, each of us was involved in many other research activities, so the joint work was never a burden. Our joint work influenced and enriched our individual research, as well as research with other colleagues, and led to cross-fertilization and influence.

Now, after going over all our papers as a reader, I look back with satisfaction and gratitude. This was a fruitful and enjoyful period which I hope will continue.

References

[1] H. Dym, I. Gohberg, Extensions of matrix valued functions with rational polynomial inverses. Integral Equations Operator Theory, Vol. 2 503–528 (1979).

[2] H. Dym, I. Gohberg, On an extension problem, generalized Fourier analysis and an entropy formula. Integral Equations Operator Theory, 3, 143–215 (1980).

[3] H. Dym, I. Gohberg, Extensions of band matrices with band inverses. Linear Algebra and its Applicatiobns, 36, 1–14 (1981).

[4] H. Dym, I. Gohberg, Extensions of kernels of Fredholm operators. Journal d'Analyse Mathématique, 42, 51–97 (1982/83).

[5] H. Dym, I. Gohberg, Extensions of triangular operators and matrix functions. Indiana University Mathematics Journal, Vol. 31, no. 4, 579–606 (1982).

[6] H. Dym, I. Gohberg, Extensions of matrix valued functions and block matrices. Indiana University Mathematics Journal, Vol. 31, no. 5, 733–765 (1982)

[7] H. Dym, I. Gohberg, Unitary interpolants, factorization indices and infinite Hankel block matrices. Journal of Functional Analysis, Vol. 54, no. 3, 229–289 (1983).

[8] H. Dym, I. Gohberg, Hankel integral operators and isometric interpolants on the line. Journal of Functional Analysis, Vol. 54, no. 3, 290–307 (1983).

[9] H. Dym, I. Gohberg, On unitary interpolants and Fredholm infinite block Toeplitz matrices. Integral Equations Operator Theory, 6, 863–878 (1983).

[10] H. Dym, I. Gohberg, A maximum entropy principle for contractive interpolants. Journal of Functional Analysis 65, 83–125 (1986).

[11] H. Dym, I. Gohberg, A new class of contractive interpolants and maximum entropy principles. Operator Theory: Advances and Applications, 29, 117–150, Birkhäuser, Basel (1988).

[12] H. Dym, I. Gohberg, On maximum entropy interpolants and maximum determinant completions of associated Pick matrices. Integral Equations Operator Theory, Vol. 23, 61–88 (1995).

[13] H. Dym, I. Gohberg (editors), Topics in Operator Theory, Systems and Networks. Workshop on Applications of Linear Operator Theory to Systems and Networks, Rehovot (Israel), June 13–16 (1983), Operator Theory: Advances and Applications, vol. 12, Birkhäuser Verlag (1984).

Israel Gohberg
School of Mathematical Sciences
Raymond and Beverly Sackler Faculty of Exact Sciences
Tel Aviv University
Ramat Aviv
69978 Tel Aviv, Israel

Operator Theory:
Advances and Applications, Vol. 134, 31–66

Methods of Kreĭn Space Operator Theory

James Rovnyak

Abstract. This paper is a survey of old and recent methods of Kreĭn space operator theory centering around Julia operators, extension problems for contraction operators, Hermitian kernels, and uniqueness questions. Examples related to coefficient problems for univalent functions are briefly discussed.

1. Introduction

The author was originally led to Kreĭn space operator theory by a problem of L. de Branges concerning the coefficients of univalent functions. The particular question was resolved in the negative, but the operator methods used to show this are related to other areas which remain currently active, such as the study of generalized Schur and Nevanlinna functions. The methods are of a general nature and based on familiar Hilbert space concepts, including contraction operators, their dilations, and reproducing kernel spaces. Today the Kreĭn space counterparts of many of these ideas are complete to a high degree. As always, there are difficulties and new issues in the indefinite theory. For example, it turns out that uniqueness questions play a more important role in the indefinite theory than in the definite case. In this paper we survey some old and recent results in these areas, with an aim to show that tools which have found wide applicability in Hilbert space problems are also available in Kreĭn space operator theory.

In outline, the contents are as follows:

§2. *Examples from function theory*
Generalizations of the Dirichlet space yield interesting examples, including contraction operators on indefinite inner product spaces defined by substitution by normalized univalent functions. Multiplication by the independent variable on similar spaces gives examples of indefinite two-isometries as studied by Agler, Richter, and others.

This article is an expanded version of the author's Toeplitz Lectures, which were given at Tel Aviv University in March 1999. Special thanks are given to Israel Gohberg for organizing the series of Toeplitz Lectures in commemoration of the impact of Otto Toeplitz, and also to D. Alpay and V. Vinnikov for their efficient work organizing the Toeplitz Lectures 1999 and Workshop in Operator Theory in honor of Harry Dym. The author is indebted to D. Alpay, V. Bolotnikov, T. Constantinescu, A. Dijksma, M.A. Dritschel, and H.S.V. de Snoo for many conversations on the material of this survey. The author is supported by NSF Grant DMS-9801016.

§3. *Definitions and basic notions*
Basic ideas are discussed here in order to make the paper self-contained.

§4. *Three useful tools of Kreĭn space operator theory*
Our goal is to adapt Hilbert space methods to Kreĭn space operators, but some elementary constructions break down when positivity is abandoned. Here we show that there are simple replacements in the indefinite theory. For example, the replacement for the Hilbert space construction of a nonnegative square root of a nonnegative operator is a factorization of any selfadjoint operator C on a Kreĭn space $\mathfrak{H}$ in the form $C = AA^*$ where $A \in \mathfrak{L}(\mathfrak{A}, \mathfrak{H})$ for some Kreĭn space $\mathfrak{A}$ and $\ker A = \{0\}$. Factorizations of this type are one of the main themes of this survey. Though elementary, they are extremely useful.

§5. *Julia operators and extension problems*
In §5.1 and §5.2, we discuss Julia operators and the most basic kinds of row, column, and matrix completions. In §5.3, we contrast several forms of commutant lifting in the indefinite setting.

§6. *Uniqueness questions*
A selfadjoint operator $C \in \mathfrak{L}(\mathfrak{H})$ is said to have the unique factorization property if the representation $C = AA^*$, $A \in \mathfrak{L}(\mathfrak{A}, \mathfrak{H})$, described above can only be changed by replacing the Kreĭn space $\mathfrak{A}$ by an isomorphic copy. We give necessary and sufficient conditions for uniqueness and identify situations in which uniqueness is automatic.

§7. *Kolmogorov decompositions of Hermitian kernels*
L. Schwarz introduced a number of elegant ideas into Kreĭn space operator theory in a 1964 paper, but they have become mainstream only more recently. Here we present the ideas in the form of the theory of Hermitian kernels. Particular cases include finite and infinite block operator matrices and reproducing kernels. A Hermitian kernel is a collection of Kreĭn space operators $K_{ij} = K_{ji}^* \in \mathfrak{L}(\mathfrak{H}_j, \mathfrak{H}_i)$, $i, j \in J$. A Kolmogorov decomposition is a representation in the form

$$K_{ij} = V_i^* V_j, \qquad i, j \in J,$$

where $V_j \in \mathfrak{L}(\mathfrak{H}_j, \mathfrak{K})$, $j \in J$, for some Kreĭn space $\mathfrak{K}$ such that $\mathfrak{K} = \bigvee_{j \in J} V_j \mathfrak{H}_j$. The general theory is concerned with criteria for existence and uniqueness. Our account is expository and follows recent work of Constantinescu and Gheondea.

§8. *Examples of Hermitian kernels*
The theory of Kolmogorov decompositions is illustrated with reproducing kernel spaces and holomorphic kernels. Another special case yields criteria for existence and uniqueness of completions of pre-Kreĭn spaces, which behave differently from pre-Hilbert spaces.

§9. *The contractive substitution property*
We return to the coefficient problems discussed in §2 and show, by numerical evidence, that the contractive substitution property, while not sufficient to characterize coefficients, nevertheless does an excellent job constraining low order coefficients. Some open questions are stated.

Related topics appear in the six lectures of Dritschel and Rovnyak [37]. Definitive accounts of the general theory of operators on indefinite inner product spaces, along with authoritative literature notes, are given in the books by Azizov and Iokhvidov [12], Bognár [14], and Iokhvidov, Kreĭn, and Langer [47]. Azizov, Ginzburg, and Langer [11] discuss M. G. Kreĭn's vision and contributions in this area. These and other sources should be consulted to see the great diversity of Kreĭn space operator theory and something of the many topics that are omitted here.

2. Examples from function theory

We give some examples which arise from coefficient problems for univalent functions. For the author personally, these examples were a compelling reason to undertake learning the indefinite theory. A deeper understanding of them is a long-range goal and challenge for the subject.

A holomorphic function $f(z)$ is **univalent** if it takes distinct values at distinct points. Coefficient problems play a central role in the theory of univalent functions which are defined on the unit disk $\mathbf{D} = \{z : |z| < 1\}$. A highlight of the theory is de Branges' proof [16] of the Bieberbach conjecture: *Let $f(z)$ be univalent on $\mathbf{D}$ and normalized so that $f(0) = 0$ and $f'(0) > 0$. If $f(z) = a_1 z + a_2 z^2 + \cdots$, then $|a_n| \leq n a_1$ for all $n \geq 2$.* The inequality, however, is satisfied by many functions which are not univalent. Ideally we would like to find stronger conditions which are more characteristic of univalent functions. We restrict attention to the subclass of functions which are bounded by one in $\mathbf{D}$. The following problem is classical.

Coefficient Interpolation Problem: *For any positive integer r, characterize all complex numbers $B_1, \ldots, B_r$ ($B_1 > 0$) such that there exists a univalent and normalized function $B(z)$ satisfying $|B(z)| \leq 1$ on $\mathbf{D}$ and such that $B(z) = B_1 z + \cdots + B_r z^r + \mathcal{O}(z^{r+1})$.*

Necessary conditions follow from a generalized form of the area theorem. Assume that such a function $B(z)$ exists for given numbers $B_1, \ldots, B_r$ ($B_1 > 0$). For any real number ν, consider an arbitrary generalized power series

$$h(z) = a_1 z^{\nu+1} + a_2 z^{\nu+2} + \cdots \tag{2.1}$$

with complex coefficients (constants terms, which arise when ν is a negative integer, are identified to zero). Define $h(B(z)) = b_1 z^{\nu+1} + b_2 z^{\nu+2} + \cdots$ by formal substitution. Then

$$\sum_{n=1}^{r} (\nu + n)|b_n|^2 \leq \sum_{n=1}^{r} (\nu + n)|a_n|^2.$$

Equivalently,

$$\langle h(B(z)), h(B(z)) \rangle_{\mathfrak{D}_r^\nu} \leq \langle h(z), h(z) \rangle_{\mathfrak{D}_r^\nu}, \tag{2.2}$$

where $\mathfrak{D}_r^\nu$ is the linear space of series (2.1) in the inner product

$$\langle h(z), h(z)\rangle_{\mathfrak{D}_r^\nu} = \sum_{n=1}^{r} (\nu+n)|a_n|^2.$$

The inequality (2.2) is proved by de Branges [17] when $\nu \geq -r-1$, and the restriction on ν is removed by Li and Rovnyak [51]; for a proof, see [65, Section 7.5]. Nikolskii and Vasyunin [59, 60] give another view of these inequalities and explain their connection with subordination (see Section P45, p. 1202, in the English translation of [60]); see also Ghosechowdhury [43, 44] and Rovnyak [67]. The conditions (2.2) depend only on $B_1, \ldots, B_r$ and are thus necessary conditions on these numbers for the existence of an interpolating function $B(z)$. It is natural to ask if the necessary conditions are sufficient:

Problem (de Branges [17, 19]). *Let $B_1, \ldots, B_r$ be complex numbers with $B_1 > 0$ such that* (2.2) *holds for all real numbers ν and all generalized power series* (2.1). *Does it follow that $B(z) = B_1 z + \cdots + B_r z^r + \mathcal{O}(z^{r+1})$ where $B(z)$ is univalent and $|B(z)| \leq 1$ on* **D**?

The simple answer is negative (see §9).

The main point here, however, is that we obtain a large class of examples of contraction operators. Namely, by (2.2) the operator

$$T : h(z) \to h(B(z)) \tag{2.4}$$

is a contraction on the space $\mathfrak{D}_r^\nu$ for any positive integer r, any real number ν, and any function $B(z)$ which is univalent, normalized, and bounded by one in **D**. The space $\mathfrak{D}_r^\nu$ is indefinite when $\nu < -1$. In the same way, (2.4) acts as a contraction in the infinite-dimensional space $\mathfrak{D}^\nu$ of series (2.1) such that $\sum_{n=1}^{\infty}(\nu+n)|a_n|^2 < \infty$ in the inner product

$$\langle h(z), h(z)\rangle_{\mathfrak{D}^\nu} = \sum_{n=1}^{\infty} (\nu+n)|a_n|^2,$$

and this inner product is indefinite when $\nu < -1$. Another interesting example in $\mathfrak{D}^\nu$ is multiplication by z:

$$S\colon f(z) \to zf(z).$$

In the classical case ($\nu = 0$), this is the Dirichlet shift. In general, S is a two-isometry in the sense that $S^{*2}S^2 - 2S^*S + 1 = 0$, or in terms of inner products,

$$\langle z^2 f(z), z^2 f(z)\rangle_{\mathfrak{D}^\nu} - 2\langle zf(z), zf(z)\rangle_{\mathfrak{D}^\nu} + \langle f(z), f(z)\rangle_{\mathfrak{D}^\nu} = 0$$

for all $f(z)$ in $\mathfrak{D}^\nu$. Two-isometries and more general operators on Hilbert spaces are studied by Agler and Stankus [1]. A two-isometry is called analytic if the intersection of the ranges of its powers is zero. Richter [63] constructed a model theory for cyclic analytic two-isometries on a Hilbert space, the Dirichlet shift being the motivating example [62, 64]. The beginnings of an indefinite theory have been made by Chris Hellings [46]. See also McCullough and Rodman [54, 55], who earlier proposed to extend Agler's ideas into the indefinite domain.

Such examples suggest the need for an approach that emphasizes the analogies with the Hilbert space case, and our purpose here is to outline such a viewpoint.

3. Definitions and basic notions

Inner products are assumed to be linear and symmetric. The antispace of an inner product space $(\mathfrak{H}, \langle\cdot,\cdot\rangle)$ is $(\mathfrak{H}, -\langle\cdot,\cdot\rangle)$.

As we use the term, a **Kreĭn space** is an inner product space which is expressible as an orthogonal direct sum $\mathfrak{H} = \mathfrak{H}_+ \oplus \mathfrak{H}_-$ of a Hilbert space $\mathfrak{H}_+$ and the antispace $\mathfrak{H}_-$ of a Hilbert space (for simplicity, Hilbert spaces are assumed to be separable). Any such representation is a **fundamental decomposition**. The induced Hilbert space topology is the **strong topology** of $\mathfrak{H}$. The dimensions of $\mathfrak{H}_\pm$ are the **indices** of $\mathfrak{H}$. A Kreĭn space is also called a **Pontryagin space** if it has finite negative index. These definitions do not depend on the choice of fundamental decomposition. When nothing is said, underlying spaces are assumed to be Kreĭn spaces (which might be Pontryagin spaces or finite-dimensional).

Spaces $\mathfrak{L}(\mathfrak{H})$ and $\mathfrak{L}(\mathfrak{H}, \mathfrak{K})$ of continuous operators and adjoint operators are defined for Kreĭn spaces in the same way as for Hilbert spaces. Thus if $A \in \mathfrak{L}(\mathfrak{H}, \mathfrak{K})$, then $A^* \in \mathfrak{L}(\mathfrak{K}, \mathfrak{H})$ and $\langle Af, g\rangle = \langle f, A^*g\rangle$ for all f in $\mathfrak{H}$ and g in $\mathfrak{K}$. An operator $A \in \mathfrak{L}(\mathfrak{H})$ is

selfadjoint if $A^* = A$,
a **projection** if A is selfadjoint and $A^2 = A$, and
nonnegative if $\langle Af, f\rangle \geq 0$ for every $f \in \mathfrak{H}$.

If $A \in \mathfrak{L}(\mathfrak{H})$ is selfadjoint, let $\mathrm{ind}_+ A$ ($\mathrm{ind}_- A$) be the supremum of all r such that there exists an r-dimensonal subspace of $\mathfrak{H}$ which is a Hilbert space (antispace of a Hilbert space) in the inner product $\langle f, g\rangle_A = \langle Af, g\rangle$, $f, g \in \mathfrak{H}$. An operator $B \in \mathfrak{L}(\mathfrak{H}, \mathfrak{K})$ is

isometric if $B^*B = 1_{\mathfrak{H}}$,
partially isometric if $BB^*B = B$,
unitary if both B and B^* are isometric,
a **contraction** if $B^*B \leq 1_{\mathfrak{H}}$, and
a **bicontraction** if both B and B^* are contractions.

An **isomorphism** of inner product spaces is a one-to-one and onto linear mapping which preserves inner products. As in the Hilbert space case, the class of isomorphisms between two Kreĭn spaces $\mathfrak{H}$ and $\mathfrak{K}$ coincides with the set of unitary operators between the spaces.

Orthogonality is defined for Kreĭn spaces as for Hilbert spaces. The relation $\mathfrak{H} = \mathfrak{M} \oplus \mathfrak{M}^\perp$ is not always true for all closed subspaces $\mathfrak{M}$ of a Kreĭn space $\mathfrak{H}$, however. It is true for an important subclass of subspaces. A linear subspace $\mathfrak{M}$ of a Kreĭn space $\mathfrak{H}$ is a **Kreĭn subspace**, or a **regular subspace**, if $\mathfrak{M}$ is closed and a

Kreĭn space in the inner product of $\mathfrak{H}$. If $\mathfrak{M}$ is a linear subspace of $\mathfrak{H}$, the following assertions are equivalent:

(1) $\mathfrak{M}$ is a Kreĭn subspace;
(2) $\mathfrak{H} = \mathfrak{M} \oplus \mathfrak{M}^{\perp}$;
(3) $\mathfrak{M} = \operatorname{ran} P$, where $P \in \mathfrak{L}(\mathfrak{H})$ is a projection operator.

In this case, restriction of the strong topology of $\mathfrak{H}$ to $\mathfrak{M}$ coincides with the strong topology of $\mathfrak{M}$ as a Kreĭn space. For details and other basic notions, see [12, 14, 36, 47].

4. Three useful tools of Kreĭn space operator theory

Kreĭn space operator theory is much like the Hilbert space special case despite failure of some of the most basic notions in the indefinite situation. The explanation is that there are effective substitutes for the missing Hilbert space results.

Tool #1: a factorization theorem for selfadjoint operators.

One of the cornerstones of Hilbert space operator theory is that every nonnegative operator has a nonnegative square root. The Kreĭn space counterpart is a factorization theorem for any selfadjoint operator. The result is old, but its systematic use is more recent [26, 37, 36].

Theorem 4.1. *Every selfadjoint operator $C \in \mathfrak{L}(\mathfrak{H})$, $\mathfrak{H}$ a Kreĭn space, can be written $C = AA^*$ where $A \in \mathfrak{L}(\mathfrak{A}, \mathfrak{H})$ for some Kreĭn space $\mathfrak{A}$ and* $\ker A = \{0\}$.

The first step in the proof, reduction to the Hilbert space case, is worth separate notice:

> *Every selfadjoint operator on a Kreĭn space is congruent to a selfadjoint operator on a Hilbert space.*

That is, if $\mathfrak{H}$ is a Kreĭn space and $C \in \mathfrak{L}(\mathfrak{H})$ is a selfadjoint operator, there is a Hilbert space $\mathfrak{K}$, a selfadjoint operator $B \in \mathfrak{L}(\mathfrak{K})$, and an invertible operator $X \in \mathfrak{L}(\mathfrak{H}, \mathfrak{K})$ such that

$$C = X^*BX.$$

In fact, let X be any invertible operator from $\mathfrak{H}$ onto any Hilbert space $\mathfrak{K}$, and take $B = X^{*-1}CX^{-1}$.

Proof of Theorem 4.1. It is sufficient to prove the theorem when $\mathfrak{H}$ is a Hilbert space. In this case, we can decompose $\mathfrak{H}$ into spectral subspaces for C for the sets $(0, \infty)$, $\{0\}$, $(-\infty, 0)$, say $\mathfrak{H} = \mathfrak{H}_+ \oplus \mathfrak{H}_0 \oplus \mathfrak{H}_-$. Define $\mathfrak{A} = \mathfrak{H}_+ \oplus \mathfrak{H}_-$ in the inner product

$$\langle f, g \rangle_{\mathfrak{A}} = \pm \langle f, g \rangle_{\mathfrak{H}}, \qquad f, g \in \mathfrak{H}_{\pm}.$$

We easily check that the operator A defined by $Af = |C|^{1/2} f$, $f \in \mathfrak{A}$, has the required properties. □

Tool #2: extension theorems for densely defined operators.

A different factorization occurs in Hilbert space operator theory. In a typical situation, we are given Hilbert space operators $A \in \mathfrak{L}(\mathfrak{H}, \mathfrak{A})$ and $B \in \mathfrak{L}(\mathfrak{H}, \mathfrak{B})$ with $B^*B \leq A^*A$. If A has dense range, then the partially defined operator

$$C_0 : Af \to Bf, \qquad f \in \mathfrak{H},$$

has a contractive (hence continuous) extension $C \in \mathfrak{L}(\mathfrak{A}, \mathfrak{B})$ such that $B = CA$. When the underlying spaces are Kreĭn spaces, C_0 may not be well defined (that is, $Af_1 = Af_2$ and $Bf_1 \neq Bf_2$ for some $f_1, f_2 \in \mathfrak{H}$), and even if it is it may not have a continuous extension. See [6, p. 429] for examples.

What is needed is a means to define continuous contraction operators by specifying their action on dense sets. An index condition resolves the difficulties. A **linear relation** from $\mathfrak{H}$ to $\mathfrak{K}$ is a linear subspace $\mathbf{R}$ of $\mathfrak{H} \times \mathfrak{K}$. The domain of $\mathbf{R}$ is the set of all first elements f of the pairs (f, g) in $\mathbf{R}$.

Theorem 4.2. *Let $\mathfrak{H}$ and $\mathfrak{K}$ be Pontryagin spaces such that* $\operatorname{ind}_- \mathfrak{H} = \operatorname{ind}_- \mathfrak{K}$. *Let* $\mathbf{R}$ *be a linear relation such that*

(1) $\mathbf{R}$ *has dense domain,*
(2) $\langle g, g \rangle_{\mathfrak{K}} \leq \langle f, f \rangle_{\mathfrak{H}}$ *for all* $(f, g) \in \mathbf{R}$.

Then the closure of $\mathbf{R}$ *is the graph of a contraction* $C \in \mathfrak{L}(\mathfrak{H}, \mathfrak{K})$.

See [5, 6] for two different proofs ofTheorem 4.2. The known Kreĭn space generalizations of Theorem 4.2 require strong hypotheses which are difficult to verify in applications (Shmul'yan [70], Dritschel and Rovnyak [37, Theorem 1.4.4] and [36, Supplement]). An exception here is the following nice result which is given in Constantinescu and Gheondea [25, Lemma 2.3].

Theorem 4.3. *Let $\mathfrak{H}$ and $\mathfrak{K}$ be Kreĭn spaces. Let* $\mathbf{R}$ *be a linear relation such that*

(1) $\mathbf{R}$ *has dense domain and dense range,*
(2) $\langle g, g \rangle_{\mathfrak{K}} = \langle f, f \rangle_{\mathfrak{H}}$ *for all* $(f, g) \in \mathbf{R}$,
(3) *the domain of* $\mathbf{R}$ *contains one of the subspaces* $\mathfrak{H}_\pm$ *in some fundamental decomposition* $\mathfrak{H} = \mathfrak{H}_+ \oplus \mathfrak{H}_-$.

Then the closure of $\mathbf{R}$ *is the graph of a unitary operator* $U \in \mathfrak{L}(\mathfrak{H}, \mathfrak{K})$.

A finite-dimensional example in [6, p. 429] shows that Theorem 4.2 is not valid if $\operatorname{ind}_- \mathfrak{H} \neq \operatorname{ind}_- \mathfrak{K}$. The same example shows that the conclusion of Theorem 4.3 can fail if all conditions are met except the range of $\mathbf{R}$ is not dense.

Typical applications of Theorems 4.2 and 4.3 arise from inequalities $B^*B \leq A^*A$, where $A \in \mathfrak{L}(\mathfrak{H}, \mathfrak{A})$ and $B \in \mathfrak{L}(\mathfrak{H}, \mathfrak{B})$ are Kreĭn space operators. Under suitable conditions, the linear relation

$$\mathbf{R} = \{(Af, Bf) : f \in \mathfrak{H}\}.$$

satisfies the hypotheses of the theorems. Then we obtain a factorization $B = CA$ with $C \in \mathfrak{L}(\mathfrak{A}, \mathfrak{B})$ a contraction operator or unitary operator, as in the Hilbert space case.

Tool #3: continuous isometries and partial isometries.

Recall that a partial isometry is defined as a Kreĭn space operator $A \in \mathfrak{L}(\mathfrak{H}, \mathfrak{K})$ such that $AA^*A = A$. Such operators have properties much the same as in the Hilbert space case.

Theorem 4.4. *If $A \in \mathfrak{L}(\mathfrak{H}, \mathfrak{K})$, $\mathfrak{H}$ and $\mathfrak{K}$ Kreĭn spaces, the following assertions are equivalent:*

(1) *A is a partial isometry;*
(2) *A^*A is a projection operator and* $\ker A^*A = \ker A$*;*
(3) *AA^* is a projection operator and* $\ker AA^* = \ker A^*$*;*
(4) *there exist Kreĭn subspaces $\mathfrak{M}$ of $\mathfrak{H}$ and $\mathfrak{N}$ of $\mathfrak{K}$ such that A maps $\mathfrak{M}$ in a one-to-one way onto $\mathfrak{N}$ with $\langle Af, Ag\rangle_{\mathfrak{K}} = \langle f, g\rangle_{\mathfrak{H}}$ for all $f, g \in \mathfrak{M}$, and $Af = 0$ for all $f \in \mathfrak{M}^{\perp}$.*

In this case, A^*A and AA^* are the projections onto $\mathfrak{M}$ and $\mathfrak{N}$. If, in fact, A is an isometry, then in addition

(5) *A maps closed subspaces of $\mathfrak{H}$ onto closed subspaces of $\mathfrak{K}$;*
(6) *A maps Kreĭn subspaces of $\mathfrak{H}$ onto Kreĭn subspaces of $\mathfrak{K}$.*

In particular, the range of an isometry $A \in \mathfrak{L}(\mathfrak{H}, \mathfrak{K})$ is a Kreĭn subspace of $\mathfrak{K}$.

The conditions on kernels in parts (2) and (3) of Theorem 4.4 do not appear in the Hilbert space case because they hold automatically when $\mathfrak{H}$ and $\mathfrak{K}$ are Hilbert spaces. For proofs of the assertions in Theorem 4.4, see the *Supplement and errata* cited in [36, pp. 156–57].

Theorem 4.4 plays a greater role in the indefinite theory than in the special case of Hilbert spaces. It can only be appreciated in the light of pathological examples of "isometries" on Kreĭn spaces: if $\mathfrak{H}$ is an infinite-dimensional Hilbert space and $\mathfrak{K}$ is an infinite dimensional Pontryagin space with $\operatorname{ind}_- \mathfrak{K} = 1$, there exists an everywhere defined linear transformation V on $\mathfrak{H}$ into $\mathfrak{K}$ such that $\langle Vf, Vg\rangle_{\mathfrak{K}} = \langle f, g\rangle_{\mathfrak{H}}$ for all f and g in $\mathfrak{H}$, yet V is not continuous with respect to the strong topologies of $\mathfrak{H}$ and $\mathfrak{K}$ (for example, see [36, Supplement]). Obviously all manner of bad behavior is to be expected in such a situation, and the point of Theorem 4.4 is that order is restored with the hypothesis of continuity. While our definition of an "isometry" presumes continuity, this practice is not universal, and in other sources the meaning of the term should be verified.

5. Julia operators and extension problems

5.1 Defect and Julia operators

Much of the theory of contraction operators on Hilbert spaces in Sz.-Nagy and Foias [72] carries over to the indefinite setting. Dilation properties and model theory are discussed in Davis [28], Davis and Foias [29] and McEnnis [56, 57, 58]. We focus on more recent developments in the Kreĭn space theory that include notions of defect and Julia operators, matrix extension theorems, and the commutant lifting theorem. In the definite case, the history of results in this area is long and complex and closely connected with interpolation theory; for example, see Foias and Frazho [40]; a recent sequel to this standard source is given in Foias, Frazho, Kaashoek, and Gohberg [41]. The indefinite theory for these areas originates with Constantinescu and Gheondea [22, 24] and Dritschel [32].

Defect and Julia operators play an even greater role in Kreĭn space operator theory than in the Hilbert space case. The first constructions are due to Arsene, Constantinescu, and Gheondea [10]. Let $T \in \mathfrak{L}(\mathfrak{H},\mathfrak{K})$, where $\mathfrak{H}$ and $\mathfrak{K}$ are Kreĭn spaces. By a **defect operator** for T we mean any operator $\tilde{D} \in \mathfrak{L}(\tilde{\mathfrak{D}},\mathfrak{H})$, where $\tilde{\mathfrak{D}}$ is a Kreĭn space, such that $\ker \tilde{D} = \{0\}$ and the operator

$$V = \begin{pmatrix} T \\ \tilde{D}^* \end{pmatrix} \in \mathfrak{L}(\mathfrak{H}, \mathfrak{K}\oplus\tilde{\mathfrak{D}}) \tag{5.1}$$

is an isometry, that is, $T^*T + \tilde{D}\tilde{D}^* = 1$. A **Julia operator** for T is any unitary operator

$$U = \begin{pmatrix} T & D \\ \tilde{D}^* & -L^* \end{pmatrix} \in \mathfrak{L}(\mathfrak{H}\oplus\mathfrak{D}, \mathfrak{K}\oplus\tilde{\mathfrak{D}}), \tag{5.2}$$

where $\mathfrak{D}$ and $\tilde{\mathfrak{D}}$ are Kreĭn spaces, such that the operators $D \in \mathfrak{L}(\mathfrak{D},\mathfrak{H})$ and $\tilde{D} \in \mathfrak{L}(\tilde{\mathfrak{D}},\mathfrak{H})$ have zero kernels. Julia operators are also called **elementary rotations** in the literature.

The preceding definitions of defect and Julia operators apply to any operator $T \in \mathfrak{L}(\mathfrak{H},\mathfrak{K})$, and they do not presume that T is a contraction operator. So even when $\mathfrak{H}$ and $\mathfrak{K}$ are Hilbert spaces, the definitions are more general than the standard definitions which are given in the Hilbert space case.

Theorem 5.1. *Let $T \in \mathfrak{L}(\mathfrak{H},\mathfrak{K})$, where $\mathfrak{H}$ and $\mathfrak{K}$ are Kreĭn spaces.*
(1) *A defect operator $\tilde{D} \in \mathfrak{L}(\tilde{\mathfrak{D}},\mathfrak{H})$ for T exists, and for any such operator*

$$\operatorname{ind}_\pm \tilde{\mathfrak{D}} = \operatorname{ind}_\pm (1 - T^*T).$$

(2) *A Julia operator $U \in \mathfrak{L}(\mathfrak{H}\oplus\mathfrak{D}, \mathfrak{K}\oplus\tilde{\mathfrak{D}})$ for T exists, and for any such operator*

$$\operatorname{ind}_\pm \mathfrak{D} = \operatorname{ind}_\pm (1 - TT^*) \qquad \text{and} \qquad \operatorname{ind}_\pm \tilde{\mathfrak{D}} = \operatorname{ind}_\pm (1 - T^*T).$$

Proof. We obtain (1) by applying Theorem 4.1 to $C = 1 - T^*T$. To prove (2), apply Theorem 4.1 a second time to $C = 1 - VV^*$, where V is given by (5.1). For details, see Dritschel and Rovnyak [36, Theorem 2.3]. □

We give an elementary illustration how Theorem 5.1, combined with the good behavior of isometric and unitary operators, can be used to obtain information about general operators. The result itself is old and has a simple direct proof [10].

Theorem 5.2. *If $T \in \mathfrak{L}(\mathfrak{H},\mathfrak{K})$ for any Kreĭn spaces $\mathfrak{H}$ and $\mathfrak{K}$, then*

$$\operatorname{ind}_\pm \mathfrak{H} + \operatorname{ind}_\pm (1 - TT^*) = \operatorname{ind}_\pm \mathfrak{K} + \operatorname{ind}_\pm (1 - T^*T).$$

In particular, if $\operatorname{ind}_- \mathfrak{H} = \operatorname{ind}_- \mathfrak{K} < \infty$, *then* $T^*T \le 1$ *implies* $TT^* \le 1$.

Proof. Choose a Julia operator (5.2) for T. By the unitarity of U and Theorem 5.1(2), $\operatorname{ind}_{\pm} \mathfrak{H} + \operatorname{ind}_{\pm}(1 - TT^*) = \operatorname{ind}_{\pm} \mathfrak{H} + \operatorname{ind}_{\pm} \mathfrak{D} = \operatorname{ind}_{\pm} \mathfrak{K} + \operatorname{ind}_{\pm} \tilde{\mathfrak{D}} = \operatorname{ind}_{\pm} \mathfrak{K} + \operatorname{ind}_{\pm}(1 - T^*T)$. □

Another basic problem is to describe all contractive row, column, and matrix extensions

$$\begin{pmatrix} T & F \end{pmatrix} \in \mathfrak{L}(\mathfrak{H} \oplus \mathfrak{F}, \mathfrak{K}),$$
$$\begin{pmatrix} T \\ G \end{pmatrix} \in \mathfrak{L}(\mathfrak{H}, \mathfrak{K} \oplus \mathfrak{G}),$$
$$\begin{pmatrix} T & F \\ G & H \end{pmatrix} \in \mathfrak{L}(\mathfrak{H} \oplus \mathfrak{F}, \mathfrak{K} \oplus \mathfrak{G}),$$

of a given contraction operator $T \in \mathfrak{L}(\mathfrak{H}, \mathfrak{K})$, where $\mathfrak{H}$ and $\mathfrak{K}$ are Kreĭn spaces. The problem has several variants, such as dropping the hypothesis that T is a contraction. We can alternatively consider operators T such that $\operatorname{ind}_{-}(1 - T^*T) < \infty$ and ask for contractive extensions or extensions which also satisfy index conditions.

5.2 Basic extension theorems

Let $T \in \mathfrak{L}(\mathfrak{H}, \mathfrak{K})$, where $\mathfrak{H}$ and $\mathfrak{K}$ are Kreĭn spaces. Choose a Julia operator

$$\begin{pmatrix} T & D_T \\ \tilde{D}_T^* & -L_T^* \end{pmatrix} \in \mathfrak{L}(\mathfrak{H} \oplus \mathfrak{D}_T, \mathfrak{K} \oplus \tilde{\mathfrak{D}}_T) \tag{5.4}$$

for T. This is, of course, a particular extension of T. When $\mathfrak{H}, \mathfrak{K}, \mathfrak{F}, \mathfrak{G}$ are Hilbert spaces and T is a contraction, it is a well-known result that all contractive row, column, and matrix extensions are given by

$$\begin{pmatrix} T & D_T X \end{pmatrix} \in \mathfrak{L}(\mathfrak{H} \oplus \mathfrak{F}, \mathfrak{K}), \tag{5.5}$$
$$\begin{pmatrix} T \\ Y^* \tilde{D}_T^* \end{pmatrix} \in \mathfrak{L}(\mathfrak{H}, \mathfrak{K} \oplus \mathfrak{G}), \tag{5.6}$$

and

$$\begin{pmatrix} T & D_T X \\ Y^* \tilde{D}_T^* & -Y^* L_T^* X + \tilde{D}_Y Z \tilde{D}_X^* \end{pmatrix} \in \mathfrak{L}(\mathfrak{H} \oplus \mathfrak{F}, \mathfrak{K} \oplus \mathfrak{G}), \tag{5.7}$$

where X, Y, Z are contraction operators on appropriate spaces as required to make the formulas meaningful and $\tilde{D}_X$ and $\tilde{D}_Y$ are defect operators for X and Y.

The next result describes the situation when T is a contraction.

Theorem 5.3. *Assume that $T \in \mathfrak{L}(\mathfrak{H}, \mathfrak{K})$ is a contraction, $\mathfrak{H}, \mathfrak{K}, \mathfrak{F}$ are Kreĭn spaces, and $\mathfrak{G}$ is a Hilbert space. Then all contractive row, column, and matrix extensions of T are given by* (5.5), (5.6), *and* (5.7) *again where X, Y, Z are contraction operators on appropriate spaces as required to make the formulas meaningful and $\tilde{D}_X$ and $\tilde{D}_Y$ are defect operators for X and Y.*

The asymmetry in Theorem 5.3 is due to the fact that the adjoint of a contraction operator on Kreĭn spaces is not necessarily a contraction. Thus, for example, the row extension theorem cannot be deduced by applying the column extension to T^*; the row and column extensions need separate proofs. When $\mathfrak{G}$ is a Kreĭn space, the conclusions can fail [36, p. 172]. Nevertheless, a more general result holds and provides another illustration of the role played by index conditions in Kreĭn space operator theory.

When T is not necessarily a contraction, or $\mathfrak{G}$ is not a Hilbert space, similar conclusions hold but with other hypotheses in the form of index conditions.

Theorem 5.4 (Row extensions). *Assume that $\mathfrak{H}, \mathfrak{K}, \mathfrak{F}$ are Kreĭn spaces. Let $T \in \mathfrak{L}(\mathfrak{H}, \mathfrak{K})$ (not necessarily a contraction), and let $D_T \in \mathfrak{L}(\mathfrak{D}_T, \mathfrak{H})$ be a defect operator for T^*. Let $R = \begin{pmatrix} T & F \end{pmatrix} \in \mathfrak{L}(\mathfrak{H} \oplus \mathfrak{F}, \mathfrak{K})$ be a row extension of T satisfying at least one of the conditions*

$$\operatorname{ind}_- (1 - RR^*) + \operatorname{ind}_- \mathfrak{F} = \operatorname{ind}_- (1 - TT^*) < \infty, \tag{5.8}$$

$$\operatorname{ind}_- (1 - R^*R) = \operatorname{ind}_- (1 - T^*T) < \infty. \tag{5.9}$$

Then R has the form (5.5), *where $X \in \mathfrak{L}(\mathfrak{F}, \mathfrak{D}_T)$ is a contraction. Conversely, every such operator* (5.5) *satisfies both of the equalities in* (5.8) *and* (5.9) (*with possibly infinite values*).

Theorem 5.5 (Column extensions). *Let $\mathfrak{H}, \mathfrak{K}, \mathfrak{G}$ be Kreĭn spaces. Assume that $T \in \mathfrak{L}(\mathfrak{H}, \mathfrak{K})$ and that $\tilde{D}_T \in \mathfrak{L}(\tilde{\mathfrak{D}}_T, \mathfrak{H})$ is a defect operator for T. Let*

$$C = \begin{pmatrix} T \\ G \end{pmatrix} \in \mathfrak{L}(\mathfrak{H}, \mathfrak{K} \oplus \mathfrak{G})$$

be a column extension of T satisfying at least one of the conditions

$$\operatorname{ind}_- (1 - C^*C) + \operatorname{ind}_- \mathfrak{G} = \operatorname{ind}_- (1 - T^*T) < \infty, \tag{5.10}$$

$$\operatorname{ind}_- (1 - CC^*) = \operatorname{ind}_- (1 - TT^*) < \infty. \tag{5.11}$$

Then C has the form (5.6), *where $Y \in \mathfrak{L}(\mathfrak{G}, \tilde{\mathfrak{D}}_T)$ is a contraction. Conversely, every such operator* (5.6) *satisfies both of the equalities in* (5.10) *and* (5.11) (*with possibly infinite values*).

A similar result holds for matrix extensions of the form (5.7). Dritschel [33] has given a beautiful method of proof of such theorems. The results are first proved in the special case when the given operators are isometries; in this simple case we are able to use what are essentially Hilbert space methods, and these methods work for Kreĭn spaces because by Theorem 4.4 the properties of continuous partial isometries on Kreĭn spaces are much the same as in the Hilbert space case. The second step is to reduce the general results to the case of isometries by means of extensions using defect and Julia operators. It is only necessary to prove Theorems 5.4 and 5.5 and the counterpart for (5.7), as these imply Theorem 5.3; for example, the row and column statements in Theorem 5.3 follow when the equalities in (5.9) and (5.10) hold with the value zero. Full details are given in [36, Lecture 3].

5.3 Commutant lifting

Commutant lifting provides operator extensions with additional properties. Already in the definite case, the commutant lifting theorem has a number of formulations, but the different versions have essentially the same content. In the case of Kreĭn space operators, there are several natural extensions of the commutant lifting theorem. While obviously related, however, they are not easily compared. A survey of this area by itself would be a sizable undertaking, and we limit this discussion to several results and some citations to other sources.

One result simply says that the theorem of Sz.-Nagy and Foias [72] remains true if Hilbert spaces are replaced by Kreĭn spaces. If $\mathfrak{K}$ is a Kreĭn space with Kreĭn subspace $\mathfrak{H}$, let $P_{\mathfrak{H}}$ be the projection operator on $\mathfrak{K}$ with range $\mathfrak{H}$. A **minimal isometric dilation** of an operator $A \in \mathfrak{L}(\mathfrak{H})$, $\mathfrak{H}$ a Kreĭn space, is an isometric operator $U \in \mathfrak{L}(\mathfrak{K})$, where $\mathfrak{K}$ is a Kreĭn space containing $\mathfrak{H}$ as a Kreĭn subspace, such that $A^n = P_{\mathfrak{H}} U^n|_{\mathfrak{H}}$ for all $n = 1, 2, \ldots$, and $\bigvee_{n=0}^{\infty} U^n \mathfrak{H} = \mathfrak{K}$. A minimal isometric dilation exists for any Kreĭn space operator $A \in \mathfrak{L}(\mathfrak{H})$; if A is a contraction, it is essentially unique as in the Hilbert space case [37].

Commutant Lifting Theorem I (Dritschel [32]). *Let $\mathfrak{H}_1$ and $\mathfrak{H}_2$ be Kreĭn spaces, and let $T \in \mathfrak{L}(\mathfrak{H}_1, \mathfrak{H}_2)$ be a contraction operator such that $TA_1 = A_2 T$ for some contraction operators $A_1 \in \mathfrak{L}(\mathfrak{H}_1)$ and $A_2 \in \mathfrak{L}(\mathfrak{H}_2)$. Let $U_1 \in \mathfrak{L}(\mathfrak{K}_1)$ and $U_2 \in \mathfrak{L}(\mathfrak{K}_2)$ be minimal isometric dilations of A_1 and A_2. Then there is a contraction $\hat{T} \in \mathfrak{L}(\mathfrak{K}_1, \mathfrak{K}_2)$ such that $U_2 \hat{T} = \hat{T} U_1$ and $P_{\mathfrak{H}_2} \hat{T} = T P_{\mathfrak{H}_1}$.*

The proof is an application of Theorems 5.4 and 5.5. It is simplified in Dritschel and Rovnyak [37]. For different proofs, see Dijksma, Dritschel, Marcantognini, and de Snoo [30], and Marcantognini [52]. A module formulation has been given by Dritschel [35]. Earlier results in the same direction were obtained by Constantinescu and Gheondea; see [22, 24].

Another version of the commutant lifting theorem also starts with the Sz.-Nagy and Foias theorem and weakens the hypothesis that the intertwining operator T is a contraction. In its original form, the underlying spaces are again Hilbert spaces.

Commutant Lifting Theorem II (Ball and Helton [13]). *Let $\mathfrak{H}_1$ and $\mathfrak{H}_2$ be Hilbert spaces, $A_1 \in \mathfrak{L}(\mathfrak{H}_1)$ and $A_2 \in \mathfrak{L}(\mathfrak{H}_2)$ contractions with minimal isometric dilations $U_1 \in \mathfrak{L}(\mathfrak{K}_1)$ and $U_2 \in \mathfrak{L}(\mathfrak{K}_2)$. Assume that $T \in \mathfrak{L}(\mathfrak{H}_1, \mathfrak{H}_2)$ is a contraction operator such that*

$$\operatorname{ind}_-(1 - T^*T) \le \kappa$$

for some nonnegative integer κ. Then there is a U_1-invariant subspace $\hat{\mathfrak{K}}_1$ of $\mathfrak{K}_1$ of codimension at most κ and a contraction operator $\hat{T}\colon \hat{\mathfrak{K}}_1 \to \mathfrak{K}_2$ such that $U_2 \hat{T} = \hat{T} U_1|_{\hat{\mathfrak{K}}_1}$ and $P_{\mathfrak{H}_2} \hat{T} = T P_{\mathfrak{H}_1}|_{\hat{\mathfrak{K}}_1}$.

Independently, Gheondea [42] and Arocena, Azizov, Dijksma, and Marcantognini [7, 8] have extended the Ball and Helton theorem to allow $\mathfrak{H}_1$ and $\mathfrak{H}_2$ to be Kreĭn spaces. In the generalization, the subspace $\hat{\mathfrak{K}}_1$ is not necessarily a Kreĭn subspace, but with a natural interpretation of contraction operator the statement is otherwise identical. The formulation in [8] is more general in another direction, namely, a broader notion of isometric dilation is adopted.

Canonical models also provide a setting for commutant lifting [72]. In Alpay [3] and de Branges [20], generalizations of the commutant lifting theorem in canonical model spaces are constructed. Concerning canonical models in Kreĭn spaces, see also Yang [74].

6. Uniqueness questions

6.1 General results

While factorizations as in Theorem 4.1 always exist, they are not in general unique even up to appropriate notions of isomorphism. Indices of the underlying Kreĭn space, at least, are unique [37, Theorem 1.2.1]:

Theorem 6.1. *Let $C \in \mathfrak{L}(\mathfrak{H})$ be a selfadjoint operator on a Kreĭn space $\mathfrak{H}$. In any way, factor C in the form $C = AA^*$ where $A \in \mathfrak{L}(\mathfrak{A}, \mathfrak{H})$ for some Kreĭn space $\mathfrak{A}$ and $\ker A = \{0\}$ as in Theorem 4.1. Then*

$$\operatorname{ind}_\pm \mathfrak{A} = \operatorname{ind}_\pm C.$$

In particular, the indices $\operatorname{ind}_\pm \mathfrak{A}$ do not depend on the choice of factorization.

We turn to conditions which imply that a factorization $C = AA^*$, $A \in \mathfrak{L}(\mathfrak{A}, \mathfrak{H})$, $\ker A = \{0\}$ is unique up to replacement of $\mathfrak{A}$ by an isomorphic copy. Examples show that this is not always the case (see [34, p. 217] and [38, p. 891]). Such a notion of uniqueness is of interest in its own right and also because some applications use special properties of the particular factorization which is constructed in the proof of Theorem 4.1; see Dritschel and Rovnyak [38, Lecture 6].

Definition 6.2. *A selfadjoint operator $C \in \mathfrak{L}(\mathfrak{H})$ is said to have the* **unique factorization property** *if for any two factorizations*

$$C = A_j A_j^*, \qquad A_j \in \mathfrak{L}(\mathfrak{A}_j, \mathfrak{H}), \quad \ker A_j = \{0\}, \quad j = 1, 2, \tag{6.1}$$

there is an isomorphism $U \in \mathfrak{L}(\mathfrak{A}_1, \mathfrak{A}_2)$ such that $A_1 = A_2 U$.

This property holds in many naturally occurring situations. In fact, it is possible to completely characterize when the property holds.

Theorem 6.3. *Let $\mathfrak{H}$ be a Kreĭn space, and let $C \in \mathfrak{L}(\mathfrak{H})$ be a selfadjoint operator. The following conditions are equivalent:*

(1) *C has the unique factorization property;*
(2) *for some Hilbert space selfadjoint operator B congruent to C, $\sigma(B)$ omits an interval of the form $(-\epsilon, 0)$ or $(0, \epsilon)$ with $\epsilon > 0$;*
(3) *for some factorization $C = AA^*$ as in Theorem 4.1, $\operatorname{ran} A^*$ contains one of the subspaces $\mathfrak{A}_+$ or $\mathfrak{A}_-$ in some fundamental decomposition $\mathfrak{A} = \mathfrak{A}_+ \oplus \mathfrak{A}_-$.*

In this case, (2) holds for any selfadjoint operator congruent to C, and (3) holds for any factorization of C as in Theorem 4.1.

For a proof see [26, Theorem 2.8]. Condition (2) in Theorem 6.3 was given by Constantinescu and Gheondea [23, 25], Ćurgus and Langer [27], and Hara [45]. Condition (3) is given in a different form in Dritschel [34] and Dritschel and Rovnyak [38].

Theorem 6.4. *Let $\mathfrak{H}$ be a Kreĭn space, and let $C \in \mathfrak{L}(\mathfrak{H})$ be a selfadjoint operator. Each of the following conditions is sufficient for C to have the unique factorization property:*

(1) $C \geq 0$;
(2) *one of the indices* $\operatorname{ind}_{\pm} C$ *is finite;*
(3) $C^2 \leq C$.

Sketch of proof. (1), (2) Assume that $\operatorname{ind}_{-} C < \infty$. We check condition (2) in Theorem 6.3. Suppose that B is a selfadjoint operator on a Hilbert space $\mathfrak{K}$ which is congruent to C. Then $\sigma(B) \cap (-\infty, 0)$ is a finite set, and so (2) holds. We obtain (1) as a special case of (2).

(3) We deduce this from Theorems 6.5 and 6.6 below. Assume that $C^2 \leq C$. Suppose that we have two factorizations $C = A_j A_j^*$, $A_j \in \mathfrak{L}(\mathfrak{A}_j, \mathfrak{H})$, $\ker A_j = \{0\}$, $j = 1, 2$. For $j = 1, 2$, let $\mathfrak{G}_j$ be the range of A_j in the inner product that makes A_j an isomorphism from $\mathfrak{A}_j$ onto $\mathfrak{G}_j$. Then $\mathfrak{G}_j$ is a Kreĭn space which is contained continuously in $\mathfrak{H}$, and $C = E_j E_j^*$, where $E_j : \mathfrak{G}_j \to \mathfrak{H}$ is the inclusion mapping. The inequality $C^2 \leq C$ implies that the inclusion operators E_j are contractions. Applying Theorems 6.5 and 6.6 with $P = C$, we see that C has the unique factorization property. □

Alternatively, to prove Theorem 6.4(2) we can verify condition (3) in Theorem 6.3 with the aid of

Pontryagin's Theorem: *Let $\mathfrak{D}$ be a dense linear subspace of a Pontryagin space $\mathfrak{G}$. Then $\mathfrak{D}$ contains the negative subspace $\mathfrak{G}_-$ in some fundamental decomposition $\mathfrak{G} = \mathfrak{G}_+ \oplus \mathfrak{G}_-$.*

Suppose again that $\operatorname{ind}_{-} C < \infty$, and let $C = AA^*$ be any factorization as in Theorem 4.1. By Theorem 6.1, $\operatorname{ind}_{-} \mathfrak{A} = \operatorname{ind}_{-} C < \infty$. Since $\ker A = \{0\}$, $\operatorname{ran} A^*$ is dense in $\mathfrak{A}$, and so (3) follows from Pontryagin's theorem.

6.2 Examples of uniqueness results

(i) **Continuous inclusion of Kreĭn spaces and complementation in the sense of de Branges**. The simplest case here comes from a Kreĭn subspace $\mathfrak{G}$ of a Kreĭn space $\mathfrak{H}$. The inclusion mapping $E : \mathfrak{G} \to \mathfrak{H}$ is a continuous isometry in this case. The operator $P = EE^*$ is the projection on $\mathfrak{H}$ with range $\mathfrak{G}$. These notions have far-reaching generalizations in the work of de Branges [18]. We follow the operator range view in [38] in which P can be any selfadjoint operator on a Kreĭn space.

A Kreĭn space $\mathfrak{G}$ is said to be **contained continuously** in a Kreĭn space $\mathfrak{H}$ if $\mathfrak{G}$ is a linear subspace of $\mathfrak{H}$ and the inclusion mapping $E : \mathfrak{G} \to \mathfrak{H}$ is continuous. In this situation $P = EE^*$ is a selfadjoint operator on $\mathfrak{H}$. It is not hard to see that the range of P is contained in $\mathfrak{G}$ as a dense subspace, and

$$\langle Pf, Pg \rangle_{\mathfrak{G}} = \langle Pf, g \rangle_{\mathfrak{H}}, \qquad f, g \in \mathfrak{H}. \tag{6.2}$$

We call P the **generalized projection operator** for the inclusion of $\mathfrak{G}$ in $\mathfrak{H}$.

It is easy to see that every selfadjoint operator $P \in \mathfrak{L}(\mathfrak{H})$ arises as a generalized projection operator. In fact, if $P \in \mathfrak{L}(\mathfrak{H})$ is a given selfadjoint operator, write $P = AA^*$, $A \in \mathfrak{L}(\mathfrak{A}, \mathfrak{H})$, $\ker A = \{0\}$, as in Theorem 4.1. Let $\mathfrak{G}$ be the range of A in the inner product which makes A an isomorphism. It is not hard to see that $\mathfrak{G}$ is a Kreĭn space which is contained continuously in $\mathfrak{H}$, and $P = EE^*$ where E is the inclusion mapping.

Uniqueness questions arise. In the preceding situation, the indices $\operatorname{ind}_\pm \mathfrak{G}$ are determined by P. However, $\mathfrak{G}$ itself is not necessarily determined by P: it may occur that P is the generalized projection operator for distinct Kreĭn spaces $\mathfrak{G}_1$ and $\mathfrak{G}_2$ which are contained continuously in $\mathfrak{H}$; that is, $P = E_1E_1^* = E_2E_2^*$, where $E_1 : \mathfrak{G}_1 \to \mathfrak{H}$ and $E_2 : \mathfrak{G}_2 \to \mathfrak{H}$ are the inclusion mappings.

Theorem 6.5. *Let $\mathfrak{H}$ be a Kreĭn space, and let $P \in \mathfrak{L}(\mathfrak{H})$ be a selfadjoint operator. The following conditions are equivalent:*

(1) *P is the generalized projection operator for a unique Kreĭn space which is contained continuously in $\mathfrak{H}$;*
(2) *P has the unique factorization property.*

Uniqueness is automatic in some cases. Suppose that $\mathfrak{G}$ is contained continuously in $\mathfrak{H}$. We say that the inclusion is **contractive** if

$$\langle g, g \rangle_{\mathfrak{H}} \leq \langle g, g \rangle_{\mathfrak{G}}, \qquad g \in \mathfrak{G},$$

that is, the inclusion mapping is contractive; by (6.2), this occurs if and only if the associated generalized projection operator P satisfies $P^2 \leq P$. The notion of an **isometric** inclusion is defined similarly but with equality in the preceding inequalities.

Theorem 6.6. *Conditions* (1) *and* (2) *in Theorem* 6.5 *are satisfied if P is the generalized projection operator for some Kreĭn space $\mathfrak{G}$ which is contained continuously and contractively in $\mathfrak{H}$. In particular, such a space $\mathfrak{G}$ is unique.*

Let $\mathfrak{H}_1, \mathfrak{H}_2$ be Kreĭn spaces which are contained continuously and contractively in a Kreĭn space $\mathfrak{H}$. We say that $\mathfrak{H}_1$ and $\mathfrak{H}_2$ are **complementary in the sense of de Branges** or simply **complementary** if the mapping $(h_1, h_2) \to h_1 + h_2$ is a contractive partial isometry from $\mathfrak{H}_1 \times \mathfrak{H}_2$ onto $\mathfrak{H}$. In this case, for every $h \in \mathfrak{H}$,

$$\langle h, h \rangle_{\mathfrak{H}} = \min_{h = h_1 + h_2} \left(\langle h_1, h_1 \rangle_{\mathfrak{H}_1} + \langle h_2, h_2 \rangle_{\mathfrak{H}_2} \right),$$

and $\operatorname{ind}_- \mathfrak{H} = \operatorname{ind}_- \mathfrak{H}_1 + \operatorname{ind}_- \mathfrak{H}_2$. Examples appear in the theory of reproducing kernel spaces (see §8). The general theory is given in [5, 18, 38].

(ii) **Defect and Julia operators**. Defect and Julia operators can be changed by replacing the underlying Kreĭn spaces by isomorphic spaces. It is of interest to know if any two defect or Julia operators for a given operator $T \in \mathfrak{L}(\mathfrak{H}, \mathfrak{K})$ are related in this way.

Definition 6.7. *Let $T \in \mathfrak{L}(\mathfrak{H}, \mathfrak{K})$, where $\mathfrak{H}$ and $\mathfrak{K}$ are Kreĭn spaces.*

(1) *We say that T has an* **essentially unique defect operator** *if any two defect operators $\tilde{D}_j \in \mathfrak{L}(\tilde{\mathfrak{D}}_j, \mathfrak{H})$, $j = 1, 2$, are related by $\tilde{D}_1 = \tilde{D}_2 \tilde{V}$, where $\tilde{V}$ is an isomorphism from $\tilde{\mathfrak{D}}_1$ onto $\tilde{\mathfrak{D}}_2$.*

(2) *We say that T has an* **essentially unique Julia operator** *if any two Julia operators*

$$\begin{pmatrix} T & D_j \\ \tilde{D}_j^* & -L_j^* \end{pmatrix} \in \mathfrak{L}(\mathfrak{H} \oplus \mathfrak{D}_j, \mathfrak{K} \oplus \tilde{\mathfrak{D}}_j), \qquad j = 1, 2,$$

are related by

$$\begin{pmatrix} T & D_1 \\ \tilde{D}_1^* & -L_1^* \end{pmatrix} = \begin{pmatrix} 1_{\mathfrak{K}} & 0 \\ 0 & \tilde{V}^* \end{pmatrix} \begin{pmatrix} T & D_2 \\ \tilde{D}_2^* & -L_2^* \end{pmatrix} \begin{pmatrix} 1_{\mathfrak{H}} & 0 \\ 0 & V \end{pmatrix}$$

where $\tilde{V}$ is an isomorphism from $\tilde{\mathfrak{D}}_1$ onto $\tilde{\mathfrak{D}}_2$ and V is an isomorphism from $\mathfrak{D}_1$ onto $\mathfrak{D}_2$

A complete analysis of these conditions is given in Dritschel [34]. The following result probably covers the most important special cases

Theorem 6.8. *Let $T \in \mathfrak{L}(\mathfrak{H}, \mathfrak{K})$, where $\mathfrak{H}$ and $\mathfrak{K}$ are Kreĭn spaces. Each of the following conditions is sufficient for T to have essentially unique defect and Julia operators:*

(1) *T is a contraction;*
(2) *T^* is a contraction;*
(3) *one of the four indices* $\operatorname{ind}_\pm (1 - T^*T)$, $\operatorname{ind}_\pm (1 - TT^*)$ *is finite.*

Conditions (1) and (2) in Theorem 6.8 are included for emphasis, but they are special cases of (3). In the case of Julia operators, Theorem 6.8 is given in Dritschel and Rovnyak [37, p. 298]. The result for defect operators can be deduced from this and the fact that a Julia operator (5.2) can be constructed with any prescribed defect operator $\tilde{D}$ for T.

7. Kolmogorov decompositions of Hermitian kernels

The theory of Hermitian kernels provides a unified environment for common constructions that appear in a number of areas including the study of reproducing kernels, inner products, and selfadjoint operator matrices. The indefinite theory originates with Schwartz [69]. We follow the approach of Constantinescu and Gheondea [26]. The form of the uniqueness result in Theorem 7.3 is implicit in [26] and was communicated privately by the authors.

A **(Hermitian) kernel** is an indexed collection

$$K = \{K_{ij}\}_{i,j \in J}, \qquad K_{ij} \in \mathfrak{L}(\mathfrak{H}_j, \mathfrak{H}_i), \tag{7.1}$$

of operators satisfying $K_{ij} = K_{ji}^*$ for all $i, j \in J$. Here J is an index set, and the underlying spaces $\mathfrak{H}_j$, $j \in J$, are Kreĭn spaces. We say that K has a **Kolmogorov decomposition** if there exist a Kreĭn space $\mathfrak{K}$ and operators $V_j \in \mathfrak{L}(\mathfrak{H}_j, \mathfrak{K})$, $j \in J$, such that

$$K_{ij} = V_i^* V_j, \qquad i, j \in J, \tag{7.2}$$

and $\mathfrak{K} = \bigvee_{j \in J} V_j \mathfrak{H}_j$. The term "Kolmogorov decomposition" is derived from a theorem of Kolmogorov [48] as it appears, for example, in Martin and Putinar [53, p. 34]. Two Kolmogorov decompositions with operators $V_{1j} \in \mathfrak{L}(\mathfrak{H}_j, \mathfrak{K}_1)$ and $V_{2j} \in \mathfrak{L}(\mathfrak{H}_j, \mathfrak{K}_2)$, $j \in J$, are called **equivalent** if there is an isomorphism $W \in \mathfrak{L}(\mathfrak{K}_1, \mathfrak{K}_2)$ such that $V_{2j} = W V_{1j}$ for all $j \in J$. If any two Kolmogorov decompositions are equivalent, we say that K has an **essentially unique Kolmogorov decomposition**.

Sums and differences of kernels are defined in the obvious way when the underlying spaces are the same, and the set of such kernels has the structure of a linear space. Given a Hermitian kernel (7.1), let $\mathfrak{F}$ be the linear space of all finitely nonzero indexed sets $f = \{f_j\}_{j \in J}$ of vectors $f_j \in \mathfrak{H}_j$, $j \in J$. Define a **K-inner product** on $\mathfrak{F}$ by

$$\langle f, g \rangle_K = \sum_{i,j \in J} \langle K_{ij} f_j, g_i \rangle_{\mathfrak{H}_i}, \qquad f, g \in \mathfrak{F}.$$

We call K **nonnegative** and write $K \geq 0$ if the K-inner product (7.3) is nonnegative. The inequality $K_1 \leq K_2$ for two Hermitian kernels means that $K_2 - K_1 \geq 0$

A **nonnegative majorant** for a Hermitian kernel K is a Hermitian kernel L having the same underlying spaces such that $L \geq 0$ and $-L \leq K \leq L$. In this situation, we associate a Hilbert space $\mathfrak{H}_L$ with L by a standard construction. A dense set in $\mathfrak{H}_L$ is the quotient space $\mathfrak{F}/\mathfrak{N}_L$, where $\mathfrak{F}$ is as above and $\mathfrak{N}_L$ the subspace of elements which are orthogonal to all of $\mathfrak{F}$ in the L-inner product. If $f \in \mathfrak{F}$, let $[f]$ be the corresponding coset in $\mathfrak{F}/\mathfrak{N}_L$. The inner product in $\mathfrak{H}_L$ is given on the dense set by

$$\langle [f], [g] \rangle_{\mathfrak{H}_L} = \langle f, g \rangle_L, \qquad f, g \in \mathfrak{F}.$$

Arguments in [26, p. 929] show that there is a unique operator $G \in \mathfrak{L}(\mathfrak{H}_L)$ such that

$$\langle G[f], [g] \rangle_{\mathfrak{H}_L} = \langle f, g \rangle_K, \qquad f, g \in \mathfrak{F}.$$

The operator G is selfadjoint and satisfies $\|G\| \leq 1$. It is called the **Gram operator** of the kernel K for the majorant L.

To avoid repetitive statements, throughout §7 the underlying spaces for Hermitian kernels are assumed to be as in (7.1), and $\mathfrak{F}$ has the same meaning as above. We likewise use the same notation $\mathfrak{F}/\mathfrak{N}_L$ for the dense set in the Hilbert space $\mathfrak{H}_L$ as in the definition of a Gram operator.

Theorem 7.1. *If K is a Hermitian kernel, the following assertions are equivalent:*

(1) *K has a Kolmogorov decomposition;*
(2) *K has a nonnegative majorant;*
(3) *$K = K_+ - K_-$ for some Hermitian kernels $K_+ \geq 0$ and $K_- \geq 0$.*

In this case, the decomposition in (3) *can be chosen such that the only Hermitian kernel M such that $0 \le M \le K_\pm$ is $M = 0$.*

Proof. (1) $\Leftrightarrow$ (2) Assume that a Kolmogorov decomposition (7.2) exists. In any way construct a Hilbert space $\mathfrak{M}$ and an invertible operator $X \in \mathfrak{L}(\mathfrak{K}, \mathfrak{M})$ such that

$$|\langle k, k\rangle_{\mathfrak{K}}| \le \langle Xk, Xk\rangle_{\mathfrak{M}}, \qquad k \in \mathfrak{K}.$$

Define $L = \{L_{ij}\}_{i,j\in J}$ by $L_{ij} = V_i^* X^* X V_j \in \mathfrak{L}(\mathfrak{H}_j, \mathfrak{H}_i)$, $i, j \in J$. It is easy to see that L is a nonnegative majorant for K.

Conversely, let K have a nonnegative majorant L, and let $G \in \mathfrak{L}(\mathfrak{H}_L)$ be the associated Gram operator. Using Theorem 4.1, factor $G = AA^*$ with $A \in \mathfrak{L}(\mathfrak{K}, \mathfrak{H}_L)$ and $\ker A = \{0\}$. For each $j \in J$, there is a natural continuous embedding operator E_j from $\mathfrak{H}_j$ into $\mathfrak{H}_L$, namely, $E_j u = [f_u]$, where f_u is the element of $\mathfrak{F}$ whose j-th component is u and all other components are zero. Then $V_j = A^* E_j$, $j \in J$, defines a Kolmogorov decomposition.

(2) $\Leftrightarrow$ (3) If $K = K_+ - K_-$ as in (3), $L = K_+ + K_-$ is a nonnegative majorant for K.

Conversely, suppose that K has a nonnegative majorant L, and let G be the corresponding Gram operator. In terms of the embedding operators E_j, $j \in J$, defined above, we have

$$K_{ij} = E_i^* G E_j, \qquad i, j \in J.$$

Let $P_\pm, P_0$ be the spectral projections for $(0, \infty)$, $(-\infty, 0)$, $\{0\}$ for G. Then the formula

$$K_{\pm ij} = E_i^*(\pm P_\pm) G E_j, \qquad i, j \in J.$$

defines kernels $K_\pm$ such that $K_\pm \ge 0$ and $K = K_+ - K_-$.

The kernels $K_\pm$ constructed in this way have the property in the last statement of the theorem. For assume that $0 \le M \le K_\pm$. Since $\|G\| \le 1$, $K^\pm \le L$, and thus $0 \le M \le L$. The Gram operator $H \in \mathfrak{L}(\mathfrak{H}_L)$ of M relative to L satisfies

$$0 \le \langle H[f], [f]\rangle_{\mathfrak{H}_L} \le \langle \pm P_\pm G[f], [f]\rangle_{\mathfrak{H}_L}, \qquad f \in \mathfrak{F}.$$

Since P_+G and P_-G are supported on orthogonal subspaces of $\mathfrak{H}_L$, $H = 0$. Hence $\langle f, g\rangle_M = \langle H[f], [g]\rangle_{\mathfrak{H}_L} = 0$ for all $f, g \in \mathfrak{F}$, and so $M = 0$. □

Let K be a Hermitian kernel with nonnegative majorant L. A Kolmogorov decomposition (7.2) for K is said to be **L-continuous** if the mapping $[f]$ into $\sum_{j\in J} V_j f_j$ on $\mathfrak{F}/\mathfrak{N}_L$ extends to a continuous operator from on $\mathfrak{H}_L$ into $\mathfrak{K}$.

Lemma 7.2. *Let K be a Hermitian kernel.*

(1) *If K has a Kolmogorov decomposition* (7.2), *the decomposition is L-continuous with respect to the nonnegative majorant L constructed in the proof of Theorem* 7.1, (1) *implies* (2).

(2) *If K has a nonnegative majorant L, the Kolmogorov decomposition of K constructed in the proof of Theorem* 7.1, (2) *implies* (1), *is L-continuous.*

Proof. (1) In the notation of Theorem 7.1, (1) implies (2),

$$\langle [f],[g]\rangle_{\mathfrak{H}_L} = \left\langle X\textstyle\sum_{j\in J} V_j f_j, X\sum_{i\in J} V_i g_i\right\rangle_{\mathfrak{M}}, \qquad f,g\in\mathfrak{F}.$$

Thus the mapping $[f]$ into $X\sum_{j\in J} V_j f_j$ is a Hilbert space isometry from $\mathfrak{H}_L$ into $\mathfrak{M}$; the mapping $[f]$ into $\sum_{j\in J} V_j f_j$ on $\mathfrak{F}/\mathfrak{N}_L$ into $\mathfrak{K}$ is the composite of this isometry and X^{-1} and hence is continuous.

(2) We wish to show that, in the proof of Theorem 7.1, (2) implies (1), the mapping $[f]$ into $\sum_{j\in J} V_j f_j$ on $\mathfrak{F}/\mathfrak{N}_L$ extends to a continuous operator from $\mathfrak{H}_L$ into $\mathfrak{K}$. In fact, we show that the mapping is A^*. Since $\mathfrak{K}$ is the closed span of the ranges of the operators V_j, it is sufficient to show that for any $f,g\in\mathfrak{F}$,

$$\langle [f], A\textstyle\sum_{i\in J} V_i g_i\rangle_{\mathfrak{H}_L} = \left\langle \sum_{j\in J} V_j f_j, \sum_{i\in J} V_i g_i\right\rangle_{\mathfrak{K}}.$$

Since $V_i = A^*E_i$ for each $i\in J$ and $\sum_{i\in J} E_i g_i = [g]$, it is the same thing to show that

$$\langle [f], AA^*[g]\rangle_{\mathfrak{H}_L} = \left\langle \textstyle\sum_{j\in J} V_j f_j, \sum_{i\in J} V_i g_i\right\rangle_{\mathfrak{K}}.$$

This holds because $AA^* = G$, and so both sides are equal to $\langle f,g\rangle_K$. □

Thus L-continuous Kolmogorov decompositions always exist. Uniqueness depends on the Gram operator.

Theorem 7.3. *Let K be a Hermitian kernel with nonnegative majorant L and Gram operator G. Any two L-continuous Kolmogorov decompositions are equivalent if and only if G has the unique factorization property.*

Proof. Assume that G has the unique factorization property. Let

$$K_{ij} = V_{1i}^* V_{1j}, \quad V_{1j}\in\mathfrak{L}(\mathfrak{H}_j,\mathfrak{K}_1), \quad i,j\in J, \tag{7.6}$$
$$K_{ij} = V_{2i}^* V_{2j}, \quad V_{2j}\in\mathfrak{L}(\mathfrak{H}_j,\mathfrak{K}_2), \quad i,j\in J, \tag{7.7}$$

be two L-continuous Kolmogorov decompositions. By hypothesis, the mapping $[f]$ into $\sum_{j\in J} V_{1j} f_j$ on $\mathfrak{F}/\mathfrak{N}_L$ extends to a continuous operator from $\mathfrak{H}_L$ into $\mathfrak{K}_1$. Denote its adjoint $A_1\in\mathfrak{L}(\mathfrak{K}_1,\mathfrak{H}_L)$. Since $\mathfrak{K}_1 = \bigvee_{j\in J} V_{1j}\mathfrak{H}_j$, $\ker A_1 = \{0\}$. For all $f,g\in\mathfrak{F}$,

$$\langle [f], A_1\textstyle\sum_{i\in J} V_{1i} g_i\rangle_{\mathfrak{H}_L} = \left\langle \sum_{j\in J} V_{1j} f_j, \sum_{i\in J} V_{1i} g_i\right\rangle_{\mathfrak{K}_1}.$$

Thus

$$\langle G[f],[g]\rangle_{\mathfrak{H}_L} = \langle f,g\rangle_K = \left\langle \textstyle\sum_{j\in J} V_{1j} f_j, \sum_{i\in J} V_{1i} g_i\right\rangle_{\mathfrak{K}_1} = \langle [f], A_1A_1^*[g]\rangle_{\mathfrak{H}_L},$$

and so $G = A_1A_1^*$. Construct a factorization $G = A_2A_2^*$ in a similar way from (7.7). Since G has the unique factorization property, there is a unitary operator $W\in\mathfrak{L}(\mathfrak{K}_1,\mathfrak{K}_2)$ such that $A_1 = A_2W$. For all $j\in J$,

$$V_{2j} = A_2^*E_j = WA_1^*E_j = WV_{1j},$$

and thus the two Kolmogorov decompositions are equivalent.

Assume that any two L-continuous Kolmogorov decompositions of K are equivalent. Let

$$G = A_1 A_1^* = A_2 A_2^*$$

with $A_1 \in \mathfrak{L}(\mathfrak{K}_1, \mathfrak{H}_L)$, $A_2 \in \mathfrak{L}(\mathfrak{K}_2, \mathfrak{H}_L)$, $\ker A_1 = \{0\}$, and $\ker A_2 = \{0\}$. By Lemma 7.2, we can construct L-continuous Kolmogorov decompositions (7.6) and (7.7) by setting $V_{1j} = A_1^* E_j$ and $V_{2j} = A_2^* E_j$ for all $j \in J$. By hypothesis, there is a unitary operator $W \in \mathfrak{L}(\mathfrak{K}_1, \mathfrak{K}_2)$ such that $V_{2j} = W V_{1j}$ for all $j \in J$. Using the properties $\mathfrak{K}_1 = \bigvee_{j \in J} V_{1j} \mathfrak{H}_j$ and $\mathfrak{K}_2 = \bigvee_{j \in J} V_{2j} \mathfrak{H}_j$ of the Kolmogorov decompositions, we obtain $W A_1^* = A_2^*$ and hence $A_1 = A_2 W$, and thus G has the unique factorization property. □

A stronger uniqueness result holds with a stronger hypothesis.

Theorem 7.4. *A Hermitian kernel K has an essentially unique Kolmogorov decomposition if and only if the Gram operators for all nonnegative majorants have the unique factorization property.*

Proof. If some Gram operator does not have the unique factorization property, Theorem 7.3 implies that there exist nonequivalent Kolmogorov decompositions, which proves necessity.

Conversely, assume that every Gram operator has the essential uniqueness property. Let (7.6) and (7.7) be any two Kolmogorov decompositions of K. By Lemma 7.2, the decompositions are continuous relative to some nonnegative majorants L_1 and L_2 for K. Then $L = L_1 + L_2$ is a nonnegative majorant for K. Since $L_1 \leq L$, the "identity mapping" on $\mathfrak{F}/\mathfrak{N}_L$ to $\mathfrak{F}/\mathfrak{N}_{L_1}$ is a densely defined contraction from $\mathfrak{H}_L$ into $\mathfrak{H}_{L_1}$. Since these are Hilbert spaces, the mapping $[f]$ into $\sum_{j \in J} V_{1j} f_j$ on $\mathfrak{F}/\mathfrak{N}_L$ into $\mathfrak{K}_1$ is a composition of continuous operators, and so (7.6) is L-continuous. Similarly, (7.7) is L-continuous. By Theorem 7.3 and our hypothesis on Gram operators, the two Kolmogorov decompositions are equivalent. □

The following sufficient condition for essential uniqueness is given in [26, Theorem 4.3].

Theorem 7.5. *Let K be a Hermitian kernel, and assume that there exists a Kolmogorov decomposition* (7.2) *such that the linear span of the subspaces $V_j \mathfrak{H}_j$, $j \in J$, contains contains one of the subspaces $\mathfrak{K}_\pm$ in some fundamental decomposition $\mathfrak{K} = \mathfrak{K}_+ \oplus \mathfrak{K}_-$. Then K has an essentially unique Kolmogorov decomposition.*

Proof. Suppose that the given Kolmogorov decomposition is relabeled as (7.6), and let (7.7) be any second Kolmogorov decomposition. Define a linear relation $\mathbf{R}$ from $\mathfrak{K}_1$ into $\mathfrak{K}_2$ by

$$\mathbf{R} = \left\{ \left(\textstyle\sum_{j \in J} V_{1j} f_j, \sum_{j \in J} V_{2j} f_j \right) : f \in \mathfrak{F} \right\}.$$

By the definition of a Kolmogorov decomposition, $\mathbf{R}$ has dense domain and dense range. For all $f \in \mathfrak{F}$,

$$\left\langle \sum_{j\in J} V_{1j} f_j, \sum_{i\in J} V_{1i} f_i \right\rangle_{\mathfrak{K}_1} = \langle f, f\rangle_K = \left\langle \sum_{j\in J} V_{2j} f_j, \sum_{i\in J} V_{2i} f_i \right\rangle_{\mathfrak{K}_2} .$$

By hypothesis, the domain of $\mathbf{R}$ contains one of the subspaces $\mathfrak{K}_{1\pm}$ in some fundamental decomposition $\mathfrak{K}_1 = \mathfrak{K}_{1+} \oplus \mathfrak{K}_{1-}$. Hence by Theorem 4.3 the closure of $\mathbf{R}$ is the graph of a unitary operator $W \in \mathfrak{L}(\mathfrak{K}_1, \mathfrak{K}_2)$. By construction, $V_{2j} = WV_{1j}$ for all $j \in J$, and so the two Kolmogorov decompositions are equivalent. □

Corollary 7.6. *If a Hermitian kernel K has a Kolmogorov decomposition* (7.2) *such that $\mathfrak{K}$ is either a Pontryagin space or the antispace of a Pontryagin space, then K has an essentially unique Kolmogorov decomposition.*

Proof. Since $\mathfrak{K}$ is a Pontryagin space for the given Kolmogorov decomposition, the hypotheses of Theorem 7.5 are satisfied by Pontryagin's theorem (see §6). □

8. Examples of Hermitian kernels

8.1 Reproducing kernel Kreĭn spaces

The definite theory is classical and has many applications. In addition to the standard source of Aronszajn [9], see also, for example, Dym [39] and Saitoh [68]. The indefinite theory is due to Schwartz [69] and Sorjonen [71] and also also owes much to a series of papers in the 1970's by Kreĭn and Langer including [49, 50] and the thesis of Alpay [2]. The theory of §7 allows a quick derivation of the main results.

Consider a Hermitian kernel $K(s,t)$, $s,t \in \Omega$, with values in $\mathfrak{L}(\mathfrak{F})$ for some fixed Kreĭn space $\mathfrak{F}$ and nonempty set Ω. We call $K(s,t)$ a **reproducing kernel** for a Kreĭn space $\mathfrak{H}_K$ of $\mathfrak{F}$-valued functions on Ω if

(1) for each $s \in \Omega$ and $f \in \mathfrak{F}$, $K(s,\cdot)f$ belongs to $\mathfrak{H}_K$, and
(2) $\langle h(\cdot), K(s,\cdot)f\rangle_{\mathfrak{H}_K} = \langle h(s), f\rangle_{\mathfrak{F}}$ for every $h(\cdot)$ in $\mathfrak{H}_K$.

These conditions are equivalent to the existence of a Kolmogorov decomposition with a Kreĭn space $\mathfrak{H}_K$ and operators $V_s \colon \mathfrak{F} \to \mathfrak{H}_K$ such that

$$V_s f = K(s,\cdot)f, \qquad f \in \mathfrak{F}, \tag{8.1}$$

for all $s \in \Omega$. In other words, $V_s = E(s)^*$, where $E(s)\colon h(\cdot) \to h(s)$ is evaluation at any point $s \in \Omega$ (the evaluation mappings are continuous by the closed graph theorem).

Conversely, one can start with a Kreĭn space of functions:

Theorem 8.1. *Let $\mathfrak{H}$ be a Kreĭn space of functions on a set Ω with values in a Kreĭn space $\mathfrak{F}$. Then $\mathfrak{H}$ has a reproducing kernel if and only if all evaluation mappings $E(s)$, $s \in \Omega$, belong to $\mathfrak{L}(\mathfrak{H}, \mathfrak{F})$. The reproducing kernel is uniquely determined by the space and given by $K(s,t) = E(t)E(s)^*$, $s,t \in \Omega$.*

Notions of nonnegative kernels and nonnegative majorants have the same meaning as in the general case. The definite case is well known: a nonnegative kernel $L(s,t)$ is the reproducing kernel for a unique Hilbert space $\mathfrak{H}_L$ (this also follows from the results of §7).

Existence and uniqueness are separate issues in the indefinite case. A reproducing kernel for a Kreĭn space, when it exists, is uniquely determined by the space. However, unlike the Hilbert space case, two distinct Kreĭn spaces can have the same reproducing kernel. The uniqueness of a Kreĭn space with given reproducing kernel can be restored in a restricted sense if suitable conditions are met.

Theorem 8.2. *If $K(s,t)$, $s,t \in \Omega$, is a Hermitian kernel with values in $\mathfrak{L}(\mathfrak{F})$ for some Kreĭn space $\mathfrak{F}$, the following assertions are equivalent:*

(1) *$K(s,t)$ is the reproducing kernel for some Kreĭn space $\mathfrak{H}_K$ of functions on Ω;*
(2) *$K(s,t)$ has a nonnegative majorant $L(s,t)$ on $\Omega \times \Omega$;*
(3) *$K(s,t) = K_+(s,t) - K_-(s,t)$ for some nonnegative kernels $K_\pm(s,t)$ on $\Omega \times \Omega$.*

When these conditions hold, then moreover:

(4) *For a given nonnegative majorant $L(s,t)$ for $K(s,t)$, there is a Kreĭn space $\mathfrak{H}_K$ with reproducing kernel $K(s,t)$ which is contained continuously in the Hilbert space $\mathfrak{H}_L$ with reproducing kernel $L(s,t)$.*
(5) *In the same situation, there is a continuous selfadjoint operator G on $\mathfrak{H}_L$ such that $G\colon L(s,\cdot)f \to K(s,\cdot)f$, $s \in \Omega$, $f \in \mathfrak{F}$. The space $\mathfrak{H}_K$ in (4) is unique if and only if G has the unique factorization property.*

When the equivalent conditions in Theorem 8.2 hold, then for any space $\mathfrak{H}_K$ as in (1), the decomposition in (3) can be chosen so that $\pm K_\pm(s,t)$ are reproducing kernels for the spaces $\mathfrak{H}_K^\pm$ in a fundamental decomposition $\mathfrak{H}_K = \mathfrak{H}_K^+ \oplus \mathfrak{H}_K^-$. In fact, we need only choose $\pm K_\pm(s,t)$ to be the reproducing kernels for the spaces in a fundamental decomposition.

Proof. The equivalence of (1)–(3) follows from Theorem 7.1.

(4) We use $L(s,t)$ to construct a reproducing kernel Kreĭn space $\mathfrak{H}_K$ for $K(s,t)$ as in the proof of Theorem 7.1. The reproducing kernel Hilbert space $\mathfrak{H}_L$ is naturally identified with the abstract space denoted in the same way in §7, and the associated Gram operator G has the action described in (5). By Lemma 7.2(2) there is a continuous operator A^* on $\mathfrak{H}_L$ into $\mathfrak{H}_K$ such that

$$A^*\colon L(s,\cdot)f \to K(s,\cdot)f, \qquad s \in \Omega,\ f \in \mathfrak{F}.$$

The adjoint of this operator is the inclusion mapping A from $\mathfrak{H}_K$ into $\mathfrak{H}_L$. Thus $G = AA^*$ and $\mathfrak{H}_K$ is contained continuously in $\mathfrak{H}_L$.

(5) Assume that G has the unique factorization property, and let $\mathfrak{H}_K'$ and $\mathfrak{H}_K''$ be two Kreĭn spaces with reproducing kernel $K(s,t)$ which are contained continuously in $\mathfrak{H}_L$. The two Kolmogorov decompositions induced as in (8.1) are equivalent by Theorem 7.3. It follows that the identity mapping on the linear span

$\mathfrak{H}_0$ of all functions $K(s,\cdot)f$, $s \in \Omega$, and $f \in \mathfrak{F}$, extends to a unitary operator from $\mathfrak{H}'_K$ onto $\mathfrak{H}''_K$. The continuity of evaluation mappings for any reproducing kernel Kreĭn space implies that whenever $W\colon h_1(\cdot) \to h_1(\cdot)$, then $h(1(s) = h_2(s)$ for all $s \in \Omega$. Thus $\mathfrak{H}'_K$ and $\mathfrak{H}''_K$ are identical.

In the other direction, the existence of distinct Kreĭn spaces $\mathfrak{H}'_K$ and $\mathfrak{H}''_K$ with reproducing kernel $K(s,t)$ contained continuously in $\mathfrak{H}_L$ implies the existence of two nonequivalent L-continuous Kolmogorov decompositions. By Theorem 7.3, the Gram operator G does not have the essential uniqueness property in this situation. □

Suppose that Ω is an open set in the complex plane. A Hermitian kernel $K(w,z)$ on $\Omega \times \Omega$ with values in $\mathfrak{L}(\mathfrak{F})$ for some Kreĭn space $\mathfrak{F}$ is **holomorphic** if it is a holomorphic function of z and $\bar{w}$.

Theorem 8.3. *Let Ω be an open set in the complex plane, and let $K(w,z)$, $w,z \in \Omega$, be a holomorphic Hermitian kernel with values in $\mathfrak{L}(\mathfrak{F})$ for some Kreĭn space $\mathfrak{F}$. The following assertions are equivalent:*

(1) *$K(w,z)$ is the reproducing kernel for some Kreĭn space $\mathfrak{H}_K$ of holomorphic functions on Ω;*
(2) *$K(w,z)$ has a nonnegative holomorphic majorant $L(w,z)$ on $\Omega \times \Omega$;*
(3) *$K(w,z) = K_+(w,z) - K_-(w,z)$ for some nonnegative holomorphic kernels $K_\pm(w,z)$ on $\Omega \times \Omega$.*

When these conditions hold, then moreover:

(4) *For a given nonnegative holomorphic majorant $L(w,z)$ for $K(w,z)$, there is a Kreĭn space $\mathfrak{H}_K$ of holomorphic functions with reproducing kernel $K(w,z)$ which is contained continuously in the Hilbert space $\mathfrak{H}_L$ with reproducing kernel $L(w,z)$.*
(5) *In the same situation, there is a continuous selfadjoint operator G on $\mathfrak{H}_L$ such that $G\colon L(w,\cdot)f \to K(w,\cdot)f$, $w \in \Omega$, $f \in \mathfrak{F}$. The space $\mathfrak{H}_K$ in (4) is unique if and only if G has the unique factorization property.*

Proof. It is well known that the reproducing kernel Hilbert space associated with a nonnegative holomorphic Hermitian kernel consists of holomorphic functions, and conversely. Given this, we proceed as in the proof of Theorem 8.2 to obtain the result. □

The conditions for existence of a reproducing kernel Kreĭn space are automatically satisfied in many cases of interest. Suppose that Ω is an open set in the complex plane.

Theorem 8.4 (Alpay [4]). *Let $K(w,z)$ be a holomorphic Hermitian kernel on $\Omega \times \Omega$ with values in $\mathfrak{L}(\mathfrak{F})$ for some Kreĭn space $\mathfrak{F}$. Assume that Ω is a disk or half-plane, and that $K(w,z)$ is bounded relative to some and hence any norm which determines the strong topology of $\mathfrak{F}$. Then there exist nonnegative holomorphic Hermitian kernels $K_\pm(w,z)$ such that $K(w,z) = K_+(w,z) - K_-(w,z)$, $w,z \in \Omega$. In particular, $K(w,z)$ satisfies the equivalent conditions (1)–(3) of Theorem 8.2.*

Proof. Without loss of generality, take $\Omega = \mathbf{D}$. It is also sufficient to prove the result when $\mathfrak{F}$ is a Hilbert space, since otherwise we need only consider $X^*K(w,z)X$, where X is an invertible continuous operator on $\mathfrak{F}$ onto a Hilbert space. Let $H^2_{\mathfrak{F}}$ be the Hardy space of holomorphic $\mathfrak{F}$-valued functions on $\mathbf{D}$.

For any polynomial $p(z)$ with coefficients in $\mathfrak{F}$ and $0 < r < 1$, the formula

$$h_p(z) = \frac{1}{2\pi}\int_0^{2\pi} K(re^{it}, z)p(r^{-1}e^{it})\,dt$$

defines a holomorphic function on the disk $|z| < r$. This function is independent of r. It is enough to show this for a monomial $p(z) = fz^m$, $f \in \mathfrak{F}$. If $K(w,z) = \sum_0^\infty A_n(z)\bar{w}^n$, then in this case

$$h_p(z) = \frac{1}{2\pi}\int_0^{2\pi} \sum_0^\infty A_n(z)r^n e^{-int} r^{-m} e^{imt} f\,dt = A_m(z)f,$$

which is independent of r.

Let $p(z)$ be any polynomial with coefficients in $\mathfrak{F}$. We show that the function h_p belongs to $H^2_{\mathfrak{F}}$ and $\|h_p\|_{H^2_{\mathfrak{F}}} \le M\,\|p\|_{H^2_{\mathfrak{F}}}$, where $M > 0$ is a constant. Given any $\rho \in (0,1)$ and $\rho < r < 1$,

$$\begin{aligned}
\frac{1}{2\pi}\int_0^{2\pi} \left\|h_p(\rho e^{i\theta})\right\|_{\mathfrak{F}}^2 d\theta &\le \frac{1}{2\pi}\int_0^{2\pi} \left\|\frac{1}{2\pi}\int_0^{2\pi} K(re^{it}, \rho e^{i\theta})p(r^{-1}e^{it})\,dt\right\|_{\mathfrak{F}}^2 d\theta \\
&\le \frac{M^2}{2\pi}\int_0^{2\pi}\left[\frac{1}{2\pi}\int_0^{2\pi} \left\|p(r^{-1}e^{it})\right\|_{\mathfrak{F}}\right]^2 d\theta \\
&\le \frac{M^2}{2\pi}\int_0^{2\pi}\left[\frac{1}{2\pi}\int_0^{2\pi} 1\cdot\,dt\right]\left[\frac{1}{2\pi}\int_0^{2\pi}\left\|p(r^{-1}e^{it})\right\|_{\mathfrak{F}}^2 dt\right] d\theta \\
&= \frac{M^2}{2\pi}\int_0^{2\pi}\left\|p(r^{-1}e^{it})\right\|_{\mathfrak{F}}^2 dt.
\end{aligned}$$

The constant M in the estimate is any bound for $\|K(w,z)\|_{\mathfrak{L}(\mathfrak{F})}$ on $\mathbf{D}\times\mathbf{D}$. The assertion follows on letting $r \uparrow 1$ and using the arbitrariness of ρ. It follows that there is a bounded operator P on $H^2_{\mathfrak{F}}$ such that $P\colon p \to h_p$ on polynomials. The operator P is selfadjoint by the symmetry of the kernel $K(w,z)$. Applying P to monomials of the form $f\bar{w}^n z^n$ and summing over $n \ge 0$, we find that

$$P\colon (1-\bar{w}z)^{-1}f \to K(w,z)f, \qquad w \in \mathbf{D},\ f \in \mathfrak{F},$$

and so for any $f, g \in \mathfrak{F}$ and $\alpha, \beta \in \mathbf{D}$,

$$\langle K(\alpha,\beta)f, g\rangle_{\mathfrak{F}} = \langle P\{(1-\bar{\alpha}z)^{-1}f\}, (1-\bar{\beta}z)^{-1}g\rangle_{H^2_{\mathfrak{F}}}\,.$$

In any way write $P = P_+ - P_-$, where $P_\pm$ are nonnegative selfadjoint operators on $H^2_{\mathfrak{F}}$. Then kernels $K_\pm(w,z)$ of the required type are defined by requiring

$$\langle K_\pm(\alpha,\beta)f,g\rangle_{\mathfrak{F}} = \left\langle P_\pm\{(1-\bar{\alpha}z)^{-1}f\},(1-\bar{\beta}z)^{-1}g\right\rangle_{H^2_{\mathfrak{F}}}\,.$$

for all $f,g \in \mathfrak{F}$ and $\alpha,\beta \in \mathbf{D}$. □

Theorem 8.4 can be used for unbounded as well as bounded kernels.

Corollary 8.5. *Let $S(z)$ be a holomorphic function on $\Omega = \mathbf{D}$, with possibly an isolated set of points Z omitted. Assume that there is a bounded scalar-valued function $u(z)$ on $\mathbf{D}$ which is nonvanishing on Ω and such that $u(z)S(z)$ is bounded relative to some and hence any norm which determines the strong topology of $\mathfrak{F}$. Then the kernels*

$$\frac{1_{\mathfrak{F}} - S(z)S(w)^*}{1-z\bar{w}} \qquad \textit{and} \qquad \frac{S(z)S(w)^*}{1-z\bar{w}} \tag{8.3}$$

satisfy the equivalent conditions (1)–(3) *of Theorem* 8.2 *on $\Omega\times\Omega$. In particular, they are reproducing kernels for Kreĭn spaces $\mathfrak{H}(S)$ and $\mathfrak{M}(S)$ of functions on Ω.*

Proof. Each of the kernels has the form $K(w,z) = L(w,z)/(1-\bar{w}z)$, where $u(z)L(w,z)\overline{u(w)}$ is a bounded holomorphic Hermitian kernel on $\mathbf{D}\times\mathbf{D}$ (any removable singularities of $u(z)S(z)$ are presumed to be removed). By Theorem 8.4, $u(z)L(w,z)\overline{u(w)} = M_+(w,z) - M_-(w,z)$, where $M_\pm(w,z)$ are nonnegative holomorphic kernels on $\mathbf{D}\times\mathbf{D}$. Thus

$$\begin{aligned} K(w,z) &= \frac{L(w,z)}{1-\bar{w}z} \\ &= \sum_{n=0}^{\infty} \frac{z^n}{u(z)} M_+(w,z)\left(\frac{w^n}{u(w)}\right)^{-} - \sum_{n=0}^{\infty} \frac{z^n}{u(z)} M_-(w,z)\left(\frac{w^n}{u(w)}\right)^{-}. \end{aligned}$$

The two sums on the right define nonnegative holomorphic kernels $K_\pm(w,z)$ such that $K(w,z) = K_+(w,z) - K_-(w,z)$ on $\Omega\times\Omega$. This verifies condition (3) in Theorem 8.2, and the each of the kernels (8.3) is the reproducing kernel for some Kreĭn space of functions on Ω. □

8.2 Reproducing kernel Pontryagin spaces

Again consider a Hermitian kernel $K(s,t)$, $s,t\in\Omega$, with values in $\mathfrak{L}(\mathfrak{F})$ for some fixed Kreĭn space $\mathfrak{F}$ and nonempty set Ω. Stronger results than those above hold when $K(s,t)$ has κ **negative squares**, that is, the maximum number of negative eigenvalues of all matrices

$$\left(\langle K(s_j,s_i)f_j,f_i\rangle_{\mathfrak{F}}\right)_{i,j=1}^{n}, \qquad s_1,\dots,s_n\in\Omega,\ f_1,\dots,f_n\in\mathfrak{F},\ n\ge 1,$$

is a nonnegative integer κ. In this case, we write $\mathrm{sq}_-\,K = \kappa$. An associated reproducing kernel space $\mathfrak{H}_K$ automatically exists. It is unique and a Pontryagin space of negative index κ. Conversely, the reproducing kernel of any given reproducing kernel Pontryagin space is a Hermitian kernel which has κ negative squares [71].

The classical Aronszajn theory of sums and differences of kernels has a natural generalization in the present setting.

Theorem 8.6. *Let $K(s,t), K_1(s,t), K_2(s,t)$ be Hermitian kernels on $\Omega \times \Omega$ with values in $\mathfrak{L}(\mathfrak{F})$ for some Kreĭn space $\mathfrak{F}$. If $K(s,t) = K_1(s,t) + K_2(s,t)$, then*

$$\mathrm{sq}_- K \leq \mathrm{sq}_- K_1 + \mathrm{sq}_- K_2.$$

Suppose these numbers are finite, and let $\mathfrak{H}_1, \mathfrak{H}_2, \mathfrak{H}$ be the associated reproducing kernel Pontryagin spaces. Then the following conditions are equivalent:

(1) $\mathrm{sq}_- K = \mathrm{sq}_- K_1 + \mathrm{sq}_- K_2$;

(2) *$\mathfrak{H}_1$ and $\mathfrak{H}_2$ are contained continuously and contractively as complementary spaces in $\mathfrak{H}$;*

(3) *the intersection $\mathfrak{R} = \mathfrak{H}_1 \cap \mathfrak{H}_2$ is a Hilbert space in the inner product*

$$\langle h, k\rangle_{\mathfrak{R}} = \langle h, k\rangle_{\mathfrak{H}_1} + \langle h, k\rangle_{\mathfrak{H}_2}, \qquad h, k \in \mathfrak{R}.$$

Theorem 8.7. *Let $\mathfrak{H}, \mathfrak{H}_1$ be Pontryagin spaces of functions defined on a set Ω with values in a Kreĭn space $\mathfrak{F}$ and such that $\mathfrak{H}_1$ is contained continuously and contractively in $\mathfrak{H}$. If the spaces have reproducing kernels $K(s,t), K_1(s,t)$, then*

$$K_2(s,t) = K(s,t) - K_1(s,t)$$

defines a Hermitian kernel such that $\mathrm{sq}_- K_2 = \mathrm{sq}_- K - \mathrm{sq}_- K_1$. The associated reproducing kernel Pontryagin space $\mathfrak{H}_2$ is also contained continuously and contractively in $\mathfrak{H}$, and $\mathfrak{H}_1$ and $\mathfrak{H}_2$ are complementary spaces in $\mathfrak{H}$.

Theorem 8.8. *Let $K(w,z)$ be a holomorphic kernel on $\Omega \times \Omega$ with values in $\mathfrak{L}(\mathfrak{F})$ for some Kreĭn space $\mathfrak{F}$ and region Ω in the complex plane. Let Ω_0 be a subregion of Ω, and assume that the restriction of $K(w,z)$ to $\Omega_0 \times \Omega_0$ has κ negative squares. Then $K(w,z)$ has κ negative squares on $\Omega \times \Omega$.*

Proofs. See [5, Theorems 1.1.4, 1.5.5, 1.5.6]. □

8.3 On pre-Kreĭn spaces

Another special case of the general theory of Kolmogorov decompositions gives results on completions of inner product spaces. We again follow [26].

If an inner product space $\mathfrak{H}_0$ is nonnegative, a standard quotient-completion construction produces an essentially unique Hilbert space. More generally, let $\mathfrak{H}_0$ be any linear and symmetric inner product space. Define a quotient space $\mathfrak{H}_0/\mathfrak{N}$, where $\mathfrak{N}$ is the set of elements of $\mathfrak{H}_0$ which are orthogonal to the full space. If $[f]$ is the coset determined by any $f \in \mathfrak{H}_0$, we obtain an inner product on $\mathfrak{H}_0/\mathfrak{N}$ by writing

$$\langle [f], [g]\rangle_{\mathfrak{H}_0/\mathfrak{N}} = \langle f, g\rangle_{\mathfrak{H}_0}, \qquad f, g \in \mathfrak{H}_0.$$

The quotient space is **nondegenerate**: the only element which is orthogonal to the full space is the zero element. In the nonnegative case, this means that the inner product is strictly positive, and therefore $\mathfrak{H}_0$ has an essentially unique completion to a Hilbert space. In general, we are interested to know, under what conditions does a nondegenerate inner product space have a "completion" to a Kreĭn space, and when is such a completion unique?

We first give formal definitions. Let $\mathfrak{H}_0$ be a nondegenerate inner product space with inner product $\langle\cdot,\cdot\rangle_{\mathfrak{H}_0}$. A **completion** of $\mathfrak{H}_0$ is a Kreĭn space $\mathfrak{H}$ which contains $\mathfrak{H}_0$ isometrically as a dense subspace (that is, $\mathfrak{H}_0$ is a dense linear subspace of $\mathfrak{H}$ and $\langle f,g\rangle_{\mathfrak{H}} = \langle f,g\rangle_{\mathfrak{H}_0}$ for all $f,g \in \mathfrak{H}_0$). By a **nonnegative majorant** for $\langle\cdot,\cdot\rangle_{\mathfrak{H}_0}$ we mean a nonnegative inner product $\langle\cdot,\cdot\rangle_+$ on $\mathfrak{H}_0$ such that

$$-\langle f,f\rangle_+ \leq \langle f,f\rangle_{\mathfrak{H}_0} \leq \langle f,f\rangle_+, \qquad f \in \mathfrak{H}_0.$$

Since we assume that $\mathfrak{H}_0$ is nondegenerate, such a majorant is strictly positive [26, p.929]; thus $\mathfrak{H}_0$ has a completion to a Hilbert space $\mathfrak{H}_+$ relative to $\langle\cdot,\cdot\rangle_+$, and there is a **Gram operator** $G \in \mathfrak{L}(\mathfrak{H}_+)$ such that

$$\langle f,g\rangle_{\mathfrak{H}_0} = \langle Gf,g\rangle_+, \qquad f,g \in \mathfrak{H}_0.$$

The Gram operator G is selfadjoint and satisfies $\|G\| \leq 1$. Two completions $\mathfrak{H}_1$ and $\mathfrak{H}_2$ of $\mathfrak{H}_0$ are **equivalent** if the identity mapping on $\mathfrak{H}_0$ extends to an isomorphism from $\mathfrak{H}_1$ onto $\mathfrak{H}_2$. We call $\mathfrak{H}_0$ a **pre-Kreĭn** space if it has a completion and any two completions are equivalent.

These notions correspond to their counterparts in §7 for an associated Hermitian kernel. Namely, given a nondegenerate inner product space $\mathfrak{H}_0$ as above, we define a Hermitian kernel by setting

$$K_{gf} = \langle f,g\rangle_{\mathfrak{H}_0}, \qquad f,g \in \mathfrak{H}_0.$$

The index set for the kernel is $\mathfrak{H}_0$ itself, and the underlying spaces are all chosen to be **C**, the complex numbers in the Euclidean metric. It is immediate from the definitions that $\mathfrak{H}_0$ has a completion if and only if the Hermitian kernel has a Kolmogorov decomposition. The definitions of nonnegative majorant and Gram operator for the inner product correspond to the same notions for the Hermitian kernel.

Theorem 8.9. *If $\mathfrak{H}_0$ is a nondegenerate inner product space, the following are equivalent:*

(1) *$\mathfrak{H}_0$ has a completion to a Kreĭn space;*
(2) *the inner product of $\mathfrak{H}_0$ has a nonnegative majorant $\langle\cdot,\cdot\rangle_+$;*
(3) *the inner product of $\mathfrak{H}_0$ is a difference of nonnegative inner products.*

If these conditions hold and $\langle\cdot,\cdot\rangle_+$ is a nonnegative majorant as in (2), there is a completion $\mathfrak{H}$ of $\mathfrak{H}_0$ which is contained continuously in the Hilbert space completion $\mathfrak{H}_+$ of $\mathfrak{H}_0$ in the (necessarily strictly positive) inner product $\langle\cdot,\cdot\rangle_+$. Any two such completions are equivalent if and only if the associated Gram operator for the nonnegative majorant has the unique factorization property.

Theorem 8.9 is a special case of Theorems 7.1 and 7.3. It can also be proved directly by repeating arguments in this special case. In a similar way, Theorem 7.4 yields:

Theorem 8.10. *A nondegenerate inner product space $\mathfrak{H}_0$ is a pre-Kreĭn space if and only if* (i) *it satisfies the conditions* (1)–(3) *of Theorem* 8.9, *and* (ii) *the Gram operator for every nonnegative majorant has the unique factorization property.*

9. The contractive substitution property

We return to the problem of de Branges on the coefficients of univalent functions (see §2): do the conditions (2.2) characterize initial segments $B_1, \ldots, B_r$ of the coefficients of a normalized univalent function which is bounded by one in the unit disk? The answer, as already indicated, is negative, but there is more to the story. It is not hard to see that the conditions are sufficient for $r = 1, 2$ [51, Theorem 3.4], and they are also sufficient in the limit as $r \to \infty$ [15]:

Theorem 9.1 (de Branges). *Let $B(z) = B_1 z + B_2 z^2 + B_3 z^3 + \cdots$ be a formal power series such that $B_1 > 0$ and* (2.2) *holds for all $r = 1, 2, \ldots$, every $\nu = -1, -2, \ldots$, and every generalized power series* (2.1). *Then $B(z)$ represents a univalent function which is bounded by one in the unit disk.*

See [66] for an account of the original proof by de Branges; a different proof is due to Nikolskii and Vasyunin in their work on coefficient problems and functional analytic aspects of the proof of the Bieberbach conjecture [59, 60] (see Theorem D180, p. 1219, and the remark D270, p. 1225, in the English translation of [60]).

The conditions (2.2) are reduced to a procedure which is analogous to the Schur algorithm in Christner, Li, and Rovnyak [21]. The classical Schur algorithm does not make sense in the present context, but the operator transcription in Foias and Frazho [40] can be adapted to the indefinite situation using properties of Julia operators as discussed in §5. The outcome is that if $B_1, \ldots, B_r$, $B_1 > 0$, are given numbers satisfying (2.2) for all real numbers ν, then the set of numbers B_{r+1} such that $B_1, \ldots, B_r, B_{r+1}$, satisfy the same conditions with r replaced by $r+1$ is the intersection of a family of closed disks $\Delta_r(\nu)$, $-\infty < \nu < \infty$. The centers and radii of the disks $\Delta_r(\nu)$ are functions of $B_1, \ldots, B_r$ which are given by recursive formulas. The formulas were implemented in a *Mathematica* program by an undergraduate student, A. Pitsillides [61]. A typical run is shown in the Figures 1–5 below. In each case, the white oval region inside the system of circles shows the possible values of B_{r+1} for the given numbers $B_1, \ldots, B_r$. The same formulas produced a counterexample when $r = 3$ in another undergraduate project by D. Dreibelbis [31]: the numbers

$$B_1 = \frac{1}{6}, \qquad B_2 = \frac{1}{4}, \qquad B_3 = \frac{4}{15} + \frac{i}{18} \tag{9.1}$$

satisfy (2.2) with $r = 3$ for all real ν, but there is no univalent function which has the form

$$B(z) = \frac{1}{6}\, z + \frac{1}{4}\, z^2 + \left(\frac{4}{15} + \frac{i}{18}\right) z^3 + \mathcal{O}(z^4)$$

and is bounded by one in the unit disk.

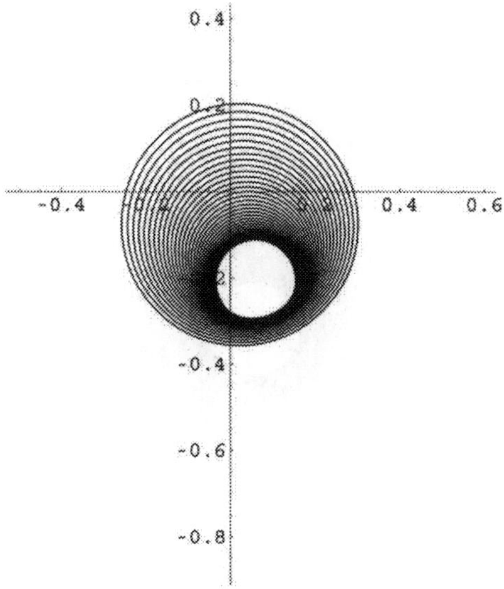

FIGURE 1. Possible values of B_3 if $B_1 = 0.2$ and $B_2 = -0.2 + 0.15\,i$

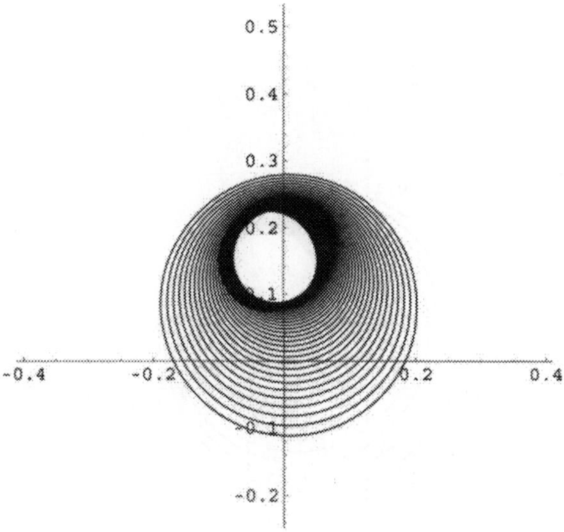

FIGURE 2. Possible values of B_4 if $B_3 = 0.1 - 0.2\,i$

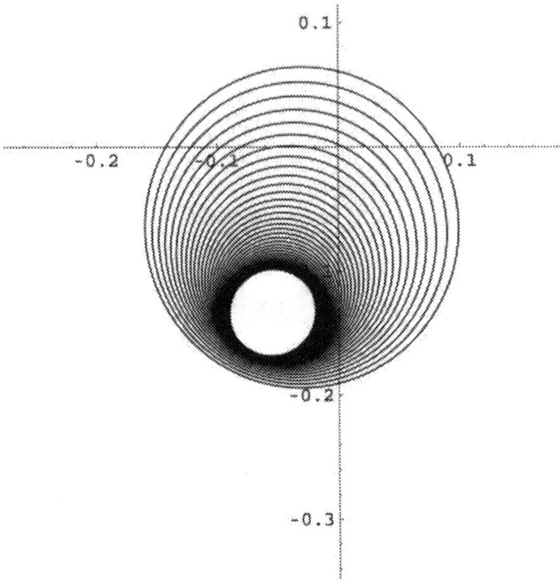

FIGURE 3. Possible values of B_5 if $B_4 = -0.02 + 0.2\,i$

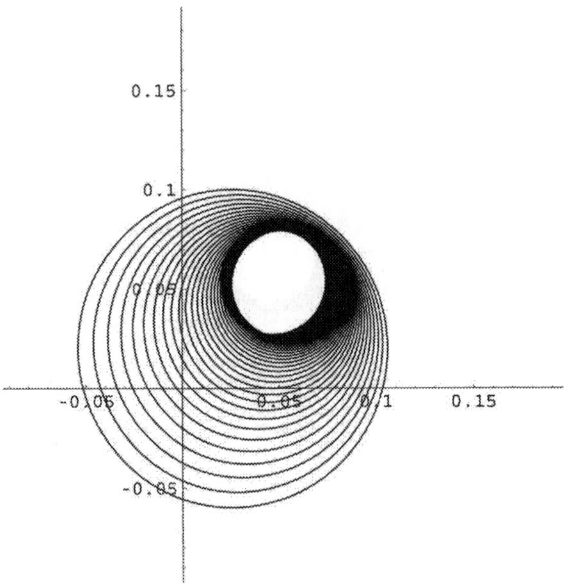

FIGURE 4. Possible values of B_6 if $B_5 = -0.05 - 0.15\,i$

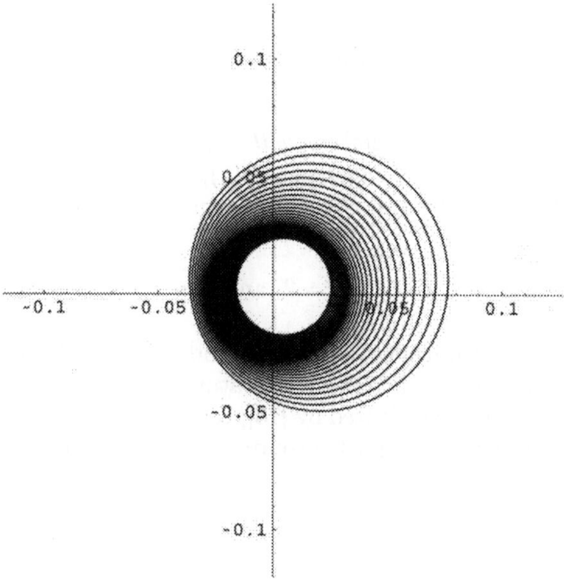

FIGURE 5. Possible values of B_7 if $B_6 = 0.06 + 0.05\,i$

Nevertheless, there is numerical evidence in favor of something like (2.2). The most basic example of a normalized univalent function which is bounded by one in the unit disk is a **bounded Koebe mapping**, by which we mean a solution $B(z) = B_{b,a,u}(z)$ of the functional equation $f_{a,u}(z) = f_{b,u}(B(z))$, where $0 < a \leq b < \infty$, $|u| = 1$, and

$$f_{t,u}(z) = tz/(1 - uz)^2, \qquad 0 < t < \infty,$$

is a Koebe function [65, §8.1]; that is,

$$B_{b,a,u}(z) = f_{b,u}^{-1}(f_{a,u}(z)). \tag{9.2}$$

Compositions of bounded Koebe mappings provide many data sets for numerical experiments. It appears that counterexamples such as (9.1) are only possible when numbers are chosen very close to the boundaries of the regions predicted by (2.2). In private discussions, M.A. Dritschel and the author have considered possible variations of (2.2), including:

Problem. *Let B_1, B_2, B_3, B_4 be given numbers with $B_1 > 0$. If (2.2) holds for $r = 4$, all real numbers ν, and all generalized power series (2.1), are B_1, B_2, B_3 the coefficients of a normalized univalent function which is bounded by one in the unit disk?*

The numbers (9.1) are not a counterexample because there is no way to choose B_4 to meet the conditions. More generally, if (2.2) holds for some numbers $B_1, \ldots, B_r$ $(B_1 > 0)$, are some of these numbers the coefficients of a normalized univalent function which is bounded by one in the unit disk? A related problem asks if, in some sense, the bounded Koebe mappings (9.2) play a role analogous to single Blaschke factors.

Problem. *Let $B_1, \ldots, B_r$ be numbers with $B_1 > 0$. If there exists a univalent and normalized function $B(z)$ satisfying $|B(z)| \leq 1$ on $\mathbf{D}$ and such that $B(z) = B_1 z + \cdots + B_r z^r + \mathcal{O}(z^{r+1})$, can $B(z)$ be chosen to be a composition of r bounded Koebe mappings?*

The answer is affirmative for $r = 2$, and numerical evidence for $r = 3$ seems strong.

The coefficients of univalent functions, and in particular bounded univalent functions, are extensively studied in the literature on classical function theory. The first four coefficients are investigated by [73]; connections with the problems stated above may be present but are not transparent to the author.

References

[1] J. Agler and M. Stankus, *m-isometric transformations of Hilbert space. I, II, III*, Integral Equations and Operator Theory **21** (1995), no. 4, 383–429, ibid. **23** (1995), no. 1, 1–48, ibid. **24** (1996), no. 4, 379–421.

[2] D. Alpay, *Reproducing kernel Kreĭn spaces of analytic functions and inverse scattering*, Ph.D. thesis, Weizmann Institute of Science, 1985.

[3] ———, *Dilatations des commutants d'opérateurs pour des espaces de Kreĭn de fonctions analytiques*, Ann. Inst. Fourier (Grenoble) **39** (1989), no. 4, 1073–1094.

[4] ———, *Some remarks on reproducing kernel Kreĭn spaces*, Rocky Mountain J. Math. **21** (1991), no. 4, 1189–1205.

[5] D. Alpay, A. Dijksma, J. Rovnyak, and H.S.V. de Snoo, *Schur functions, operator colligations, and reproducing kernel Pontryagin spaces*, Oper. Theory Adv. Appl., vol. 96, Birkhäuser, Basel, 1997.

[6] ———, *Reproducing kernel Pontryagin spaces*, Holomorphic Spaces, MSRI Publications, vol. 33, Cambridge University Press, Cambridge, 1998, pp. 425–444.

[7] R. Arocena, T.Ya. Azizov, A. Dijksma, and S.A.M. Marcantognini, *On commutant lifting with finite defect*, J. Operator Theory **35** (1996), no. 1, 117–132.

[8] ———, *On commutant lifting with finite defect. II*, J. Funct. Anal. **144** (1997), no. 1, 105–116.

[9] N. Aronszajn, *Theory of reproducing kernels*, Trans. Amer. Math. Soc. **68** (1950), 337–404.

[10] Gr. Arsene, T. Constantinescu, and A. Gheondea, *Lifting of operators and prescribed numbers of negative squares*, Michigan Math. J. **34** (1987), no. 2, 201–216.

[11] T.Ya. Azizov, Yu.P. Ginzburg, and H. Langer, *On the work of M. G. Kreĭn in the theory of spaces with an indefinite metric*, Ukraïn. Mat. Zh. **46** (1994), no. 1–2, 5–17.

[12] T.Ya. Azizov and I.S. Iokhvidov, *Linear operators in spaces with an indefinite metric*, John Wiley & Sons Ltd., Chichester, 1989.

[13] J.A. Ball and J.W. Helton, *A Beurling-Lax theorem for the Lie group $U(m, n)$ which contains most classical interpolation theory*, J. Operator Theory **9** (1983), no. 1, 107–142.

[14] J. Bognár, *Indefinite inner product spaces*, Springer-Verlag, New York, 1974, Ergebnisse der Mathematik und ihrer Grenzgebiete, Band 78.

[15] L. de Branges, *Square summable power series*, Unpublished book ms., 1985.

[16] ———, *A proof of the Bieberbach conjecture*, Acta Math. **154** (1985), no. 1-2, 137–152.

[17] ———, *Unitary linear systems whose transfer functions are Riemann mapping functions*, Operator theory and systems (Amsterdam, 1985), Oper. Theory Adv. Appl., vol. 19, Birkhäuser, Basel, 1986, pp. 105–124.

[18] ———, *Complementation in Kreĭn spaces*, Trans. Amer. Math. Soc. **305** (1988), no. 1, 277–291.

[19] ———, *Underlying concepts in the proof of the Bieberbach conjecture*, Proceedings of the International Congress of Mathematicians (Berkeley, California, 1986), Amer. Math. Soc., Providence, RI, 1988, pp. 25–42.

[20] ———, *A construction of Kreĭn spaces of analytic functions*, J. Funct. Anal. **98** (1991), no. 1, 1–41.

[21] G. Christner, Kin Y. Li, and J. Rovnyak, *Julia operators and coefficient problems*, Nonselfadjoint operators and related topics (Beer Sheva, 1992), Oper. Theory Adv. Appl., vol. 73, Birkhäuser, Basel, 1994, pp. 138–181.

[22] T. Constantinescu and A. Gheondea, *Minimal signature in lifting of operators. I*, J. Operator Theory **22** (1989), no. 2, 345–367.

[23] ———, *Extending factorizations and minimal negative signatures*, J. Operator Theory **28** (1992), no. 2, 371–402.

[24] ———, *Minimal signature in lifting of operators. II*, J. Funct. Anal. **103** (1992), no. 2, 317–351.

[25] ———, *Elementary rotations of linear operators in Kreĭn spaces*, J. Operator Theory **29** (1993), no. 1, 167–203.

[26] ———, *Representations of Hermitian kernels by means of Kreĭn spaces*, Publ. Res. Inst. Math. Sci. **33** (1997), no. 6, 917–951.

[27] B. Ćurgus and H. Langer, *On a paper of de Branges*, preprint, 1990.

[28] Ch. Davis, *J-unitary dilation of a general operator*, Acta Sci. Math. (Szeged) **31** (1970), 75–86.

[29] Ch. Davis and C. Foias, *Operators with bounded characteristic function and their J-unitary dilation*, Acta Sci. Math. (Szeged) **32** (1971), 127–139.

[30] A. Dijksma, M. Dritschel, S.A.M. Marcantognini, and H.S.V. de Snoo, *The commutant lifting theorem for contractions on Kreĭn spaces*, Operator extensions, interpolation of functions and related topics (Timişoara, 1992), Oper. Theory Adv. Appl., vol. 61, Birkhäuser, Basel, 1993, pp. 65–83.

[31] D. Dreibelbis, J. Rovnyak, and Kin Y. Li, *Coefficients of bounded univalent functions*, Proceedings of the Conference on Complex Analysis (Tianjin, 1992) (Cambridge, MA), Internat. Press, 1994, pp. 45–58.

[32] M.A. Dritschel, *Extension theorems for operators on Kreĭn spaces*, Ph.D. thesis, University of Virginia, 1989.

[33] ______, *Commutant lifting when the intertwining operator is not necessarily a contraction*, unpublished paper, 1993.

[34] ______, *The essential uniqueness property for operators on Kreĭn spaces*, J. Funct. Anal. **118** (1993), no. 1, 198–248.

[35] ______, *A module approach to commutant lifting on Kreĭn spaces*, Operator theory, system theory and related topics. The Moshe Livšic anniversary volume, Oper. Theory Adv. Appl., vol. 123, Birkhäuser, Basel, 2001, pp. 195–206.

[36] M.A. Dritschel and J. Rovnyak, *Operators on indefinite inner product spaces*, Lectures on operator theory and its applications (Waterloo, ON, 1994), Amer. Math. Soc., Providence, RI, pp. 141–232, 1996. *Supplement and errata*, `www.people.virginia.edu/~jlr5m/papers/papers.html`.

[37] ______, *Extension theorems for contraction operators on Kreĭn spaces*, Extension and interpolation of linear operators and matrix functions, Oper. Theory Adv. Appl., vol. 47, Birkhäuser, Basel, 1990, pp. 221–305.

[38] ______, *Julia operators and complementation in Kreĭn spaces*, Indiana Univ. Math. J. **40** (1991), no. 3, 885–901.

[39] H. Dym, *J contractive matrix functions, reproducing kernel Hilbert spaces and interpolation*, CBMS Regional Conference Series in Math., vol. **71**, Amer. Math. Soc., Providence, RI, 1989.

[40] C. Foias and A.E. Frazho, *The commutant lifting approach to interpolation problems*, Oper. Theory Adv. Appl., vol. 44, Birkhäuser, Basel, 1990.

[41] C. Foias, A.E. Frazho, I. Gohberg, and M.A. Kaashoek, *Metric constrained interpolation, commutant lifting and systems*, Oper. Theory Adv. Appl., vol. 100, Birkhäuser, Basel, 1998.

[42] A. Gheondea, *Contractive intertwining dilations of quasi-contractions*, Z. Anal. Anwendungen **15** (1996), no. 1, 31–44.

[43] S. Ghosechowdhury, *An expansion theorem for state space of unitary linear system whose transfer function is a Riemann mapping function*, Reproducing kernels and their applications (Newark, DE, 1997), Kluwer Acad. Publ., Dordrecht, 1999, pp. 81–95.

[44] ______, *Löwner expansions*, Math. Nachr. **210** (2000), 111–126.

[45] T. Hara, *Operator inequalities and construction of Kreĭn spaces*, Integral Equations and Operator Theory **15** (1992), no. 4, 551–567.

[46] C. Hellings, *Two-isometries on Pontryagin spaces*, Ph.D. thesis, University of Virginia, 2000.

[47] I.S. Iokhvidov, M.G. Kreĭn, and H. Langer, *Introduction to the spectral theory of operators in spaces with an indefinite metric*, Akademie-Verlag, Berlin, 1982.

[48] A.N. Kolmogorov, *Stationary sequences in Hilbert spaces*, Vestnik Moskov. Univ. Ser. I, Mat. Mech. **6** (1941), 1–40.

[49] M.G. Kreĭn and H. Langer, *Über die verallgemeinerten Resolventen und die charakteristische Funktion eines isometrischen Operators im Raume Π_κ*, Hilbert space operators and operator algebras (Proc. Internat. Conf., Tihany, 1970), North-Holland, Amsterdam, 1972, pp. 353–399. Colloq. Math. Soc. János Bolyai, 5.

[50] ———, *Über einige Fortsetzungsprobleme, die eng mit der Theorie hermitescher Operatoren im Raume Π_κ zusammenhängen. I. Einige Funktionenklassen und ihre Darstellungen*, Math. Nachr. **77** (1977), 187–236.

[51] Kin Y. Li and J. Rovnyak, *On the coefficients of Riemann mappings of the unit disk into itself*, Contributions to operator theory and its applications, Oper. Theory Adv. Appl., vol. 62, Birkhäuser, Basel, 1993, pp. 145–163.

[52] S.A.M. Marcantognini, *The commutant lifting theorem in the Kreĭn space setting: a proof based on the coupling method*, Indiana Univ. Math. J. **41** (1992), no. 4, 1303–1314.

[53] M. Martin and M. Putinar, *Lectures on hyponormal operators*, Oper. Theory Adv. Appl., vol. 39, Birkhäuser, Basel, 1989.

[54] S.A. McCullough and L. Rodman, *Two-selfadjoint operators in Kreĭn spaces*, Integral Equations and Operator Theory **26** (1996), no. 2, 202–209.

[55] ———, *Hereditary classes of operators and matrices*, Amer. Math. Monthly **104** (1997), no. 5, 415–430.

[56] B.W. McEnnis, *Characteristic functions and dilations of noncontractions*, J. Operator Theory **3** (1980), no. 1, 71–87.

[57] ———, *Models for operators with bounded characteristic function*, Acta Sci. Math. (Szeged) **43** (1981), no. 1-2, 71–90.

[58] ———, *Shifts on Kreĭn spaces*, Operator theory: operator algebras and applications, Part 2 (Durham, NH, 1988), Amer. Math. Soc., Providence, RI, 1990, pp. 201–211.

[59] N.K. Nikolskii and V.I. Vasyunin, *Quasi-orthogonal decompositions with respect to complementary metrics, and estimates for univalent functions*, Algebra i Analiz **2** (1990), no. 4, 1–81, Engl. transl., Leningrad Math. J. 2 (1991), no. 4, 691–764.

[60] ———, *Operator-valued measures and coefficients of univalent functions*, Algebra i Analiz **3** (1991), no. 6, 1–75 (1992), Engl. transl., St. Petersburg Math. J. 3 (1992), no. 6, 1199-1270.

[61] A. Pitsillides, *Mathematica program*, REU project, available along with the web version of this survey at `www.people.virginia.edu/~jlr5m/papers/papers.html`.

[62] S. Richter, *Invariant subspaces of the Dirichlet shift*, J. Reine Angew. Math. **386** (1988), 205–220.

[63] ———, *A representation theorem for cyclic analytic two-isometries*, Trans. Amer. Math. Soc. **328** (1991), no. 1, 325–349.

[64] S. Richter and C. Sundberg, *Multipliers and invariant subspaces in the Dirichlet space*, J. Operator Theory **28** (1992), no. 1, 167–186.

[65] M. Rosenblum and J. Rovnyak, *Topics in Hardy classes and univalent functions*, Birkhäuser, Basel, 1994.

[66] J. Rovnyak, *An extension problem for the coefficients of Riemann mappings*, Seminar lecture, `www.people.virginia.edu/~jlr5m/papers/papers.html`.

[67] ———, *Coefficient estimates for Riemann mapping functions*, J. Analyse Math. **52** (1989), 53–93.

[68] S. Saitoh, *Theory of reproducing kernels and its applications*, Longman Scientific & Technical, Harlow, 1988.

[69] L. Schwartz, *Sous-espaces hilbertiens d'espaces vectoriels topologiques et noyaux associés (noyaux reproduisants)*, J. Analyse Math. **13** (1964), 115–256.

[70] Yu.L. Shmul'yan, *Division in the class of J-expansive operators*, Mat. Sb. (N. S.) **74 (116)** (1967), 516–525; Engl. transl.: Math. USSR-Sbornik **3** (1967), 471–479.

[71] P. Sorjonen, *Pontryaginräume mit einem reproduzierenden Kern*, Ann. Acad. Sci. Fenn. Ser. A I Math. (1975), no. 594, 30 pages.

[72] B.Sz.-Nagy and C. Foias, *Harmonic analysis of operators on Hilbert space*, North-Holland Publishing Co., Amsterdam, 1970.

[73] O. Tammi, *Extremum problems for bounded univalent functions. I, II*, Lecture Notes in Mathematics, Springer-Verlag, Berlin, vol. 646, 1978; ibid. vol. 913, 1982.

[74] A.M. Yang, *A construction of unitary linear systems*, Integral Equations and Operator Theory **19** (1994), no. 4, 477–499.

James Rovnyak
Department of Mathematics
University of Virginia
P. O. Box 400137
Charlottesville, VA 22903-3199, USA
e-mail: rovnyak@Virginia.EDU

Operator Theory:
Advances and Applications, Vol. 134, 67–97

Notes on Interpolation in the Generalized Schur Class. I. Applications of Realization Theory

D. Alpay, T. Constantinescu, A. Dijksma, and J. Rovnyak

To Harry Dym: teacher, colleague and friend, in appreciation and with best wishes.

Abstract. Realization theory is used to study Nevanlinna-Pick and Carathéodory-Fejér interpolation problems for generalized Schur classes. In the first part of the paper, conditions are given for the existence of a solution of a factorization problem that includes Nevanlinna-Pick interpolation and factorization problems of Leech type for operator-valued functions. In the second part, an analysis is made of the numbers of positive and negative eigenvalues of classical matrices which arise in coefficient problems. The complete solution of an indefinite Carathéodory-Fejér problem is obtained.

Introduction

The classical approach to functions $S(z)$ in the generalized Schur class $\mathbf{S}_\kappa$ is by means of the Schwarz-Pick kernels on the unit disk $\mathbf{D}$. In the scalar case, which for the moment we assume, recall that these are defined by

$$\begin{cases} K_S(w,z) = \dfrac{1 - S(z)\overline{S(w)}}{1 - z\bar{w}}, \qquad K_{\tilde{S}}(w,z) = \dfrac{1 - \tilde{S}(z)\overline{\tilde{S}(w)}}{1 - z\bar{w}}, \\ D_S(w,z) = \begin{pmatrix} K_S(w,z) & \dfrac{S(z) - S(\bar{w})}{z - \bar{w}} \\ \dfrac{\tilde{S}(z) - \tilde{S}(\bar{w})}{z - \bar{w}} & K_{\tilde{S}}(w,z) \end{pmatrix}, \end{cases} \tag{1}$$

where $\tilde{S}(z) = \overline{S(\bar{z})}$. If $S(z)$ is analytic on an open subset Ω of $\mathbf{D}$ and one of the three kernels has κ negative squares, then all three kernels have κ negative squares [3, Theorem 2.5.2]. In this case $S(z)$ has an extension to a meromorphic function on $\mathbf{D}$, and $\mathbf{S}_\kappa$ is defined as the set of all such functions. The general theory and interpolation properties of generalized Schur functions are developed

A. Dijksma is grateful to Mr. Harry T. Dozor for supporting his research through a Dozor Fellowship at the Ben-Gurion University of the Negev, Beer-Sheva, Israel. J. Rovnyak is supported by NSF Grant DMS-9801016.

in well-known papers including Adamjan, Arov, and Kreĭn [1], Ball and Helton [7], Kreĭn and Langer [17, 18], Nudel′man [19], and Takagi [21], to name a few. The Schwarz-Pick kernels also arise in the realization theory of co-isometric, isometric, and unitary colligations [3], where they are reproducing kernels for state spaces. In this paper we study problems of the Nevanlinna-Pick and Carathéodory-Fejér types in which realization theory and kernels such as (1) and their generalizations play an important role.

In Part I we consider a form of the Nevanlinna-Pick interpolation problem which can also be viewed as a factorization problem: given functions $A(z)$ and $B(z)$ on a subset Ω of the unit disk, we seek a function $S(z)$ in $\mathbf{S}_\kappa$ such that

$$B(z) = A(z)S(z) \tag{2}$$

on Ω. Nothing is assumed about the set Ω, and so the factorization problem includes Nevanlinna-Pick interpolation. When Ω is an open set and $A(z)$ and $B(z)$ are holomorphic, (2) is a factorization problem of Leech type [3, 5]. Set

$$K(w,z) = \frac{A(z)\overline{A(w)} - B(z)\overline{B(w)}}{1 - z\bar{w}}, \qquad w, z \in \Omega.$$

If a solution $S(z)$ of (2) exists, then $K(w,z) = A(z)K_S(w,z)\overline{A(w)}$. Therefore a necessary condition for the existence of a solution is that the kernel $K(w,z)$ has κ negative squares. In Part I we show that when this necessary condition is satisfied and additional properties hold, it is possible to construct a solution of (2) in the form of a characteristic function

$$S(z) = H + wG(1 - wT)^{-1}F$$

of a partially isometric operator colligation $V = \{T, F, G, H\}$. We note that such expressions appear in many places in interpolation theory and applications in systems theory (for example, see Ball, Gohberg, and Rodman [6]). In systems language, we may think of the function $A(z)$ as an input, and then the problem (2) is to find a system whose transfer function produces the output $B(z) = A(z)S(z)$. The results of Part I are presented for operator-valued functions.

In Part II we discuss coefficient problems and their connection with the kernels (1). In this case also realization theory plays a role in establishing analyticity (see Theorem 8). Necessary conditions for the existence of solutions derived from the three kernels (1) are shown to be equivalent (Theorem 7). Using results of Iokhvidov [13], we provide a complete solution to an indefinite form of the Carathéodory-Fejér problem analogous to results of Woracek [22, 23] for the Nevanlinna-Pick problem. The solution is obtained by means of an equivalence of two matrix extension problems, one involving lower triangular Toeplitz matrices and the other Hermitian Toeplitz matrices. Scalar-valued functions are assumed in Part II.

An Appendix is devoted to a result that identifies the number of negative squares of a holomorphic kernel $K(w,z)$, $|w| < R$, $|z| < R$, in terms of the coefficients in the Taylor expansion $K(w,z) = \sum_{m,n=0}^{\infty} C_{mn} z^m \bar{w}^n$.

I. Interpolation and factorization

The most natural setting for the factorization problem (2) uses operator-valued functions. Let $\mathfrak{F}$ and $\mathfrak{G}$ be Pontryagin spaces having the same negative index (in most of our applications these are Hilbert spaces). For any integer $\kappa \geq 0$, the **generalized Schur class** $\mathbf{S}_\kappa(\mathfrak{F},\mathfrak{G})$ is the set of functions $S(z)$ with values in $\mathfrak{L}(\mathfrak{F},\mathfrak{G})$ which are holomorphic on some subregion Ω of the open unit disk $\mathbf{D}$ such that the kernel

$$K_S(w,z) = \frac{1 - S(z)S(w)^*}{1 - z\bar{w}} \tag{3}$$

has κ negative squares. If $S(z)$ belongs to $\mathbf{S}_\kappa(\mathfrak{F},\mathfrak{G})$, $\mathfrak{H}(S)$ is the Pontryagin space with reproducing kernel (3). Terminology and notation follow [3] (the definition of $\mathbf{S}_\kappa(\mathfrak{F},\mathfrak{G})$ in [3] requires the functions to be holomorphic at the origin, and we do not require this now). In particular, $\mathrm{sq}_- H$ is the number of negative squares of a Hermitian kernel H. In the scalar case, that is, when $\mathfrak{F} = \mathfrak{G} = \mathbf{C}$ is the space of complex numbers in the Euclidean metric, we write $\mathbf{S}_\kappa$ instead of $\mathbf{S}_\kappa(\mathfrak{F},\mathfrak{G})$.

Our first result contains the main construction, and it is in some sense the most general possible for the method. In Theorem 2 we recast the conditions in a more geometrical form in a particular case.

Theorem 1. *Let $\mathfrak{F}$, $\mathfrak{G}$, $\mathfrak{K}$ be Hilbert spaces, and let Ω be a subset of the unit disk containing the point w_0. Let $A(z)$ and $B(z)$ be functions on Ω with values in $\mathfrak{L}(\mathfrak{G},\mathfrak{K})$ and $\mathfrak{L}(\mathfrak{F},\mathfrak{K})$. Assume that the kernel*

$$K(w,z) = \frac{A(z)A(w)^* - B(z)B(w)^*}{1 - z\bar{w}} \tag{4}$$

has κ negative squares on $\Omega \times \Omega$, and let $\mathfrak{H}_K$ be the associated reproducing kernel Pontryagin space. Let $\mathfrak{M}$ be the subspace of $\mathfrak{H}_K \oplus \mathfrak{G}$ consisting of all elements $k(z) \oplus g$ such that

$$A(w_0)g = 0 \quad \text{and} \quad \frac{z - w_0}{\sqrt{1-|w_0|^2}}\, k(z) + [A(z) - A(w_0)]g \equiv 0 \quad \text{on} \quad \Omega.$$

Let $\mathfrak{N}$ be the subspace of $\mathfrak{H}_K \oplus \mathfrak{F}$ consisting of all elements $h(z) \oplus f$ such that

$$\frac{1 - z\bar{w}_0}{\sqrt{1-|w_0|^2}}\, h(z) + B(z)f \equiv 0 \quad \text{on} \quad \Omega.$$

Assume that $\mathfrak{M}$ and $\mathfrak{N}$ are Hilbert spaces in the inner products of the larger spaces. Then there is a function $S(z) \in \mathbf{S}_{\kappa'}(\mathfrak{F},\mathfrak{G})$ for some $\kappa' \leq \kappa$ such that

$$B(z) = A(z)S(z)$$

for $z = w_0$ and for all but at most κ points z of $\Omega \setminus \{w_0\}$. In this case, $\kappa' = \kappa$ if and only if the elements h of $\mathfrak{H}(S)$ such that $A(z)h(z) \equiv 0$ on Ω form a Hilbert subspace of $\mathfrak{H}(S)$.

The function $S(z)$ which is constructed in the proof is holomorphic at w_0. The subspaces $\mathfrak{M}$ and $\mathfrak{N}$ defined in the statement of the theorem are automatically closed by the continuity of function values in a reproducing kernel space [3, Theorem 1.1.2].

Proof. It is sufficient to prove the result when $0 \in \Omega$ and $w_0 = 0$. For suppose that the result is known in this case, and consider the general situation. Let φ be the linear fractional mapping of $\mathbf{D}$ onto itself given by $\varphi(z) = (w_0 - z)/(1 - z\bar{w}_0)$. Thus $\varphi(w_0) = 0$ and $\varphi^{-1} = \varphi$. Put $\Omega' = \varphi(\Omega)$, $w_0' = 0$, and

$$\begin{aligned} A'(z) &= A(\varphi^{-1}(z)), \qquad z \in \Omega', \\ B'(z) &= B(\varphi^{-1}(z)), \qquad z \in \Omega'. \end{aligned}$$

Define $K'(w,z)$ on $\Omega' \times \Omega'$ by (10) using $A'(z)$ and $B'(z)$ in place of $A(z)$ and $B(z)$. A short calculation shows that

$$K'(w,z) = \frac{1 - |w_0|^2}{(1 - z\bar{w}_0)(1 - \bar{w}w_0)} K(\varphi^{-1}(w), \varphi^{-1}(z)), \qquad w, z \in \Omega',$$

and so sq_$K' = \kappa$; write $\mathfrak{H}_{K'}$ for the associated reproducing kernel Pontryagin space. The preceding reproducing kernel identity may be used to show that the mapping

$$V' \colon f(z) \to \frac{\sqrt{1 - |w_0|^2}}{1 - z\bar{w}_0} f(\varphi^{-1}(z))$$

acts as an isometry from $\mathfrak{H}_K$ onto $\mathfrak{H}_{K'}$. Writing $\mathfrak{M}$ and $\mathfrak{N}$ for the subspaces defined in the theorem for the original functions $A(z)$ and $B(z)$ and point $w_0 \in \Omega$, and $\mathfrak{M}'$ and $\mathfrak{N}'$ for the corresponding subspaces relative to $A'(z)$ and $B'(z)$ and point $w_0' \in \Omega'$, we find that

$$(V' \oplus -1_{\mathfrak{G}})\,\mathfrak{M} = \mathfrak{M}' \quad \text{and} \quad (V' \oplus 1_{\mathfrak{F}})\,\mathfrak{N} = \mathfrak{N}.$$

Since we assume the result when $0 \in \Omega$ and $w_0 = 0$, we can find a function $S'(z) \in \mathbf{S}_{\kappa'}(\mathfrak{F}, \mathfrak{G})$ for some $\kappa' \le \kappa$ such that $S'(z)$ is holomorphic at $w_0' = 0$ and $B'(z) = A'(z)S'(z)$ for $z = w_0'$ and for all but at most κ points z of $\Omega' \setminus \{w_0'\}$. Then $S(z) = S'(\varphi(z))$ has the required properties.

Thus without loss of generality, we may assume that $0 \in \Omega$ and $w_0 = 0$. Define a linear relation $\mathbf{R}$ in $(\mathfrak{H}_K \oplus \mathfrak{G}) \times (\mathfrak{H}_K \oplus \mathfrak{F})$ as the span of all pairs

$$\left(\begin{pmatrix} K(\alpha,\cdot)u_1 \\ \dfrac{A(\alpha)^* - A(0)^*}{\bar{\alpha}} u_1 + A(0)^* u_2 \end{pmatrix}, \begin{pmatrix} \dfrac{K(\alpha,\cdot) - K(0,\cdot)}{\bar{\alpha}} u_1 + K(0,\cdot)u_2 \\ \dfrac{B(\alpha)^* - B(0)^*}{\bar{\alpha}} u_1 + B(0)^* u_2 \end{pmatrix} \right) \tag{5}$$

with $\alpha \in \Omega \setminus \{0\}$ and $u_1, u_2 \in \mathfrak{K}$. A direct calculation shows that $\mathbf{R}$ is isometric. In fact, consider a second pair with α replaced by β and u_1, u_2 replaced by v_1, v_2. Expand and simplify the inner products of the first members in $\mathfrak{H}_K \oplus \mathfrak{G}$ and second members in $\mathfrak{H}_K \oplus \mathfrak{F}$. After simplification, in both cases we obtain

$$\langle K(\alpha,\beta)u_1, v_1\rangle_{\mathfrak{K}} + \left\langle \frac{A(\beta)A(\alpha)^* - A(\beta)A(0)^* - A(0)A(\alpha)^* + A(0)A(0)^*}{\bar{\alpha}\beta} u_1, v_1 \right\rangle_{\mathfrak{K}}$$
$$+ \left\langle \frac{A(0)A(\alpha)^* - A(0)A(0)^*}{\bar{\alpha}} u_1, v_2 \right\rangle_{\mathfrak{K}} + \left\langle \frac{A(\beta)A(0)^* - A(0)A(0)^*}{\beta} u_2, v_1 \right\rangle_{\mathfrak{K}}$$
$$+ \langle A(0)A(0)^* u_2, v_2\rangle_{\mathfrak{K}}, \tag{6}$$

and this verifies the assertion. The orthogonal complement of the domain of $\mathbf{R}$ is $\mathfrak{M}$, and the orthogonal complement of the range of $\mathbf{R}$ is $\mathfrak{N}$. Since these are Hilbert spaces, it follows from [3, Theorem 1.4.2] (or [4, Theorem 2.2]) that there is a continuous partial isometry

$$V = \begin{pmatrix} T & F \\ G & H \end{pmatrix} : \begin{pmatrix} \mathfrak{H}_K \\ \mathfrak{F} \end{pmatrix} \to \begin{pmatrix} \mathfrak{H}_K \\ \mathfrak{G} \end{pmatrix}$$

such that V^* has initial space $\overline{\operatorname{dom}}\,\mathbf{R}$ and final space $\overline{\operatorname{ran}}\,\mathbf{R}$ and

$$V^* : \begin{pmatrix} K(\alpha,\cdot)u_1 \\ \dfrac{A(\alpha)^* - A(0)^*}{\bar{\alpha}} u_1 + A(0)^* u_2 \end{pmatrix} \to \begin{pmatrix} \dfrac{K(\alpha,\cdot) - K(0,\cdot)}{\bar{\alpha}} u_1 + K(0,\cdot)u_2 \\ \dfrac{B(\alpha)^* - B(0)^*}{\bar{\alpha}} u_1 + B(0)^* u_2 \end{pmatrix}$$

for all $\alpha \in \Omega \setminus \{0\}$ and $u_1, u_2 \in \mathfrak{K}$. Calculating as in [3, p. 51], we find that

$$\begin{aligned}
(Th)(z) &= \frac{h(z) - A(z)Gh}{z}, && z \in \Omega \setminus \{0\}, \\
(Ff)(z) &= \frac{B(z) - A(z)H}{z} f, && z \in \Omega \setminus \{0\}, \\
A(0)Gh &= h(0), \\
A(0)Hf &= B(0)f,
\end{aligned}$$

for all $h \in \mathfrak{H}_K$ and $f \in \mathfrak{F}$.

Since V is a partial isometry whose kernel is a Hilbert space, V is a contraction. The embedding mappings $E_{\mathfrak{F}}$ and $E_{\mathfrak{G}}$ from $\mathfrak{H}_K$ into $\mathfrak{H}_K \oplus \mathfrak{F}$ and $\mathfrak{H}_K \oplus \mathfrak{G}$ are contractions (in fact isometries), and their adjoints act as projections. The adjoints are also contractions because we assume that $\mathfrak{F}$ and $\mathfrak{G}$ are Hilbert spaces. Therefore

$$T = E_{\mathfrak{G}}^* V E_{\mathfrak{F}}$$

is a contraction on the Pontryagin space $\mathfrak{H}_K$. By [15, Lemma 11.1 (p. 75)], the part of the spectrum of T that lies in $|\lambda| > 1$ consists of normal eigenvalues. By [15, Theorem 11.2 (p. 84)], the span of root manifolds for eigenvalues in $|\lambda| > 1$ is contained in a nonpositive subspace, and hence the number of such eigenvalues is at most $\mathrm{sq}_-\mathfrak{H}_K = \kappa$. It follows that $1 - zT$ is invertible for all but at most κ points in $\mathbf{D}$. Since these exceptional points obviously do not include 0, $1 - zT$ is invertible for all $z \in \Omega \setminus \{\lambda_1, \dots, \lambda_q\}$ for some nonzero numbers $\lambda_1, \dots, \lambda_q$ in $\mathbf{D}$; here $q \le \kappa$ and possibly $q = 0$ when there are no exceptional points.

Claim 1: *If* $w \in \Omega \setminus \{\lambda_1, \ldots, \lambda_q\}$, $h \in \mathfrak{H}_K$, *and* $(1 - wT)^{-1}h = g$, *then*

$$g(z) = \frac{zh(z) - wA(z)Gg}{z - w}, \qquad z \in \Omega \setminus \{w\}, \tag{7}$$

and $h(w) = A(w)Gg$.

Since this is trivially true if $w = 0$, assume that $w \neq 0$. Then

$$h(z) = g(z) - w\,\frac{g(z) - A(z)Gg}{z}, \qquad z \in \Omega \setminus \{0\}. \tag{8}$$

Since $w \neq 0$, we can take $z = w$ in (8) to get $h(w) = A(w)Gg$. Again by (8),

$$(z - w)g(z) = zh(z) - wA(z)Gg$$

for $z \in \Omega \setminus \{0\}$. Trivially the last identity holds for $z = 0$ as well, and we obtain (7).

Claim 2: *Define* $S(w) = H + wG(1 - wT)^{-1}F$ *for all* $w \in \mathbf{D} \setminus \{\lambda_1, \ldots, \lambda_q\}$. *Then*

$$B(w) = A(w)S(w)$$

for all $w \in \Omega \setminus \{\lambda_1, \ldots, \lambda_q\}$.

The case $w = 0$ is clear. Assume $w \in \Omega \setminus \{\lambda_1, \ldots, \lambda_q\}$ and $w \neq 0$. Fix $f \in \mathfrak{F}$. We use Claim 1 with $g = (1 - wT)^{-1}h$, $h = Ff$. Thus

$$wA(w)G(1 - wT)^{-1}Ff = wA(w)Gg = wh(w) = w\,(Ff)(w) = B(w)f - A(w)Hf.$$

Claim 2 follows.

Claim 3: $S \in \mathbf{S}_{\kappa'}$ *for some* $\kappa' \leq \kappa$.

It is clear from the definition of $S(z)$ that it is a holomorphic function on $\mathbf{D} \setminus \{\lambda_1, \ldots, \lambda_q\}$. For all $w, z \in \mathbf{D} \setminus \{\lambda_1, \ldots, \lambda_q\}$, by the identity [3, (1.2.9)],

$$\begin{aligned}
1 - S(z)S(w)^* &= \begin{pmatrix} G(1 - zT)^{-1} & 1 \end{pmatrix} \begin{pmatrix} (1 - \bar{w}T^*)^{-1}G^* \\ 1 \end{pmatrix} \\
&\quad - \begin{pmatrix} zG(1 - zT)^{-1} & 1 \end{pmatrix} VV^* \begin{pmatrix} \bar{w}(1 - \bar{w}T^*)^{-1}G^* \\ 1 \end{pmatrix} \\
&= \begin{pmatrix} G(1 - zT)^{-1} & 1 \end{pmatrix} \begin{pmatrix} (1 - \bar{w}T^*)^{-1}G^* \\ 1 \end{pmatrix} \\
&\quad - \begin{pmatrix} zG(1 - zT)^{-1} & 1 \end{pmatrix} \begin{pmatrix} \bar{w}(1 - \bar{w}T^*)^{-1}G^* \\ 1 \end{pmatrix} \\
&\quad + \begin{pmatrix} zG(1 - zT)^{-1} & 1 \end{pmatrix} (1 - VV^*) \begin{pmatrix} \bar{w}(1 - \bar{w}T^*)^{-1}G^* \\ 1 \end{pmatrix} \\
&= (1 - z\bar{w})G(1 - zT)^{-1}(1 - \bar{w}T^*)^{-1}G^* \\
&\quad + \begin{pmatrix} zG(1 - zT)^{-1} & 1 \end{pmatrix} (1 - VV^*) \begin{pmatrix} \bar{w}(1 - \bar{w}T^*)^{-1}G^* \\ 1 \end{pmatrix}.
\end{aligned}$$

Since $1 - VV^* \geq 0$ in the partial ordering of selfadjoint operators, $1 - VV^* = MM^*$ for some operator $M \in \mathfrak{L}(\mathfrak{D}, \mathfrak{H}_K \oplus \mathfrak{G})$, where $\mathfrak{D}$ is a Hilbert space (see, for example,

[12, Theorem 2.1]; we can choose M so that it has zero kernel, but this property is not needed. Therefore

$$K_S(w,z) = G(1-zT)^{-1}(1-\bar{w}T^*)^{-1}G^* + \frac{\Phi(z)\Phi(w)^*}{1-z\bar{w}}, \quad w,z \in \mathbf{D}\setminus\{\lambda_1,\dots,\lambda_q\}, \tag{9}$$

where

$$\Phi(z) = \begin{pmatrix} zG(1-zT)^{-1} & 1 \end{pmatrix} M, \qquad z \in \mathbf{D}\setminus\{\lambda_1,\dots,\lambda_q\},$$

is a holomorphic function with values in $\mathfrak{L}(\mathfrak{D},\mathfrak{G})$. The first summand on the right of (9) has κ'' negative squares for some $\kappa'' \le \kappa$ by [3, Lemma 1.1.1′], and the second summand is nonnegative because $\mathfrak{D}$ is a Hilbert space. Thus by [3, Theorem 1.5.5] the kernel (9) has κ' negative squares, where $\kappa' \le \kappa'' \le \kappa$. Hence $S \in \mathbf{S}_{\kappa'}$, which proves Claim 3.

The function $S(z)$ has the required properties by Claims 2 and 3. The last statement, which gives the condition for $\kappa' = \kappa$, follows from [3, Theorem 1.5.7]. □

The next result identifies a case in which the conditions in Theorem 1 can be verified. Namely, we assume that the values of $A(z)$ are "square" in the sense that $\mathfrak{K} = \mathfrak{G}$ and so the values of $A(z)$ are in $\mathfrak{L}(\mathfrak{G})$. We also assume that one of these values is invertible, and we take this to be $1_{\mathfrak{G}}$.

Theorem 2. *Let $\mathfrak{F}$ and $\mathfrak{G}$ be Hilbert spaces, and let $A(z)$ and $B(z)$ be functions which are defined on a subset Ω of $\mathbf{D}$ with values in $\mathfrak{L}(\mathfrak{G})$ and $\mathfrak{L}(\mathfrak{F},\mathfrak{G})$. Assume that the kernel*

$$K(w,z) = \frac{A(z)A(w)^* - B(z)B(w)^*}{1-z\bar{w}} \tag{10}$$

has κ negative squares on $\Omega\times\Omega$, and let $\mathfrak{H}_K$ be the associated reproducing kernel Pontryagin space. Assume that there is a point $w_0 \in \Omega$ such that

(1) *$A(w_0) = 1_{\mathfrak{G}}$, and*
(2) *the set of elements of $\mathfrak{H}_K$ which vanish on $\Omega\setminus\{w_0\}$ is a Hilbert subspace of $\mathfrak{H}_K$.*

Then there is a function $S(z) \in \mathbf{S}_{\kappa'}(\mathfrak{F},\mathfrak{G})$ for some $\kappa' \le \kappa$ such that

$$B(z) = A(z)S(z)$$

for $z = w_0$ and for all but at most κ points z of $\Omega\setminus\{w_0\}$. In this case, $\kappa' = \kappa$ if and only if the elements h of $\mathfrak{H}(S)$ such that $A(z)h(z) \equiv 0$ on Ω form a Hilbert subspace of $\mathfrak{H}(S)$.

The function $S(z)$ constructed in the proof is holomorphic at w_0.

Proof. The last statement follows from [3, Theorem 1.5.7]. It is convenient to assume that $0 \in \Omega$ and $w_0 = 0$. If the result is known in this case, then as in the proof of Theorem 1, define $A'(z)$ and $B'(z)$ on $\Omega' = \varphi(\Omega)$, where $\varphi(z) = (w_0 - z)/(1 - z\bar{w}_0)$. As in the same proof, introduce the kernel $K'(w,z)$ and isomorphism V' from $\mathfrak{H}_K$ onto $\mathfrak{H}_{K'}$. Under V', the functions in $\mathfrak{H}_K$ which vanish on $\Omega \setminus \{w_0\}$ correspond to the functions in $\mathfrak{H}_{K'}$ which vanish on $\Omega' \setminus \{w_0'\}$, where $w_0' = 0$. Then as before, the special case implies the general result.

In what follows, we assume that $0 \in \Omega$ and $w_0 = 0$. We apply Theorem 1 in this situation and also with $\mathfrak{K} = \mathfrak{G}$. It is easy to see that the subspace $\mathfrak{M}$ in Theorem 1 coincides with the the set of elements of $\mathfrak{H}_K$ which vanish on $\Omega \setminus \{0\}$ and is thus a Hilbert space by hypothesis. We show that the subspace $\mathfrak{N}$ in Theorem 1 is a Hilbert space. By the first part of the proof of Theorem 1, $\mathfrak{N}$ is the orthogonal complement of the range of the relation $\mathbf{R}$ in $\mathfrak{H}_K \oplus \mathfrak{F}$, and therefore it is the same thing to show that the range of $\mathbf{R}$ contains a strictly negative subspace of dimension κ. By [3, Lemma 1.1.1′], it is sufficient to show that some Gram matrix of elements of the range of $\mathbf{R}$ has κ negative eigenvalues. In fact, consider two of the second members of the pairs (5) that define $\mathbf{R}$, say

$$\begin{pmatrix} \dfrac{K(\alpha,\cdot)-K(0,\cdot)}{\bar\alpha}u_1 + K(0,\cdot)u_2 \\ \dfrac{B(\alpha)^*-B(0)^*}{\bar\alpha}u_1 + B(0)^*u_2 \end{pmatrix} \quad \text{and} \quad \begin{pmatrix} \dfrac{K(\beta,\cdot)-K(0,\cdot)}{\bar\beta}v_1 + K(0,\cdot)v_2 \\ \dfrac{B(\beta)^*-B(0)^*}{\bar\beta}v_1 + B(0)^*v_2 \end{pmatrix}.$$

By (6), since now $A(0) = 1_{\mathfrak{G}}$, the inner product of these elements in $\mathfrak{H}_K \oplus \mathfrak{F}$ is equal to

$$\begin{aligned} &\langle K(\alpha,\beta)u_1, v_1\rangle_{\mathfrak{G}} + \left\langle \frac{A(\beta)A(\alpha)^* - A(\beta) - A(\alpha)^* + 1_{\mathfrak{G}}}{\bar\alpha\beta} u_1, v_1 \right\rangle_{\mathfrak{G}} \\ &\quad + \left\langle \frac{A(\alpha)^* - 1_{\mathfrak{G}}}{\bar\alpha} u_1, v_2 \right\rangle_{\mathfrak{G}} + \left\langle \frac{A(\beta) - 1_{\mathfrak{G}}}{\beta} u_2, v_1 \right\rangle_{\mathfrak{G}} + \langle u_2, v_2\rangle_{\mathfrak{G}} \\ &= \langle K(\alpha,\beta)u_1, v_1\rangle_{\mathfrak{G}} + \left\langle \frac{A(\alpha)^* - 1_{\mathfrak{G}}}{\bar\alpha} u_1 + u_2, \frac{A(\beta)^* - 1_{\mathfrak{G}}}{\bar\beta} v_1 + v_2 \right\rangle_{\mathfrak{G}}. \end{aligned}$$

Here we can choose α, β and u_1, u_2 arbitrarily, and then choose v_1, v_2 so that

$$\frac{A(\alpha)^* - 1_{\mathfrak{G}}}{\bar\alpha} u_1 + u_2 = \frac{A(\beta)^* - 1_{\mathfrak{G}}}{\bar\beta} v_1 + v_2 = 0.$$

Since we assume that $\mathrm{sq}_- K = \kappa$, it follows that some Gram matrix of elements of the range of $\mathbf{R}$ has κ negative eigenvalues, as was to be shown. This completes the proof that $\mathfrak{N}$ is a Hilbert space.

The hypotheses of Theorem 1 are thus met, and Theorem 1 yields a function $S(z) \in \mathbf{S}_{\kappa'}(\mathfrak{F}, \mathfrak{G})$, $\kappa' \le \kappa$, such that $B(z) = A(z)S(z)$ for $z = 0$ and for all but at most κ points z of $\Omega \setminus \{0\}$. □

We give another condition for interpolation. Suppose that $S(z)$ belongs to $\mathbf{S}_\kappa$ and is holomorphic at the origin. Then $zS(z)$ also belongs to $\mathbf{S}_\kappa$, and thus both kernels

$$\frac{1 - S(z)\overline{S(w)}}{1 - z\bar w} \quad \text{and} \quad \frac{1 - z\bar w S(z)\overline{S(w)}}{1 - z\bar w}$$

have κ negative squares (see [3, Example 1 on p. 132]). In the other direction, a condition on two kernels is sufficient for interpolation from an arbitrary set Ω with at most a finite number of exceptional points.

Theorem 3. *Let $A(z)$ and $B(z)$ be functions defined on a subset Ω of the unit disk* $\mathbf{D}$ *with values in $\mathfrak{L}(\mathfrak{G},\mathfrak{K})$ and $\mathfrak{L}(\mathfrak{F},\mathfrak{K})$, where $\mathfrak{F},\mathfrak{G},\mathfrak{K}$ are Hilbert spaces. Assume that both*

$$K_1(w,z) = \frac{A(z)A(w)^* - B(z)B(w)^*}{1 - z\bar{w}}$$

and

$$K_2(w,z) = \frac{A(z)A(w)^* - z\bar{w}B(z)B(w)^*}{1 - z\bar{w}}$$

have κ negative squares on $\Omega \times \Omega$. Then there is a function $S(z)$ in $\mathbf{S}_{\kappa'}(\mathfrak{F},\mathfrak{G})$, $\kappa' \leq \kappa$, such that

$$B(z) = A(z)S(z)$$

for all but at most κ points z of Ω. In this case, $\kappa' = \kappa$ if and only if the elements h of $\mathfrak{H}(S)$ such that $A(z)h(z) \equiv 0$ on Ω form a Hilbert subspace of $\mathfrak{H}(S)$.

The proof uses a different colligation from that of Theorem 1. It is adapted from the work of V.E. Katsnelson, A. Kheifets, and P.M. Yuditskiĭ; see Kheifets [16] for an account and references to earlier works. The idea is used by Ball and Trent [8], who extend it to a several variable setting and apply it in a form for reproducing kernel functions that is close to our situation.

Theorem 3 is a non-holomorphic analog of [5, Theorem 11]: there the coefficient spaces are indefinite, but we have the stronger hypothesis that Ω is a neighborhood of the origin and $A(z)$ and $B(z)$ are holomorphic. Now the functions $A(z)$ and $B(z)$ are not assumed to be holomorphic, but in compensation $\mathfrak{F}$ and $\mathfrak{G}$ are required to be Hilbert spaces (for simplicity we have taken $\mathfrak{K}$ to be a Hilbert space also, but this plays no role in the argument). The proof of Theorem 3 runs along the same lines.

Proof. Write $\mathfrak{H}(K_1)$ and $\mathfrak{H}(K_2)$ for the Pontryagin spaces with reproducing kernels $K_1(w,z)$ and $K_2(w,z)$. Define a relation

$$\mathbf{R} = \operatorname{span}\left\{ \left(\begin{pmatrix} K_1(w,\cdot)k \\ B(w)^*k \end{pmatrix}, \begin{pmatrix} \bar{w}K_1(w,\cdot)k \\ A(w)^*k \end{pmatrix} \right) : w \in \Omega,\ k \in \mathfrak{K} \right\}$$
$$\subseteq \begin{pmatrix} \mathfrak{H}(K_1) \\ \mathfrak{F} \end{pmatrix} \times \begin{pmatrix} \mathfrak{H}(K_1) \\ \mathfrak{G} \end{pmatrix}.$$

It is easy to see that $\mathbf{R}$ is isometric. We show that the domain $\mathfrak{M}$ of $\mathbf{R}$ contains a maximal uniformly negative subspace of $\mathfrak{H}(K_1) \oplus \mathfrak{F}$. To this end, consider a Gram matrix of the form

$$M = \left(\left\langle \begin{pmatrix} K_1(w_j,\cdot)k_j \\ B(w_j)^*k_j \end{pmatrix}, \begin{pmatrix} K_1(w_i,\cdot)k_i \\ B(w_i)^*k_i \end{pmatrix} \right\rangle_{\mathfrak{H}(K_1)\oplus\mathfrak{F}} \right)_{i,j=1}^{n},$$

where $w_1, \ldots, w_n$ are any points in Ω and $k_1, \ldots, k_n$ are arbitrary vectors in $\mathfrak{K}$. Thus

$$M = \left(\langle \left[K_1(w_j, w_i) + B(w_i)B(w_j)^* \right] k_j, k_i \rangle_{\mathfrak{K}} \right)_{i,j=1}^{n} = \left(\langle K_2(w_j, w_i) k_j, k_i \rangle_{\mathfrak{K}} \right)_{i,j=1}^{n}.$$

Since we assume that $K_2(w, z)$ has κ negative squares, M has at most κ negative eigenvalues no matter how $w_1, \ldots, w_n$ and $k_1, \ldots, k_n$ are chosen, and some such Gram matrix has exactly κ negative eigenvalues. By [3, Lemma 1.1.1′], $\mathfrak{M}$ contains a κ-dimensional subspace which is the antispace of a Hilbert space in the inner product of $\mathfrak{H}(K_1) \oplus \mathfrak{F}$. Since

$$\mathrm{sq}_-(\mathfrak{H}(K_1) \oplus \mathfrak{F}) = \kappa,$$

this verifies the assertion. It follows that the closure of $\mathfrak{M}$ in $\mathfrak{H}(K_1) \oplus \mathfrak{F}$ is a regular subspace whose orthogonal complement $\mathfrak{M}^\perp$ is a Hilbert space.

By [3, Theorem 1.4.2], the closure of the range of $\mathbf{R}$ is likewise a regular subspace $\mathfrak{N}$ of $\mathfrak{H}(K_1) \oplus \mathfrak{G}$, and we can construct a partial isometry

$$V = \begin{pmatrix} T & F \\ G & H \end{pmatrix} : \begin{pmatrix} \mathfrak{H}(K_1) \\ \mathfrak{F} \end{pmatrix} \to \begin{pmatrix} \mathfrak{H}(K_1) \\ \mathfrak{G} \end{pmatrix}$$

with initial space $\mathfrak{M}$ and final space $\mathfrak{N}$ such that

$$V^* = \begin{pmatrix} T^* & G^* \\ F^* & H^* \end{pmatrix} : \begin{pmatrix} \bar{w} K_1(w, \cdot)k \\ A(w)^* k \end{pmatrix} \to \begin{pmatrix} K_1(w, \cdot)k \\ B(w)^* k \end{pmatrix}$$

for all $k \in \mathfrak{K}$ and all $w \in \Omega$. Thus for $w \in \Omega$,

$$T^* \{\bar{w} K_1(w, \cdot)k\} + G^* \{A(w)^* k\} = K_1(w, \cdot)k, \tag{11}$$

and

$$F^* \{\bar{w} K_1(w, \cdot)k\} + H^* \{A(w)^* k\} = B(w)^* k. \tag{12}$$

Hence

$$(1 - \bar{w} T^*) \{K_1(w, \cdot) A(w)^* k\} = G^* \{A(w)^* k\}. \tag{13}$$

Since $\ker V$ is a Hilbert space, V is a contraction. As in the proof of Theorem 1, because we assume that $\mathfrak{F}$ and $\mathfrak{G}$ are Hilbert spaces, T is a contraction, and the part of the spectrum of T that lies in $|\lambda| > 1$ consists of at most κ normal eigenvalues.

Let $\Omega' = \Omega \setminus \{\lambda_1, \ldots, \lambda_q\}$, where $\lambda_1, \ldots, \lambda_q$ are the points λ of the unit disk at which $1 - \lambda T$ is not invertible ($q \leq \kappa$). For all $w \in \Omega'$ and all $k \in \mathfrak{K}$,

$$K_1(w, \cdot)k = (1 - \bar{w} T^*)^{-1} G^* \{A(w)^* k\}$$

by (13). Define

$$S(z) = H + zG(1 - zT)^{-1} F, \qquad z \in \mathbf{D} \setminus \{\lambda_1, \ldots, \lambda_q\}.$$

Then

$$B(w) = A(w) S(w), \qquad w \in \Omega',$$

by (11) and (12). The proof that $\hat{S} \in \mathbf{S}_{\kappa'}$ for some $\kappa' \leq \kappa$ is the same as in the proof of Theorem 1. The last statement follows from [3, Theorem 1.5.7]. □

In the next theorem, we allow $\mathfrak{F}, \mathfrak{G}, \mathfrak{K}$ to be indefinite, but the functions $A(z)$ and $B(z)$ are required to be holomorphic. This yields a new result of Leech type factorization theorems as a companion to those of [5].

Theorem 4. *Let $\mathfrak{F}$, $\mathfrak{G}$, $\mathfrak{K}$ be Kreĭn spaces with* $\mathrm{sq}\_\mathfrak{F} = \mathrm{sq}\_\mathfrak{G} < \infty$. *Let Ω be a subregion of the unit disk containing the origin. Let $A(z)$ and $B(z)$ be holomorphic functions on Ω with values in $\mathfrak{L}(\mathfrak{G}, \mathfrak{K})$ and $\mathfrak{L}(\mathfrak{F}, \mathfrak{K})$. Assume that the kernel*

$$K(w,z) = \frac{A(z)A(w)^* - B(z)B(w)^*}{1 - z\bar{w}} \tag{14}$$

has κ negative squares on $\Omega \times \Omega$, and let $\mathfrak{H}_K$ be the associated reproducing kernel Pontryagin space. Let $\mathfrak{M}$ be the subspace of $\mathfrak{H}_K \oplus \mathfrak{G}$ consisting of all elements $k(z) \oplus g$ such that

$$A(0)g = 0 \quad \text{and} \quad zk(z) + [A(z) - A(0)]g \equiv 0 \quad \text{on} \quad \Omega.$$

Let $\mathfrak{N}$ be the subspace of $\mathfrak{H}_K \oplus \mathfrak{F}$ consisting of all elements $h(z) \oplus f$ such that

$$h(z) + B(z)f \equiv 0 \quad \text{on} \quad \Omega.$$

Assume that $\mathfrak{M}$ and $\mathfrak{N}$ are Hilbert spaces in the inner products of the larger spaces. Then there is a function $S(z) \in \mathbf{S}_{\kappa'}(\mathfrak{F}, \mathfrak{G})$ for some $\kappa' \leq \kappa$ which is holomorphic at the origin and such that

$$B(z) = A(z)S(z)$$

for all but at most κ points z of Ω. In this case, $\kappa' = \kappa$ if and only if the elements h of $\mathfrak{H}(S)$ such that $A(z)h(z) \equiv 0$ on Ω form a Hilbert subspace of $\mathfrak{H}(S)$.

Proof. We repeat the constructions in the proof of Theorem 1. The partial isometry V is again a contraction in the present situation. In general, the operator T is not a contraction, but it is a bounded operator and so $(1 - wT)^{-1}$ is defined for $|w|$ sufficiently small. The argument goes through if we restrict attention to a suitable neighborhood of the origin. At the end, the identity $B(z) = A(z)S(z)$ extends to all but at most κ points of Ω by analytic continuation. □

II. Coefficient and moment problems

Let $z_1, \dots, z_n$ be points in the unit disk, and let $w_1, \dots, w_n$ be any complex numbers. If we specialize Part I to the scalar case and set $\Omega = \{z_1, \dots, z_n\}$, $A(z_j) = 1$, and $B(z_j) = w_j$ for all $j = 1, \dots, n$, then the interpolation problem in Part I reduces to the Nevanlinna-Pick problem. The indefinite form of interpolation was introduced by Takagi [21], and it has been studied by Adamjan, Arov, and Kreĭn [1], Kreĭn and Langer [18], and others. A rather complete picture of the solution of the indefinite Nevanlinna-Pick problem emerged from this work. A remaining issue concerning the degenerate case was recently settled. Namely, one can ask, for which nonnegative integers κ can the Nevanlinna-Pick problem be solved in $\mathbf{S}_\kappa$ for given data $z_1, \dots, z_n$ and $w_1, \dots, w_n$? A more precise question can be posed. Define $\mathbf{S}_{\nu,\pi}$ as the class of all meromorphic functions $S(z)$ on the unit disk for which the kernel

$K_S(w,z)$ has ν negative squares and π positive squares (thus $\mathbf{S}_{\nu,\pi}$ is a subclass of $\mathbf{S}_\nu$). For which nonnegative integers ν and π can the Nevanlinna-Pick problem be solved in $\mathbf{S}_{\nu,\pi}$ for given data $z_1,\dots,z_n$ and $w_1,\dots,w_n$? These questions were answered by Woracek [22, 23] (with the disk replaced by the upper half-plane), yielding a complete solution of the Nevanlinna-Pick problem in the scalar case.

We consider analogous questions for the indefinite Carathéodory-Fejér problem and obtain a complete solution in the scalar case. The solution depends on results of Iokhvidov [13] on a related trigonometric moment problem. In the positive definite case this connection is well known. We refer to [13] for references to the original papers (some jointly with M.G. Kreĭn) pertaining to this problem. A key step involves another application of the characteristic function of a partially isometric operator colligation, which was the principal tool in Part I.

Problem I (Carathéodory-Fejér problem). *Let $a_0, a_1, \dots, a_{n-1}$ be n complex numbers. For which nonnegative integers κ is there a function $S(z)$ in $\mathbf{S}_\kappa$ which is holomorphic at the origin and such that $S(z) = a_0 + a_1 z + \cdots + a_{n-1}z^{n-1} + \mathcal{O}(z^n)$ in a neighborhood of the origin? For which ν and π do there exist solutions in $\mathbf{S}_{\nu,\pi}$?*

Necessary conditions on coefficients are obtained from the series expansions of standard kernel functions. Suppose that $S(z)$ is a holomorphic (scalar-valued) function defined in a neighborhood of the origin. Let $S(z) = a_0 + a_1 z + a_2 z^2 + \cdots$ be its Taylor series expansion, and write

$$T_r = \begin{pmatrix} a_0 & 0 & 0 & \cdots & 0 \\ a_1 & a_0 & 0 & \cdots & 0 \\ & & \cdots & & \\ a_{r-1} & a_{r-2} & a_{r-3} & \cdots & a_0 \end{pmatrix}, \qquad \overline{T}_r = \begin{pmatrix} \bar a_0 & 0 & 0 & \cdots & 0 \\ \bar a_1 & \bar a_0 & 0 & \cdots & 0 \\ & & \cdots & & \\ \bar a_{r-1} & \bar a_{r-2} & \bar a_{r-3} & \cdots & \bar a_0 \end{pmatrix}, \tag{15}$$

$$Q_r = \begin{pmatrix} a_1 & a_2 & a_3 & \cdots & a_r \\ a_2 & a_3 & a_4 & \cdots & a_{r+1} \\ & & \cdots & & \\ a_r & a_{r+1} & a_{r+2} & \cdots & a_{2r-1} \end{pmatrix}, \tag{16}$$

$r = 1, 2, \dots$. Set $\tilde S(z) = \overline{S(\bar z)}$. Straightforward calculations yield the expansions

$$K_S(w,z) = \frac{1 - S(z)\overline{S(w)}}{1 - z\bar w} = \sum_{p,q=0}^{\infty} C_{pq} z^p \bar w^q,$$

$$K_{\tilde S}(w,z) = \frac{1 - \tilde S(z)\overline{\tilde S(w)}}{1 - z\bar w} = \sum_{p,q=0}^{\infty} \overline{C}_{pq} z^p \bar w^q,$$

$$D_S(w,z) = \begin{pmatrix} K_S(w,z) & \dfrac{S(z) - S(\bar w)}{z - \bar w} \\ \dfrac{\tilde S(z) - \tilde S(\bar w)}{z - \bar w} & K_{\tilde S}(w,z) \end{pmatrix} = \sum_{p,q=0}^{\infty} D_{pq} z^p \bar w^q,$$

where

$$\left[C_{pq}\right]_{p,q=0}^{n-1} = I_n - T_n T_n^*, \qquad \left[\bar{C}_{pq}\right]_{p,q=0}^{n-1} = I_n - \bar{T}_n \bar{T}_n^*,$$

$$\left[D_{pq}\right]_{p,q=0}^{r-1} = \begin{pmatrix} I_r - T_r T_r^* & Q_r \\ Q_r^* & I_r - \bar{T}_r \bar{T}_r^* \end{pmatrix}, \qquad 1 \le r \le n/2.$$

Thus the coefficients $a_0, a_1, \ldots$ of $S(z)$ give rise to three families of matrices:

$$I_n - T_n T_n^*, \qquad I_n - \bar{T}_n \bar{T}_n^*, \qquad \begin{pmatrix} I_r - T_r T_r^* & Q_r \\ Q_r^* & I_r - \bar{T}_r \bar{T}_r^* \end{pmatrix}, \qquad 1 \le r \le n/2, \tag{17}$$

$n = 0, 1, \ldots$. For fixed n, the matrices (17) depend only on $a_0, \ldots, a_{n-1}$.

If $S(z)$ belongs to $\mathbf{S}_\kappa$, then the three kernels each have κ negative squares [3, Theorem 2.5.2]. It follows that the number of negative eigenvalues of each of the matrices in (17) is a nondecreasing function of the order of the matrix, and this number is ultimately equal to κ in each case (see the result in the Appendix at the end of the paper).

If $S(z)$ belongs to $\mathbf{S}_{\nu,\pi}$, similar remarks apply not only to the number of negative squares but also to the number of positive squares. For simplicity, suppose that $S(0) \neq 0$, and note the identities

$$K_S(w,z) = -S(z) K_{1/S}(w,z) \overline{S(w)},$$

$$K_{\tilde{S}}(w,z) = -\tilde{S}(z) K_{1/\tilde{S}}(w,z) \overline{\tilde{S}(w)},$$

$$D_S(w,z) = -\begin{pmatrix} S(z) & 0 \\ 0 & \tilde{S}(z) \end{pmatrix} D_{1/S}(w,z) \begin{pmatrix} \overline{S(w)} & 0 \\ 0 & \overline{\tilde{S}(w)} \end{pmatrix}.$$

The numbers of positive squares of

$$K_S(w,z), K_{\tilde{S}}(w,z), D_S(w,z)$$

thus coincide with the numbers of negative squares of

$$K_{1/S}(w,z), K_{1/\tilde{S}}(w,z), D_{1/S}(w,z),$$

respectively. Hence if one of the three kernels has π positive squares, then all do. In this case, applying the previous assertions concerning negative squares, we see that the number of positive eigenvalues of each of the matrices in (17) is a nondecreasing function of the order of the matrix, and this number is ultimately equal to π in each case.

This raises questions concerning the general behavior of the numbers of negative and positive eigenvalues for the matrices (17) whenever (15) and (16) are defined for any complex numbers $a_0, a_1, \ldots$, whether these numbers are the Taylor coefficients of a holomorphic function or not. We show that the behavior is indeed always similar to the special cases noted above: the numbers of negative (positive) eigenvalues for the three types are nondecreasing functions of the order, and if one eventually has some constant value, then all have the same constant value eventually. These questions are purely algebraic. There is a separate convergence question, namely, under what conditions are the given numbers $a_0, a_1, \ldots$ the Taylor coefficients of a holomorphic function $S(z)$ in $\mathbf{S}_\kappa$ or $\mathbf{S}_{\nu,\pi}$? Finally, if we only define (15), (16), and (17) as far as we can go with a finite sequence $a_0, \ldots, a_{n-1}$, what are the possible extensions to an infinite sequence $a_0, a_1, \ldots$?

To answer such questions, we relate given complex numbers $a_0, \ldots, a_{n-1}$ to a trigonometric moment problem. Define $c_0 = 1, c_1, \ldots, c_n$ by

$$\begin{cases} c_0 &= 1, \\ c_1 &= c_0 a_0, \\ c_2 &= c_0 a_1 + c_1 a_0, \\ & \ldots \\ c_n &= c_0 a_{n-1} + c_1 a_{n-2} + \cdots + c_{n-1} a_0, \end{cases} \tag{18}$$

This correspondence is one-to-one and has the property that if $a_0, \ldots, a_{n-1}$ corresponds to $c_0 = 1, c_1, \ldots, c_n$ then for each $1 \le k \le n$, $a_0, \ldots, a_{k-1}$ corresponds to $c_0 = 1, c_1, \ldots, c_k$ also via (18) with n replaced by k. We consider the associated matrix

$$M_n = \begin{pmatrix} c_0 & \bar{c}_1 & \bar{c}_2 & \ldots & \bar{c}_n \\ c_1 & c_0 & \bar{c}_1 & \ldots & \bar{c}_{n-1} \\ c_2 & c_1 & c_0 & \ldots & \bar{c}_{n-2} \\ & & \ldots & & \\ c_n & c_{n-1} & c_{n-2} & \ldots & c_0 \end{pmatrix} \tag{19}$$

In the sequel J_n stands for the selfadjoint and unitary $n \times n$ matrix

$$J_n = \begin{pmatrix} 0 & 0 & \ldots & 0 & 1 \\ 0 & 0 & \ldots & 1 & 0 \\ & & \ldots & & \\ 1 & 0 & \ldots & 0 & 0 \end{pmatrix}.$$

Also define

$$B_r = \begin{pmatrix} c_0 & 0 & \ldots & 0 \\ c_1 & c_0 & \ldots & 0 \\ & & \ldots & \\ c_r & c_{r-1} & \ldots & c_0 \end{pmatrix}, \quad C_r = \begin{pmatrix} I_r & 0 \\ 0 & B_r \end{pmatrix} \begin{pmatrix} 0 & B_r^* J_{r+1} \\ I_r & 0 \end{pmatrix}.$$

Theorem 5. *Let $a_0, a_1, \ldots, a_{n-1}$ be complex numbers and define $c_0 = 1, c_1, \ldots, c_n$ by* (18). *The following equalities hold:*

$$M_r = B_r \begin{pmatrix} 1 & 0 \\ 0 & I_r - T_r T_r^* \end{pmatrix} B_r^* = B_r^* \begin{pmatrix} I_r - T_r^* T_r & 0 \\ 0 & 1 \end{pmatrix} B_r, \quad 1 \le r \le n, \tag{20}$$

$$\overline{M}_r = \overline{B}_r^* \begin{pmatrix} I_r - \overline{T}_r^* \overline{T}_r & 0 \\ 0 & 1 \end{pmatrix} \overline{B}_r, \quad 1 \le r \le n, \tag{21}$$

and

$$M_{2r} = C_r \begin{pmatrix} I_r - T_r T_r^* & 0 & Q_r \\ 0 & 1 & 0 \\ Q_r^* & 0 & I_r - \overline{T}_r \overline{T}_r^* \end{pmatrix} C_r^*, \quad 1 \le r \le n/2. \tag{22}$$

In (21) as in (15), a bar on a matrix indicates that all entries of the matrix are replaced by their complex conjugates.

Proof. The first equality in (20) can be shown by induction. The second equality follows from the first. To see this, use the identities

$$J_{r+1} M_r J_{r+1} = \overline{M}_r, \qquad J_{r+1} B_r J_{r+1} = \overline{B}_r^*, \qquad J_r T_r J_r = \overline{T}_r^*,$$

and

$$\begin{pmatrix} 0 & J_r \\ 1 & 0 \end{pmatrix} \begin{pmatrix} 1 & 0 \\ 0 & I_r - T_r T_r^* \end{pmatrix} \begin{pmatrix} 0 & 1 \\ J_r & 0 \end{pmatrix} = \begin{pmatrix} I_r - \overline{T}_r^* \overline{T}_r & 0 \\ 0 & 1 \end{pmatrix}$$

to obtain

$$M_r = J_{r+1} \overline{M}_r J_{r+1} = J_{r+1} \overline{B}_r J_{r+1} \begin{pmatrix} I_r - T_r^* T_r & 0 \\ 0 & 1 \end{pmatrix} J_{r+1} \overline{B}_r^* J_{r+1}$$

$$= B_r^* \begin{pmatrix} I_r - T_r^* T_r & 0 \\ 0 & 1 \end{pmatrix} B_r,$$

which is the second equality in (20). We get (21) on replacing the entries of the matrices by their complex conjugates.

We prove (22). Assume $1 \le r \le n/2$. Then

$$M_{2r} = \begin{pmatrix} M_{r-1} & S_r^* \\ S_r & M_r \end{pmatrix}, \qquad S_r = \begin{pmatrix} c_r & c_{r-1} & \dots & c_2 & c_1 \\ c_{r+1} & c_r & \dots & c_3 & c_2 \\ & & & \dots & \\ c_{2r} & c_{2r-1} & \dots & c_{r+2} & c_{r+1} \end{pmatrix}. \tag{23}$$

In (23) we use the first equality in (20) to obtain

$$M_{2r} = \begin{pmatrix} M_{r-1} & S_r^* \\ S_r & B_r \begin{pmatrix} 1 & 0 \\ 0 & I_r - T_r T_r^* \end{pmatrix} B_r^* \end{pmatrix}$$

$$= \begin{pmatrix} I_r & 0 \\ 0 & B_r \end{pmatrix} \begin{pmatrix} M_{r-1} & S_r^* B_r^{*-1} \\ B_r^{-1} S_r & \begin{pmatrix} 1 & 0 \\ 0 & I_r - T_r T_r^* \end{pmatrix} \end{pmatrix} \begin{pmatrix} I_r & 0 \\ 0 & B_r^* \end{pmatrix}.$$

Due to the lower triangular form of B_r, we get

$$B_r^{-1} S_r = \begin{pmatrix} 1 & 0 & \dots & 0 & 0 \\ * & 1 & \dots & 0 & 0 \\ & & \dots & & \\ * & * & \dots & * & 1 \end{pmatrix} \begin{pmatrix} c_r & c_{r-1} & \dots & c_1 \\ & & * & \\ & & & \end{pmatrix} = \begin{pmatrix} c_r & c_{r-1} & \dots & c_1 \\ & & Z_r & \\ & & & \end{pmatrix}.$$

With this definition of Z_r and (20), we obtain

$$M_{2r} = \begin{pmatrix} I_r & 0 \\ 0 & B_r \end{pmatrix} \left(\begin{matrix} M_{r-1} & \begin{pmatrix} \bar{c}_r \\ \vdots \\ \bar{c}_1 \end{pmatrix} \quad Z_r^* \\ \begin{pmatrix} c_r & c_{r-1} & \dots & c_1 \\ & Z_r & & \end{pmatrix} & \begin{pmatrix} 1 & 0 & \dots & 0 \\ 0 & & & \\ \vdots & & I_r - T_r T_r^* & \\ 0 & & & \end{pmatrix} \end{matrix} \right) \begin{pmatrix} I_r & 0 \\ 0 & B_r^* \end{pmatrix}$$

$$= \begin{pmatrix} I_r & 0 \\ 0 & B_r \end{pmatrix} \begin{pmatrix} M_r & \begin{pmatrix} Z_r^* \\ 0 \ \dots \ 0 \end{pmatrix} \\ \begin{pmatrix} Z_r & \begin{matrix} 0 \\ \vdots \\ 0 \end{matrix} \end{pmatrix} & I_r - T_r T_r^* \end{pmatrix} \begin{pmatrix} I_r & 0 \\ 0 & B_r^* \end{pmatrix}$$

$$= \begin{pmatrix} I_r & 0 \\ 0 & B_r \end{pmatrix} \begin{pmatrix} B_r^* \begin{pmatrix} I_r - T_r^* T_r & 0 \\ 0 & 1 \end{pmatrix} B_r & \begin{pmatrix} Z_r^* \\ 0 \ \dots \ 0 \end{pmatrix} \\ \begin{pmatrix} Z_r & \begin{matrix} 0 \\ \vdots \\ 0 \end{matrix} \end{pmatrix} & I_r - T_r T_r^* \end{pmatrix} \begin{pmatrix} I_r & 0 \\ 0 & B_r^* \end{pmatrix}$$

$$= \begin{pmatrix} I_r & 0 \\ 0 & B_r \end{pmatrix} \begin{pmatrix} B_r^* & 0 \\ 0 & I_r \end{pmatrix} \begin{pmatrix} \begin{pmatrix} I_r - T_r^* T_r & 0 \\ 0 & 1 \end{pmatrix} & B_r^{*-1} \begin{pmatrix} Z_r^* \\ 0 \ \dots \ 0 \end{pmatrix} \\ \begin{pmatrix} Z_r & \begin{matrix} 0 \\ \vdots \\ 0 \end{matrix} \end{pmatrix} B_r^{-1} & I_r - T_r T_r^* \end{pmatrix} \cdot$$

$$\cdot \begin{pmatrix} B_r & 0 \\ 0 & I_r \end{pmatrix} \begin{pmatrix} I_r & 0 \\ 0 & B_r^* \end{pmatrix}$$

Here the matrix

$$C_r' = \begin{pmatrix} I_r & 0 \\ 0 & B_r \end{pmatrix} \begin{pmatrix} B_r^* & 0 \\ 0 & I_r \end{pmatrix}$$

is invertible. Note also that

$$\begin{pmatrix} & 0 \\ Z_r & \vdots \\ & 0 \end{pmatrix} B_r^{-1} = \begin{pmatrix} & 0 \\ Z_r & \vdots \\ & 0 \end{pmatrix} \begin{pmatrix} 1 & 0 & \dots & 0 \\ * & 1 & \dots & 0 \\ & \dots & & \\ * & * & \dots & 1 \end{pmatrix} = \begin{pmatrix} & 0 \\ Y_r & \vdots \\ & 0 \end{pmatrix},$$

so that with Y_r defined in this way, we have

$$M_{2r} = C_r' \begin{pmatrix} I_r - T_r^* T_r & 0 & Y_r^* \\ 0 & 1 & 0 \\ Y_r & 0 & I_r - T_r T_r^* \end{pmatrix} C_r'^*, \quad 1 \le r \le n/2. \tag{24}$$

We now identify Y_r as

$$Y_r = \begin{pmatrix} a_r & a_{r-1} & \dots & a_1 \\ a_{r+1} & a_r & \dots & a_2 \\ & \dots & & \\ a_{2r-1} & a_{2r-2} & \dots & a_r \end{pmatrix}. \tag{25}$$

From the definition of B_r we find that

$$B_r^{-1} = \begin{pmatrix} 1 & 0 & \dots & 0 \\ -a_0 & 1 & \dots & 0 \\ & \dots & & \\ -a_{r-1} & -a_{r-2} & \dots & 1 \end{pmatrix}.$$

It follows that

$$B_r^{-1} S_r = \begin{pmatrix} (c_r \quad c_{r-1} \quad \dots \quad c_1) \\ \begin{pmatrix} a_r & a_{r-1} & \dots & a_1 \\ a_{r+1} & a_r & \dots & a_2 \\ & \dots & & \\ a_{2r-1} & a_{2r-2} & \dots & a_r \end{pmatrix} B_{r-1} \end{pmatrix}$$

and

$$Z_r = \begin{pmatrix} a_r & a_{r-1} & \dots & a_1 \\ a_{r+1} & a_r & \dots & a_2 \\ & \dots & & \\ a_{2r-1} & a_{2r-2} & \dots & a_r \end{pmatrix} B_{r-1}.$$

Finally, we obtain

$$(Y_r \quad 0) = (Z_r \quad 0)\, B_r^{-1} = \left(\begin{pmatrix} a_r & a_{r-1} & \dots & a_1 \\ a_{r+1} & a_r & \dots & a_2 \\ & \dots & & \\ a_{2r-1} & a_{2r-2} & \dots & a_r \end{pmatrix} B_{r-1} \quad 0 \right) \begin{pmatrix} B_{r-1}^{-1} & 0 \\ * & 1 \end{pmatrix}$$

$$= \left(\begin{pmatrix} a_r & a_{r-1} & \dots & a_1 \\ a_{r+1} & a_r & \dots & a_2 \\ & \dots & & \\ a_{2r-1} & a_{2r-2} & \dots & a_r \end{pmatrix} \quad 0 \right),$$

proving (25). Evidently, $Q_r = Y_r J_r$ and

$$C_r = C_r' \begin{pmatrix} 0 & J_{r+1} \\ I_r & 0 \end{pmatrix} = C_r' \begin{pmatrix} 0 & 0 & J_r \\ 0 & 1 & 0 \\ I_r & 0 & 0 \end{pmatrix}.$$

Substituting this in (24) we obtain (22). □

A number of consequences follow. For any Hermitian matrix A we write $\pi(A)$ and $\nu(A)$ for the numbers of positive and negative eigenvalues of A counting multiplicity.

Corollary 6. *Let $a_0, a_1, \dots, a_{n-1}$ be complex numbers.*

(1) *Each of the four quantities*

$$\nu(I_r - T_r T_r^*), \quad \pi(I_r - T_r T_r^*), \qquad 1 \le r \le n,$$

$$\nu \begin{pmatrix} I_r - T_r T_r^* & Q_r \\ Q_r^* & I_r - \overline{T}_r \overline{T}_r^* \end{pmatrix}, \quad \pi \begin{pmatrix} I_r - T_r T_r^* & Q_r \\ Q_r^* & I_r - \overline{T}_r \overline{T}_r^* \end{pmatrix}, \qquad 1 \le r \le n/2,$$

is a nondecreasing function of r.

(2) *For $0 \le r \le n$,*

$$\nu(I_r - T_r T_r^*) = \nu(I_r - T_r^* T_r) = \nu(I_r - \overline{T}_r \overline{T}_r^*) = \nu(I_r - \overline{T}_r^* \overline{T}_r)$$

and

$$\pi(I_r - T_r T_r^*) = \pi(I_r - T_r^* T_r) = \pi(I_r - \overline{T}_r \overline{T}_r^*) = \pi(I_r - \overline{T}_r^* \overline{T}_r).$$

(3) *If $\nu(I_n - T_n T_n^*) = \kappa$, then all of the matrices in* (17) *have at most κ negative eigenvalues.*

The condition $\nu(I_n - T_n T_n^*) = \kappa$ is necessary that $a_0, a_1, \dots, a_{n-1}$ are the first n Taylor coefficients of a function in $\mathbf{S}_\kappa$. The point of statement (3) in the preceding corollary is that no stronger necessary condition can be obtained from the other matrices in (17).

Proof. Define $c_0 = 1, c_1, \ldots, c_n$ by (18) and associated matrices M_r as in (19).

(1) By the first equality in (20),

$$\nu(I_r - T_rT_r^*) = \nu(M_r),$$
$$\pi(I_r - T_rT_r^*) = \pi(M_r) - 1.$$

By (22),

$$\nu\begin{pmatrix} I_r - T_rT_r^* & Q_r \\ Q_r^* & I_r - \overline{T}_r\overline{T}_r^* \end{pmatrix} = \nu(M_{2r}),$$
$$\pi\begin{pmatrix} I_r - T_rT_r^* & Q_r \\ Q_r^* & I_r - \overline{T}_r\overline{T}_r^* \end{pmatrix} = \pi(M_{2r}) - 1.$$

If $s < r$, then M_s is a submatrix of M_r obtained by deleting a set of rows and corresponding columns, and therefore $\nu(M_s) \le \nu(M_r)$, yielding (1).

(2) The first and third equalities hold by (20). Since $J_rT_r^* = \overline{T}_rJ_r$ and hence

$$I_r - \overline{T}_r\overline{T}_r^* = I_r - J_rT_r^*T_rJ_r = J_r(I_r - T_r^*T_r)J_r,$$

the second equality also holds.

(3) By part (2), $\nu(I_n - \overline{T}_n\overline{T}_n^*) = \nu(I_n - T_nT_n^*) = \kappa$. By the proof of (1), if $1 \le r \le n/2$, then

$$\nu\begin{pmatrix} I_r - T_rT_r^* & Q_r \\ Q_r^* & I_r - \overline{T}_r\overline{T}_r^* \end{pmatrix} = \nu(M_{2r}) \le \nu(M_n) = \nu(I_n - T_nT_n^*) = \kappa,$$

and this proves (3). □

Corollary 7. *Let $a_0, a_1, a_2, \ldots$ be complex numbers. If one of the three nondecreasing sequences*

$$\{\nu(I_r - T_rT_r^*)\}_1^\infty, \quad \left\{\nu(I_r - \overline{T}_r\overline{T}_r^*)\right\}_1^\infty, \quad \left\{\nu\begin{pmatrix} I_r - T_rT_r^* & Q_r \\ Q_r^* & I_r - \overline{T}_r\overline{T}_r^* \end{pmatrix}\right\}_1^\infty$$

has constant value κ from some point on, then all do. If one of the three nondecreasing sequences

$$\{\pi(I_r - T_rT_r^*)\}_1^\infty, \quad \left\{\pi(I_r - \overline{T}_r\overline{T}_r^*)\right\}_1^\infty, \quad \left\{\pi\begin{pmatrix} I_r - T_rT_r^* & Q_r \\ Q_r^* & I_r - \overline{T}_r\overline{T}_r^* \end{pmatrix}\right\}_1^\infty$$

has constant value κ from some point on, then all do.

Proof. Define $c_0 = 1, c_1, \ldots, c_n$ by (18) and associated matrices M_r as in (19). The corollary follows on expressing all of the quantities in terms of the sequences $\{\nu(M_r)\}_{r=1}^\infty$ and $\{\pi(M_r)\}_{r=1}^\infty$. For example, for the negative eigenvalues, if one of the quantities has constant value κ from some point on, then $\nu(M_r) = \kappa$ for all sufficiently large r, and all have constant value κ from some point on. □

We next recall a result from [11] on the convergence of power series. We include a proof for the convenience of the reader and to show the role of realization theory: the coefficients of the power series are represented as Taylor coefficients of a transfer function, which is holomorphic in a neighborhood of the origin.

Theorem 8. *Let $a_0, a_1, a_2, \dots$ be complex numbers such that the matrices $I_j - T_j T_j^*$ have κ negative eigenvalues for all sufficiently large j. Then the power series $S(z) = \sum_{j=0}^{\infty} a_j z^j$ converges in some disk $|z| < \delta$ where $\delta > 0$.*

Proof. Let $\mathfrak{F} = \mathbf{C}$ be the complex numbers viewed as a Hilbert space in the Euclidean metric. Define $c_0, c_1, c_2, \dots$ by (18). Then by (20), the matrices (19) have κ negative eigenvalues for all sufficiently large r, that is, the sequence $c_0, c_1, c_2, \dots$ belongs to $\mathfrak{P}_\kappa$. As in Iokhvidov and Kreĭn [14, pp. 312–314], construct a Naimark dilation for $c_0, c_1, c_2, \dots$; that is, we construct a Pontryagin space $\mathfrak{K}$ that contains $\mathfrak{F}$ isometrically as a regular subspace, and a unitary operator $U \in \mathfrak{L}(\mathfrak{K})$ such that

$$c_j = P_{\mathfrak{F}} U^j|_{\mathfrak{F}}, \qquad j = 0, 1, 2, \dots,$$

where $P_{\mathfrak{F}}$ is the projection on $\mathfrak{K}$ with range $\mathfrak{F}$. Since $\mathfrak{F}$ is a regular subspace of $\mathfrak{K}$, we can write $\mathfrak{K} = \mathfrak{H} \oplus \mathfrak{F}$ where $\mathfrak{H}$ is a regular subspace of $\mathfrak{K}$. Let

$$U = \begin{pmatrix} A & B \\ C & D \end{pmatrix}$$

relative to this decomposition. We show that

$$a_0 = D \qquad \text{and} \qquad a_m = CA^{m-1}B, \quad m \geq 1. \tag{26}$$

The cases $m = 0, 1$ are immediate. We prove the formula for a_m assuming it is known for $a_0, \dots, a_{m-1}$. By (18),

$$c_{m+1} = c_0 a_m + c_1 a_{m-1} + \cdots + c_m a_0,$$

so it is the same thing to show that

$$c_{m+1} = c_0 CA^{m-1}B + c_1 CA^{m-2}B + \cdots + c_{m-1}CB + c_m D. \tag{27}$$

Put

$$\begin{pmatrix} A & B \\ C & D \end{pmatrix}^j = \begin{pmatrix} A_j & B_j \\ C_j & D_j \end{pmatrix}, \qquad j \geq 0.$$

Then

$$\begin{pmatrix} A_{m+1} & B_{m+1} \\ C_{m+1} & D_{m+1} \end{pmatrix} = \begin{pmatrix} A_m & B_m \\ C_m & D_m \end{pmatrix} \begin{pmatrix} A & B \\ C & D \end{pmatrix} = \begin{pmatrix} A_m A + B_m C & A_m B + B_m D \\ C_m A + D_m C & C_m B + D_m D \end{pmatrix}.$$

Since $D_j = P_{\mathfrak{F}} U^j|_{\mathfrak{F}} = c_j$ for all $j \geq 0$, $c_{m+1} = C_m B + c_m D$. This allows us to bring (27) to the form

$$C_m B = c_0 CA^{m-1}B + c_1 CA^{m-2}B + \cdots + c_{m-1}CB. \tag{28}$$

Dropping the factor B on the right in each term, we easily verify (28) by induction: the formula is evident for $m = 1$, and the inductive step follows from the identity $C_{m+1} = C_m A + c_m C$. This completes the proof of (26). The identity (26) implies that $|a_j| \leq K\rho^j$ for some positive constants K and ρ, and therefore the power series $\sum_{j=0}^{\infty} a_j z^j$ converges in a neighborhood of the origin. $\square$

We can now relate Problem I to an indefinite form of the trigonometric moment problem.

Let $\mathfrak{P}_\kappa$ ($\mathfrak{P}_{\nu,\pi}$) be the set of all sequences $\{c_j\}_{j=0}^{\infty}$ with $\bar{c}_0 = c_0$ such that the matrix M_r has κ negative (ν negative and π positive) eigenvalues for all sufficiently large r.

Problem II (Trigonometric moment problem). *Let $c_0, c_1, \ldots, c_{n-1}$ be n complex numbers with $\bar{c}_0 = c_0$. Determine for which nonnegative integers κ there is a sequence $\{c_p\}_{p=0}^{\infty}$ in $\mathfrak{P}_\kappa$ that extends the given numbers. Determine for which nonnegative integers ν and π there is a sequence $\{c_p\}_{p=0}^{\infty}$ in $\mathfrak{P}_{\nu,\pi}$ that extends the given numbers.*

This problem is an indefinite form of the trigonometric moment problem and it was considered by Iokhvidov and Kreĭn [14, §19]. In the classical case, this concerns the Fourier coefficients, or moments,

$$c_j = \int e^{-ijt}\, d\mu(t), \qquad j = 0, \pm 1, \pm 2, \ldots,$$

of a nonnegative measure μ on $[0, 2\pi)$. In this case, the matrix $(c_{i-j})_{i,j=0}^{n}$ is nonnegative for every $n \geq 0$, since

$$\sum_{j,k=0}^{n} c_{k-j}\lambda_k\bar{\lambda}_j = \int_{[0,2\pi)} \sum_{j,k=0}^{n} \lambda_k\bar{\lambda}_j e^{-i(k-j)t}\, d\mu(t)$$

$$= \int_{[0,2\pi)} \left| \sum_{j,k=0}^{n} \lambda_k e^{-ikt} \right|^2 d\mu(t) \geq 0$$

for arbitrary numbers $\lambda_0, \ldots, \lambda_n$. When μ is a probability measure, $c_0 = 1$. The classical trigonometric moment problem is to extend given numbers $c_0, c_1, \ldots, c_{n-1}$ with $\bar{c}_0 = c_0$ to such a moment sequence. In the indefinite extension, we still speak of the "trigonometric moment problem," but the underlying function theory is not the same.

We can show now that Problem I and Problem II are equivalent.

Theorem 9 (Equivalence of Problems I and II). *Assume that the numbers $a_0, \ldots, a_{n-1}$ and $c_0 = 1, c_1, \ldots, c_{n-1}, c_n$ are connected as in (18). Then Problem I is solvable with the data $a_0, \ldots, a_{n-1}$ if and only if Problem II is solvable with the data $c_0, \ldots, c_{n-1}, c_n$.*

Proof. Suppose that Problem I with the data $a_0, \ldots, a_{n-1}$ has a solution in $\mathbf{S}_\kappa$. Let

$$S(z) = \sum_{j=0}^{\infty} a_j z^j$$

be the Taylor expansion of this solution. By the necessary conditions for Problem I discussed above, $I_j - T_j T_j^*$ has κ negative eigenvalues for all sufficiently large j. Define $c_{n+1}, c_{n+2}, \dots$ so that

$$c_j = c_0 a_{j-1} + c_1 a_{j-2} + \cdots c_{j-1} a_0$$

for all $j = 1, 2, \dots$. Then (20) implies that M_j has κ negative eigenvalues for all $j = 0, 1, 2, \dots$. Therefore $c_0, c_1, c_2, \dots$ is a solution to Problem II with the data $c_0, \dots, c_{n-1}, c_n$.

Conversely, assume that Problem II is solvable with the data $c_0, \dots, c_{n-1}, c_n$, that is, the numbers can be extended to a sequence $c_0, c_1, c_2, \dots$ in $\mathfrak{P}_\kappa$. Then the matrices (19) have κ negative eigenvalues for all sufficiently large r. Reversing the process above, we obtain a sequence $a_0, a_1, a_2, \dots$ that extends $a_0, \dots, a_{n-1}$ such that the matrices $I_j - T_j T_j^*$ have κ negative eigenvalues for all sufficiently large j. By Theorem 8, the series $S(z) = \sum_{j=0}^\infty a_j z^j$ converges in some disk $|z| < \delta$ where $\delta > 0$, and by a theorem of Kreĭn and Langer in [18, Theorem 6.3], the function $S(z)$ so defined belongs to $\mathbf{S}_\kappa$. Thus Problem I is solvable with the data $a_0, \dots, a_{n-1}$. The argument for the classes $\mathfrak{P}_{\nu,\pi}$ and $\mathbf{S}_{\nu,\pi}$ is similar. □

We use a series of propositions from [13]. The matrices $M_0, M_1, M_2, \dots$ that appear in the list below are Hermitian matrices of the form (19) defined for appropriate numbers $c_0 = \bar{c}_0, c_1, c_2, \dots,$ and n is any positive integer. Recall that for any Hermitian matrix A we write $\pi(A)$ and $\nu(A)$ for the numbers of positive and negative eigenvalues of A counting multiplicity. The **signature** of A is $\sigma(A) = \pi(A) - \nu(A)$. Write $|A|$ for the determinant of A and $\rho(A) = \pi(A) + \nu(A)$ for the rank of A.

1°) The difference $\rho(M_n) - \rho(M_{n-1})$ is either 0, 1, or 2.

2°) If $\rho(M_n) - \rho(M_{n-1}) = 0$, then $\pi(M_n) = \pi(M_{n-1})$ and $\nu(M_n) = \nu(M_{n-1})$.

3°) If $\rho(M_n) - \rho(M_{n-1}) = 1$, then either $\pi(M_n) = \pi(M_{n-1}) + 1$ and $\nu(M_n) = \nu(M_{n-1})$, or $\pi(M_n) = \pi(M_{n-1})$ and $\nu(M_n) = \nu(M_{n-1}) + 1$.

4°) If $\rho(M_n) - \rho(M_{n-1}) = 2$, then $\pi(M_n) = \pi(M_{n-1}) + 1$ and $\nu(M_n) = \nu(M_{n-1}) + 1$.

5°) If $|M_{n-1}| \neq 0$, then there are infinitely many c_n such that $\rho(M_n) = \rho(M_{n-1})$.

6°) If $|M_{n-1}| = 0$ and $|M_{\rho(M_{n-1})-1}| \neq 0$, then there is a unique c_n such that $\rho(M_n) = \rho(M_{n-1})$.

7°) The assumptions in 6°) imply that there is a unique extension $(c_j)_{j=0}^\infty$ of $(c_j)_{j=0}^{n-1}$ such that $\rho(M_j) = \rho(M_{n-1})$, $j \geq n$.

8°) There exists a c_n with $\rho(M_n) = \rho(M_{n-1})$ if and only if $|M_{\rho(M_{n-1})-1}| \neq 0$.

9°) If $|M_{r-1}| \neq 0$ and $|M_{n-1}| = \cdots = |M_r| = 0$ for some $0 \leq r < \rho(M_{n-1})$ ($|M_{-1}| = 1$ by definition), then $\rho(M_n) = \rho(M_{n-1}) + 2$.

10°) If $|M_{n-1}| \neq 0$, then for each $k = 1, 2, \dots,$ there are infinitely many c_n, $\dots$, c_{n+k-1} such that $\nu(M_{n+k-1}) = \nu(M_{n-1}) + k$ and $|M_{n+k-1}| \neq 0$.

11°) If $|M_{n-1}| \neq 0$, then for each $\ell = 1, 2, \ldots$, there are infinitely many $c_n, \ldots, c_{n+\ell-1}$ such that $\pi(M_{n+\ell-1}) = \pi(M_{n-1}) + \ell$ and $|M_{n+\ell-1}| \neq 0$.

12°) $\sigma(M_{n-1}) = \sum_{j=0}^{n-1} \operatorname{sign}(|M_{j-1}||M_j|)$, where, by definition, $|M_{-1}| = 1$ and $\operatorname{sign} 0 = 0$.

13°) If $\rho(M_j)$ is a constant ρ for all sufficiently large j, then $|M_{\rho-1}| \neq 0$.

Proofs. All of the citations below are from [13].

1°) Corollary on p. 34.

2°) Theorem 6.2, p. 36.

3°) Theorem 6.3, p. 36.

4°) Theorem 6.1, p. 35.

5°) Theorem 13.1, p. 97, and Remark 1, p. 98.

6°) Theorem 13.2, p. 100, and Remark 1, p. 102.

7°) Corollary on p. 101 and Remark 1, p. 102.

8°) The "if" part follows from 5°) and 6°), the "only if" part from Theorem 15.3, p. 119.

9°) Proposition 3°, p. 121.

10°) and 11°) It is enough to prove these statements for $k = 1$ in 10°) and $\ell = 1$ in 11°). To do this, we use the proof of Theorem 13.1, p. 97, and Remark 1, p. 98, to construct infinitely many extensions with $|M_n| > 0$ and infinitely many extensions with $|M_n| < 0$ (treat the subcases $|M_{n-2}| \neq 0$ and $|M_{n-2}| = 0$ separately using the argument on p. 99). Then 10°) and 11°) follow from 3°).

12°) Theorem 16.1, p. 129.

13°) Theorem 15.4, p. 119. □

Our solution of Problem II is presented in Theorem 10. The first parts of the statements (a), (c), and (f) can be found in Iokhvidov's book as Excercise 8 on pp. 133–134; in the interest of completeness we prove these statements as well. It is clear that a given sequence $(c_j)_{j=0}^{n-1}$ does not have any extension $(c_j)_{j=0}^{\infty}$ in $\mathfrak{P}_\nu$ if $\nu < \nu(M_{n-1})$, and there is no extension in $\mathfrak{P}_{\nu,\pi}$ if either $\nu < \nu(M_{n-1})$ or $\pi < \pi(M_{n-1})$, because by 1°)–4°), $\nu(M_j)$ and $\pi(M_j)$ are nondecreasing functions of j. If an extension $(c_j)_{j=0}^{\infty}$ belongs to the class $\mathfrak{P}_\nu$ then it is possible that $\rho(M_j)$ and hence also $\pi(M_j)$ tends to ∞ as $j \to \infty$. Such an extension does not belong to any of the classes $\mathfrak{P}_{\nu,\pi}$. According to 13°) a necessary condition for $(c_j)_{j=0}^{\infty}$ to belong to $\mathfrak{P}_{\nu,\pi}$ is that $|M_{\nu+\pi-1}| \neq 0$.

Theorem 10. *Let* $c_0 = \bar{c}_0, c_1, \ldots, c_{n-1}$ *be given numbers, and define* $M_0, \ldots, M_{n-1}$ *as in* (19).

Assume $|M_{n-1}| \neq 0$.

(a) *There exist infinitely many extensions in* $\mathfrak{P}_{\nu(M_{n-1})}$, *even infinitely many extensions in the smaller set* $\mathfrak{P}_{\nu(M_{n-1}),\pi(M_{n-1})}$.

(b) *There exist infinitely many extensions in* $\mathfrak{P}_{\nu(M_{n-1})+\nu,\pi(M_{n-1})+\pi}$ *for all* $\nu \geq 0$ *and* $\pi \geq 0$.

Assume $|M_{n-1}| = 0$ *and* $|M_{\rho(M_{n-1})-1}| \neq 0$.

(c) *There is a unique extension in* $\mathfrak{P}_{\nu(M_{n-1})}$; *it belongs to* $\mathfrak{P}_{\nu(M_{n-1}),\pi(M_{n-1})}$.

(d) *There are no extensions in* $\mathfrak{P}_\nu$ *for*

$$\nu(M_{n-1}) < \nu < \nu(M_{n-1}) + \dim\ker M_{n-1};$$

there are no extensions in $\mathfrak{P}_{\nu,\pi}$ *if*

$$\nu(M_{n-1}) < \nu < \nu(M_{n-1}) + \dim\ker M_{n-1}$$

or if

$$\pi(M_{n-1}) < \pi < \pi(M_{n-1}) + \dim\ker M_{n-1}.$$

(e) *There are infinitely many extensions in* $\mathfrak{P}_{\nu,\pi}$ *for all pairs* (ν,π) *with*

$$\nu \geq \nu(M_{n-1}) + \dim\ker M_{n-1} \quad \text{and} \quad \pi \geq \pi(M_{n-1}) + \dim\ker M_{n-1}.$$

Assume $|M_{n-1}| = 0$ *and* $|M_{\rho(M_{n-1})-1}| = 0$.

(f) *There are no extensions in* $\mathfrak{P}_{\nu(M_{n-1})}$.

(g) *There are no extensions in* $\mathfrak{P}_\nu$ *if* $\nu < \nu(M_{n-1}) + \dim\ker M_{n-1}$; *there are no extensions in* $\mathfrak{P}_{\nu,\pi}$ *if*

$$\nu < \nu(M_{n-1}) + \dim\ker M_{n-1}$$

or if

$$\pi < \pi(M_{n-1}) + \dim\ker M_{n-1}.$$

(h) *There are infinitely many extensions in* $\mathfrak{P}_{\nu,\pi}$ *for every pair* (ν,π) *with*

$$\nu \geq \nu(M_{n-1}) + \dim\ker M_{n-1} \quad \text{and} \quad \pi \geq \pi(M_{n-1}) + \dim\ker M_{n-1}.$$

Proof. For any extension of the given sequence by numbers $c_n, c_{n+1}, \ldots$, we assume that $M_n, M_{n+1}, \ldots$ are defined as in (19).

(a) According to 5°) there are infinitely many c_n such that $\rho(M_n) = \rho(M_{n-1}) = n$. For such M_n we have $|M_n| = 0$ and $|M_{\rho(M_n)-1}| \neq 0$. Hence by 7°) there is an extension $(c_j)_{j=0}^\infty$ of $(c_j)_{j=0}^{n-1}$ such that $\rho(M_j) = \rho(M_{n-1})$ for all $j \geq n-1$. Statement 2°) implies that

$$\nu(M_j) = \nu(M_{n-1}) \quad \text{and} \quad \pi(M_j) = \pi(M_{n-1}), \qquad j \geq n-1,$$

and hence $(c_j)_{j=0}^\infty$ belongs to $\mathfrak{P}_{\nu(M_{n-1}),\pi(M_{n-1})}$.

(b) By 10°) there are infinitely many numbers c_n such that $\nu(M_n) = \nu(M_{n-1})+1$ and $|M_n| \neq 0$. Therefore $\rho(M_n) = \rho(M_{n-1})+1$ and by 3°), $\pi(M_n) = \pi(M_{n-1})$. After ν steps, we obtain numbers $c_n, \ldots, c_{n+\nu-1}$ such that

$$|M_{n+\nu-1}| \neq 0, \qquad \nu(M_{n+\nu-1}) = \nu, \quad \text{and} \quad \pi(M_{n+\nu-1}) = \pi(M_{n-1}).$$

Using the same argument with 11°) instead of 10°), we obtain numbers $c_{n+\nu}, \ldots, c_{n+\nu+\pi-1}$ (each of which can be chosen in infinitely many ways) such that

$$|M_{n+\nu+\pi-1}| \neq 0, \qquad \nu(M_{n+\nu+\pi-1}) = \nu, \quad \text{and} \quad \pi(M_{n+\nu+\pi-1}) = \pi.$$

Now (b) follows from (a).

(c) According to 7°) there exists a unique extension $(c_j)_{j=0}^\infty$ of $(c_j)_{j=0}^{n-1}$ such that

$$\rho(M_j) = \rho(M_{n-1}), \qquad j \geq n.$$

It follows from 2°) that also $\pi(M_j) = \pi(M_{n-1})$ and $\nu(M_j) = \nu(M_{n-1})$ for $j \geq n$. Therefore there exists a unique extension of $(c_j)_{j=0}^{n-1}$ in the class $\mathfrak{P}_{\nu(M_{n-1})}$ and this extension belongs to $\mathfrak{P}_{\nu(M_{n-1}),\pi(M_{n-1})}$ (for the uniqueness part, note that by 3°) the equality $\nu(M_n) = \nu(M_{n-1})$ can only hold in the present situation when $\rho(M_n) = \rho(M_{n-1})$).

(d) and (e). By hypothesis

$$|M_{n-1}| = 0 \qquad \text{and} \qquad |M_{\rho(M_{n-1})-1}| \neq 0. \tag{29}$$

The unique extension described in part (c) of the theorem cannot meet any of the conditions in parts (d) and (e); since for this extension $\rho(M_{n-1}) = \rho(M_n) = \rho(M_{n+1}) = \cdots$, in parts (d) and (e) we need only consider extensions such that

$$\rho(M_{n-1}) = \cdots = \rho(M_{n+k-1}) < \rho(M_{n+k})$$

for some $k \geq 0$. In this situation (29) holds with n replaced by $n+k$, and therefore we may restrict attention to extensions satisfying

$$\rho(M_{n-1}) < \rho(M_n). \tag{30}$$

By 6°), (30) holds for all but one choice of c_n; in what follows, we assume that c_n is chosen so that (30) is satisfied. The question then is if the sequence $(c_j)_{j=0}^{n}$ can be further extended to an infinite sequence $(c_j)_{j=0}^{\infty}$ as required in (d) and (e).

Case (i): $\rho(M_n) = n+1$.

Since $\rho(M_{n-1}) < n$ by (29), by 1°) we must have $\rho(M_{n-1}) = n-1$. Thus $\dim\ker M_{n-1} = 1$, and hence part (d) holds vacuously. Part (e) also holds in this case. For by statement 4°), $\nu(M_n) = \nu(M_{n-1}) + 1$ and $\pi(M_n) = \pi(M_{n-1}) + 1$ and since M_n is invertible, part (e) follows from (a).

Case (ii): $\rho(M_n) < n+1$.

Then with $r = \rho(M_{n-1})$, in view of (29) and (30),

$$|M_{r-1}| = |M_{\rho(M_{n-1})-1}| \neq 0, \quad |M_r| = \cdots = |M_{n-1}| = |M_n| = 0.$$

Consider any extension of $(c_j)_{j=0}^{n}$ by a number c_{n+1}. By (30), $r < \rho(M_{n+1})$. Applying 9°) with n replaced by $n+1$, we obtain

$$\rho(M_{n+1}) = \rho(M_n) + 2,$$

and by 4°),

$$\nu(M_{n+1}) = \nu(M_n) + 1 \qquad \text{and} \qquad \pi(M_{n+1}) = \pi(M_n) + 1.$$

If $\rho(M_{n+1}) < n+2$, we can repeat this argument. We continue in this way for $k = 1, 2, \ldots$ and extend $(c_j)_{j=0}^{n}$ with any numbers $c_{n+1}, \ldots, c_{n+k}$, $k = 1, 2, \ldots$; by 9°) and 4°), we have $r < \rho(M_{n+k})$,

$$\begin{aligned} \rho(M_{n+k}) &= \rho(M_n) + 2k, \\ \nu(M_{n+k}) &= \nu(M_n) + k, \\ \pi(M_{n+k}) &= \pi(M_n) + k, \end{aligned}$$

and

$$|M_{r-1}| \neq 0, \quad |M_r| = \cdots = |M_{n-1}| = |M_n| = \cdots = |M_{n+k}| = 0,$$

provided $\rho(M_{n+k}) = \rho(M_n) + 2k < n + k + 1$. If equality holds, that is,

$$k = k_0 := n - \rho(M_n) + 1,$$

then M_{n+k_0} is invertible and the process stops. Hence if such an extension of $(c_j)_{j=0}^n$ can be continued to a sequence in a class $\mathfrak{P}_{\nu,\pi}$, then necessarily

$$\nu \geq \nu_0 := \nu(M_{n+k_0}) = \nu(M_n) + k_0 = \nu(M_n) + n - \rho(M_n) + 1,$$

$$\pi \geq \pi_0 := \pi(M_{n+k_0}) = \pi(M_n) + k_0 = \pi(M_n) + n - \rho(M_n) + 1,$$

and according to (a) and (b) each of the classes $\mathfrak{P}_\nu$ and $\mathfrak{P}_{\nu,\pi}$ contains infinitely many extensions. Thus the first part of (d) and (e) will follow once we show that

$$\nu_0 = \nu(M_{n-1}) + \dim\ker M_{n-1} \quad \text{and} \quad \pi_0 = \pi(M_{n-1}) + \dim\ker M_{n-1}.$$

Since $|M_{n-1}| = 0$, 12°) implies that

$$\sigma(M_n) - \sigma(M_{n-1}) = \operatorname{sign}|M_{n-1}||M_n| = 0$$

and since $\rho(M_n) > \rho(M_{n-1})$, we therefore have $\rho(M_n) = \rho(M_{n-1}) + 2$, and by 4°), $\nu(M_n) = \nu(M_{n-1}) + 1$ and $\pi(M_n) = \pi(M_{n-1}) + 1$. This implies that $\nu_0 = \nu(M_{n-1}) + \dim\ker M_{n-1}$ and also that π_0 has the desired value.

From the first part of (d) it follows that there are no extensions in $\mathfrak{P}_{\nu,\pi}$ if

$$\nu(M_{n-1}) < \nu < \nu(M_{n-1}) + \dim\ker M_{n-1},$$

whatever the value of π. By considering the sequence $(-c_j)_{j=0}^{n-1}$ and its extensions $(-c_j)_{j=0}^\infty$ and applying the results just proved (together with $\nu(-M_j) = \pi(M_j)$) we find that there are no extensions in $\mathfrak{P}_{\nu,\pi}$ if

$$\pi(M_{n-1}) < \pi < \pi(M_{n-1}) + \dim\ker M_{n-1}$$

whatever the value of ν.

(f) is part of (g).

(g) and (h). By 8°), (30) holds for any choice of c_n. This allows us to proceed by an argument which is similar to the proof of (d) and (e) above; in case (ii) there, the exact value of r is unimportant in order to obtain the conclusion. □

We can now deal with Problem I. According to Theorem 9, we must apply the previous result to the case where $c_0 = 1$, n is replaced by $n+1$, $|M_n| = |I_n - T_nT_n^*|$, $\dim\ker M_n = \dim\ker(I_n - T_nT_n^*)$, and

$$\rho(I_n - T_nT_n^*) = \rho(M_n) - 1,\ \pi(I_n - T_nT_n^*) = \pi(M_n) - 1,\ \nu(I_n - T_nT_n^*) = \nu(M_n).$$

Note that $\mathfrak{P}_{\nu,\pi}$ corresponds to the class $\mathbf{S}_{\nu,\pi'}$ with $\pi' = \pi - 1$. We obtain the following solution for the Carathéodory-Fejér problem.

Theorem 11. *Let $a_0, \ldots, a_{n-1}$ be given numbers, and define $T_1, \ldots, T_n$ as in (15). Assume $|I_n - T_nT_n^*| \neq 0$.*

(a′) *There exist infinitely many solutions of Problem I in $\mathbf{S}_{\nu(I_n - T_nT_n^*)}$, even in the smaller set $\mathbf{S}_{\nu(I_n-T_nT_n^*),\pi(I_n-T_nT_n^*)}$.*

(b′) *There exist infinitely many solutions in $\mathbf{S}_{\nu,\pi}$ for all pairs (ν, π) with $\nu \geq \nu(I_n - T_nT_n^*)$ and $\pi \geq \pi(I_n - T_nT_n^*)$.*

Assume $|I_n - T_nT_n^*| = 0$ *and* $|I_\rho - T_\rho T_\rho^*| \neq 0$, *where* $\rho = \rho(I_n - T_nT_n^*)$.

(c′) *There is a unique solution in* $\mathbf{S}_{\nu(I_n - T_nT_n^*)}$; *it belongs to*

$$\mathbf{S}_{\nu(I_n - T_nT_n^*),\pi(I_n - T_nT_n^*)}.$$

(d′) *There are no solutions in* $\mathbf{S}_\nu$ *for* $\nu(I_n - T_nT_n^*) < \nu < \nu(I_n - T_nT_n^*) +$ $\dim\ker I_n - T_nT_n^*$; *there are no solutions in* $\mathbf{S}_{\nu,\pi}$ *if*

$$\nu(I_n - T_nT_n^*) < \nu < \nu(I_n - T_nT_n^*) + \dim\ker(I_n - T_nT_n^*)$$

or if

$$\pi(I_n - T_nT_n^*) < \pi < \pi(I_n - T_nT_n^*) + \dim\ker(I_n - T_nT_n^*).$$

(e′) *There are infinitely many solutions in* $\mathbf{S}_{\nu,\pi}$ *for all pairs* (ν, π) *with*

$$\nu \geq \nu(I_n - T_nT_n^*) + \dim\ker(I_n - T_nT_n^*) \text{ and } \pi \geq \pi(I_n - T_nT_n^*) + \dim\ker(I_n - T_nT_n^*).$$

Assume $|I_n - T_nT_n^*| = 0$ *and* $|I_\rho - T_\rho T_\rho^*| = 0$.

(f′) *There are no solutions in* $\mathbf{S}_{\nu(I_n - T_nT_n^*)}$.

(g′) *There are no solutions in* $\mathbf{S}_\nu$ *if* $\nu < \nu(I_n - T_nT_n^*) + \dim\ker(I_n - T_nT_n^*)$; *there are no solutions in* $\mathbf{S}_{\nu,\pi}$ *if*

$$\nu < \nu(I_n - T_nT_n^*) + \dim\ker(I_n - T_nT_n^*)$$

or if

$$\pi < \pi(I_n - T_nT_n^*) + \dim\ker(I_n - T_nT_n^*).$$

(h′) *There are infinitely many solutions in* $\mathbf{S}_{\nu,\pi}$ *for every pair* (ν, π) *with*

$$\nu \geq \nu(I_n - T_nT_n^*) + \dim\ker(I_n - T_nT_n^*) \text{ and } \pi \geq \pi(I_n - T_nT_n^*) + \dim\ker(I_n - T_nT_n^*).$$

We mention a consequence of the solution of Problem I for the case $\nu = \nu(I_n - T_nT_n^*)$.

Corollary 12. *Let* $a_0, a_1, \ldots, a_{n-1}$ *be numbers such that* $I_n - T_nT_n^*$ *has* ν *negative eigenvalues.*

(1) *If* $I_n - T_nT_n^*$ *is invertible, Problem I has infinitely many solutions in* $\mathbf{S}_\nu$.

(2) *If* $I_n - T_nT_n^*$ *is singular and* $\rho(I_{n-1} - T_{n-1}T_{n-1}^*) = \rho(I_n - T_nT_n^*)$, *Problem I has a unique solution in* $\mathbf{S}_\nu$.

(3) *If* $I_n - T_nT_n^*$ *is singular and* $\rho(I_{n-1} - T_{n-1}T_{n-1}^*) < \rho(I_n - T_nT_n^*)$, *Problem I has no solution in* $\mathbf{S}_\nu$.

The results in [9] and [11] give a solution to the existence and uniqueness problems for the matrix versions of both the trigonometric moment problem and Carathéodory-Fejér coefficients problem in the indefinite case, but the question of obtaining a matrix analogue of (d′)-(h′) in Theorem 11 is open.

Appendix: A remark on holomorphic kernels

The result below is used in Part I and is well known in particular cases. The general result is presumably also known, but we do not know a reference. For the convenience of the reader, we sketch a proof.

Let $K(w,z) = \sum_{m,n=0}^{\infty} C_{mn} z^m \bar{w}^n$ be a holomorphic Hermitian kernel defined for $|w| < R$ and $|z| < R$, with values in $\mathfrak{L}(\mathfrak{F})$ for some Kreĭn space $\mathfrak{F}$. For any nonnegative integer r, we may alternatively view the matrix $(C_{mn})_{m,n=0}^{r}$ as a selfadjoint operator on $\mathfrak{F}^{r+1} = \mathfrak{F} \oplus \cdots \oplus \mathfrak{F}$, where there are $r+1$ summands on the right side, or as a kernel on a finite set. The number of negative eigenvalues of $(C_{mn})_{m,n=0}^{r}$ as an operator and the number of negative squares of $(C_{mn})_{m,n=0}^{r}$ as a kernel coincide.

Theorem 13. *Let κ be a nonnegative integer. Then* $\mathrm{sq}_- K = \kappa$ *if and only if*

$$\nu(C_{mn})_{m,n=0}^{r} \leq \kappa$$

for all nonnegative integers r and equality holds for all sufficiently large r.

We can formulate this result in another way. Let $\mathbb{N}_0$ be the set of nonnegative integers. Define a kernel C on $\mathbb{N}_0 \times \mathbb{N}_0$ by

$$C(m,n) = C_{mn}, \qquad m,n \in \mathbb{N}_0.$$

Then $\mathrm{sq}_- K = \mathrm{sq}_- C$. The theory of Kolmogorov decompositions [10] gives a natural approach to this result, but we base our argument on similar notions for reproducing kernel Pontryagin spaces.

Proof. Since a holomorphic Hermitian kernel has the same number of negative squares on subregions [3, Theorem 1.1.4], by a change of scale we may assume that $R > 1$. We may also assume without loss of generality that $\mathfrak{F}$ is a Hilbert space. Let $H^2_{\mathfrak{F}}$ be the Hardy class of $\mathfrak{F}$-valued functions on the unit disk $\mathbf{D}$.

Assume that $\mathrm{sq}_- K = \kappa$. By a method of Alpay [2], we define a bounded selfadjoint operator P on $H^2_{\mathfrak{F}}$ such that

$$P\colon (1 - z\bar{w})^{-1} f \to K(w,z) f, \qquad w \in \mathbf{D}, f \in \mathfrak{F},$$

and

$$P\colon z^n f \to A_n(z) f, \qquad f \in \mathfrak{F},\ n = 0,1,2,\ldots.$$

where $K(w,z) = \sum_{n=0}^{\infty} A_n(z)\bar{w}^n$, that is, $A_n(z) = \sum_{m=0}^{\infty} C_{mn} z^m$ for all $n = 0,1,2,\ldots$. For another account of the construction of P, see [20, Theorem 8.4]. By the spectral theorem, we can write

$$P = P_+ + P_0 + P_-,$$

where $P_\pm$ and P_0 are selfadjoint operators corresponding to the spectral subspaces $\mathfrak{H}_+$, $\mathfrak{H}_-$, and $\mathfrak{H}_0 = \ker P$ for the sets $(0,\infty)$, $(-\infty,0)$, and $\{0\}$. Since $\mathrm{sq}_- K = \kappa$, $\dim \mathfrak{H}_- = \kappa$. Let $\mathfrak{K}_0$ be $H^2_{\mathfrak{F}} / \ker P$. Write

$$[h] = h + \ker P$$

for the coset determined by an element h of $H^2_{\mathfrak{F}}$. Define a nondegenerate inner product on $\mathfrak{K}_0$ by

$$\langle [h],[k]\rangle_{\mathfrak{K}_0} = \langle Ph, k\rangle_{H^2_{\mathfrak{F}}}, \qquad h,k \in H^2_{\mathfrak{F}}.$$

Using [15, Theorem 2.5, p. 20], complete $\mathfrak{K}_0$ to a Pontryagin space $\mathfrak{K}$ having negative index κ. The cosets determined by the polynomials are dense in $H^2_{\mathfrak{F}}/\ker P$ by [15, statement (i) on p. 20], and therefore $\{[z^n f]\colon f \in \mathfrak{F}, n = 0,1,2,\dots\}$ is a total set in $\mathfrak{K}$. By construction,

$$\langle [z^m f_1],[z^n f_2]\rangle_{\mathfrak{K}} = \langle C_{mn} f_1, f_2\rangle_{\mathfrak{F}}, \qquad f_1, f_2 \in \mathfrak{F},\ m,n = 0,1,2,\dots.$$

Hence by [3, Lemma 1.1.1], the matrix $(C_{mn})^r_{m,n=0}$ has at most κ negative eigenvalues for all $r = 0,1,2,\dots$ and one such matrix has exactly κ negative eigenvalues. Since the number of negative eigenvalues of $(C_{mn})^r_{m,n=0}$ is a nondecreasing function of r, this number is κ for all sufficiently large r.

Conversely, assume that the matrix $(C_{mn})^r_{m,n=0}$ has at most κ negative eigenvalues for all $r = 0,1,2,\dots$ and exactly κ negative eigenvalues for all sufficiently large r. By what we showed above, if we can only show that $\operatorname{sq}_- K \le \kappa$, it will follow that $\operatorname{sq}_- K = \kappa$. Let $\mathbb{N}_0$ be the set of nonnegative integers, and define a kernel C on $\mathbb{N}_0 \times \mathbb{N}_0$ by

$$C(m,n) = C_{mn}, \qquad m,n \in \mathbb{N}_0.$$

Our hypotheses imply that $\operatorname{sq}_- C = \kappa$. By [3, Theorem 1.1.3], there is a unique Pontryagin space $\mathfrak{H}_C$ of functions $h = \{h_n\}_{n=0}^\infty$ on $\mathbb{N}_0$ with reproducing kernel C. This means that for each $m \in \mathbb{N}_0$ and $f \in \mathfrak{F}$, the sequence $C(m,\cdot)f = \{C_{mn}f\}_{n=0}^\infty$ belongs to $\mathfrak{H}_C$, and for any element $h = \{h_n\}_{n=0}^\infty$ of $\mathfrak{H}_C$,

$$\langle \{h_n\}_{n=0}^\infty, \{C_{mn}f\}_{n=0}^\infty\rangle_{\mathfrak{H}_C} = \langle h_m, f\rangle_{\mathfrak{F}}.$$

By [3, Theorem 1.1.2], we can represent the kernel C in the form

$$C_{mn} = A_n^* A_m, \qquad m,n \in \mathbb{N}_0,$$

where for each $k \in \mathbb{N}_0$, A_k^* is the evaluation mapping on $\mathfrak{H}_C$ to $\mathfrak{F}$: $A_k^*(\{h_n\}_{n=0}^\infty) = h_k$. By the Cauchy representation, the operators C_{mn} are uniformly bounded, and therefore for w and z in a suitable neighborhood of the origin,

$$K(w,z) = \sum_{m,n=0}^\infty A_n^* A_m z^m \bar{w}^n = A(w)^* A(z),$$

where $A(z) = \sum_{m=0}^\infty A_m z^m$. The values of $A(z)$ lie in the Pontryagin space $\mathfrak{H}_C$, which has negative index κ. The restriction of $K(w,z)$ to a suitable neighborhood of the origin thus has at most κ negative squares, and since the number of negative squares is independent of the domain (see [3, Theorem 1.1.4]), $\operatorname{sq}_- K \le \kappa$. As noted above, this implies that $\operatorname{sq}_- K = \kappa$. □

References

[1] V.M. Adamjan, D.Z. Arov, and M.G. Kreĭn, *Analytic properties of the Schmidt pairs of a Hankel operator and the generalized Schur-Takagi problem*, Mat. Sb. (N.S.) **86(128)** (1971), 34–75.

[2] D. Alpay, *Some remarks on reproducing kernel Kreĭn spaces*, Rocky Mountain J. Math. **21** (1991), no. 4, 1189–1205.

[3] D. Alpay, A. Dijksma, J. Rovnyak, and H.S.V. de Snoo, *Schur functions, operator colligations, and reproducing kernel Pontryagin spaces*, Oper. Theory Adv. Appl., vol. 96, Birkhäuser, Basel, 1997.

[4] ______, *Reproducing kernel Pontryagin spaces*, Holomorphic Spaces (S. Axler, J.E. McCarthy, and D. Sarason, eds.), MSRI Publications, vol. 33, Cambridge University Press, Cambridge, 1998, pp. 425–444.

[5] ______, *Realization and factorization in reproducing kernel Pontryagin spaces*, Operator theory, system theory and related topics. The Moshe Livšic anniversary volume (D. Alpay and V. Vinnikov, eds.), Oper. Theory Adv. Appl., vol. 123, Birkhäuser, Basel, 2001, pp. 43–65.

[6] J.A. Ball, I. Gohberg, and L. Rodman, *Interpolation of rational matrix functions*, Oper. Theory Adv. Appl., vol. 45, Birkhäuser, Basel, 1990.

[7] J.A. Ball and J.W. Helton, *A Beurling-Lax theorem for the Lie group $U(m,n)$ which contains most classical interpolation theory*, J. Operator Theory **9** (1983), no. 1, 107–142.

[8] J.A. Ball and T.T. Trent, *Unitary colligations, reproducing kernel Hilbert spaces, and Nevanlinna-Pick interpolation in several variables*, J. Funct. Anal. **157** (1998), no. 1, 1–61.

[9] T. Constantinescu and A. Gheondea, *On the indefinite trigonometric moment problem of I. S. Iohvidov and M. G. Kreĭn*, Math. Nachr. **171** (1995), 79–94.

[10] ______, *Representations of Hermitian kernels by means of Kreĭn spaces*, Publ. Res. Inst. Math. Sci. **33** (1997), no. 6, 917–951.

[11] ______, *On the Carathéodory type problem of M. G. Kreĭn and H. Langer*, C. R. Acad. Sci. Paris Sér. I Math. **327** (1998), no. 3, 243–247.

[12] M.A. Dritschel and J. Rovnyak, *Operators on indefinite inner product spaces*, Lectures on operator theory and its applications (Waterloo, ON, 1994), Fields Institute Monographs, vol. 3, Amer. Math. Soc., Providence, RI, 1996, pp. 141–232, Supplementary material and errata are available at
`http://wsrv.clas.virginia.edu/ jlr5m/papers/papers.html`.

[13] I.S. Iokhvidov, *Hankel and Toeplitz matrices and forms, algebraic theory*, Birkhäuser, Boston, 1982.

[14] I.S. Iokhvidov and M.G. Kreĭn, *Spectral theory of operators in spaces with indefinite metric. II*, Trudy Moskov. Mat. Obšč. **8** (1959), 413–496, English transl.: Amer. Math. Soc. Transl. (2) **34** (1963), 283–373.

[15] I.S. Iokhvidov, M.G. Kreĭn, and H. Langer, *Introduction to the spectral theory of operators in spaces with an indefinite metric*, Mathematical Research, vol. 9, Akademie-Verlag, Berlin, 1982.

[16] A. Kheifets, *The abstract interpolation problems and applications*, Holomorphic Spaces (S. Axler, J. E. McCarthy, and D. Sarason, eds.), MSRI Publications, vol. 33, Cambridge University Press, Cambridge, 1998, pp. 351–379.

[17] M.G. Kreĭn and H. Langer, *Über die verallgemeinerten Resolventen und die charakteristische Funktion eines isometrischen Operators im Raume* Π_κ, Hilbert space operators and operator algebras (Proc. Internat. Conf., Tihany, 1970), North-Holland, Amsterdam, 1972, pp. 353–399. Colloq. Math. Soc. János Bolyai, 5.

[18] ———, *Über einige Fortsetzungsprobleme, die eng mit der Theorie hermitescher Operatoren im Raume* Π_κ *zusammenhängen. I. Einige Funktionenklassen und ihre Darstellungen*, Math. Nachr. **77** (1977), 187–236.

[19] A.A. Nudel′man, *A generalization of classical interpolation problems*, Dokl. Akad. Nauk SSSR **256** (1981), no. 4, 790–793.

[20] J. Rovnyak, *Methods of Kreĭn space operator theory*, this volume.

[21] T. Takagi, *On an algebraic problem related to an analytic theorem of Carathéodory and Fejér*, Japan J. Math. **1** (1924), 83–93, ibid. **2** (1925), 13–17.

[22] H. Woracek, *An operator-theoretic approach to degenerated Nevanlinna-Pick interpolation*, Math. Nachr. **176** (1995), 335–350.

[23] ———, *Nevanlinna-Pick interpolation: the degenerated case*, Linear Algebra Appl. **252** (1997), 141–158.

D. Alpay
Department of Mathematics, Ben-Gurion University of the Negev
P. O. Box 653, 84105 Beer-Sheva, Israel
e-mail: `dany@math.bgu.ac.il`

T. Constantinescu
Programs in Mathematical Sciences, University of Texas at Dallas
Box 830688, Richardson, TX 75083-0688, U. S. A.
e-mail: `tiberiu@utdallas.edu`

A. Dijksma
Department of Mathematics, University of Groningen
P. O. Box 800, 9700 AV Groningen, The Netherlands
e-mail: `dijksma@math.rug.nl`

J. Rovnyak
Department of Mathematics, University of Virginia
P. O. Box 400137, Charlottesville, VA 22904-4137, U. S. A.
e-mail: `rovnyak@Virginia.EDU`

Operator Theory:
Advances and Applications, Vol. 134, 99–136

Stable Dissipative Linear Stationary Dynamical Scattering Systems

D.Z. Arov
(translated and with an appendix by D.Z. Arov and J. Rovnyak)

D.Z. Arov and J. Rovnyak dedicate this translation and the remarks that follow to Harry Dym, with affection and appreciation for his many contributions. D.Z. Arov wishes also to say that it has been a great pleasure to work with Harry for the past ten years.

Introduction

In the theory of passive linear electrical networks, the Darlington method is well known as a realization of a finite ideal passive 1-port with losses via a finite ideal passive lossless 2-port closed by one resistance [9]. The reflection coefficient Θ of such a 1-port is an element of the scattering matrix $\widetilde{\Theta}$ of a corresponding lossless 2-port; the lossless behavior is indicated in the property that $\widetilde{\Theta}$ has unitary values on the boundary of the physical domain (in the right or upper half-plane, or inside the unit disk). The consideration of scattering matrices allowed Belevich to generalize Darlington's result on finite ideal n-ports with losses [16]. Darlington himself did not consider Θ and $\widetilde{\Theta}$ but other frequency characteristics: the impedance Z of 1-ports and the transmission matrix $\widetilde{A}$ of 2-ports (Z and $\widetilde{A}$ have simple representations by means of Θ and $\widetilde{\Theta}$). In this way, the Darlington result was generalized to finite ideal n-ports with losses by V.P. Potapov [14] and his student E.Ya. Melamud [12].

It should be mentioned that Darlington proposed his method of network synthesis [18] as universal, which can be applied in the investigation via a frequency characteristic of systems with losses of an arbitrary physical nature. After Darlington, but independently from him, the representation of the scattering matrix Θ of a system with losses as a block of a scattering matrix $\widetilde{\Theta}$ of a lossless system was used in the physics of nuclear reactions [11].

The main part of this paper consists of a translation of the article "Устойчвые диссипативные линейные стационарные динамические системы рассеяния," *Journal of Operator Theory* **2**, no. 1 (1979), 95–126, which was prepared by the author and J. Rovnyak. Commentary and an update of the results are provided by D.Z. Arov in Appendix 1. Appendix 2 by D.Z. Arov and J. Rovnyak shows some directions for generalizations and further development; this work was supported by NSF grant DMS–9801016.

Nevertheless, simple analytical considerations demonstrate that if, for example, Θ has unitary values only on some interval of the boundary of the physical domain, the Darlington method is not applicable. In the paper [1] the author selected a natural class $B\Pi$ of scattering matrices Θ which have a Darlington representation: $\Theta \in B\Pi$ if and only if Θ is a holomorphic and contractive function in the physical domain that has in a certain sense (via boundary values almost everywhere across the boundary) a continuation into the exterior of the domain, and which has a representation here as a ratio of holomorphic bounded functions. Independently from the author, a generalization of the Darlington method for the class $B\Pi$ was obtained by Dewilde [19] inspired by the corresponding Belevich result for rational matrix-valued functions (see also Douglas and Helton [20]). The values of the function Θ can be linear operators acting from one Hilbert space to another. Starting with [1], the author considered two forms of the Darlington representation: first by consideration of the scattering matrix of a lossless system (independently from Belevich), and second via the transmission matrix. In this paper the representations in the first form are considered. They are obtained and described for operator-valued functions $\Theta(z)$ ($|z| < 1$) of a wider class than $B\Pi$. In the matrix case these classes coincide.

The description of the Darlington representation of Θ is given in the form

$$\widetilde{\Theta} = \begin{pmatrix} I & 0 \\ 0 & b_2 \end{pmatrix} \begin{pmatrix} \varphi_1 & \Theta \\ h_0 & \varphi_2 \end{pmatrix} \begin{pmatrix} b_1 & 0 \\ 0 & I \end{pmatrix},$$

where $\varphi_1^*(\bar{z})$ and $\varphi_2(z)$ are outer functions which are the solutions of the factorization problem

$$\varphi_1(\zeta)\varphi_1^*(\zeta) = I - \Theta(\zeta)\Theta^*(\zeta), \qquad \varphi_2^*(\zeta)\varphi_2(\zeta) = I - \Theta^*(\zeta)\Theta(\zeta) \qquad (|\zeta| = 1),$$

$h_0(\zeta)$ is a function which is defined by the relation

$$h_0(\zeta)\varphi_1^*(\zeta) = -\varphi_2(\zeta)\Theta^*(\zeta), \qquad (|\zeta| = 1),$$

and $b_1^*(\bar{z})$ and $b_2(z)$ are inner functions such that $b_2(\zeta)h_0(\zeta)b_1(\zeta)$ are the boundary values of a bounded holomorphic function defined for $|z| < 1$. From among the "denominators" $\{b_2, b_1\}$ of the function h_0, minimal ones are selected in a natural way; the minimal $\widetilde{\Theta}$ correspond to them. The description of Darlington representations in the second form, which is given in the paper [2], can be obtained by passing from the first form to the second.

It is known [13], [8], that an arbitrary function $\widetilde{\Theta}(z)$ which is holomorphic and contractive for $|z| < 1$ with unitary boundary values $\widetilde{\Theta}(\zeta)$ ($|\zeta| = 1$, a.e.) can be realized as the transfer function (scattering matrix)

$$\widetilde{\Theta}(z) = \widetilde{S} + z\widetilde{G}(I - z\widetilde{T})^{-1}\widetilde{F}$$

of some minimal (controllable and observable) stable conservative linear stationary dynamical scattering system $\tilde{\alpha}$ with discrete time

$$\tilde{h}(n+1) = \widetilde{T}\tilde{h}(n) + \widetilde{F}\tilde{\varphi}^-(n), \qquad \tilde{\varphi}^+(n) = \widetilde{G}\tilde{h}(n) + \widetilde{S}\tilde{\varphi}^-(n)$$

with Hilbert spaces $\widetilde{\mathfrak{N}}^-$ (input), $\widetilde{\mathfrak{N}}^+$ (output) and $\widetilde{\mathfrak{H}}$ (state space). The conservativeness condition for the scattering system $\tilde{\alpha}$ is equivalent to the requirement that $\widetilde{F}$, $\widetilde{T}$, $\widetilde{G}$, and $\widetilde{S}$, are the blocks of an operator which maps $\widetilde{\mathfrak{N}}^- \oplus \widetilde{\mathfrak{H}}$ unitarily onto $\widetilde{\mathfrak{H}} \oplus \widetilde{\mathfrak{N}}^+$; the stability condition means that

$$\text{s-}\lim_{n\to\infty} \widetilde{T}^n = 0, \qquad \text{s-}\lim_{n\to\infty} (\widetilde{T}^*)^n = 0, \qquad (\widetilde{T} \in C_{00}),$$

and the conditions of controllability and observability are

$$\widetilde{\mathfrak{H}} = \bigvee_0^\infty \widetilde{T}^n \widetilde{F} \widetilde{\mathfrak{N}}^-, \qquad \widetilde{\mathfrak{H}} = \bigvee_0^\infty (\widetilde{T}^*)^n \widetilde{G}^* \widetilde{\mathfrak{N}}^+$$

($\vee \mathfrak{D}_n$ is the smallest subspace which contains all $\mathfrak{D}_n$). The Darlington representation of Θ in the first form, in the form of a block of $\widetilde{\Theta}$, is associated in the present paper with a realization of Θ as the scattering matrix of a stable dissipative linear stationary dynamical scattering system α, which is obtained via $\tilde{\alpha}$ by losses of parts of some scattering channels: in α the input space $\mathfrak{N}^-$ and output space $\mathfrak{N}^+$ are subspaces of spaces $\widetilde{\mathfrak{N}}^-$ and $\widetilde{\mathfrak{N}}^+$, respectively, the state space $\mathfrak{H}$ coincides with $\widetilde{\mathfrak{H}}$, and the coefficients are

$$F = \widetilde{F}|\mathfrak{N}^+, \quad T = \tilde{T}, \quad G = P_{\mathfrak{N}^+}\widetilde{G}, \quad S = P_{\mathfrak{N}^+}\widetilde{S}|\mathfrak{N}^-.$$

It is proved here (Theorem 3) that the dissipative system α obtained in this way is minimal if and only if $\widetilde{\Theta}$ is minimal.

Functions Θ of the class $B\Pi$ are realized as the scattering matrices of systems α with basic operators T of the class C_0, which was introduced and investigated by B. Sz.-Nagy and C. Foias [13, 25–27]. It will be proved here (Theorems 5, 6): 1) if $T \in C_0$, then $\Theta \in B\Pi$; 2) if $\Theta \in B\Pi$, then for a minimal system α with scattering matrix Θ we have $T \in C_0$, and moreover the minimal function $m_T(z)$ of the contraction T does not depend on the choice of minimal system α, and it is in fact the minimal scalar denominator $b_\Theta(z)$ of the function $\Theta(1/z)$ ($b_\Theta(z)$ is contained in the set of the scalar inner functions $b(z)$ such that $b(z)\Theta(1/z)$ is a holomorphic and contractive function for $|z| < 1$, and $b_\Theta(z)$ is a divisor of all such $b(z)$); 3) for minimal $\widetilde{\Theta}$ we have $b_{\widetilde{\Theta}}(z) = b_\Theta(z)$. Proposition 1 (and 2) were announced in [3] (see also [4]).

With the Darlington repesentation of Θ we obtain in particular the synthesis of a stable, controllable, optimal system $\mathring{\alpha}$, i.e. such that for each other passive system α with the same scattering matrix Θ, for arbitrary $\varphi_k^- (\in \mathfrak{N}^-)$, $n \geq 0$, we have

$$\Big\| \sum_{k=0}^n \mathring{T}^k \mathring{F} \varphi_k^- \Big\| \leq \Big\| \sum_{k=0}^n T^k F \varphi_k^- \Big\|.$$

The existence of a controllable optimal system is proved in the paper for an arbitrary scattering matrix Θ.

For scattering systems with continuous time, one can obtain corresponding results via passage to scattering systems with discrete time by Cayley transform (see [7]).

For resistance and transmission systems with discrete and continuous time, one can obtain dissipative realizations of transfer functions (impedance matrices and transmission matrices) by passing to corresponding scattering systems.

The author's results on impedance matrices which are analogous to Theorems 7, 8, and 5 and obtained by the same methods will be presented in another place.

Note that the consideration of a stable dissipative scattering system differs from a scattering scheme which satisfies Lax-Phillips postulates [24] with orthogonal subspaces $\mathfrak{D}^-$ and $\mathfrak{D}^+$, "incoming and outgoing waves" (see [3], [23]), only by ignoring Lax scattering channels. This is the reason why the Darlington representation of Θ in the first form arises in the work of C. Foias [21] and Bondy [17].

A dissipative scattering system has a conservative dilation and can be obtained from it by excluding the inner Lax scattering channels from consideration (see [7]). Thus by Darlington's method one can obtain two types of realizations involving losses of part of the exterior scattering channels: a conservative (see [4]) and a dissipative. Both types in essence are considered in the physics of nuclear reactions [11] (pp. 147–148), where the dissipative realization leads to the Teichmann and Wigner method.

Finally we shall indicate an application of the description of minimal $\widetilde{\Theta}$ to the synthesis of electrical networks. For rational real matrix-functions Θ of size n, contractive for $|z| < 1$, the minimal $\widetilde{\Theta}$ are rational matrix functions of size $n + r$, where $r = \text{rank } [I - \Theta^*(1/\bar{z})\Theta(z)]$, and one can choose such $\widetilde{\Theta}$ to be real (see [2], §6).

Via these $\widetilde{\Theta}$ one can obtain the realizations of Θ as scattering matrices of ideal n-ports that have a minimal number of resistances, equal to r, and at the same time a minimal number $m_C + m_L$ of capacitors and inductors, equal to the degree of the rational matrix function $\Theta(z)$, $m_C + m_L = \deg \Theta = \deg \widetilde{\Theta}$. (The problem to obtain such a realization of Θ was formulated by Tellegen [28].) It is enough to realize $\widetilde{\Theta}$ as the scattering matrix of a lossless ideal $(n + r)$-port with $m_C + m_L = \deg \widetilde{\Theta}$, for example in the manner indicated in Helton [22], and then r corresponding exterior branches should be loaded with 1-ohm resistances. The Darlington representation in the second form with entire real transmission matrix functions $\widetilde{A}$ of size $2n$ already gives the synthesis of a non-finite ideal n-port which is obtained via loading of an ideal regular n-line on the output branches of 1-ohm resistances through transformers [5]. For $n = 1$, in this way we obtain the realization of an ideal regular string with friction on one end, and with a given coefficient $Z(\lambda)$ of dynamical pliability of the velocity on the other end (see [5], §4). Using considerations in [5], it is not difficult to show that the minimal realization should have the minimal length of string $\ell = -Z(0)\sum \text{Re } (1/\lambda_j)$ and the minimal mass $M(\ell) = -Z^{-1}(0)\sum \text{Re } (1/\mu_j)$, where λ_j and μ_j are zeros and poles of $Z(\lambda)$ counting multiplicities; precisely such strings were investigated in [5].

1. Scattering matrices of passive systems with the basic operators of the classes $C_{0\cdot}$, $C_{\cdot 0}$, and C_{00}

1. A linear stationary dynamical system (LSDS) α in the separable Hilbert spaces $\mathfrak{N}^-$, $\mathfrak{N}^+$ (exterior) and $\mathfrak{H}$ (inner) with discrete time n $(= 0, 1, 2, \dots)$ is defined as a system

$$\begin{cases} h(n+1) = Th(n) + F\varphi^-(n), \\ \varphi^+(n) = Gh(n) + S\varphi^-(n), \end{cases} \qquad (n \geq 0) \tag{1}$$

with constant coefficients T, F, G and S,

$$T \in [\mathfrak{H}, \mathfrak{H}], \quad F \in [\mathfrak{N}^-, \mathfrak{H}], \quad G \in [\mathfrak{H}, \mathfrak{N}^+], \quad S \in [\mathfrak{N}^-, \mathfrak{N}^+]$$

(by $[\mathfrak{N}_1, \mathfrak{N}_2]$ is meant the set of bounded linear operators acting from $\mathfrak{N}_1$ into $\mathfrak{N}_2$). The vectors $\varphi^-(n)$, $\varphi^+(n)$ and $h(n)$ from $\mathfrak{N}^-, \mathfrak{N}^+$ and $\mathfrak{H}$ can be interpreted as data from the input, output, and inside the system, respectively, at time n. The evolution of the inner state under zero data on the input is described by the operator T: $h(n) = T^n h(0)$ for $\varphi^-(n) \equiv 0$. The operator T is called the basic operator of the system α. The data $h(0)$ and $\{\varphi^-(n)\}_0^m$ from the system (1) uniquely determine $\{h(n)\}_1^{m+1}$ and $\{\varphi^+(n)\}_0^m$. For $h(0) = 0$, we have:

$$\varphi^+(0) = S\varphi^-(0), \qquad \varphi^+(n) = S\varphi^-(n) + \sum_{k=0}^{n-1} GT^k F\varphi^-(n-k-1), \qquad (n \geq 1).$$

Via formal power series

$$\Phi^\pm(z) = \sum_{n=0}^{\infty} \varphi^\pm(n) z^n, \qquad \Theta(z) = S + \sum_{n=1}^{\infty} GT^{n-1} F z^n$$

one can write this system of equalities in the form

$$\Phi^+(z) = \Theta(z)\Phi^-(z).$$

One can consider $\Theta(z)$ as the power series representation in a sufficiently small neighborhood of zero of the function

$$\Theta_\alpha(z) = S + zG(I - zT)^{-1}F, \tag{2}$$

which is holomorphic for $z = 0$. This function is called the transfer function of the system α.

Below the data $\varphi^-(n)$ and $\varphi^+(n)$ are interpreted as incoming and reflected waves, which bring and remove energy, and the squares of the norms of vectors in $\mathfrak{N}^\pm$ and $\mathfrak{H}$ are interpreted as energy.

We will call α a *passive scattering system* if for arbitrary data $h(0)$ and $\{\varphi^-(n)\}_0^\infty$ the condition

$$\left\|\varphi^-(n)\right\|^2 - \left\|\varphi^+(n)\right\|^2 \geq \|h(n+1)\|^2 - \|h(n)\|^2,$$

is satisfied, which means that an inner source of energy is absent. Since

$$\begin{pmatrix} h(n+1) \\ \varphi^+(n) \end{pmatrix} = \begin{pmatrix} T & F \\ G & S \end{pmatrix} \begin{pmatrix} h(n) \\ \varphi^-(n) \end{pmatrix},$$

this condition means that the operator

$$V = \begin{pmatrix} T & F \\ G & S \end{pmatrix} (\in [\mathfrak{H} \oplus \mathfrak{N}^-, \mathfrak{H} \oplus \mathfrak{N}^+])$$

is contractive ($V^*V \leq I$), i.e. that the two equivalent inequalities

$$\begin{pmatrix} I|\mathfrak{H} & 0 \\ 0 & I|\mathfrak{N}^- \end{pmatrix} - \begin{pmatrix} T & F \\ G & S \end{pmatrix}^* \begin{pmatrix} T & F \\ G & S \end{pmatrix} \geq 0,$$

$$\begin{pmatrix} I|\mathfrak{H} & 0 \\ 0 & I|\mathfrak{N}^+ \end{pmatrix} - \begin{pmatrix} T & F \\ G & S \end{pmatrix} \begin{pmatrix} T & F \\ G & S \end{pmatrix}^* \geq 0$$

hold.

If in each of the preceding two inequalities, equality "=" holds, i.e. V is a unitary operator, then we call α a *conservative scattering system*, and in the opposite case a *dissipative scattering system.* The transfer function of a passive scattering system α will be called the *scattering matrix.* Because the basic operator T in such a system is a contraction, $\Theta_\alpha(z)$ is determined by formula (2) in the unit disk (for $|z| < 1$).

2. The conservative scattering LSDS's in essence are the subject of the investigation of contractions in Hilbert space, which is developed by M.S. Livšic, and B. Sz.-Nagy and C. Foias, their followers, and others (see [13, 8, 7]).

For a system α, we denote

$$\mathfrak{H}_\alpha^c = \bigvee_0^\infty T^n F \mathfrak{N}^-, \qquad \mathfrak{H}_\alpha^o = \bigvee_0^\infty (T^*)^n G^* \mathfrak{N}^+$$

($\vee_n \mathfrak{D}_n$ is the minimal subspace which contains all $\mathfrak{D}_n$).

We will call a LSDS *simple* if

$$\mathfrak{H} = \mathfrak{H}_\alpha^c \vee \mathfrak{H}_\alpha^o.$$

For a conservative scattering system, this condition means that T has no unitary part (T is a completely nonunitary contraction). We will denote by B the class of functions $\Theta(z)$ which are holomorphic for $|z| < 1$ and have values from $[\mathfrak{N}^-, \mathfrak{N}^+]$ for some $\mathfrak{N}^-$ and $\mathfrak{N}^+$ and which have $\|\Theta(z)\| \leq 1$ ($|z| < 1$).

One of the significant results in the theory of contractions on Hilbert space (see [13], [8]) can be formulated in the following form.

Proposition 1. *The scattering matrix of an arbitrary conservative LSDS belongs to the class B. An arbitrary function $\Theta(z)$ in the class B is the scattering matrix of a simple conservative LSDS, which is determined by $\Theta(z)$ up to unitary similarity.*

Two LSDS's α and α_1 are called (unitarily) similar if there exists a (unitary) bounded and invertible operator X ($\in [\mathfrak{H}, \mathfrak{H}_1]$) such that

$$F_1 = XF, \quad T_1 X = XT, \quad G = G_1 X, \quad (S_1 = S).$$

There are functional models of a simple conservative scattering system α, which are constructed using the boundary values $\Theta_\alpha(\zeta)$ of the function $\Theta_\alpha(z)$ ($\in B$),

$$\Theta_\alpha(\zeta) = \text{s-}\lim_{r\uparrow 1} \Theta_\alpha(r\zeta) \qquad (|\zeta| = 1, \text{ a.e.})$$

(see [13], [8]). There are especially simple models for systems α for which T satisfies the condition

$$\text{s-}\lim_{n\to\infty} T^n = 0 \quad \big(\text{s-}\lim_{n\to\infty} (T^*)^n = 0\big).$$

The class of such contractions T is denoted by $C_{0\cdot}$, ($C_{\cdot 0}$), and the intersection $C_{0\cdot} \cap C_{\cdot 0}$ is denoted by C_{00}. It is known [13] that the scattering matrices of conservative systems with basic operators from the class $C_{0\cdot}$ ($C_{\cdot 0}$, C_{00}) coincide with the subclass of functions $\Theta(z)$ in B which have isometric ($*$-isometric, unitary) boundary values $\Theta(\zeta)$ almost everywhere on the circle $|\zeta| = 1$. Such functions $\Theta(z)$ are called inner ($*$-inner, bi-inner). We will write the model in the case that $\Theta_\alpha(z)$ is an inner function.

Let $L^2(\mathfrak{N})$ be the Hilbert space of weakly measurable functions $f(\zeta)$ ($|\zeta| = 1$) whose values belong to the Hilbert space $\mathfrak{N}$ for which

$$\|f\|^2 = \frac{1}{2\pi} \int_{|\zeta|=1} \|f(\zeta)\|^2 \, |d\zeta| < \infty,$$

and let $H_+^2(\mathfrak{N})$ be the subspace of functions $f(\zeta)$ in $L^2(\mathfrak{N})$ which have Fourier series representations involving only nonnegative powers of ζ. We will identify $H_+^2(\mathfrak{N})$ with the Hardy space of functions $f(z)$ which are holomorphic in the unit disk and have values in $\mathfrak{N}$ and for which $\|f\|^2 = \sup_{r<1} \frac{1}{2\pi} \int_{|\zeta|=1} \|f(r\zeta)\|^2 \, |d\zeta| < \infty$;

$$f(\zeta) = \text{s-}\lim_{r\uparrow 1} f(r\zeta) \qquad (|\zeta| = 1, \text{ a.e.}).$$

Proposition 2. [8]. *Let $\Theta(z)$ be an inner function which has values in $[\mathfrak{N}^-, \mathfrak{N}^+]$. Consider the system $\dot\alpha$ with exterior spaces $\mathfrak{N}^-$ and $\mathfrak{N}^+$ for which the inner space $\dot{\mathfrak{H}}$ and coefficients $\dot T$, $\dot F$, $\dot G$, $\dot S$ are determined by the formulas*

$$\dot{\mathfrak{H}} = H_+^2(\mathfrak{N}^+) \ominus \Theta(\zeta) H_+^2(\mathfrak{N}^-),$$

$$\dot T h = \zeta^{-1}[h(\zeta) - h(0)], \qquad \dot G h = h(0) \qquad (h = h(\zeta) \in \dot{\mathfrak{H}}),$$

$$\dot F \varphi^- = \zeta^{-1}[\Theta(\zeta) - \Theta(0)]\varphi^-, \qquad \dot S \varphi^- = \Theta(0)\varphi^- \qquad (\varphi^- \in \mathfrak{N}^-).$$

Then $\dot\alpha$ is a simple conservative system with scattering matrix $\Theta(z)$.

If α is a system with coefficients T, F, G, S, then the system with coefficients T^*, G^*, F^*, S^* is denoted by α^* and is called the adjoint of α. In the above model the coefficients of the adjoint system $\dot\alpha^*$ are defined by

$$F^* h = \frac{1}{2\pi} \int_{|\zeta|=1} \zeta \Theta^*(\zeta) h(\zeta) \, |d\zeta| \; (= p_h), \qquad \dot T^* h = \zeta h(\zeta) - \Theta(\zeta) p_h \quad (h \in \dot{\mathfrak{H}}),$$

$$G^* \varphi^+ = [I - \Theta(\zeta)\Theta^*(0)]\varphi^+, \qquad \dot S^* \varphi^+ = \Theta^*(0)\varphi^+ \quad (\varphi^+ \in \mathfrak{N}^+).$$

If $\Theta(z)$ is the scattering matrix of the system α, then $\Theta^{\sim}(z)$ $(= \Theta^*(\bar{z}))$ is the scattering matrix of the system α^*. Consequently, $\dot{\alpha}^*$ is the model of a simple conservative system, the scattering matrix of which $\Theta^{\sim}(z)$ is a $*$-inner function.

3. **Theorem 1.** *Let α be a simple passive scattering system with bi-inner function $\Theta_\alpha(z)$. Then α is a conservative system.*

Proof. Let α be a passive scattering system, and let the corresponding contraction with block-coefficients T, F, G, S of the system α be V $(\in [\mathfrak{H} \oplus \mathfrak{N}^-, \mathfrak{H} \oplus \mathfrak{N}^+])$. Consider

$$\mathring{\mathfrak{N}}^- = \overline{(I - VV^*)(\mathfrak{H} \oplus \mathfrak{N}^+)} \ (= \mathfrak{D}_{V^*}), \qquad \mathring{\mathfrak{N}}^+ = \overline{(I - V^*V)(\mathfrak{H} \oplus \mathfrak{N}^-)} \ (= \mathfrak{D}_V),$$

$$\mathcal{D}_{V^*} = (I - VV^*)^{1/2}, \qquad \mathcal{D}_V = (I - V^*V)^{1/2},$$

$$\widetilde{V} = \begin{pmatrix} \mathcal{D}_{V^*} & V \\ -V^* & \mathcal{D}_V \end{pmatrix} \in [\mathring{\mathfrak{N}}^- \oplus (\mathfrak{H} \oplus \mathfrak{N}^-), (\mathfrak{H} \oplus \mathfrak{N}^+) \oplus \mathring{\mathfrak{N}}^+];$$

$$\widetilde{\mathfrak{N}}^- = \mathring{\mathfrak{N}}^- \oplus \mathfrak{N}^-, \qquad \widetilde{\mathfrak{N}}^+ = \mathfrak{N}^+ \oplus \mathring{\mathfrak{N}}^+,$$

let $\widetilde{\alpha}$ be the system with exterior spaces $\widetilde{\mathfrak{N}}^-$ and $\widetilde{\mathfrak{N}}^+$, inner space $\mathfrak{H}$ and coefficients $\widetilde{T}$, $\widetilde{F}$, $\widetilde{G}$, $\widetilde{S}$ which are blocks of the operator $\widetilde{V}$. One can easily check that $\widetilde{\alpha}$ is a conservative system and α is a part of it, that is,

$$F = \widetilde{F}|\mathfrak{N}^-, \qquad T = \widetilde{T}, \qquad G = P_{\mathfrak{N}^+}\widetilde{G}, \qquad S = P_{\mathfrak{N}^+}\widetilde{S}|\mathfrak{N}^-. \tag{3}$$

The scattering matrix $\Theta_\alpha(z)$ is represented in the form of a block of the matrix $\Theta_{\widetilde{\alpha}}(z)$: $\Theta_\alpha(z) = P_{\mathfrak{N}^+}\Theta_{\widetilde{\alpha}}(z)|\mathfrak{N}^-$. Let $\Theta_\alpha(z)$ be a bi-inner function. Since in this case $\Theta_\alpha(\zeta)$ has unitary values for $|\zeta| = 1$ (a.e.), and $\|\Theta_{\widetilde{\alpha}}(\zeta)\| \le 1$, we have

$$\Theta_\alpha(\zeta) = \Theta_{\widetilde{\alpha}}(\zeta)|\mathfrak{N}^-, \qquad \Theta^*_\alpha(\zeta) = \Theta^*_{\widetilde{\alpha}}(\zeta)|\mathfrak{N}^+ \quad \text{(a.e.)}$$

These equalities are equivalent to the following:

$$S = \widetilde{S}|\mathfrak{N}^-, \qquad GT^nF = \widetilde{G}\widetilde{T}^n\widetilde{F}|\mathfrak{N}^- \qquad (n \ge 0);$$

$$S^* = \widetilde{S}^*|\mathfrak{N}^+, \qquad F^*(T^*)^nG^* = \widetilde{F}^*(\widetilde{T}^*)^n\widetilde{G}^*|\mathfrak{N}^+ \qquad (n \ge 0).$$

Since

$$\widetilde{T} = T, \quad \widetilde{S} = \begin{pmatrix} P_{\mathfrak{N}^+}\mathcal{D}_{V^*} & S \\ -V^* & \mathcal{D}_V \end{pmatrix} \Bigg| \mathfrak{D}_{V^*} \oplus \mathfrak{N}^-,$$

$$\widetilde{G} = \begin{pmatrix} G \\ \mathcal{D}_V|\mathfrak{H} \end{pmatrix}, \quad \widetilde{F} = \left(P_{\mathfrak{H}}\mathcal{D}_{V^*}|\mathfrak{D}_{V^*} \quad F\right),$$

we obtain

$$\mathcal{D}_V^2|\mathfrak{N}^- = 0, \quad \mathcal{D}_V^2T^nF = 0; \quad \mathcal{D}_{V^*}^2|\mathfrak{N}^+ = 0, \quad \mathcal{D}_{V^*}^2(T^*)^nG^* = 0 \quad (n \ge 0).$$

It is to be proved that α is a conservative system, i.e. that $\mathcal{D}^2_{V^*} = 0$ and $\mathcal{D}^2_V = 0$. Because of the hypothesis of the theorem that α is a simple system, and for $\mathcal{D}^2_{V^*}$ (≥ 0) and $\mathcal{D}^2_V$ (≥ 0) we already have

$$\mathcal{D}^2_V|\mathfrak{N}^- \oplus \mathfrak{H}^c_\alpha = 0, \qquad \mathcal{D}^2_{V^*}|\mathfrak{N}^+ \oplus \mathfrak{H}^o_\alpha = 0,$$

it remains to prove that

$$P_{\mathfrak{H}}\mathcal{D}^2_V|\mathfrak{H}^o_\alpha = 0, \qquad P_{\mathfrak{H}}\mathcal{D}^2_{V^*}|\mathfrak{H}^c_\alpha = 0.$$

We will prove the first equality, and the second follows analogously. Since

$$\mathcal{D}^2_V|\mathfrak{N}^- = \begin{pmatrix} -T^*F - G^*S \\ I - F^*F - S^*S \end{pmatrix} = 0, \qquad \mathcal{D}^2_{V^*}|\mathfrak{N}^+ = \begin{pmatrix} -TG^* - FS^* \\ I - GG^* - SS^* \end{pmatrix} = 0,$$

$$\mathcal{D}^2_{V^*}(T^*)^n G^* = (I - TT^* - FF^*)(T^*)^n G^* = 0 \quad (n \geq 0),$$

we have

$$T^*F = -G^*S, \quad TG^* = -FS^*,$$

$$GG^* = I - SS^*, \quad (TT^*)(T^*)^n G^* = (I - FF^*)(T^*)^n G^*.$$

Therefore

$$P_{\mathfrak{H}}\mathcal{D}^2_V G^* = (I - T^*T - G^*G)G^* = G^* - T^*(TG^*) - G^*(GG^*)$$

$$= G^* - T^*(-FS^*) - G^*(I - SS^*) = (T^*F)S^* + (G^*S)S^*$$

$$= (-G^*S)S^* + (G^*S)S^* = 0,$$

$$P_{\mathfrak{H}}\mathcal{D}^2_V(T^*)^n G^* = (I - T^*T - G^*G)(T^*)^n G^* = (T^*)^n G^* - T^*(TT^*)(T^*)^{n-1}G^*$$

$$-G^*(GT^*)(T^*)^{n-1}G^* = (T^*)^n G^* - T^*(I - FF^*)(T^*)^{n-1}G^*$$

$$-G^*(-SF^*)(T^*)^{n-1}G^* = (T^*F)F^*(T^*)^{n-1}G^* + (G^*S)F^*(T^*)^{n-1}G^* = 0.$$

Consequently, $P_{\mathfrak{H}}\mathcal{D}^2_V|\mathfrak{H}^o_\alpha = 0$. Thus, $\mathcal{D}^2_V = 0$, $\mathcal{D}^2_{V^*} = 0$, i.e. V is a unitary operator. The theorem is proved.

A simple passive scattering LSDS α will be called *lossless* if α is a conservative system with basic operator T in the class C_{00}; otherwise α is called a system with losses. As we see, for a simple passive scattering LSDS α to be lossless, it is necessary and sufficient that the scattering matrix $\Theta_\alpha(z)$ is a bi-inner function.

4. We will now investigate the properties of the scattering matrices of dissipative systems with basic operators from the classes $C_{0\cdot}$, $C_{\cdot 0}$, and C_{00}.

Proposition 3. *In order that $\Theta(z)$ is the scattering matrix of a passive system with basic operator of class $C_{0\cdot}$ ($C_{\cdot 0}$, C_{00}) it is necessary and sufficient that it has a representation in the form of a block*

$$\Theta(z) = P_{\mathfrak{N}^+}\widetilde{\Theta}(z)|\mathfrak{N}^-$$

of an inner ($*$-inner, bi-inner) function $\widetilde{\Theta}(z)$.

Proof. Let α be a passive scattering system with basic operator T of the class $C_{0\cdot}$ ($C_{\cdot 0}$, C_{00}). Let us represent α as a part of a conservative system $\widetilde{\alpha}$ as constructed in the proof of the preceding theorem. Since the basic operator $\widetilde{T}$ $(= T)$ of the system $\widetilde{\alpha}$ belongs to the class $C_{0\cdot}$ ($C_{\cdot 0}$, C_{00}), $\Theta_{\widetilde{\alpha}}(z)$ is an inner ($*$-inner, bi-inner) function. The matrix $\Theta(z)$ is a block of the matrix $\Theta_{\widetilde{\alpha}}(z)$. Necessity is proved.

Let now $\Theta(z)$ have a representation in the form of a block of some inner ($*$-inner, bi-inner) function $\widetilde{\Theta}(z)$. Let us consider a simple conservative system $\widetilde{\alpha}$ with scattering matrix $\widetilde{\Theta}(z)$. Its basic operator $\widetilde{T}$ belongs to the class $C_{0\cdot}$ ($C_{\cdot 0}$, C_{00}). Let $\Theta(z)$ have values from $[\mathfrak{N}^-, \mathfrak{N}^+]$. We consider the system α with exterior spaces $\mathfrak{N}^-$ and $\mathfrak{N}^+$, which is a part of system $\widetilde{\alpha}$ in above mentioned sense. Its basic operator T $(= \widetilde{T})$ belongs to the class $C_{0\cdot}$ ($C_{\cdot 0}$, C_{00}), and $\Theta_\alpha(z) = \Theta(z)$. The proposition is proved.

Let $\Theta(z)$ be a block of an inner function $\widetilde{\Theta}(z)$. We can suppose that

$$\widetilde{\Theta}(z) = \begin{pmatrix} \Theta_{11}(z) & \Theta_{12}(z) \\ \Theta_{21}(z) & \Theta_{22}(z) \end{pmatrix}, \quad \Theta_{12}(z) = \Theta(z).$$

Then the function $\Theta_{22}(z)$ is a solution of the factorization problem:

$$\Theta_{22}^*(\zeta)\Theta_{22}(\zeta) = I - \Theta^*(\zeta)\Theta(\zeta) \qquad (|\zeta| = 1, \text{ a.e.}), \quad \Theta_{22}(z) \in B. \tag{4}$$

Conversely, if $\Theta_{22}(z)$ is solution of this problem, then $\Theta(z) = \begin{pmatrix} \Theta(z) \\ \Theta_{22}(z) \end{pmatrix}$ is an inner function with given block $\Theta(z)$. In the same way $\Theta(z)$ is a block of a $*$-inner function if and only if the factorization problem

$$\Theta_{11}(\zeta)\Theta_{11}^*(\zeta) = I - \Theta(\zeta)\Theta^*(\zeta), \qquad (|\zeta| = 1, \text{ a.e.}), \quad \Theta_{11}(z) \in B, \tag{5}$$

has a solution. Thus we have:

Proposition 4. *A function $\Theta(z)$ is the scattering matrix of a passive system with basic operator of the class $C_{0\cdot}$ ($C_{\cdot 0}$), if and only if the factorization* (4) ((5)) *has a solution.*

There are known [15] necessary and sufficient conditions for the solvability of the factorization problem

$$h^*(\zeta)h(\zeta) = f(\zeta) \qquad (|\zeta| = 1, \text{ a.e.}), \quad h(z) \in B$$

for a nonnegative function $f(\zeta)$ $(\leq I)$, which have values in $[\mathfrak{N}, \mathfrak{N}]$. A sufficient condition is due to Devinatz: $\ln \|f^{-1}(\zeta)\| \in L^1$. If this problem is solvable, the set of all solutions is described by the formula

$$h(z) = b(z)\varphi(z),$$

where $\varphi(z)$ is an outer function with values in $[\mathfrak{N}, \mathfrak{N}_\varphi]$ ($\varphi(z) \in B$, $\overline{\varphi(\zeta)H_+^2(\mathfrak{N})} = H_+^2(\mathfrak{N}_\varphi)$), and $b(z)$ is an arbitrary inner function with values in $[\mathfrak{N}_\varphi, \mathfrak{N}_*]$, where $\mathfrak{N}_*$ is a Hilbert space with $\dim \mathfrak{N}_* \geq \dim \mathfrak{N}_\varphi$, $\mathfrak{N}_\varphi \subset \mathfrak{N}$. Under the normalization

$\varphi(0)|\mathfrak{N}_\varphi > 0$ the function $\varphi(z)$ is uniquely determined by $f(\zeta)$. The dimension of the space $\mathfrak{N}_\varphi$ is determined by the equality

$$\dim \mathfrak{N}_\varphi = \operatorname{rank} f(\zeta) \qquad (|\zeta| = 1, \text{ a.e.}),$$

so if the factorization problem is solvable, rank $f(\zeta)$ ($= \dim f(\zeta)\mathfrak{N}$) is constant (a.e., $|\zeta| = 1$).

Thus the solutions of the problems (4) and (5) can be written in the form

$$\Theta_{11}(z) = \varphi_1(z)b_1(z), \qquad \Theta_{22}(z) = b_2(z)\varphi_2(z), \tag{6}$$

where $b_2(z)$ and $\varphi_2(z)$ are inner and outer functions with values in $[\mathfrak{N}_{\varphi_2}, \overset{\circ}{\mathfrak{N}}{}^+]$ and $[\mathfrak{N}^-, \mathfrak{N}_{\varphi_2}]$, and $b_1(z)$ and $\varphi_1(z)$ are $*$-inner and $*$-outer ($\varphi_1^\sim(z) = \varphi_1^*(\bar{z})$ is outer) functions with values in $[\overset{\circ}{\mathfrak{N}}{}^-, \mathfrak{N}_{\varphi_1^\sim}]$ and $[\mathfrak{N}_{\varphi_1^\sim}, \mathfrak{N}^+]$,

$$\dim \overset{\circ}{\mathfrak{N}}{}^- \geq \operatorname{rank}\,[I - \Theta(\zeta)\Theta^*(\zeta)], \qquad \dim \overset{\circ}{\mathfrak{N}}{}^+ \geq \operatorname{rank}\,[I - \Theta^*(\zeta)\Theta(\zeta)] \tag{7}$$

($|\zeta| = 1$). The functions $b_1(z)$ and $b_2(z)$ here are bi-inner if and only if

$$\overline{\Theta_{22}(\zeta)L^2(\mathfrak{N}^-)} = L^2(\overset{\circ}{\mathfrak{N}}{}^+), \qquad \overline{\Theta_{11}^*(\zeta)L^2(\mathfrak{N}^+)} = L^2(\overset{\circ}{\mathfrak{N}}{}^-). \tag{8}$$

If this condition is satisfied, then in (7) we have, instead of the signs "$\geq$", the signs "$=$":

$$\dim \overset{\circ}{\mathfrak{N}}{}^- = \operatorname{rank}\,[I - \Theta(\zeta)\Theta^*(\zeta)], \quad \dim \overset{\circ}{\mathfrak{N}}{}^+ = \operatorname{rank}\,[I - \Theta^*(\zeta)\Theta(\zeta)] \tag{9}$$

($|\zeta| = 1$, a.e.). If here the right sides are finite, then the conditions (9) and (8) are equivalent.

5. We formulate the necessary and sufficient conditions that $\Theta(z)$ has a representation as the block $\Theta_{12}(z)$ of a bi-inner function $\widetilde{\Theta}(z) = [\Theta_{ik}(z)]_1^2$ which satisfies the condition (8), and we will describe all such $\widetilde{\Theta}(z)$.

Theorem 2. *Let $\varphi_1(z)$ and $\varphi_2(z)$ be $*$-outer and outer functions which are solutions of the factorization problems* (5) *and* (4)*:*

$$\varphi_1(\zeta)\varphi_1^*(\zeta) = I - \Theta(\zeta)\Theta^*(\zeta); \qquad \varphi_2^*(\zeta)\varphi_2(\zeta) = I - \Theta^*(\zeta)\Theta(\zeta) \qquad \text{(a.e.)} \tag{10}$$

$$\begin{pmatrix} \varphi_1(z) \in B, \quad \varphi_1^*(0)|\mathfrak{N}_{\varphi_1^\sim} > 0; & \varphi_2(z) \in B, \quad \varphi_2(0)|\mathfrak{N}_{\varphi_2} > 0, \\ \overline{\varphi_1^*(\bar{\zeta})H_+^2(\mathfrak{N}^+)} = H_+^2(\mathfrak{N}_{\varphi_1^\sim}); & \overline{\varphi_2(\zeta)H_+^2(\mathfrak{N}^-)} = H_+^2(\mathfrak{N}_{\varphi_2}) \end{pmatrix}$$

Then the equality

$$h_0^*(\zeta)\varphi_2(\zeta) = -\varphi_1^*(\zeta)\Theta(\zeta) \tag{11}$$

determines a contractive function $h_0(\zeta)$ which has values in $[\mathfrak{N}_{\varphi_1^\sim}, \mathfrak{N}_{\varphi_2}]$ (weakly measurable, $\|h_0(\zeta)\| \leq 1$ a.e.). Let $b_1(z)$ and $b_2(z)$ be bi-inner functions with values in $[\overset{\circ}{\mathfrak{N}}{}^-, \mathfrak{N}_{\varphi_1^\sim}]$ and $[\mathfrak{N}_{\varphi_2}, \overset{\circ}{\mathfrak{N}}{}^+]$ such that

$$\Theta_{21}(\zeta) = b_2(\zeta)h_0(\zeta)b_1(\zeta) \tag{12}$$

are the boundary values of a function $\Theta_{21}(z)$ of the class B. Let $\Theta_{12}(z) = \Theta(z)$, and define $\Theta_{11}(z)$ and $\Theta_{22}(z)$ by the formulas (6). Then $\widetilde{\Theta}(z) = [\Theta_{ik}(z)]_1^2$ is a bi-inner function with the given block $\Theta_{12}(z) = \Theta(z)$ which satisfies the condition (8). All $\widetilde{\Theta}(z)$ with the stated properties are obtained in this way.

Proof. Let $\widetilde{\Theta}(z) = [\Theta_{ik}(z)]_1^2$ be a bi-inner function with given block $\Theta_{12}(z) = \Theta(z)$ satisfying the condition (8). Then $\Theta_{11}(z)$ and $\Theta_{22}(z)$ are solutions of the factorization problems (5) and (4) satisfying the condition (8). Therefore they are represented in the form (6), where $b_1(z)$ and $b_2(z)$ are bi-inner functions and $\varphi_1(z)$ and $\varphi_2(z)$ are $*$-outer and outer functions which are solutions of the problems (10). It follows from the unitarity of the boundary values $\widetilde{\Theta}(\zeta) = [\Theta_{ik}(\zeta)]_1^2$ that

$$\Theta_{11}^*(\zeta)\Theta_{12}(\zeta) + \Theta_{21}^*(\zeta)\Theta_{22}(\zeta) = 0; \qquad \Theta_{21}^*(\zeta)b_2(\zeta)\varphi_2(\zeta) = -b_1^*(\zeta)\varphi_1^*(\zeta)\Theta(\zeta).$$

We conclude that the contractive function

$$h_0(\zeta) = b_2^*(\zeta)\Theta_{21}(\zeta)b_1^*(\zeta)$$

satisfies the relation (11); $\Theta_{21}(\zeta)$ is expressed in terms of $h_0(\zeta)$ by formula (12).

We now show that if $\varphi_1(z)$ and $\varphi_2(z)$ are solutions of the problems (10), then the relation (11) uniquely determines a function $h_0(\zeta)$ with values in $[\mathfrak{N}_{\varphi_1^\sim}, \mathfrak{N}_{\varphi_2}]$. By means of the equation

$$K\varphi_2(\zeta)g(\zeta) = -\varphi_1^*(\zeta)\Theta(\zeta)g(\zeta) \qquad (g(\zeta) \in L^2(\mathfrak{N}^-))$$

we define a contraction operator K on the dense lineal $\varphi_2(\zeta)L^2(\mathfrak{N}^-)$ in $L^2(\mathfrak{N}_{\varphi_2})$ which maps this lineal into $L^2(\mathfrak{N}_{\varphi_1^\sim})$. This follows from the inequality

$$\|\varphi_1^*(\zeta)\Theta(\zeta)g(\zeta)\|^2_{\mathfrak{N}_{\varphi_1^\sim}} \leq \|\varphi_2(\zeta)g(\zeta)\|^2_{\mathfrak{N}_{\varphi_2}},$$

which is easily verified.

We extend K by continuity to all of the space $L^2(\mathfrak{N}_{\varphi_2})$. In this way we obtain a contraction K $(\in [L^2(\mathfrak{N}_{\varphi_2}), L^2(\mathfrak{N}_{\varphi_1^\sim})])$, possessing the property: $K\dot{U}_2 = \dot{U}_1 K$, where $\dot{U}_1$ and $\dot{U}_2$ are the operators of multiplication by ζ on $L^2(\mathfrak{N}_{\varphi_1^\sim})$ and $L^2(\mathfrak{N}_{\varphi_2})$. Therefore K is an operator of "multiplication" by some contractive function $K(\zeta)$ with values in $[\mathfrak{N}_{\varphi_2}, \mathfrak{N}_{\varphi_1^\sim}]$ (see [13], Ch. V, proof of Lemma 3.1). It remains to put $h_0(\zeta) = K^*(\zeta)$ and to remark that $K(\zeta)$ is uniquely determined by K.

For a function $h_0(\zeta)$ determined by the relation (11) these equations hold:

$$\begin{gathered} h_0(\zeta)\varphi_1^*(\zeta) = -\varphi_2(\zeta)\Theta^*(\zeta), \\ \varphi_1^*(\zeta)\varphi_1(\zeta) + h_0^*(\zeta)h_0(\zeta) = I, \quad \varphi_2(\zeta)\varphi_2^*(\zeta) + h_0(\zeta)h_0^*(\zeta) = I. \end{gathered} \tag{13}$$

Since $\varphi_1^*(\bar{z})$ and $\varphi_2(z)$ are outer functions, for their proofs it is sufficient to show that

$$\begin{aligned} \varphi_2^*(\zeta)\left[\varphi_2(\zeta)\Theta^*(\zeta) + h_0(\zeta)\varphi_1^*(\zeta)\right] &= 0, \\ \varphi_2^*(\zeta)\left[\varphi_2(\zeta)\varphi_2^*(\zeta) + h_0(\zeta)h_0^*(\zeta) - I\right] &= 0, \\ \varphi_1(\zeta)\left[\varphi_1^*(\zeta)\varphi_1(\zeta) + h_0^*(\zeta)h_0(\zeta) - I\right] &= 0. \end{aligned}$$

These equations follow from the relations (10)–(11).

Equations (10), (11) and (13) show that the function $\widetilde{\Theta}_0(\zeta)$, defined by the formula

$$\widetilde{\Theta}_0(\zeta) = \begin{pmatrix} \varphi_1(\zeta) & \Theta(\zeta) \\ h_0(\zeta) & \varphi_2(\zeta) \end{pmatrix} \quad (= [\Theta^0_{ik}(\zeta)]_1^2),$$

has unitary values (a.e. $|\zeta| = 1$).

Now assume that for the function $h_0(\zeta)$ there exist bi-inner functions $b_1(z)$ and $b_2(z)$ such that the function $\Theta_{21}(\zeta)$ determined by formula (12) gives the boundary values of a function $\Theta_{21}(z)$ of the class B. We put $\Theta_{12}(z) = \Theta(z)$, define $\Theta_{11}(z)$ and $\Theta_{22}(z)$ by the formulas (6), and prove that the resulting function $\widetilde{\Theta}(z) = [\Theta_{ik}(z)]_1^2$ is bi-inner. Since $\Theta_{ik}(z) \in B$, $\widetilde{\Theta}(z)$ is a bounded holomorphic function in the unit disk. Therefore $\widetilde{\Theta}(z) \in B$ if only $\|\widetilde{\Theta}(z)\| \leq 1$ a.e. But

$$\widetilde{\Theta}(\zeta) = \begin{pmatrix} I & 0 \\ 0 & b_2(\zeta) \end{pmatrix} \widetilde{\Theta}_0(\zeta) \begin{pmatrix} b_1(\zeta) & 0 \\ 0 & I \end{pmatrix} \tag{14}$$

and therefore $\widetilde{\Theta}(\zeta)$ admits unitary values ($|\zeta| = 1$, a.e.). Consequently, $\widetilde{\Theta}(z)$ is a bi-inner function. It satisfies condition (8). The theorem is proved.

2. Synthesis of minimal dissipative scattering systems with basic operators of the class C_{00}

1. Suppose $\Theta(z)$ is represented in the form of a block $\Theta_{12}(z)$ of some bi-inner function $\widetilde{\Theta}(z) = [\Theta_{ik}(z)]_1^2$. Assume that $\Theta(z)$ and $\widetilde{\Theta}(z)$ take values, respectively, in $[\mathfrak{N}^-, \mathfrak{N}^+]$ and $[\widetilde{\mathfrak{N}}^-, \widetilde{\mathfrak{N}}^+]$, $\widetilde{\mathfrak{N}}^- = \mathring{\mathfrak{N}}^- \oplus \mathfrak{N}^-$, $\widetilde{\mathfrak{N}}^+ = \mathfrak{N}^+ \oplus \mathring{\mathfrak{N}}^+$. We consider a simple conservative system $\tilde{\alpha}$ with scattering matrix $\widetilde{\Theta}(z)$ and its part α with scattering matrix $\Theta(z)$. The basic operator T $(= \widetilde{T})$ of the system α belongs to the class C_{00}. The system $\tilde{\alpha}$ is unitarily similar to its model $\dot{\tilde{\alpha}}$, which is constructed from $\widetilde{\Theta}(z)$ as in Proposition 2. The part $\dot{\alpha}$ of the system $\dot{\tilde{\alpha}}$ with the same inner space as $\dot{\tilde{\alpha}}$, and with the exterior spaces $\mathfrak{N}^-$ and $\mathfrak{N}^+$, is a model for the system α. The inner space $\dot{\mathfrak{H}}$ $(= \dot{\tilde{\mathfrak{H}}})$ and coefficients $\dot{T}$, $\dot{F}$, $\dot{G}$, $\dot{S}$ of the system $\dot{\alpha}$ are defined by the formulas

$$\begin{gathered}
\dot{\mathfrak{H}} = H^2_+(\widetilde{\mathfrak{N}}^+) \ominus \widetilde{\Theta}(\zeta) H^2_+(\widetilde{\mathfrak{N}}^-), \\
\dot{T}h = \zeta^{-1}[h(\zeta) - h(0)], \quad \dot{G}h = h_1(0), \\
(h = h(\zeta) = h_1(\zeta) \oplus h_2(\zeta) \in \dot{\mathfrak{H}}, \;\; h_1(\zeta) \in H^2_+(\mathfrak{N}^+), \;\; h_2(\zeta) \in H^2_+(\mathring{\mathfrak{N}}^+), \\
\dot{F}\varphi^- = \zeta^{-1}[\widetilde{\Theta}(\zeta) - \widetilde{\Theta}(0)]\varphi^-, \quad \dot{S}\varphi^- = \Theta(0)\varphi^- \quad (\varphi^- \in \mathfrak{N}^-).
\end{gathered}$$

For the coefficients of the adjoint system $\dot{\alpha}^*$, we have:

$$\dot{F}^* h = P_{\mathfrak{N}^-} p_h, \quad \dot{T}^* h = \zeta h(\zeta) - \widetilde{\Theta}(\zeta) p_h \quad (h \in \dot{\mathfrak{H}}),$$

$$p_h = \frac{1}{2\pi} \int_{|\zeta|=1} \zeta \widetilde{\Theta}(\zeta) h(\zeta)\, |d\zeta|,$$

$$\dot{G}^* \varphi^+ = [I - \widetilde{\Theta}(\zeta)\widetilde{\Theta}^*(0)]\varphi^+, \quad \dot{S}^* \varphi^+ = \Theta^*(0)\varphi^+ \quad (\varphi^+ \in \mathfrak{N}^+).$$

The system α ($\dot{\alpha}$) is conservative if and only if $\widetilde{\Theta}(0)$ maps $\mathring{\mathfrak{N}}^-$ unitarily onto $\mathring{\mathfrak{N}}^+$, and this occurs if and only if $\Theta_{11}(z) \equiv 0$, $\Theta_{22}(z) \equiv 0$, $\Theta_{21}(z) \equiv \Theta_{21}(0)$, $\Theta_{21}(0)$ is a unitary operator, and $\Theta_{12}(z) = \Theta(z)$ is a bi-inner function. This case is not of interest here and is excluded from consideration in what follows. Thus $\dot{\alpha}$ is a model for a dissipative system with basic operator of class C_{00}, constructed as part of a conservative system with this basic operator.

We call a representation of $\Theta(z)$ in the form of a block $\Theta_{12}(z)$ of a bi-inner function $\widetilde{\Theta}(z) = [\Theta_{ik}(z)]_1^2$ a $\mathfrak{D}$-representation.

2. It will be an interesting problem to ask, for which $\widetilde{\Theta}(z)$ constructed with the aid of a $\mathfrak{D}$-representation, the dissipative system α ($\dot{\alpha}$) is *minimal*, i.e. simultaneously *controllable* ($\mathfrak{H} = \mathfrak{H}_\alpha^c$) and *observable* ($\mathfrak{H} = \mathfrak{H}_\alpha^o$).

We call a $\mathfrak{D}$-representation *minimal* if $\widetilde{\Theta}(z)$ does not have nontrivial bi-inner left and right divisors in the class B of the forms $\begin{pmatrix} I & 0 \\ 0 & u(z) \end{pmatrix}$ and $\begin{pmatrix} v(z) & 0 \\ 0 & I \end{pmatrix}$, respectively, i.e. if $\widetilde{\Theta}(z)$ has no representation in the form

$$\widetilde{\Theta}(z) = \begin{pmatrix} I|\mathfrak{N}^+ & 0 \\ 0 & u(z) \end{pmatrix} \mathring{\widetilde{\Theta}}(z) \begin{pmatrix} v(z) & 0 \\ 0 & I|\mathfrak{N}^- \end{pmatrix}, \tag{15}$$

where $\mathring{\widetilde{\Theta}}(z) \in B$, $u(z)$ and $v(z)$ are bi-inner functions, and at least one of these is nonconstant.

Theorem 3. *Suppose that $\Theta(z)$ has a $\mathfrak{D}$-representation determined by the function $\widetilde{\Theta}(z)$, and α is a dissipative system constructed (by formula (3)) as part of the lossless system $\tilde{\alpha}$ with scattering matrix $\widetilde{\Theta}(z)$. The system α with scattering matrix $\Theta(z)$ is minimal if and only if the associated $\mathfrak{D}$-representation is minimal.*

Proof. We observe that $\mathfrak{H} \ominus \mathfrak{H}_\alpha^c$ is the largest of all subspaces $\mathfrak{D}_*$ such that $F^*\mathfrak{D}_* = \{0\}$, $T^*\mathfrak{D}_* \subset \mathfrak{D}_*$, and $\mathfrak{H} \ominus \mathfrak{H}_\alpha^o$ is the largest of all subspaces $\mathfrak{D}$ such that $G\mathfrak{D} = \{0\}$, $T\mathfrak{D} \subset \mathfrak{D}$. Therefore α is a minimal system if and only if there are no nonzero subspeces $\mathfrak{D}_*$ and $\mathfrak{D}$ with these properties.

We show that $\mathfrak{H} = \mathfrak{H}_\alpha^o$ if and only if $\widetilde{\Theta}(z)$ has no nontrivial left bi-inner divisor of the form $\begin{pmatrix} I & 0 \\ 0 & u(z) \end{pmatrix}$. Suppose that such a divisor exists. We consider the subspace $\dot{\mathfrak{D}}$ in $H_+^2(\widetilde{\mathfrak{N}}^+)$ consisting of all $h(\zeta) = h_1(\zeta) \oplus h_2(\zeta)$ for which $h_1(\zeta) = 0$

and $h_2(\zeta) \in H^2_+(\overset{\circ}{\mathfrak{N}}{}^+) \ominus u(\zeta) H^2_+(\overset{\circ}{\mathfrak{N}}{}^+)$. It is obvious that $\dot{\mathfrak{D}} \neq \{0\}$, $\dot{\mathfrak{D}} \subset \dot{\mathfrak{H}}$, $\dot{G}\dot{\mathfrak{D}} = \{0\}$, $\dot{T}\dot{\mathfrak{D}} \subset \dot{\mathfrak{D}}$. Therefore $\dot{\mathfrak{H}} \neq \mathfrak{H}^o_{\dot{\alpha}}$, and consequently $\mathfrak{H} \neq \mathfrak{H}^o_\alpha$.

Now suppose it is known that $\mathfrak{H} \neq \mathfrak{H}^o_\alpha$, i.e. $\dot{\mathfrak{H}} \neq \mathfrak{H}^o_{\dot{\alpha}}$. We consider $\dot{\mathfrak{D}} = \dot{\mathfrak{H}} \ominus \mathfrak{H}^o_{\dot{\alpha}}$. From the formulas defining $\dot{T}$ and $\dot{G}$, it is clear that $\dot{\mathfrak{D}}$ is the subspace of all $h(\zeta)$ $(= h_1(\zeta) \oplus h_2(\zeta))$ in $\dot{\mathfrak{H}}$ for which $h_1(\zeta) = 0$. Consequently, we can assert that $\dot{\mathfrak{D}} \subset H^2_+(\overset{\circ}{\mathfrak{N}}{}^+)$ (identifying $0 \oplus h_2(\zeta)$ with $h_2(\zeta)$). The subspace $\dot{\mathfrak{D}}$ is invariant with respect to the operator $\dot{T}$. Therefore $H^2_+(\overset{\circ}{\mathfrak{N}}{}^+) \ominus \dot{\mathfrak{D}}$ is a subspace which is invariant with respect to the operator multiplication by ζ. Therefore there exists an inner function $u(z)$ with values in $[\mathfrak{N}, \overset{\circ}{\mathfrak{N}}{}^+]$ $(\mathfrak{N} \subset \overset{\circ}{\mathfrak{N}}{}^+)$ such that

$$H^2_+(\overset{\circ}{\mathfrak{N}}{}^+) \ominus \dot{\mathfrak{D}} = u(\zeta) H^2_+(\mathfrak{N}), \qquad \dot{\mathfrak{D}} = H^2_+(\overset{\circ}{\mathfrak{N}}{}^+) \ominus u(\zeta) H^2_+(\mathfrak{N}).$$

We show that $u(z)$ is a bi-inner function.

In fact, since

$$(\dot{\mathfrak{D}} =) H^2_+(\overset{\circ}{\mathfrak{N}}{}^+) \ominus u(\zeta) H^2_+(\mathfrak{N}) \subset H^2_+(\widetilde{\mathfrak{N}}^+) \ominus \widetilde{\Theta}(\zeta) H^2_+(\widetilde{\mathfrak{N}}^-) (= \dot{\mathfrak{H}}),$$

we obtain the other inclusions:

$$\begin{aligned} \widetilde{\Theta}(\zeta) H^2_+(\widetilde{\mathfrak{N}}^-) &\subset H^2_+(\mathfrak{N}^+) \oplus u(\zeta) H^2_+(\mathfrak{N}), \\ \widetilde{\Theta}(\zeta) L^2(\widetilde{\mathfrak{N}}^-) &\subset L^2(\mathfrak{N}^+) \oplus u(\zeta) L^2(\mathfrak{N}). \end{aligned} \tag{16}$$

On the other hand, $\widetilde{\Theta}(\zeta)$ has unitary values, and therefore

$$\widetilde{\Theta}(\zeta) L^2(\widetilde{\mathfrak{N}}^-) = L^2(\widetilde{\mathfrak{N}}^+) = L^2(\mathfrak{N}^+) \oplus L^2(\overset{\circ}{\mathfrak{N}}{}^+) \supset L^2(\mathfrak{N}^+) \oplus u(\zeta) L^2(\mathfrak{N}).$$

We obtain that $u(\zeta) L^2(\mathfrak{N}) = L^2(\overset{\circ}{\mathfrak{N}}{}^+)$, and this is possible for an inner function $u(z)$ only when it is a bi-inner function. We now consider

$$\overset{\circ}{\widetilde{\Theta}}(\zeta) = \begin{pmatrix} I|\mathfrak{N}^+ & 0 \\ 0 & u^*(\zeta) \end{pmatrix} \widetilde{\Theta}(\zeta).$$

The function $\overset{\circ}{\widetilde{\Theta}}(\zeta)$ admits unitary values ($|\zeta| = 1$ a.e.), and thus in view of the inclusions (16), we have for it:

$$\overset{\circ}{\widetilde{\Theta}}(\zeta) H^2_+(\widetilde{\mathfrak{N}}^-) \subset H^2_+(\mathfrak{N}^+ \oplus \mathfrak{N}).$$

Therefore $\overset{\circ}{\widetilde{\Theta}}(z)$ is a bi-inner function. We obtain a representation of $\widetilde{\Theta}(z)$ in the form (15) with $v(z) = I$ and $u(z)$ a nonconstant bi-inner function. Consequently, the $\mathfrak{D}$-representation determined by $\overset{\circ}{\widetilde{\Theta}}(z)$ is not minimal.

We note now that: 1) $\mathfrak{H}^c_\alpha = \mathfrak{H}^o_{\alpha^*}$, 2) α^* is constructed from $\tilde{\alpha}^*$ with the scattering matrix $\widetilde{\Theta}^*(\bar{z})$ in the same way as α is constructed from $\tilde{\alpha}$, 3) $\begin{pmatrix} v(z) & 0 \\ 0 & I \end{pmatrix}$ is a right bi-inner divisor of $\widetilde{\Theta}(z)$ if and only if $\begin{pmatrix} v^*(\bar{z}) & 0 \\ 0 & I \end{pmatrix}$ is a left bi-inner divisor of the function $\widetilde{\Theta}^*(\bar{z})$. Therefore from what has already been shown it follows that $\mathfrak{H} \neq \mathfrak{H}_\alpha^c$ if and only if $\widetilde{\Theta}(z)$ has a nontrivial right bi-inner divisor of the form $\begin{pmatrix} v(z) & 0 \\ 0 & I \end{pmatrix}$. The theorem follows.

3. The $\mathfrak{D}$-representation of $\Theta(z)$ described in Theorem 2 satisfies condition (8). Serving as "parameters" is the ordered pair $\{b_2(z), b_1(z)\}$ of bi-inner functions for which $b_2(\zeta)h_0(\zeta)b_1(\zeta)$ are the boundary values of a function of class B. We call this pair a *denominator* of the function $h_0(\zeta)$. In a natural way we introduce a notion of divisibility for the denominator functions of $h_0(\zeta)$: $\{b_2(z), b_1(z)\}$ is divided by $\{\mathring{b}_2(z), \mathring{b}_1(z)\}$ if $b_2(z) = u(z)\mathring{b}_2(z)$ and $b_1(z) = \mathring{b}_1(z)v(z)$, where $u(z)$ and $v(z)$ are bi-inner functions. The denominator $\{b_2(z), b_1(z)\}$ is called *minimal* if it has no nontrivial divisors.

Proposition 5. *A $\mathfrak{D}$-representation of a function $\Theta(z)$ satisfying condition* (8) *is minimal if and only if the corresponding denominator $\{b_2(z), b_1(z)\}$ of the function $h_0(\zeta)$ is minimal.*

Proof. Suppose that the $\mathfrak{D}$-representation of $\Theta(z)$ with bi-inner function $\widetilde{\Theta}(z)$, satisfying the condition (8), corresponds to the denominator $\{b_2(z), b_1(z)\}$ of $h_0(\zeta)$. Assume that this representation is not minimal, i.e. a relation (15) holds, where $u(z)$, $\mathring{\widetilde{\Theta}}(z)$ and $v(z)$ are bi-inner functions, and $u(z)$ or $v(z)$ is nonconstant. Then the function $\mathring{\widetilde{\Theta}}(z)$ determines a $\mathfrak{D}$-representation which satisfies condition (8). Suppose that it corresponds to the denominator $\{\mathring{b}_2(z), \mathring{b}_1(z)\}$ of the function $h_0(\zeta)$. From the uniqueness of the representation of the function $\widetilde{\Theta}(\zeta)$ in the form (14), it follows that $b_2(z) = u(z)\mathring{b}_2(z)$ and $b_1(z) = \mathring{b}_1(z)v(z)$, i.e. $\{\mathring{b}_2(z), \mathring{b}_1(z)\}$ is a nontrivial divisor of the denominator $\{b_2(z), b_1(z)\}$. Conversely, if we have such a divisor, we obtain a relation (15) in which one of the bi-inner functions $u(z)$ and $v(z)$ is not a constant.

Proposition 6. *For an arbitrary denominator $\{b_2(z), b_1(z)\}$ of the function $h_0(\zeta)$ there exists a divisor $\{\mathring{b}_2(z), \mathring{b}_1(z)\}$ which is a minimal denominator of this function.*

Proof. Let $\mathfrak{L}_1$ denote the set of functions $h(\zeta)$ in $H_+^2(\mathfrak{N}_{\widetilde{\varphi_1}})$ for which

$$b_2(\zeta)h_0(\zeta)h(\zeta) \in H_+^2(\mathring{\mathfrak{N}}^+).$$

Since $\mathfrak{L}_1$ is a subspace of $H_+^2(\mathfrak{N}_{\widetilde{\varphi_1}})$ which is invariant under the operator multiplication by ζ, then $\mathfrak{L}_1 = \mathring{b}_1(\zeta)H_+^2(\mathfrak{N}_1^-)$, where $\mathring{b}_1(z)$ is an inner function with values in $[\mathfrak{N}_1^-, \mathfrak{N}_{\widetilde{\varphi_1}}]$. From the inclusion $b_1(\zeta)H_+^2(\mathring{\mathfrak{N}}^-) \subset \mathring{b}_1(\zeta)H_+^2(\mathfrak{N}_1^-)$ it follows that $\mathring{b}_1(z)$ is a bi-inner function and $b_1(z) = \mathring{b}_1(z)v(z)$, where $v(z)$ is a bi-inner function with values in $[\mathring{\mathfrak{N}}^-, \mathfrak{N}_1^-]$. Now let $\mathfrak{L}_2$ denote the subspace of functions

$h(\zeta)$ in $H^2_+(\mathfrak{N}_{\varphi 2})$ for which $\overset{\circ}{b}{}^*_1(\bar\zeta)h^*_0(\bar\zeta)h(\zeta) \in H^2_+(\mathfrak{N}^-_1)$. This is a subspace of $H^2_+(\mathfrak{N}_{\varphi 2})$ which is invariant under the operator multiplication by ζ. Therefore $\mathfrak{L}_2 = \overset{\circ}{b}{}^*_2(\bar\zeta)H^2_+(\mathfrak{N}^+_1)$, where $\overset{\circ}{b}{}^*_2(\bar z)$ is an inner function with values in $[\mathfrak{N}^+_1, \mathfrak{N}_{\varphi 2}]$. Using the inclusion $b^*_2(\bar\zeta)H^2_+(\mathfrak{N}^+) \subset \overset{\circ}{b}{}^*_2(\bar\zeta)H^2_+(\mathfrak{N}^+_1)$, we obtain that $\overset{\circ}{b}_2(z)$ is a bi-inner function and $b_2(z) = u(z)\overset{\circ}{b}_2(z)$, where $u(z)$ is a bi-inner function with values in $[\mathfrak{N}^+_1, \mathfrak{N}^+]$. Since $\overset{\circ}{b}{}^*_1(\bar\zeta)h^*_0(\bar\zeta)\overset{\circ}{b}{}^*_2(\bar\zeta)H^2_+(\mathfrak{N}^+_1) \subset H^2_+(\mathfrak{N}^-_1)$, $\{\overset{\circ}{b}{}^*_1(\bar z), \overset{\circ}{b}{}^*_2(\bar z)\}$ is a denominator of the function $h^*_0(\bar\zeta)$, and therefore $\{\overset{\circ}{b}_2(z), \overset{\circ}{b}_1(z)\}$ is a denominator of the function $h_0(\zeta)$. The latter is a divisor of the denominator $\{b_2(z), b_1(z)\}$. It remains to show that $\{\overset{\circ}{b}_2(z), \overset{\circ}{b}_1(z)\}$ is a minimal denominator. Assume that $\{d_2(z), d_1(z)\}$ is a divisor of it. Then

$$d^*_1(\bar\zeta)h^*_0(\bar\zeta)d^*_2(\bar\zeta)H^2_+(\mathfrak{N}^+_2) \subset H^2_+(\mathfrak{N}^-_2), \qquad \overset{\circ}{b}{}^*_1(\bar\zeta)h^*_0(\bar\zeta)d^*_2(\bar\zeta)H^2_+(\mathfrak{N}^+_2) \subset H^2_+(\mathfrak{N}^-_1)$$

and therefore

$$d^*_2(\bar\zeta)H^2_+(\mathfrak{N}^+_2) \subset \overset{\circ}{b}{}^*_2(\bar\zeta)H^2_+(\mathfrak{N}^+_1)\ (= \mathfrak{L}_2), \qquad \overset{\circ}{b}_2(\bar\zeta)d^*_2(\bar\zeta)H^2_+(\mathfrak{N}^+_2) \subset H^2_+(\mathfrak{N}^+_1).$$

It follows from the last inclusion that $\overset{\circ}{b}_2(z)$ is a right divisor of $d_2(z)$, i.e. $d_2(z) = \overset{\circ}{u}(z)\overset{\circ}{b}_2(z)$, where $\overset{\circ}{u}$ is a bi-inner function. Since, on the other hand, $d_2(z)$ is a right divisor of $\overset{\circ}{b}_2(z)$, we obtain that $\overset{\circ}{u}(z) = \overset{\circ}{u}$ is a constant. In the same way, from the inclusions

$$b_2(\zeta)h_0(\zeta)d_1(\zeta)H^2_+(\mathfrak{N}^+_2) \subset H^2_+(\overset{\circ}{\mathfrak{N}}{}^+), \qquad d_1(\zeta)H^2_+(\mathfrak{N}^+_2) \subset \overset{\circ}{b}_1(\zeta)H^2_+(\mathfrak{N}^-_1)$$

it follows that $d_1(z) = \overset{\circ}{b}_1(z)\overset{\circ}{v}$, where $\overset{\circ}{v}$ is a constant. We conclude that $\{\overset{\circ}{b}_2(z), \overset{\circ}{b}_1(z)\}$ is a minimal denominator of the function $h_0(\zeta)$ which is a divisor of the denominator $\{b_2(z), b_1(z)\}$. The theorem follows.

3. Scattering matrices in the classes $B\widetilde{\Pi}$ and $B\Pi$

1. We now consider the following class of scattering matrices, for which there exist $\mathfrak{D}$-representations satisfying condition (8).

We say that $\Theta(z)$ is in the class $B\widetilde{\Pi}$ if: 1) $\Theta(z) \in B$, 2) there exist $c(z)$ and $d(z)$ in the class B such that $c(\zeta)\Theta^*(\zeta)$ and $\Theta^*(\zeta)d(\zeta)$ are the boundary values of a function in the class B, and moreover $\ker c(\zeta) = \{0\}$ and $\ker d^*(\zeta) = \{0\}$, $|\zeta| = 1$ a.e. (By $\ker c$ we mean the kernel of the operator c.) If

$$c^*(\bar z) = b_1(z)\varphi(z), \qquad d(z) = b_2(z)\psi(z)$$

are representations of the functions $c^*(\bar z)$ and $d(z)$ in the form of products of inner and outer functions, then $b^*_1(\bar\zeta)\Theta^*(\zeta)$ and $\Theta^*(\zeta)b_2(\zeta)$ are the boundary values of functions in the class B. The conditions $\ker c(\zeta) = \{0\}$ and $\ker d^*(\zeta) = \{0\}$ (a.e.)

mean that $b_1(z)$ and $b_2(z)$ are bi-inner functions. For $\Theta(z)$ ($\in B\widetilde{\Pi}$) there exist bi-inner functions $b_\Pi(z)$ and $b_\Omega(z)$ such that:

1) $b_\Pi(\zeta)\Theta^*(\zeta)$ and $\Theta^*(\zeta)b_\Omega(\zeta)$ are the boundary values of functions of the class B,
2) $b_\Pi(\zeta)$ and $b_\Omega(\zeta)$ are common divisors from the right and left, respectively, of functions $c(z)$ and $d(z)$ with the above properties, i.e. $c(\zeta)b_\Pi^{-1}(\zeta)$ and $b_\Omega(\zeta)^{-1}d(\zeta)$ are the boundary values of functions of the class B.

We set

$$b_\Pi(\zeta)\Theta^*(\zeta) = \Theta_\Pi(\zeta), \qquad \Theta^*(\zeta)b_\Omega(\zeta) = \Theta_\Omega(\zeta).$$

Then

$$\Theta(\zeta) = \Theta_\Pi^*(\zeta)[b_\Pi^*(\zeta)]^{-1}, \qquad \Theta(\zeta) = [b_\Omega^*(\zeta)]^{-1}\Theta_\Omega^*(\zeta),$$

and $\Theta_\Pi^*(\zeta)$, $\Theta_\Omega^*(\zeta)$, $b_\Pi^*(\zeta)$, $b_\Omega^*(\zeta)$ are the boundary values of functions $\Theta_\Pi^*(1/\bar{z})$, $\Theta_\Omega^*(1/\bar{z})$, $b_\Pi^*(1/\bar{z})$, $b_\Omega^*(1/\bar{z})$ which are holomorphic and contractive for $|z| > 1$. We conclude that a function $\Theta(z)$ ($\in B$) belongs to the class $B\widetilde{\Pi}$ if and only if its boundary values are represented in the form of "right" and "left" quotients of the boundary values of functions which are holomorphic and contractive for $|z| > 1$; $b_\Pi^*(1/\bar{z})$ and $b_\Omega^*(1/\bar{z})$ are the least right and left denominators for $\Theta(z)$ and common divisors of all other right and left denominators for $\Theta(z)$. The functions $b_\Pi(z)$ and $b_\Omega(z)$ are determined by $\Theta(z)$ to within (left and right) constant unitary factors. If $\Theta(z) \in B\widetilde{\Pi}$, then $\Theta^*(\bar{z}) \in B\widetilde{\Pi}$: $b_\Omega(1/z)$ and $b_\Pi(1/z)$ are the least denominators for $\Theta^*(\bar{z})$.

In what follows we shall need the following proposition of V. I. Matsaev (verbal communication, Summer Mathematical School, Novgorod, 1976) which generalizes a sufficient condition of Rosenblum and Rovnyak for the solvability of a factorization problem[1].

Lemma. *Assume that, for the nonnegative contractive function $f(\zeta)$ ($|\zeta| = 1$), there exists a contractive function $c(\zeta)$ such that $\ker c(\zeta) = \{0\}$ (a.e.) and $c(\zeta)f(\zeta)$ are the boundary values of some function of the class B. Then the factorization problem for $f(\zeta)$,*

$$f(\zeta) = \varphi^*(\zeta)\varphi(\zeta) \qquad (\text{a.e., } |\zeta| = 1),\ \varphi(z) \in B,$$

is solvable.

Proof. If $f(\zeta)$ (≤ 1) has values in $[\mathfrak{N}, \mathfrak{N}]$, it is known [15] that for the solvability of the factorization problem it is necessary and sufficient that

$$\bigcap_{n \geq 0} \overline{\zeta^n f^{1/2}(\zeta) H_+^2(\mathfrak{N})} = \{0\},$$

[1]See the account by A. S. Markus, *Introduction to the Spectral Theory of Polynomial Operator Pencils*, Transl. Math. Monographs, vol. 71, Amer. Math. Soc., 1988; the result is Theorem 34.3 on p. 199, and short comments are located on p. 227.

where the bar "—" indicates closure in the metric of $L^2(\mathfrak{N})$. Denote the left side of this equation by $\mathfrak{L}$. We have

$$\mathfrak{L} \subset \overline{f^{1/2}(\zeta)H_+^2(\mathfrak{N})}.$$

Therefore, if $c(\zeta)f^{1/2}(\zeta)\mathfrak{L} = \{0\}$ and $\ker c(\zeta) = \{0\}$ (a.e.), then $\mathfrak{L} = \{0\}$. In fact since $c(\zeta)f(\zeta)$ are the boundary values of a function of class B,

$$c(\zeta)f^{1/2}(\zeta)\mathfrak{L} = c(\zeta)f^{1/2}(\zeta)\bigcap_{n\geq 0}\overline{\zeta^n f^{1/2}(\zeta)H_+^2(\mathfrak{N})} \subset$$

$$\subset \bigcap_{n\geq 0}\zeta^n\,\overline{c(\zeta)f(\zeta)H_+^2(\mathfrak{N})} \subset \bigcap_{n\geq 0}\zeta^n H_+^2(\mathfrak{N}_1) = \{0\}.$$

Thus if the condition of the lemma is satisfied, then $\mathfrak{L} = \{0\}$ and consequently the factorization problem is solvable for $f(\zeta)$.

3. We now show that any scattering matrix of the class $B\widetilde{\Pi}$ has a Darlington synthesis.

Theorem 4. *Suppose $\Theta(z)$ is in the class $B\widetilde{\Pi}$. Then $\Theta(z)$ admits a $\mathfrak{D}$-representation satisfying condition* (8).

Proof. Assume $\Theta(z)$ is in the class $B\widetilde{\Pi}$ and $b_\Pi^*(1/\bar{z})$ and $b_\Omega^*(1/\bar{z})$ are its least right and left denominators. Then the functions

$$b_\Omega^*(\bar{\zeta})\big[I - \Theta(\bar{\zeta})\Theta^*(\bar{\zeta})\big], \qquad b_\Pi(\zeta)\big[I - \Theta^*(\zeta)\Theta(\zeta)\big]$$

are the boundary values of functions of class B. By the lemma applied to the functions $I - \Theta(\bar{\zeta})\Theta^*(\bar{\zeta})$ and $I - \Theta^*(\zeta)\Theta(\zeta)$, the factorization problem is solvable, i.e. there exist solutions $\varphi_1(z)$ and $\varphi_2(z)$ for the problem (10). We consider the contractive function $h_0(\zeta)$ determined by the relation (11), and we show that there exist bi-inner functions $b_1(z)$ and $b_2(z)$ such that the function $\Theta_{21}(\zeta)$ determined by formula (12) gives the boundary values of a function of class B. Below we identify a function of class B with its boundary values. We put

$$\psi(\zeta) = b_\Pi(\zeta)\varphi_2^*(\zeta).$$

This function is of class B, since

$$\psi(\zeta)\varphi_2(\zeta) = b_\Pi(\zeta)\big[I - \Theta^*(\zeta)\Theta(\zeta)\big] \in B$$

and $\varphi_2(\zeta)$ is an outer function. If $b_\Pi(\zeta)$ takes values in $[\mathfrak{N}^-, \mathfrak{N}^-]$, then

$$\overline{\psi^*(\bar{\zeta})\mathfrak{N}^-} = \mathfrak{N}_{\varphi_2}. \qquad \text{Therefore} \quad \psi(z) = \varphi(z)b(z),$$

where $\varphi(z)$ and $b(z)$ are $*$-outer and bi-inner functions, respectively. Since

$$\psi(\zeta)h_0(\zeta) = -b_\Pi(\zeta)\Theta^*(\zeta)\varphi_1(\zeta) \in B,$$

we obtain that $b(\zeta)h_0(\zeta) \in B$, i.e. $\{b(z), I\}$ is a denominator for the function $h_0(\zeta)$. Thus we may put

$$\Theta_{21}(\zeta) = b(\zeta)h_0(\zeta).$$

By Theorem 2, the diagonal blocks of our bi-inner function $\widetilde{\Theta}(z) = [\Theta_{ik}(z)]_1^2$ satisfy condition (8), provided that we set

$$\Theta_{11}(z) = \varphi_1(z), \qquad \Theta_{22}(z) = b(z)\varphi_2(z).$$

The theorem is proved.

4. We now consider the case where, for a function $\Theta(z)$ of class B, there exists a scalar denominator — a scalar inner function $b_1(z)$ such that $b_1^*(\bar{\zeta})\Theta^*(\zeta) \in B$. This occurs if and only if $\Theta(\zeta)$ are the boundary values of a function which is meromorphic outside the circle and represented in the form of a ratio of two bounded holomorphic functions $\Theta_2^{-1}(z)\Theta_1(z)$, where $\Theta_1(z)$ is operator valued and $\Theta_2(z)$ is scalar valued.

The class of such $\Theta(z)$ is denoted $B\Pi$. Among the scalar denominators $b_1(z)$ of a function $\Theta(z)$ of class $B\Pi$ there exists a greatest common divisor of each. We denote it by $b_\Theta(z)$. The pseudo-continuation (by boundary values) to the exterior of the unit circle is represented in the form

$$\Theta(z) = b_\Theta^{-1}\left(\frac{1}{z}\right)\Theta_1(z) \qquad (|z| > 1),$$

where $\Theta_1(1/z) \in B$. The function $b_\Theta(z)$ accounts for the singularities of $\Theta(z)$: a point z is a pole of $\Theta(z)$ for $|z| > 1$ if and only if $1/z$ is a zero of $b_\Theta(z)$.

We remark that a bi-inner function $\Theta(z)$ belongs to the class $B\Pi$ if and only if it possesses a scalar multiple in the sense of [13], i.e. when there exists a scalar function $b_1(z)$ such that $b_1(z)\Theta^{-1}(z) \in B$. The function $b_\Theta(z)$ is the best minorant for $\Theta(z)$.

It is known [13] that $\Theta(z)$ is a bi-inner function with a scalar multiple if and only if the basic operator T of a simple conservative system α with scattering matrix $\Theta(z)$ belongs to the class C_0 ($\subset C_{00}$), introduced and studied by B. Sz.-Nagy and C. Foias [13, 123–126]. We recall that for a completely nonunitary contraction T the inclusion $T \in C_0$ means that there exists a function $m(z)$ ($\not\equiv 0$) of class B such that $m(T) = 0$. Among such $m(z)$ there exists a greatest common divisor. Such a function is denoted $m_T(z)$ and called the minimal function of the contraction T. For T on a finite-dimensional space, $m_T(z) = p(z)/z^n\overline{p(1/\bar{z})}$, where $p(z)$ is the minimal polynomial of the operator T and n is the degree of $p(z)$. The spectrum $\sigma(T)$ of the contraction T is completely determined by $m_T(z)$: $\sigma(T) = S_T$, where S_T is the set consisting of the zeros of $m_T(z)$ in the disk $|z| < 1$, together with the relative complement in the circle $|\zeta| = 1$ of the union of all of the arcs on which $m_T(z)$ is analytic. Every zero z_0 of the function $m_T(z)$ ($|z_0| < 1$) is an eigenvalue for T and the order of the eigenvalue z_0 is equal to the multiplicity of z_0 as a zero of $m_T(z)$. In order that the root vectors of T corresponding to points in the spectrum in the disk $|z| < 1$ generate all of the space $\mathfrak{H}$, it is necessary and sufficient that $m_T(z)$ is a Blaschke product. As shown in [10], $T \in C_0$ if and only if $T \in C_{00}$ and $(zI - T)^{-1}$ is a meromorphic function with poles satisfying the Blaschke condition and such that

$$\sup_{r<1} \int_{|\zeta|=1} \ln^+(1-r) \left\|r\zeta I - T)^{-1}\right\| \, |d\zeta| < \infty.$$

Theorem 5. *A scattering matrix of an arbitrary passive scattering LSDS with basic operator T of class C_0 belongs to the class $B\Pi$, and moreover the minimal function $m_T(z)$ of the operator T is a scalar denominator of the function $\Theta(1/z)$, i.e. $m_T(z)\Theta(1/z) \in B$. If $\Theta(z)$ is an arbitrary function of the class $B\Pi$, then the basic operator T of any minimal passive scattering LSDS α with scattering matrix $\Theta(z)$ belongs to the class C_0; in this connection, the minimal function $m_T(z)$ does not depend on the choice of minimal LSDS α and is equal to the least scalar denominator of the function $\Theta(1/z)$.*

Proof. Suppose that $\Theta(z)$ is a function of class $B\Pi$ taking values in $[\mathfrak{N}^-, \mathfrak{N}^+]$, and that $b(z)$ is a scalar denominator for $\Theta(1/z)$. Then for any f in $\mathfrak{N}^-$ and g in $\mathfrak{N}^+$, we have

$$\int_{|\zeta|=1} (\bar{\zeta}^n \overline{b(\zeta)} g, \Theta(\bar{\zeta}) f) \, |d\zeta| = 0 \qquad (n \geq 1),$$

since $\overline{b(\bar{\zeta})}\Theta^*(\zeta)$ are the boundary values of a function of class B. If we use Parseval's identity in $L^2(\mathfrak{N}^+)$, then for $b(z) = \sum_0^\infty b_k z^k$ and $\Theta(z) = \sum_0^\infty \Theta_k z^k$ we obtain

$$\sum_0^\infty (\bar{b}_k g, \Theta_{k+n} f) = 0, \qquad (n \geq 1).$$

Suppose that α is a passive scattering LSDS with coefficients T, F, G, S and transfer function $\Theta(z)$. Then $\Theta_{k+n} = GT^{k+n-1}F$ $(n \geq 1, k \geq 0)$ and therefore we have

$$\sum_0^\infty (g, b_k G T^{k+n} F f) = 0, \qquad (n \geq 0).$$

By Abel's theorem, we obtain

$$\lim_{r\uparrow 1} \left(g, G\Big(\sum_0^\infty b_k r^k T^k\Big) T^n F f\right) = 0, \qquad (n \geq 0).$$

But by definition,

$$\lim_{r\uparrow 1} \sum_0^\infty b_k r^k T^k = b(T).$$

Thus

$$(g, G\, b(T) T^n F f) = 0 \qquad (n \geq 0,\ f \in \mathfrak{N}^-,\ g \in \mathfrak{N}^+).$$

For a minimal system α it hence follows that $b(T) = 0$, since

$$\mathfrak{H} = \bigvee_0^\infty T^n F \mathfrak{N}^- = \bigvee_0^\infty (T^*)^n G^* \mathfrak{N}^+.$$

In this case, we have also shown that $T \in C_0$ and $b(z)$ is divided by $m_T(z)$.

Conversely, suppose that $T \in C_0$. We assume that $b(z) = m_T(z)$. If $\Theta(z)$ is the scattering matrix of a system α with basic operator T, then reversing the previous reasoning, we obtain that $\overline{b(\zeta)}\Theta^*(\zeta)$ are the boundary values of some function of class B, that is, $\Theta(z) \in B\Pi$ and $b(z)$ is a scalar denominator for the function $\Theta(1/z)$.

It follows from what has been shown that if $\Theta(z) \in B\Pi$ and T is a basic operator of a minimal passive scattering LSDS with transfer function $\Theta(z)$, then $T \in C_0$ and $m_T(z)$ is the least scalar denominator of the function $\Theta(1/z)$. The theorem is proved.

Theorem 6. *Assume that $\Theta(z)$ is a function of class $B\Pi$ and the function $\widetilde{\Theta}(z)$ is defined by a minimal $\mathfrak{D}$-representation of $\Theta(z)$. Then $\widetilde{\Theta}(z) \in B\Pi$ and $b_\Theta(z) = b_{\widetilde{\Theta}}(z)$.*

Proof. We consider a conservative minimal scattering system $\tilde{\alpha}$ with scattering matrix $\widetilde{\Theta}(z)$ and part α of the system $\tilde{\alpha}$ having scattering matrix $\Theta(z)$. By Theorem 3 the system α is minimal. Since $\Theta(z) \in B\Pi$, by Theorem 5 the basic operator T of the system α belongs to the class C_0 and $b_\Theta(z) = m_T(z)$. But the basic operator $\widetilde{T}$ for the system $\tilde{\alpha}$ coincides with T, hence $\widetilde{T} \in C_0$, and by Theorem 5, taking into account that the system $\tilde{\alpha}$ is minimal, we obtain that $\widetilde{\Theta}(z) \in B\Pi$ and $b_{\widetilde{\Theta}}(z) = m_{\widetilde{T}}(z) = m_T(z) = b_\Theta(z)$. The theorem is proved.

Proposition 7. *If $\Theta(z)$ is the scattering matrix (2) of a system α with basic operator T of class C_0, then*

$$\Theta\left(\frac{1}{z}\right) = S + G(zI - T)^{-1}F, \qquad (|z| < 1),$$

i.e. the pseudo-continuation of $\Theta(z)$ ($\in B\Pi$) to the exterior of the unit circle can be written as in formula (2).

In fact, the system α can be considered as a part of a conservative system $\tilde{\alpha}$ with basic operator $\widetilde{T} = T$. In this connection, $\Theta(z)$ can be written in the form of a block of the scattering matrix $\widetilde{\Theta}(z)$ of the system $\tilde{\alpha}$. Since $\widetilde{T} \in C_0$, $\widetilde{\Theta}(z)$ is a bi-inner function with scalar multiple. The pseudo-continuation of

$$\widetilde{\Theta}(z)$$

into the exterior of the unit circle can be written by the symmetry principle as

$$\widetilde{\Theta}(z) = \left[\widetilde{\Theta}^*\left(\frac{1}{\bar{z}}\right)\right]^{-1}, \qquad (|z| > 1).$$

It is immediate to show that $\left[\widetilde{\Theta}^*(1/\bar{z})\right]^{-1}$ can be expressed in terms of the coefficients $\widetilde{T}$, $\widetilde{F}$, $\widetilde{G}$, $\widetilde{S}$ of the system $\tilde{\alpha}$ in the same way as $\widetilde{\Theta}(z)$, i.e.

$$\widetilde{\Theta}(z) = \widetilde{S} + z\widetilde{G}(I - z\widetilde{T})^{-1}\widetilde{F}, \qquad (|z| > 1).$$

It remains now to take advantage of the relation (3), and by that $\Theta(1/z)$ is a block of $\widetilde{\Theta}(1/z)$ ($|z| < 1$).

4. Optimal passive scattering systems

1. It is easy to see that an LSDS α is passive if and only if for its coefficients the following system of equations

$$I-T^*T-G^*G=M^*M, \quad -T^*F-G^*S=M^*N, \quad I-F^*F-S^*S=N^*N, \tag{17}$$

with unknown operators N and M, is solvable. The system can be written in the form

$$(I-V^*V=)\begin{pmatrix} I & 0 \\ 0 & I \end{pmatrix}-\begin{pmatrix} T & F \\ G & S \end{pmatrix}^*\begin{pmatrix} T & F \\ G & S \end{pmatrix}=\begin{pmatrix} M & N \end{pmatrix}^*\begin{pmatrix} M & N \end{pmatrix}$$

(here $M\in[\mathfrak{H},\mathfrak{N}]$, $N\in[\mathfrak{N}^-,\mathfrak{N}]$, and $\mathfrak{N}$ is a Hilbert space). In fact, if these equations hold, then $I-V^*V\geq 0$, i.e. α is a passive system. Conversely, if $I-V^*V\geq 0$, we obtain a solution of the system of equations (17) if we put

$$\mathfrak{N}=\overline{(I-V^*V)(\mathfrak{H}\oplus\mathfrak{N}^-)}, \qquad M=(I-V^*V)^{1/2}|\mathfrak{H}, \qquad N=(I-V^*V)^{1/2}|\mathfrak{N}^-.$$

Let α be a passive scattering system, M and N solutions of the equations (17), and

$$\psi(z)=N+zM(I-zT)^{-1}F, \qquad K_\alpha(z)=(I-zT)^{-1}F. \tag{18}$$

Using equations (17), we obtain after simple calculations

$$I-\Theta_\alpha^*(\eta)\Theta_\alpha(\zeta)=\psi^*(\eta)\psi(\zeta)+(1-\bar{\eta}\zeta)K_\alpha^*(\eta)K_\alpha(\zeta). \tag{19}$$

It is obvious from this identity, in particular, that $I-\Theta_\alpha^*(z)\Theta_\alpha(z)\geq 0$ for $|z|<1$. Thus we have

Proposition 8. [7] *The scattering matrix of any passive LSDS belongs to the class B.*

For the boundary values of a function $\Theta_\alpha(z)$ ($\in B$) and $\psi(z)$ ($\in B$) we obtain for any h in $\mathfrak{N}$ and $|\zeta|=1$ (a.e.)

$$\left\|[I-\Theta_\alpha^*(\zeta)\Theta_\alpha(\zeta)]^{1/2}h\right\|^2=\|\psi(\zeta)h\|^2+\lim_{r\uparrow 1}(1-r^2)\|K_\alpha(r\zeta)h\|^2.$$

Therefore the following two equations are equivalent:

$$I-\Theta_\alpha^*(\zeta)\Theta_\alpha(\zeta)=\psi^*(\zeta)\psi(\zeta), \qquad \text{s-}\lim_{r\uparrow 1}(1-r^2)^{1/2}K_\alpha(r\zeta)=0. \tag{20}$$

But $\|\psi(\zeta)h\|\leq\left\|[I-\Theta_\alpha^*(\zeta)\Theta_\alpha(\zeta)]^{1/2}h\right\|$ and therefore the equations (20) hold if and only if

$$\|\psi(\zeta)h\|_{L^2(\mathfrak{N})}=\left\|[I-\Theta_\alpha^*(\zeta)\Theta_\alpha(\zeta)]^{1/2}h\right\|_{L^2(\mathfrak{N}^-)} \qquad (\forall h, h\in\mathfrak{N}^-).$$

Taking into account that $\psi(\zeta)h \in H^2_+(\mathfrak{N})$ and $\Theta_\alpha(\zeta)h \in H^2_+(\mathfrak{N}^+)$, we obtain

$$\|\psi(\zeta)h\|^2_{L^2(\mathfrak{N})} = \|Nh\|^2 + \sum_0^\infty \left\|MT^kFh\right\|^2,$$

$$\left\|[I - \Theta^*_\alpha(\zeta)\Theta_\alpha(\zeta)]^{1/2}h\right\|^2_{L^2(\mathfrak{N}^-)} = \|h\|^2 - \|\Theta_\alpha(\zeta)h\|^2_{L^2(\mathfrak{N}^+)} =$$

$$= \|h\|^2 - \|Sh\|^2 - \sum_0^\infty \left\|GT^kFh\right\|^2.$$

Now we use equation (17) and the identity

$$I - \operatorname{s-}\lim_{n\to\infty}(T^*)^nT^n = \sum_{k=0}^\infty (T^*)^k(I - T^*T)T^k.$$

As a result we arrive at the relation

$$\left\|[I - \Theta^*_\alpha(\zeta)\Theta_\alpha(\zeta)]^{1/2}h\right\|^2_{L^2(\mathfrak{N}^-)} = \|\psi(\zeta)h\|^2_{L^2(\mathfrak{N})} + \lim_{n\to\infty} \|T^nFh\|^2.$$

Consequently, condition (20) is equivalent to

$$\operatorname{s-}\lim_{n\to\infty} T^nF = 0.$$

For a controllable system α this limit relation means that $T \in C_{0\cdot}$.

2. We remark that $h(n) = \sum_{k=0}^{n-1} T^kF\varphi^-(n-k-1)$ for $h(0) = 0$. Therefore it is natural to call a passive system $\mathring{\alpha}$ with scattering matrix $\Theta(z)$ *optimal* if for any other passive system α with the same scattering matrix $\Theta(z)$ we have

$$\left\|\sum_{k=0}^n \mathring{T}^k\mathring{F}h_k\right\| \le \left\|\sum_{k=0}^n T^kFh_k\right\|$$

for arbitrary h_k in $\mathfrak{N}^-$ and $n \ge 0$. It follows from this definition that an optimal controllable passive system is determined by its scattering matrix to within unitary similarity. It is also easy to see that such a system is automatically observable and hence minimal.

Theorem 7. *An arbitrary function $\Theta(z)$ of class B is the scattering matrix of some optimal and minimal passive system.*

Proof. Given $\Theta(z)$ ($\in B$) there exists an outer function $\varphi(z)$ ($\in B$) such that

1) $\varphi^*(\zeta)\varphi(\zeta) \le I - \Theta^*(\zeta)\Theta(\zeta)$ (a.e.),
2) if $\psi(z) \in B$ and $\psi^*(\zeta)\psi(\zeta) \le I - \Theta^*(\zeta)\Theta(\zeta)$, then $\psi^*(\zeta)\psi(\zeta) \le \varphi^*(\zeta)\varphi(\zeta)$ (see [13], Ch. V, Proposition 4.2).

The function $\varphi(z)$ is defined by $\Theta(z)$ to within a constant left unitary factor. It is not excluded that $\varphi(z) \equiv 0$. If the factorization problem (4) is solvable for $\Theta(z)$, then $\varphi(z)$ is equal to the previously introduced function $\varphi_2(z)$. We consider the function $\widetilde{\Theta}(z) = \begin{pmatrix}\Theta(z)\\ \varphi(z)\end{pmatrix}$, which is holomorphic and bounded and has values in

$[\mathfrak{N}^-, \mathfrak{N}^+ \oplus \overset{\circ}{\mathfrak{N}}]$. Since $\widetilde{\Theta}^*(\zeta)\widetilde{\Theta}(\zeta) \leq I$, by the maximum principle $\widetilde{\Theta}(z) \in B$. Let $\tilde{\alpha}$ be a simple conservative system with scattering matrix $\widetilde{\Theta}(z)$ and $\overset{\circ}{\alpha}$ a part of the system $\tilde{\alpha}$ with outer spaces $\mathfrak{N}^-$, $\mathfrak{N}^+$ and coefficients $\overset{\circ}{F} = \widetilde{F}$, $\overset{\circ}{T} = \widetilde{T}$, $\overset{\circ}{G} = P_{\mathfrak{N}^+}\widetilde{G}$, $\overset{\circ}{S} = P_{\mathfrak{N}^+}\widetilde{S}$. It is obvious that $\Theta_{\overset{\circ}{\alpha}}(z) = \Theta(z)$ and

$$\varphi(z) = \overset{\circ}{N} + z\overset{\circ}{M}(I - z\overset{\circ}{T})^{-1}\overset{\circ}{F}, \qquad \overset{\circ}{N} = P_{\overset{\circ}{\mathfrak{N}}}\widetilde{S}, \qquad \overset{\circ}{M} = P_{\overset{\circ}{\mathfrak{N}}}\widetilde{G}.$$

The operators $\overset{\circ}{N}$ and $\overset{\circ}{M}$ are solutions of equations (17) for the coefficients of the system $\overset{\circ}{\alpha}$, hence according to (19) we have

$$I - \Theta^*(\eta)\Theta(\xi) = \varphi^*(\eta)\varphi(\xi) + (1 - \bar{\eta}\xi)K^*_{\overset{\circ}{\alpha}}(\eta)K_{\overset{\circ}{\alpha}}(\xi).$$

Using the definition of $\varphi(z)$ and this equation, we show that $\overset{\circ}{\alpha}$ is an optimal system. In fact, suppose that α is any other passive system with scattering matrix $\Theta(z)$. For the functions $\psi(z)$ and $K_\alpha(z)$ corresponding to it by the formulas (18) we have the identity (19), in which $\Theta_\alpha(z) = \Theta(z)$. Therefore $\psi^*(\zeta)\psi(\zeta) \leq I - \Theta^*(\zeta)\Theta(\zeta)$ and

$$K^*_\alpha(\eta)K_\alpha(\xi) - K^*_{\overset{\circ}{\alpha}}(\eta)K_{\overset{\circ}{\alpha}}(\xi) = (1 - \bar{\eta}\xi)^{-1}\varphi^*(\eta)\varphi(\xi) - (1 - \bar{\eta}\xi)^{-1}\psi^*(\eta)\psi(\xi),$$

We verify that the right side of this equation is a nonnegative definite kernel of two variables ξ and η ($|\xi| < 1$, $|\eta| < 1$). Denote by π_- the orthogonal projection from $L^2(\mathfrak{N})$ onto $H^2_-(\mathfrak{N}) = L^2(\mathfrak{N}) \ominus H^2_+(\mathfrak{N})$. Let $h_i \in \mathfrak{N}^-$, $|z_i| < 1$, $(1 \leq i \leq n)$. We consider:

$$h(\zeta) = \sum_{i=1}^{n} (\zeta - z_i)^{-1}h_i \; (\in H^2_-(\mathfrak{N}^-)), \qquad \pi_-\psi h = \sum_{i=1}^{n} (\zeta - z_i)^{-1}\psi(z_i)h_i,$$

$$\|\pi_-\psi h\|^2_{L^2(\mathfrak{N})} = \sum_{i=1}^{n}\sum_{j=1}^{n} (1 - \bar{z}_j z_i)^{-1}\big(\psi^*(z_j)\psi(z_i)h_i, h_j\big).$$

We can write an analogous equation for $\varphi(z)$. But from the definition of the function $\varphi(z)$, it follows that $\psi(z) = b(z)\varphi(z)$, where $b(z) \in B$. Therefore

$$\|\pi_-\psi h\|^2_{L^2(\mathfrak{N})} = \|\pi_- b\varphi h\|^2_{L^2(\mathfrak{N})} \leq \|\pi_-\varphi h\|^2_{L^2(\mathfrak{N})}\,.$$

Here by b we denote the operator "multiplication" by $b(\zeta)$ on $L^2(\mathfrak{N})$, so that $\|\pi_- b\| \leq 1$. Thus in turn we obtain

$$\sum_{i=1}^{n}\sum_{j=1}^{n} (1 - \bar{z}_j z_i)^{-1}\big(\psi^*(z_j)\psi(z_i)h_i, h_j\big) \leq \sum_{i=1}^{n}\sum_{j=1}^{n} (1 - \bar{z}_j z_i)^{-1}\big(\varphi^*(z_j)\varphi(z_i)h_i, h_j\big),$$

$$\sum_{i=1}^{n}\sum_{j=1}^{n} \big(K^*_\alpha(z_j)K_\alpha(z_i)h_i, h_j\big) \geq \sum_{i=1}^{n}\sum_{j=1}^{n} \big(K^*_{\overset{\circ}{\alpha}}(z_j)K_{\overset{\circ}{\alpha}}(z_i)h_i, h_j\big),$$

$$\left\|\sum_{i=1}^{n} (I - z_i T)^{-1}F h_i\right\|^2 \geq \left\|\sum_{i=1}^{n} (I - z_i\overset{\circ}{T})^{-1}\overset{\circ}{F} h_i\right\|^2.$$

We note now that

$$(I - zT)^{-1}F = \sum_0^\infty z^n T^n F, \quad r^n T^n F = \frac{1}{2\pi}\int_{|\zeta|=1} \bar{\zeta}^n (I - r\zeta T)^{-1} F\,|d\zeta|,$$

$(0 < r < 1)$. Therefore from the inequalities previously obtained, it follows that for all h_i $(\in \mathfrak{N}^-)$ and z_i $(|z_i| < 1)$,

$$\left\| \sum_{k=0}^n T^k F h_k \right\|^2 \geq \left\| \sum_{k=0}^n \overset{\circ}{T}{}^k \overset{\circ}{F} h_k \right\|^2,$$

i.e. that $\overset{\circ}{\alpha}$ is an optimal system. If it is not controllable, then we can pass from it to a new system α_0 with inner space $\mathfrak{H}_0 = \mathfrak{H}^c_{\overset{\circ}{\alpha}}$ $(= \mathfrak{H}^c_{\tilde{\alpha}})$ and coefficients

$$F_0 = \overset{\circ}{F}, \quad T_0 = \overset{\circ}{T}|\mathfrak{H}_0, \quad G_0 = \overset{\circ}{G}|\mathfrak{H}_0, \quad S_0 = \overset{\circ}{S}.$$

Since $\Theta_{\alpha_0}(z) = \Theta(z)$ and $T_0^k F_0 = \overset{\circ}{T}{}^k \overset{\circ}{F}$ $(k \geq 0)$, the system α_0 is also optimal. It is controllable and hence minimal. The theorem is proved.

Remark 1. In the proof, in place of $\begin{pmatrix}\Theta(z)\\ \varphi(z)\end{pmatrix}$ we may choose an arbitrary function $\widetilde{\Theta}(z) = [\Theta_{ik}(z)]_1^2$ of class B, which satisfies $\Theta_{12}(z) = \Theta(z)$ and $\Theta_{22}(z) = \varphi(z)$, and with the part $\overset{\circ}{\alpha}$ of a simple conservative system $\tilde{\alpha}$ with $\Theta_{\tilde{\alpha}}(z) = \widetilde{\Theta}(z)$ determined by the formulas (3). In this connection $\widetilde{\Theta}(z)$ may be chosen such that the optimal system $\overset{\circ}{\alpha}$ is controllable.

Remark 2. For an outer function $\varphi(z)$ and $\psi(z)$ $(\in B)$, from the conditions $\psi^*(z)\psi(z) \leq \varphi^*(z)\varphi(z)$ and $\psi^*(0)\psi(0) = \varphi^*(0)\varphi(0)$ it follows that $\psi(z) = b\varphi(z)$, where b is an isometric operator which does not depend on z. Therefore, if we use the identity (19), we conclude that an optimal system $\overset{\circ}{\alpha}$ is characterized by the condition $\overset{\circ}{F}{}^*\overset{\circ}{F} \leq F^*F$, where $\overset{\circ}{F}$ and F are coefficients respectively of an optimal and arbitrary passive system with one and the same scattering matrix $\Theta(z)$. In place of $z = 0$ we may choose any point ξ $(|\xi| < 1)$. The characterizing condition of an optimal system in this situation can be written in the form

$$\overset{\circ}{F}{}^*(I - \bar{\xi}\overset{\circ}{T}{}^*)^{-1}(I - \xi\overset{\circ}{T}{}^*)^{-1}\overset{\circ}{F} \leq F^*(I - \bar{\xi}T^*)^{-1}(I - \xi T)^{-1}F.$$

3. Let $\overset{\circ}{T}$ be the basic operator of an optimal controllable system $\overset{\circ}{\alpha}$ with scattering matrix $\Theta(z)$. From Proposition 4 and the definition of $\overset{\circ}{\alpha}$, it follows that $\overset{\circ}{T} \in C_{0\cdot}$ if and only if the factorization problem (4) is solvable for $\Theta(z)$.

Theorem 8. *For the inclusion $\overset{\circ}{T} \in C_{00}$ it is necessary and sufficient that for $\Theta(z)$ the factorization problems* (4), (5) *are solvable, and that for the corresponding contractive function $h_0(\zeta)$, defined by formula* (11), *there exists a bi-inner function $b(z)$ such that $h_0(\zeta)b(\zeta) \in B$. If these conditions are satisfied, the synthesis of $\overset{\circ}{\alpha}$*

may be achieved as indicated in §2 with the aid of a minimal $\mathfrak{D}$-representation of $\Theta(z)$ in the form

$$\widetilde{\Theta} = \begin{pmatrix} \varphi_1 & \Theta \\ h_0 & \varphi_2 \end{pmatrix} \begin{pmatrix} b & 0 \\ 0 & I \end{pmatrix}. \tag{21}$$

Proof. Suppose $\overset{\circ}{T} \in C_{00}$. Then $\Theta(z)$ has a $\mathfrak{D}$-representation $\widetilde{\Theta}(z) = [\Theta_{ik}(z)]_1^2$ ($\Theta_{12} = \Theta$) with $\Theta_{22} = \varphi = \varphi_2$ (see the proofs of Proposition 3 and Theorem 2). Also, as in the proof of Theorem 2, we obtain formula (21) for $\widetilde{\Theta}$, where $\{I, b(z)\}$ is a denominator for $h_0(\zeta)$. By the same token the necessity of the conditions for the inclusion $\overset{\circ}{T} \in C_{00}$ are obtained. Now suppose that the conditions are satisfied. We consider a minimal denominator $\{I, b(z)\}$ for $h_0(\zeta)$ and corresponding $\mathfrak{D}$-representation (21). With its help we construct a system $\dot{\alpha}$ as indicated in §2. Since $\Theta_{22}(z) = \varphi_2(z) = \varphi(z)$, $\dot{\alpha}$ is an optimal system. According to Theorem 3, it is minimal. For it, $\Theta_{\dot{\alpha}}(z) = \Theta(z)$, $\dot{T} \in C_{00}$ and therefore $\overset{\circ}{T} \in C_{00}$. The theorem is proved.

The bi-inner function $b(z)$ in formula (21) is determined by $\Theta(z)$ up to a constant right unitary factor by the conditions: 1) $h_0(\zeta)b(\zeta) \in B$, 2) if $b_1(z)$ is a bi-inner function and $h_0(\zeta)b_1(\zeta) \in B$, then $b_1(z)$ is divided on the left by $b(z)$, i.e. $b^*(\zeta)b_1(\zeta) \in B$.

Corollary. *If $\Theta(z) \in B\widetilde{\Pi}$, then $\overset{\circ}{T} \in C_{00}$.*

In fact, the conditions formulated in Theorem 8 are satisfied for $\Theta(z)$ of class $B\widetilde{\Pi}$. The existence of a right denominator $\{I, b(z)\}$ is proved in the same way as the existence of a left denominator is shown in Theorem 4.

According to Theorem 5, the inclusion $\overset{\circ}{T} \in C_0$ and $\Theta(z) \in B\Pi$ are equivalent. The problem remain open: is the inclusion $\overset{\circ}{T} \in C_{00}$ and $\Theta(z) \in B\widetilde{\Pi}$ equivalent?

For $\overset{\circ}{T}$ of class C_{00} the characteristic function $\Theta_{\overset{\circ}{T}}(z)$ coincides with the so-called pure part of the function $\widetilde{\Theta}(z)$, determined by formula (21) (see the appropriate definitions in [13]). Therefore in a well-known way we know the location of the spectrum of the operator $\overset{\circ}{T}$ from $\widetilde{\Theta}(z)$ (see [13], Ch. VI, Theorem 4.1). It is possible as well to use other known results from the theory of characteristic functions for the investigation of $\overset{\circ}{T}$ from $\widetilde{\Theta}(z)$. We remark that in the case where appropriate determinants are defined,

$$\det \widetilde{\Theta}(\zeta) = \det b(\zeta) \det \varphi(\zeta) / \overline{\det \varphi(\zeta)} \qquad (|\zeta| = 1),$$

where $\varphi(s) = \varphi_2(z)$.

References

(The Russian sources for items [1]–[15] have been replaced by English equivalents.)

[1] D.Z. Arov, *Darlington's method in the study of dissipative systems*, Dokl. Akad. Nauk SSSR **201** (1971), no. 3, 559–562.

[2] ______, *Realization of matrix-valued functions according to Darlington*, Izv. Akad. Nauk SSSR Ser. Mat. **37** (1973), 1299–1331.

[3] ______, *Scattering theory with dissipation of energy*, Dokl. Akad. Nauk SSSR **216** (1974), 713–716.

[4] ______, *Unitary couplings with losses (a theory of scattering with losses)*, Funkcional. Anal. i Priložen. **8** (1974), no. 4, 5–22.

[5] ______, *Realization of a canonical system with a dissipative boundary condition at one end of the segment in terms of the coefficient of dynamical compliance*, Sibirsk. Mat. Ž. **16** (1975), no. 3, 440–463.

[6] ______, *An approximation characteristic of functions of the class $B\Pi$*, Funkcional. Anal. i Priložen. **12** (1978), no. 2, 70–71.

[7] ______, *Passive linear steady-state dynamical systems*, Sibirsk. Mat. Zh. **20** (1979), no. 2, 211–228.

[8] V.M. Brodskiĭ and Ja.S. Švarcman, *Invariant subspaces of a contraction, and factorization of the characteristic function*, Teor. Funkciĭ Funkcional. Anal. i Priložen. **22** (1975), 15–35, 160.

[9] E.A. Guillemin, *Synthesis of passive networks. Theory and methods appropriate to the realization and approximation problems*, John Wiley and Sons, Inc., New York, 1958.

[10] Ju.P. Ginzburg, *The spectral subspaces of contractions with a slowly increasing resolvent*, Mat. Issled. **5** (1970), no. 4 (18), 45–62.

[11] A.M. Lane and R.G. Thomas, *R-matrix theory of nuclear reactions*, Rev. Mod. Phys. **30** (1958), 257–353.

[12] E.Ya. Malamud, *On a generalization of Darlington's theorem*, Izvestia Akad. Nauk, Arm. SSR **7** (1972), 183–195.

[13] B. Sz.-Nagy and C. Foias, *Harmonic analysis of operators on Hilbert space*, North-Holland Publishing Co., Amsterdam, 1970.

[14] V.P. Potapov, *Multiplicative representation of analytic matrix functions*, Short abstracts of scientific announcements at the International Mathematical Congress, Section 4, Moscow, 1966.

[15] Yu.A. Rozanov, *Innovation processes*, V. H. Winston & Sons, Washington, D. C., 1977.

[16] Y. Belevitch, *Elementary applications of the scattering formalism in network design*, IRE Trans. on circuit theory, CT-3, June (1956), 97–107, Ann. Télécommun. **6** (1951), 302.

[17] D.A. Bondy, *An application of functional operator models to dissipative scattering theory*, Trans. Amer. Math. Soc. **223** (1976), 1–43.

[18] S. Darlington, *Synthesis of reactance 4-poles which produce prescribed insertion loss characteristics including special applications to filter design*, J. Math. Phys. Mass. Inst. Tech. **18** (1939), 257–353.

[19] P. Dewilde, *Roomy scattering matrix synthesis*, Technical Report, 1971.

[20] R.G. Douglas and J.W. Helton, *Inner dilations of analytic matrix functions and Darlington synthesis*, Acta Sci. Math. (Szeged) **34** (1973), 61–67.

[21] C. Foias, *On the Lax-Phillips nonconservative scattering theory*, J. Functional Analysis **19** (1975), 273–301.

[22] J.W. Helton, *The characteristic functions of operator theory and electrical network realization*, Indiana Univ. Math. J. **22** (1972/73), 403–414.

[23] ———, *Discrete time systems, operator models, and scattering theory*, J. Functional Analysis **16** (1974), 15–38.

[24] P.D. Lax and R.S. Phillips, *Scattering theory for dissipative hyperbolic systems*, J. Functional Analysis **14** (1973), 172–235.

[25] B. Sz.-Nagy and C. Foias, *Opérateurs sans multiplicité*, Acta Sci. Math. (Szeged) **30** (1969), 1–18.

[26] ———, *Modèle de Jordan pour une classe d'opérateurs de l'espace de Hilbert*, Acta Sci. Math. (Szeged) **31** (1970), 91–115.

[27] ———, *Compléments à l'étude des opérateurs de classe* C_0, Acta Sci. Math. (Szeged) **31** (1970), 287–296.

[28] B.D.H. Tellegen, *Synthesis of* $2n$*-poles by networks containing the minimum number of elements*, J. Math. Physics **32** (1953), 1–18.

Appendix 1: Some further developments

D.Z. Arov

In this appendix the reader can find references to some works that are related to the paper[2] [4] and that have appeared subsequently. In the papers [8], [9], and [10] one can find: a different and purely geometrical proof of Theorem 7; a dual theorem on the existence of a minimal $*$-optimal linear discrete time stationary dissipative scattering system with a given operator function $\Theta(z)$ of the Schur class $\mathcal{S}(= B)$ as the scattering matrix of this system; generalizations of these theorems to the time-variant case; a model of the minimal optimal dissipative scattering discrete time-variant system that has been constructed by a synthesis of the methods of L. de Branges and R. Kalman; the investigation of the extremal factorization problem

$$\varphi^*\varphi \leq I - \Theta^*\Theta ,$$

where Θ is a given contractive block lower triangular operator acting from $\ell_2(\{\mathcal{U}_k\}_{-\infty}^{\infty})$ into $\ell_2(\{\mathcal{Y}_k\}_{-\infty}^{\infty})$ and φ is an unknown block lower triangular contractive operator; and the connection of the latter problem with the minimal optimal dissipative scattering realization problem for a given Θ. The theorems on D-representations and Darlington method were generalized to the time-variant case in [18]. Results have been obtained in [12] on simple conservative, minimal optimal, and minimal $*$-optimal linear stationary scattering realizations of a given holomorphic contractive operator function in the right half plane Θ as the scattering matrix of the corresponding continuous-time systems.

The minimal optimal and minimal $*$-optimal dissipative scattering realizations of a given function $\Theta \in \mathcal{S}$ were used to obtain criteria (in terms of the properties of Θ) that all of the minimal dissipative scattering realizations of Θ are unitarily equivalent [13] and that they are all similar [7], [11]. Analogous results are obtained for operator functions $\Theta(z)$ of the Carathéodory class, that is, which are holomorphic in the unit disk with Re $\Theta(z) \geq 0$ [11]. Theorems on the existence and uniqueness of minimal optimal and minimal $*$-optimal dissipative scattering realizations are obtained in [20] for operator functions $\Theta(z)$ in the generalized Schur class $\mathcal{S}_\varkappa^0$ that are meromorphic in the unit disk, for which the corresponding kernel

$$\frac{I - \Theta(z)\Theta^*(\omega)}{1 - z\overline{\omega}}$$

has $\varkappa$ negative squares ($\Theta \in \mathcal{S}_\varkappa$), and that are holomorphic at the point $z = 0$. In these generalizations the dissipative scattering systems are considered with Pontryagin state spaces $\Pi_\varkappa$ that have indefinite scalar product with $\varkappa$ negative squares. For conservative scattering realizations of a function $\Theta \in \mathcal{S}_\varkappa^0$ and related problems, see the book [1] and the references in [1]. Conservative scattering realizations of a function $\Theta(z) \in \mathcal{S}_\varkappa$ that may have a pole at the point $z = 0$ ($\Theta \notin \mathcal{S}_\varkappa^0$)

[2]The citations in Appendix 1 and Appendix 2 are to the *Supplementary References* that follow. An English translation of [4] is the main body of the present work.

and related problems on the unitary extension of an isometrical relation in the Pontryagin space $\Pi_\varkappa$ were considered in [16] and [17]. The D-representations, the Darlington method for the functions Θ of the class $\mathcal{S}_\varkappa$ and related problems, are considered below in Appendix 2. Two surveys of investigations connected with the results of the translated paper are given in [5] and [6].

Appendix 2: on Darlington representations of generalized Schur functions

D.Z. Arov and J. Rovnyak

The results of [4] have generalizations in which the state spaces are Pontryagin spaces. Such a situation arises, for example, in the analysis of circuits which have active as well as passive elements. Here we give two results which illustrate the possibilities. We plan to discuss this topic in a more systematic way in a future work.

For simplicity, we take the input and output spaces to be finite dimensional and given by $\mathfrak{N}^- = \mathbf{C}^q$ and $\mathfrak{N}^+ = \mathbf{C}^p$ in the standard Euclidean metrics. Operators on $\mathfrak{N}^-$ to $\mathfrak{N}^+$ are identified with $p \times q$ matrices in the usual way.

By $\mathcal{S}_\varkappa^{p\times q}$ we mean the **generalized Schur class** of $p\times q$ matrix-valued functions which are meromorphic on the unit disk $\mathbf{D} = \{z\colon |z| < 1\}$ such that the kernel $[I_p - S(z)S(\omega)^*]/(1 - z\bar{\omega})$ has $\varkappa$ negative squares. The basic properties of generalized Schur functions are given in a series of papers by Kreĭn and Langer. Mainly, we shall need the Kreĭn-Langer factorization. This says that $\mathcal{S}_\varkappa^{p\times q}$ coincides with the class of functions of the form

$$S(z) = B_\ell(z)^{-1} S_\ell(z),$$

where $B_\ell(z)$ is a Blaschke-Potapov product of degree $\varkappa$ whose values are $p \times p$ matrices, $S_\ell(z)$ belongs to $\mathcal{S}_0^{p\times q}$, and $B_\ell(z)$ and $S_\ell(z)$ are left co-prime in the sense that they have no common nonconstant inner left-divisor of size $p \times p$. This well-known result first appeared in [14], and there are other accounts and different proofs; for example, see §4.2 of [1]. A function $S(z)$ in $\mathcal{S}_0^{p\times q}$ is called a **(classical) Schur function.**

By the Kreĭn-Langer factorization, the boundary behavior of generalized Schur functions is immediately deducible from that of classical Schur functions. In particular, the boundary function $S(\zeta)$, $\zeta \in \partial\mathbf{D}$, of a function $S(z) \in \mathcal{S}_\varkappa^{p\times q}$ has contractive values. In some situations, it is convenient to identify $S(z)$ with its boundary function $S(\zeta)$ and write, for example, $S(\zeta) \in \mathcal{S}_\varkappa^{p\times q}$. The symbol ζ is always used for a point on $\partial\mathbf{D}$: $|\zeta| = 1$.

A scalar-valued meromorphic function is said to be of **bounded type** on a region if it is the quotient of bounded holomorphic functions with a denominator which does not vanish identically. Measure theoretic notions on the unit circle are assumed to be relative to normalized Lebesgue measure. Let $\Pi^{p\times q}$ be the class of

$p \times q$ matrix-valued functions $F(z)$ which are meromorphic separately on $\mathbf{D}$ and $\mathbf{D}^e = \{z\colon |z| > 1\} \cup \{\infty\}$ such that

(i) the entries of $F(z)$ are of bounded type on $\mathbf{D}$ and $\mathbf{D}^e$, and

(ii) the two radial limits of $F(z)$ on the unit circle $\partial\mathbf{D}$, taken from inside and outside the circle, coincide:

$$F(\zeta) \stackrel{def}{=} \lim_{r\uparrow 1} F(r\zeta) = \lim_{r\downarrow 1} F(r\zeta) \qquad \text{a.e. on } \partial\mathbf{D}.$$

We write $S(z) \in \mathcal{S}_\varkappa \Pi^{p\times q}$ if $S(z) \in \mathcal{S}_\varkappa^{p\times q}$ and $S(z)$ is the restriction to $\mathbf{D}$ of some function $F(z)$ in $\Pi^{p\times q}$. The class $\mathcal{S}_\varkappa \Pi^{p\times q}$ is a generalization of the class $B\Pi$ in [4]: $\mathcal{S}_0 \Pi^{p\times q}$ reduces to $B\Pi$ when operators are identified with matrices as above. The Darlington representation of functions in $B\Pi$ in [4] is generalized in the following result.

Theorem A. (1) *If $S(z) \in \mathcal{S}_\varkappa \Pi^{p\times q}$, there is an $m \le p+q$ and a function $\widetilde{S}(z) \in \mathcal{S}_\varkappa^{m\times m}$ whose boundary values are unitary a.e. on ∂D, such that $\widetilde{S}(z)$ has a decomposition*

$$\widetilde{S}(z) = \begin{pmatrix} S_{11}(z) & S_{12}(z) \\ S_{21}(z) & S_{22}(z) \end{pmatrix} \tag{1}$$

with $S_{12}(z) = S(z)$. Moreover, $\widetilde{S}(z)$ can be chosen such that $S_{11}(z)$, $S_{21}(z)$, and $S_{22}(z)$ are classical Schur functions.

(2) *Conversely, let $\widetilde{S}(z) \in \mathcal{S}_\varkappa^{m\times m}$ for some integer m, and assume that $\widetilde{S}(z)$ has unitary boundary values a.e. on ∂D. Decompose $\widetilde{S}(z)$ as in (1) with $S_{12}(z)$ of size $p \times q$. Then $S(z) = S_{12}(z)$ belongs to $\mathcal{S}_{\varkappa'} \Pi^{p\times q}$ for some $\varkappa' \le \varkappa$.*

An analogous problem was considered in [3] when the given function $S(z)$ of the class $\Pi^{p\times q}$ with $\|S(\zeta)\| \le 1$ a.e. on $\partial\mathbf{D}$ is not necessarily in the Schur class, but without the condition that $S(z)$ is a generalized Schur function. The present result obtains a stronger conclusion from a stronger hypothesis. The method of proof generalizes arguments in [4] for the case $\varkappa = 0$.

Proof. (1) Let $S(z) \in \mathcal{S}_\varkappa \Pi^{p\times q}$ be given. We first observe that there exists a scalar inner function $c(z)$ such that $c(\zeta)S^*(\zeta) \in \mathcal{S}_0^{q\times p}$, that is, $c(\zeta)S^*(\zeta)$ is the boundary function of a Schur function $G(z)$. In fact, since $S(z) \in \Pi^{p\times q}$, there is a function $F(z)$ which is meromorphic and of bounded type on $\mathbf{D}$ and $\mathbf{D}^e$ whose restriction to $\mathbf{D}$ is $S(z)$ and such that

$$\lim_{r\uparrow 1} F(r\zeta) = \lim_{r\downarrow 1} F(r\zeta) = S(\zeta) \qquad \text{a.e. on } \partial\mathbf{D}.$$

Choose a scalar inner function $c(z)$ such that

$$G(z) \stackrel{def}{=} c(z)F^*(1/\bar{z}), \qquad z \in \mathbf{D},$$

defines a function in $\mathcal{S}_0^{q\times p}$. Then $G(\zeta) = c(\zeta)S^*(\zeta)$ a.e. on $\partial\mathbf{D}$, and so $c(z)$ has the required property. By the Kreĭn-Langer factorization, we may also choose a scalar inner function $d(z)$ such that $d(z)S(z) \in \mathcal{S}_0^{p\times q}$. Set $b(z) = c(z)d(z)$.

The functions

$$\Delta_{S^*}(\zeta) = I_p - S(\zeta)S^*(\zeta) \qquad \text{and} \qquad \Delta_S(\zeta) = I_q - S^*(\zeta)S(\zeta)$$

satisfy $0 \le \Delta_{S^*}(\zeta) \le I_p$ and $0 \le \Delta_S(\zeta) \le I_q$ a.e. on $\partial\mathbf{D}$. If $b(z)$ is the scalar inner function defined above, then $b(\zeta)\Delta_{S^*}(\zeta)$ and $b(\zeta)\Delta_S(\zeta)$ are Schur functions. Proceeding as in the proof of Theorem 4 of [4] we apply the factorization theorem of [19] to find an outer function $\varphi_2(z)$ of size $p_1 \times q$ and a $*$-outer function $\varphi_1(z)$ of size $p \times q_1$ ($p_1 \le q$ and $q_1 \le p$) such that

$$\Delta_{S^*}(\zeta) = I_p - S(\zeta)S^*(\zeta) = \varphi_1(\zeta)\varphi_1^*(\zeta),$$
$$\Delta_S(\zeta) = I_q - S^*(\zeta)S(\zeta) = \varphi_2^*(\zeta)\varphi_2(\zeta),$$

a.e. on $\partial\mathbf{D}$. Notice that the values of $\varphi_2(z)$ and $\varphi_1(z)$ are contractions. We shall construct the required function (1) with $S_{12}(z) = S(z)$ and

$$S_{11}(z) = \varphi_1(z)b_1(z) \quad \text{and} \quad S_{22}(z) = b_2(z)\varphi_2(z), \tag{2}$$

where $b_1(z)$ and $b_2(z)$ are inner functions which will be chosen below. The entry $S_{21}(z)$ will be determined by a relation

$$S_{21}(\zeta) = b_2(\zeta)h_0(\zeta)b_1(\zeta) \qquad \text{a.e. on } \partial\mathbf{D}, \tag{3}$$

where $h_0(\zeta)$ is a measurable contractive-valued function on $\partial\mathbf{D}$ such that

$$h_0^*(\zeta)\varphi_2(\zeta) = -\varphi_1^*(\zeta)S(\zeta) \qquad \text{a.e. on } \partial\mathbf{D}. \tag{4}$$

The existence of $h_0(\zeta)$ follows as in the proof of Theorem 2 of [4]. Briefly, we first define a contraction operator $K : L^2(\mathbf{C}^{p_1}) \to L^2(\mathbf{C}^{q_1})$ by its action on a dense set:

$$K\colon \varphi_2(\zeta)g(\zeta) \to -\varphi_1^*(\zeta)S(\zeta)g(\zeta), \qquad g(\zeta) \in L^2(\mathbf{C}^q).$$

Since K intertwines multiplication by ζ on $L^2(\mathbf{C}^{p_1})$ and multiplication by ζ on $L^2(\mathbf{C}^{q_1})$, K is multiplication by a measurable contractive-valued function $h_0^*(\zeta)$, and the existence of $h_0(\zeta)$ follows.

We show that $\psi(\zeta) = b(\zeta)\varphi_2^*(\zeta)$ belongs to $\mathcal{S}_0^{q\times p_1}$. In fact,

$$\psi(\zeta)\varphi_2(\zeta) = b(\zeta)\varphi_2^*(\zeta)\varphi_2(\zeta) = b(\zeta)\Delta_S(\zeta)$$

is a Schur function, and since $\varphi_2(z)$ is an outer function, $\psi(\zeta)$ belongs to $\mathcal{S}_0^{q\times p_1}$. Let $\psi(z) = \psi_0(z)b_2(z)$, where $b_2(z)$ is inner and $\psi_0(z)$ is $*$-outer. Since

$$\psi_0(\zeta)b_2(\zeta)h_0(\zeta) = \psi(\zeta)h_0(\zeta) = b(\zeta)[\varphi_2^*(\zeta)h_0(\zeta)] = -b(\zeta)S^*(\zeta)\varphi_1(\zeta)$$

is a Schur function and $\psi_0(z)$ is $*$-outer, $b_2(\zeta)h_0(\zeta)$ is a Schur function.

Now define $S_{11}(z)$, $S_{22}(z)$, and $S_{21}(z)$ by (2) and (3) with the preceding choice of $b_2(z)$ and $b_1(z) \equiv I_{q_1}$. We show that the function $\widetilde{S}(z)$ given by (1) has the required properties. By construction, $S_{11}(z)$, $S_{22}(z)$, and $S_{21}(z)$ are Schur functions. The proof that $\widetilde{S}(z)$ has unitary boundary values a.e. on $\partial\mathbf{D}$ is the same as in the case $\varkappa = 0$. It follows that the values of $\widetilde{S}(z)$ are square matrices, and we take $m = p + p_1 = q_1 + q$. If $S(z) = b_\ell(z)^{-1} S_\ell(z)$ is a Kreĭn-Langer factorization, the representation

$$\widetilde{S}(z) = \begin{pmatrix} b_\ell(z)^{-1} & 0 \\ 0 & I \end{pmatrix} \begin{pmatrix} b_\ell(z) S_{11}(z) & S_\ell(z) \\ S_{21}(z) & S_{22}(z) \end{pmatrix} \tag{5}$$

shows that $\widetilde{S}(z) \in \mathcal{S}_{\varkappa'}^{m\times m}$ for some $\varkappa' \leq \deg b_\ell = \varkappa$. Since

$$S(z) = \begin{pmatrix} I_p & 0 \end{pmatrix} \widetilde{S}(z) \begin{pmatrix} 0 \\ I_q \end{pmatrix},$$

$\varkappa \leq \varkappa'$. It follows that $\varkappa' = \varkappa$ (in fact, (5) is a Kreĭn-Langer factorization).

(2) Assume that $\widetilde{S}(z) \in \mathcal{S}_{\varkappa}^{m\times m}$ and $\widetilde{S}(z)$ has unitary boundary values a.e. on ∂D. Then $\det \widetilde{S}(z) \not\equiv 0$. Decompose $\widetilde{S}(z)$ as in (1) with $S_{12}(z)$ of size $p \times q$, and let $S(z) = S_{12}(z)$. Using the Kreĭn-Langer factorization it is easy to see that

$$F(z) = \begin{cases} \widetilde{S}(z) & \text{on } \mathbf{D}, \\ \widetilde{S}^*(1/\bar{z})^{-1} & \text{on } \mathbf{D}^e, \end{cases}$$

defines a function which is meromorphic and of bounded type separately on $\mathbf{D}$ and $\mathbf{D}^e$. Since the boundary values of $\widetilde{S}(z)$ are unitary a.e., $F(z)$ has the same radial limits from inside and outside the circle:

$$F(\zeta) = \widetilde{S}(\zeta)$$

a.e. on $\partial\mathbf{D}$, and so $F(z) \in \Pi^{m\times m}$. Thus

$$F_{12}(z) = \begin{pmatrix} I_p & 0 \end{pmatrix} F(z) \begin{pmatrix} 0 \\ I_q \end{pmatrix}$$

belongs to $\Pi^{p\times q}$, and the restriction of $F_{12}(z)$ to $\mathbf{D}$ is $S(z) = S_{12}(z)$. Since $\widetilde{S}(z) \in \mathcal{S}_{\varkappa}^{m\times m}$, $S(z) \in \mathcal{S}_{\varkappa'}^{p\times q}$ for some $\varkappa' \leq \varkappa$. It follows that $S(z) \in \mathcal{S}_{\varkappa'}\Pi^{p\times q}$ with $\varkappa' \leq \varkappa$, as was to be shown. □

In the passive case of Darlington synthesis ($\varkappa = 0$), there is a notion of minimal losses which requires that the function $\widetilde{S}(z) \in \mathcal{S}_{\varkappa}^{m\times m}$ in Theorem A(1) be constructed with a choice of m as small as possible. We remark that the choice $m = p + p_1$ in the proof of the theorem has this property for all $\varkappa \geq 0$.

A different form of the Darlington representation may be given. Let p and q be nonnegative integers. Set $n = p + q$, and let

$$J = \begin{pmatrix} I_p & 0 \\ 0 & -I_q \end{pmatrix}.$$

An $n\times n$ matrix U is called ***J*-unitary** if $U^*JU = J$ and ***J*-contractive** if $U^*JU \leq J$. As is well known, these relations are equivalent to $UJU^* = J$ and $UJU^* \leq J$, respectively. Define the **generalized Potapov class** $\mathcal{P}_{\varkappa}^{n\times n}(J)$ as the set of meromorphic $n \times n$ matrix-valued functions $\Theta(z)$ on $\mathbf{D}$ such that the kernel $[J -$

$\Theta(z)J\Theta^*(\omega)]/(1 - z\bar{\omega})$ has $\varkappa$ negative squares. Every $\Theta(z)$ in $\mathcal{P}_\varkappa^{n\times n}(J)$ has a decomposition

$$\Theta(z) = \begin{pmatrix} \Theta_{11}(z) & \Theta_{12}(z) \\ \Theta_{21}(z) & \Theta_{22}(z) \end{pmatrix}, \tag{6}$$

where $\Theta_{22}(z)$ is of size $q \times q$ and invertible except at isolated points (see Theorem 4.3.3(1) in [1]). Hence we may define the **Potapov-Ginzburg transform**

$$S(z) = PG(\Theta(z))$$

of $\Theta(z)$ by

$$\begin{aligned} S(z) &= \begin{pmatrix} S_{11}(z) & S_{12}(z) \\ S_{21}(z) & S_{22}(z) \end{pmatrix} \\ &= \begin{pmatrix} \Theta_{11}(z) - \Theta_{12}(z)\Theta_{22}(z)^{-1}\Theta_{21}(z) & \Theta_{12}(z)\Theta_{22}(z)^{-1} \\ -\Theta_{22}(z)^{-1}\Theta_{21}(z) & \Theta_{22}(z)^{-1} \end{pmatrix}. \end{aligned}$$

Then

$$\Theta(z) = PG(S(z)),$$

that is,

$$\begin{aligned} \Theta(z) &= \begin{pmatrix} \Theta_{11}(z) & \Theta_{12}(z) \\ \Theta_{21}(z) & \Theta_{22}(z) \end{pmatrix} \\ &= \begin{pmatrix} S_{11}(z) - S_{12}(z)S_{22}(z)^{-1}S_{21}(z) & S_{12}(z)S_{22}(z)^{-1} \\ -S_{22}(z)^{-1}S_{21}(z) & S_{22}(z)^{-1} \end{pmatrix}. \end{aligned}$$

A straightforward calculation verifies the identity

$$\frac{J - \Theta(z)J\Theta^*(\omega)}{1 - z\bar{\omega}} = \Phi(z)\,\frac{I_n - S(z)S^*(\omega)}{1 - z\bar{\omega}}\,\Phi^*(\omega), \tag{7}$$

where

$$\Phi(z) = \begin{pmatrix} I_p & -\Theta_{12}(z) \\ 0 & -\Theta_{22}(z) \end{pmatrix}.$$

In particular, $S(z) \in \mathcal{S}_\varkappa^{n\times n}$. It follows that the entries of $\Theta(z)$ are of bounded type on $\mathbf{D}$, and that radial boundary limits $\Theta(\zeta)$ exist and are J-contractive a.e. on $\partial\mathbf{D}$.

Definition. *Let $\mathcal{U}_\varkappa^{n\times n}(J)$ be the set of all functions $\Theta(z)$ in $\mathcal{P}_\varkappa^{n\times n}(J)$ such that the boundary values $\Theta(\zeta)$ are J-unitary a.e. on $\partial\mathbf{D}$.*

If $\Theta(z) \in \mathcal{U}_\varkappa^{n\times n}(J)$, then $\Theta(z) \in \Pi^{n\times n}$. Indeed, the entries of the function

$$F(z) = \begin{cases} \Theta(z) & \text{on } \mathbf{D}, \\ J\Theta^*(1/\bar{z})^{-1}J & \text{on } \mathbf{D}^e, \end{cases}$$

are of bounded type on $\mathbf{D}$ and $\mathbf{D}^e$, and

$$\lim_{r\downarrow 1} F(r\zeta) = J\Theta^*(\zeta)^{-1}J = \Theta(\zeta) = \lim_{r\uparrow 1} F(r\zeta)$$

a.e. on $\partial\mathbf{D}$.

Let $\Theta(z) \in \mathcal{U}_\varkappa^{n\times n}(J)$, and decompose $\Theta(z)$ as in (6). If $\mathcal{E}$ is a constant $p \times q$ contractive matrix, then $\det[\Theta_{21}(z)\mathcal{E} + \Theta_{22}(z)] \not\equiv 0$; in fact, since the boundary values of $\Theta(z)$ are J-unitary, $\Theta_{21}(\zeta)\mathcal{E}+\Theta_{22}(\zeta)$ is invertible a.e. on $\partial\mathbf{D}$ by Theorem 1.1 in [15]. Therefore

$$S(z) = T_\Theta[\mathcal{E}] \overset{def}{=} [\Theta_{11}(z)\mathcal{E} + \Theta_{12}(z)][\Theta_{21}(z)\mathcal{E} + \Theta_{22}(z)]^{-1} \tag{8}$$

defines a function in $\Pi^{p\times q}$. An expression of the type (8) is another form of the Darlington representation. For example, see [2].

Theorem B. (1) *Assume that $\Theta(z) \in \mathcal{U}_\varkappa^{n\times n}(J)$ is decomposed as in (6) and that $\mathcal{E}$ is a constant $p \times q$ contractive matrix. Then $S(z) = T_\Theta[\mathcal{E}]$ belongs to $\mathcal{S}_{\varkappa'}\Pi^{p\times q}$ for some $\varkappa' \leq \varkappa$.*

(2) *Conversely, suppose $S(z) \in \mathcal{S}_\varkappa\Pi^{p\times q}$ and $I_q - S^*(\zeta)S(\zeta)$ is invertible a.e. on $\partial\mathbf{D}$. Then $S(z)$ has a representation $S(z) = T_\Theta[\mathcal{E}]$ as in (1), and moreover this representation can be chosen such that $\mathcal{E} = 0$.*

Proof. (1) By the discussion preceding the statement of the theorem, $S(z) \in \Pi^{p\times q}$, and what remains is to show that $S(z) \in \mathcal{S}_{\varkappa'}^{p\times q}$ for some $\varkappa' \leq \varkappa$. Straightforward algebra verifies the identity

$$\begin{aligned}
&\frac{I_q - S^*(\bar z)S(\bar\omega)}{1 - z\bar\omega} \\
&\qquad = [\mathcal{E}^*\Theta_{21}^*(\bar z) + \Theta_{22}^*(\bar z)]^{-1} \begin{pmatrix} \mathcal{E}^* & I_q \end{pmatrix} \frac{J - \Theta^*(\bar z)J\Theta(\bar\omega)}{1 - z\bar\omega} \\
&\qquad\qquad \cdot \begin{pmatrix} \mathcal{E} \\ I_q \end{pmatrix} [\Theta_{21}(\bar\omega)\mathcal{E} + \Theta_{22}(\bar\omega)]^{-1} \\
&\qquad + [\mathcal{E}^*\Theta_{21}^*(\bar z) + \Theta_{22}^*(\bar z)]^{-1} \frac{I_q - \mathcal{E}^*\mathcal{E}}{1 - z\bar\omega} [\Theta_{21}(\bar\omega)\mathcal{E} + \Theta_{22}(\bar\omega)]^{-1}.
\end{aligned} \tag{9}$$

Since we assume that $\Theta(z) \in \mathcal{U}_\varkappa^{n\times n}(J)$, the kernel $[J - \Theta(z)J\Theta^*(\omega)]/(1 - z\bar\omega)$ has $\varkappa$ negative squares. Hence so does the kernel $[J - \Theta^*(\bar z)J\Theta(\bar\omega)]/(1 - z\bar\omega)$ by Theorem 2.5.2 of [1]. Therefore the first term on the right side of (9) has at most κ negative squares. The second term on the right side of (9) is a nonnegative kernel because $\mathcal{E}$ is a contraction. Hence the kernel on the left side of (9) has at most $\varkappa$ negative squares. By another application of Theorem 2.5.2 of [1], $S(z) \in \mathcal{S}_{\varkappa'}^{p\times q}$ for some $\varkappa' \leq \varkappa$, which proves (1).

(2) Assume that $S(z) \in \mathcal{S}_\varkappa\Pi^{p\times q}$ and that $I_q - S^*(\zeta)S(\zeta)$ is invertible a.e. on $\partial\mathbf{D}$. Choose a function $\widetilde{S}(z)$ for $S(z)$ as the proof of Theorem A(1). In the same

notation, $S_{22}(z) = b_2(z)\varphi_2(z)$ where $\varphi_2(z)$ has size $p_1 \times q$ with $p_1 \leq q$. Since $\widetilde{S}(\zeta)$ is unitary a.e. on $\partial\mathbf{D}$,

$$I_q - S^*(\zeta)S(\zeta) = I_q - S_{12}^*(\zeta)S_{12}(\zeta) = S_{22}^*(\zeta)S_{22}(\zeta) = \varphi_2^*(\zeta)\varphi_2(\zeta)$$

a.e. on $\partial\mathbf{D}$. By our nondegeneracy hypothesis, it follows that $p_1 = q$ and $\det S_{22}(z) \not\equiv 0$. Define $\Theta(z) = PG(\widetilde{S}(z))$. Using the identity (7) and the fact that $\widetilde{S}(z) \in \mathcal{S}_\varkappa^{n\times n}$ and has unitary boundary values a.e. on $\partial\mathbf{D}$, we see that $\Theta(z) \in \mathcal{U}_\varkappa^{n\times n}(J)$. By construction,

$$S(z) = \Theta_{12}(z)\Theta_{22}(z)^{-1} = T_\Theta[0],$$

as was to be shown. □

Supplementary References

[1] D. Alpay, A. Dijksma, J. Rovnyak, and H.S.V. de Snoo, *Schur functions, operator colligations, and reproducing kernel Pontryagin spaces*, Oper. Theory Adv. Appl., vol. 96, Birkhäuser Verlag, Basel, 1997.

[2] D.Z. Arov, *Realization of matrix-valued functions according to Darlington*, Izv. Akad. Nauk SSSR Ser. Mat. **37** (1973), no. 6, 1299–1331, English transl.: *Math. USSR Izvestija*, 7(6): 1295–1326, 1973.

[3] ———, *Unitary couplings with losses (a theory of scattering with losses)*, Funkcional. Anal. i Priložen. **8** (1974), no. 4, 5–22, English transl.: *Funct. Anal. Appl.*, 8(4): 280–294, 1974.

[4] ———, *Stable dissipative linear stationary dynamical scattering systems*, J. Operator Theory **2** (1979), no. 1, 95–126.

[5] ———, *A survey on passive networks and scattering systems which are lossless or have minimal losses*, AEÜ Archiv für Elektronik und Übertragungstechnik. International Journal of Electronics and Communication **49** (1995), no. 5/6, 252–265.

[6] ———, *Passive linear systems and scattering theory*, Dynamical systems, control, coding, computer vision (Padova, 1998), Birkhäuser, Basel, 1999, pp. 27–44.

[7] ———, *Conditions for the similarity of all minimal passive scattering systems with given scattering matrix*, Funkcional. Anal. i Priložen. **34** (2000), no. 4, 71–74, English transl.: *Funct. Anal. Appl.*, 34(4): 293–295, 2000.

[8] D.Z. Arov, M.A. Kaashoek, and D.R. Pik, *Minimal and optimal linear discrete time-invariant dissipative scattering systems*, Integral Equations Operator Theory **29** (1997), no. 2, 127–154.

[9] ———, *Optimal time-variant systems and factorization of operators. I. Minimal and optimal systems*, Integral Equations Operator Theory **31** (1998), no. 4, 389–420.

[10] ———, *Optimal time-variant systems and factorization of operators. II. Factorization*, J. Operator Theory **43** (2000), no. 2, 263–294.

[11] D.Z. Arov and M.A. Nudel′man, *The conditions of similarity of all minimal passive realizations of a given transfer function (scattering or resistence)*, submitted.

[12] ———, *Passive linear stationary dynamical scattering systems with continuous time*, Integral Equations Operator Theory **24** (1996), no. 1, 1–45.

[13] ———, *A criterion for the unitary similarity of minimal passive systems of scattering with a given transfer function*, Ukraïn. Mat. Zh. **52** (2000), no. 2, 147–156.

[14] M.G. Kreĭn and H. Langer, *Über die verallgemeinerten Resolventen und die charakteristische Funktion eines isometrischen Operators im Raume Π_κ*, Hilbert space operators and operator algebras (Proc. Internat. Conf., Tihany, 1970), North-Holland, Amsterdam, 1972, pp. 353–399. Colloq. Math. Soc. János Bolyai, 5.

[15] M.G. Kreĭn and Ju.L. Šmul′jan, *Fractional linear transformations with operator coefficients*, Mat. Issled **2** (1967), no. 3, 64–96, English transl.: *Amer. Math. Soc. Transl.* (2), vol. 103, pages 125–152, 1974.

[16] Oleg Nitz, *Generalized resolvents of isometric linear relations in Pontryagin spaces. I. Foundations*, Operator theory and related topics, Vol. II (Odessa, 1997), Birkhäuser, Basel, 2000, pp. 303–319.

[17] ———, *Generalized resolvents of isometric linear relations in Pontryagin spaces. II. Kreĭn-Langer formula*, Methods Funct. Anal. Topology **6** (2000), no. 3, 72–96.

[18] D.R. Pik, *Block lower triangular operators and optimal contractive systems*, Ph.D. thesis, Free University of Amsterdam, 1999.

[19] M. Rosenblum and J. Rovnyak, *Hardy classes and operator theory*, Oxford University Press, New York, 1985, Dover republication, New York, 1997.

[20] S.M. Saprikin, *The theory of linear stationary passive scattering systems with Pontryagin state spaces*, Dokl. Ukrainian Akad. Nauk, submitted.

D.Z. Arov
Department of Physics and Mathematics
Division of Mathematical Analysis
South-Ukrainian Pedagogical University
Staroportofrankovskaya 26
Odessa 270020, Ukraine
e-mail: `aspect@farlep.net`

J. Rovnyak
Department of Mathematics
University of Virginia
Charlottesville, VA 22903-3199, U. S. A.
e-mail: `rovnyak@Virginia.EDU`

Operator Theory:
Advances and Applications, Vol. 134, 137–156

Concrete Interpolation of Meromorphic Matrix Functions on Riemann Surfaces

Joseph A. Ball, Kevin F. Clancey, and Victor Vinnikov

To our friend Harry Dym on the occasion of his 60th birthday

Abstract. This work investigates concrete problems of interpolating matrix pole-zero data with multiple-valued meromorphic matrix functions on closed Riemann surfaces. In the case of genus $g > 1$, a condition sufficient for the existence of a solution having constant factor of automorphy is presented. Necessary and sufficient conditions are presented in the case where $g = 1$. A necessary and sufficient condition for single-valued matrix function interpolation in arbitrary genus is also established.

This paper deals with the problem of interpolating matrix pole-zero data by regular meromorphic matrix functions on a closed Riemann surface M of genus greater than zero. In classical formulations of such interpolation problems, the data is given as a matrix divisor. A matrix divisor Θ is a section of the sheaf of germs of regular $r \times r$-meromorphic matrix functions on M modulo one-side equivalence by invertible analytic matrix functions. This notion of divisor was introduced by [15]. In general, there will exist a multiple-valued $r \times r$-meromorphic matrix function G interpolating the divisor in the sense that the matrix function germ determined by G belongs to the value of Θ at points of M. This last result is a consequence of the triviality of vector bundles on the universal cover $\rho : \widetilde{M} \to M$ of M. In fact, if $\mathcal{G}$ is the group of covering transformations for $\rho : \widetilde{M} \to M$, then the vector bundle on M determined by Θ corresponds to a holomorphic matrix factor of automorphy $\xi : \mathcal{G} \times \widetilde{M} \to GL(r, \mathbb{C})$. By definition this matrix factor of automorphy satisfies

$$\xi(ST, u) = \xi(S, Tu)\xi(T, u), \; S, T \in \mathcal{G}, \; u \in \widetilde{M}.$$

There is an $r \times r$-meromorphic matrix function $G = G(u)$ on $\widetilde{M}$ satisfying

$$G(Tu) = \xi(T, u)G(u), \; (T, u) \in \mathcal{G} \times \widetilde{M}$$

such that the germs of G agree with the values of the divisor $\rho^*\Theta$ at points of $\widetilde{M}$. This abstract solution of the interpolation problem is a nice existence theorem; however, to understand the multiple-valued nature of the solution to the interpolation problem, one must determine the factor of automorphy ξ that is associated with the divisor data Θ. One result in this direction appears in a classic paper of Weil [15]. Weil gives finite dimensional necessary and sufficient conditions for the existence of a solution to the above interpolation problem that has constant factor of automorphy, i.e., associated to a representation of $\mathcal{G}$. The existence of such a solution is equivalent to the condition that the vector bundle associated with Θ is flat.

The goal of this paper is to study the abstract interpolation problem described above in a more concrete context. In many applications, it is important to present the interpolation data in a form that explicitly displays the matrix pole-zero data at the interpolation nodes. A linear algebra description of the pole-zero data of a meromorphic matrix function has recently been given using the concept of a *null-pole triple* as described in [4]. This triple will be described in greater detail below. For now, we are content with an indication of the nature of null-pole triples. Suppose that $F = F(z)$ is a regular $r \times r$-meromorphic matrix function defined in a neighborhood of $z = 0$ in $\mathbb{C}$. The (right) null-pole triple of F at $z = 0$ has the form

$$\Upsilon = ((B_\zeta, A_\zeta), (A_\pi, C_\pi), S).$$

In this triple, the pair of matrices (A_π, C_π), where A_π is $n_\pi \times n_\pi$ and C_π is $n_\pi \times r$, captures the pole behavior of F, in the sense that for some matrix $\widetilde{B}$ the difference

$$F(z) - \widetilde{B}(zI - A_\pi)^{-1}C_\pi$$

is analytic at $z = 0$ and the pair of matrices (B_ζ, A_ζ), where A_ζ is $n_\zeta \times n_\zeta$ and B_ζ is $r \times n_\zeta$, captures the zero behavior of F in the sense that for some matrix $\widetilde{C}$ the difference

$$F^{-1}(z) - B_\zeta(zI - A_\zeta)^{-1}\widetilde{C}$$

is analytic at $z = 0$. The $n_\pi \times n_\zeta$ matrix S satisfies $A_\pi S - S A_\zeta = C_\pi B_\zeta$ and is called the coupling matrix [9]. In a somewhat imprecise sense, the matrix S accounts for any pole-zero cancellation in $\det F$ at $z = 0$. There are natural concepts of minimality and similarity for null-pole triples such that the similarity orbits of minimal null-pole triples are in a one-to-one correspondence with matrix divisors at $z = 0$. As a consequence of this last fact, the problem of interpolating matrix divisors is equivalent to the more concrete problem of interpolating null-pole triples. The concrete formulation of this interpolation problem comes with a price. Namely, the prescription of a matrix null-pole triple at a point on the Riemann surface must be done in specific local coordinates. (Below, we will offer a coordinate free method of describing null-pole triples.) If one exploits uniformization, then the dependence of the interpolation data on local coordinates is somewhat diminished (or at least hidden). Let $z_1, \ldots, z_K$ be points on M. We will specify interpolation data on M of the form

$$\mathcal{D} : \{(\Upsilon_1, z_1), \ldots, (\Upsilon_K, z_K)\},$$

where Υ_j is an admissible null-pole triple prescribed in local coordinates (s_j, V_j) at the point z_j, with $s_j(z_j) = 0$, $j = 1, \ldots, K$. One obvious question is the following: Given the data $\mathcal{D}$, find necessary and sufficient conditions in terms of the data $\mathcal{D}$ for the existence of a global meromorphic matrix function G such that near z_j, $G(p) = F_j(s_j(p))$, where Υ_j is a null-pole triple of the matrix function F_j at zero, $j = 1, \ldots, K$. We will give an answer to this question. See, e.g., Theorem 9. A related question is to find necessary and sufficient conditions involving the data $\mathcal{D}$ for the existence of a multiple-valued meromorphic matrix function G that solves the interpolation problem such that G has a constant matrix factor of automorphy. We will give a complete solution to this problem in the case of genus one (see, Theorem 1) and a sufficient condition for a solution to this problem in higher genus (see, Corollary 7). The genus one result will take advantage of Atiyah's [1] description of vector bundles over an elliptic curve. In the case where M is realized in the form $M = \mathbf{C}/\mathbb{Z} + \tau\mathbb{Z}$, where $\operatorname{Im}\tau > 0$, the results of Atiyah [1] imply that a flat bundle E over M admits a representation

$$E = \mathbb{L}_1 \otimes F_{h_1} \oplus \cdots \oplus \mathbb{L}_s \otimes F_{h_s}.$$

In this representation $\mathbb{L}_1, \ldots, \mathbb{L}_s$ are degree zero line bundles and F_h denotes a rank h flat bundle which corresponds to the representation $\xi_h : \mathbb{Z} + \tau\mathbb{Z} \to GL(h, \mathbb{C})$ given on the generators of $\mathbb{Z} + \tau\mathbb{Z}$ by

$$\xi_h(1) = I_h \ : \ \xi_h(\tau) = \begin{bmatrix} 1 & 1 & \cdots & 0 & 0 \\ 0 & 1 & 1 & \cdots & 0 \\ \vdots & \vdots & \ddots & \ddots & 0 \\ 0 & 0 & \cdots & 1 & 1 \\ 0 & 0 & \cdots & 0 & 1 \end{bmatrix}.$$

Perhaps a more natural question from the point of view of the structure of vector bundles is which interpolation data sets give rise to flat unitary vector bundles, or more generally, to "stable" or "semistable" bundles (see [14] for definitions and a full account of these concepts); we will also discuss some results on this problem in Section 3.

We close the introduction by describing a sample of our results in the simplest case on a torus where all poles and zeros are first order and occur at separate points. Suppose that M is the complex torus $M = \mathbb{C}/\mathbb{Z} + \tau\mathbb{Z}$, where $\operatorname{Im}\tau > 0$. Consider the interpolation data

$$\mathcal{D} : \{(\underline{b}_1, z_1), \ldots, (\underline{b}_N, z_N) : (\mathbf{c}_1, w_1), \ldots, (\mathbf{c}_N, w_N)\},$$

where, $\underline{b}_1, \ldots, \underline{b}_N$ are r-dimensional column vectors, $\mathbf{c}_1, \ldots, \mathbf{c}_N$ are r-dimensional row vectors and $z_1, \cdots, z_N; w_1, \cdots, w_N$ are $2N$ distinct points on M. We look for multiple valued $r \times r$-meromorphic matrix functions G on M such that the poles of entries are at most simple poles at the points $w_1, \ldots, w_N$, the only zeros of $\det G$ are simple zeros at $z_1, \ldots, z_N$ with $\underline{b}_j$ spanning the (right-)kernel of $G(z_j)$ $(j = 1, \ldots, N)$ and at w_i, G^{-1} is analytic with $\mathbf{c}_i$ spanning the left-kernel of $G^{-1}(w_i)$, $i = 1, \ldots, N$. The data set $\mathcal{D}$ can be lifted to the data set $\rho^*\mathcal{D}$ on $\mathbb{C}$. Any single-valued $r \times r$ -meromorphic matrix function $G = G(u)$ interpolating $\rho^*\mathcal{D}$ is to be considered a multiple-valued solution to the interpolation problem with data set $\mathcal{D}$. We introduce the classical theta function with characteristics $\{\frac{1}{2}, \frac{1}{2}\}$

$$\theta_*(u) = \sum_{n \in \mathbb{Z}} \exp\left\{2\pi i[\frac{1}{2}(n+\frac{1}{2})\tau(n+\frac{1}{2}) + (n+\frac{1}{2})(u+\frac{1}{2})]\right\},$$

associated with the lattice $\mathbb{Z} + \tau\mathbb{Z}$.

Theorem 1. *In order that there exist a multiple-valued $r \times r$-meromorphic matrix function $G = G(u)$ solving the interpolation problem with data $\mathcal{D}$ that has automorphic behavior*

$$G(u+1) = G(u), \ G(u+\tau) = CG(u+\tau),$$

where C is an invertible $r \times r$-matrix, it is necessary and sufficient that for some $\lambda \notin \mathbb{Z} + \tau\mathbb{Z}$ the matrix

$$\Gamma_0^\lambda = \left[\frac{\theta_*(w_i - z_j + \lambda)\mathbf{c}_i \underline{b}_j}{\theta_*(w_i - z_j)}\right]_{N \times N}$$

is invertible.

The proof of this theorem provides additional information. There is a natural vector bundle $E_{\mathcal{D}}$ associated with the interpolation data $\mathcal{D}$.

Corollary 2. *The zeros of the determinant of the matrix function Γ_0^λ, $\lambda \notin \mathbb{Z} + \tau\mathbb{Z}$, correspond to the non-trivial line bundles providing the decomposition $E_{\mathcal{D}} = \mathbb{L}_1 \otimes F_{h_1} \oplus \cdots \oplus \mathbb{L}_s \otimes F_{h_s}$. The bundle $E_{\mathcal{D}}$ is equivalent to a direct sum of non-trivial line bundles if and only if the determinant of Γ_0^λ, $\lambda \notin \mathbb{Z} + \tau\mathbb{Z}$, has precisely r zeros (counting multiplicity) in a fundamental domain for $\mathbb{Z} + \tau\mathbb{Z}$. The bundle $E_{\mathcal{D}}$ is equivalent to a direct sum of the bundles F_h if and only if the determinant of Γ_0^λ, $\lambda \notin \mathbb{Z} + \tau\mathbb{Z}$, doesn't vanish.*

The genus zero version of Theorem 1 (where the torus is replaced by the Riemann sphere and the only flat bundle is the trivial bundle) goes back to [9]; see [4] for a complete treatment. Indeed, in the case where the zeros $z_1, \ldots, z_N$ and poles $w_1, \ldots, w_N$ are in the complex plane the invertibility of the $N \times N$-matrix $\Gamma = \left[\frac{\mathbf{c}_i \underline{b}_j}{w_i - z_j}\right]$ is necessary and sufficient for the existence of a rational matrix function solving the corresponding interpolation problem on the Riemann sphere. In the sequel, for arbitrary genus $g > 0$, using a matrix analogous to Γ_0^λ, necessary and sufficient conditions will be given for the interpolation problem with data $\mathcal{D}$ to have a single valued solution. A result in this direction was given earlier in [2].

It develops that every flat vector bundle on a closed Riemann surface is equivalent to an interpolation vector bundle associated with data of the simple form $\mathcal{D}$ [3]. Thus the result in Theorem 1 gives conditions for realizing any flat vector bundle on a Riemann surface of genus $g = 1$ through global meromorphic functions on $\mathbb{C}$ having a simple pole-zero structure.

1. Local divisors and local null-pole triples

Let M be a Riemann surface. The sheaf of germs of holomorphic (respectively, meromorphic) functions on M will be denoted by $\mathcal{O}$ (respectively, $\mathcal{M}$). The notation $\mathcal{O}_p$ (respectively, $\mathcal{M}_p$) will be used for the stalk at $p \in M$ of $\mathcal{O}$ (respectively, $\mathcal{M}$). Further, we will denote $\mathcal{O}\otimes\mathbb{C}^r$ (respectively, $\mathcal{M}\otimes\mathbb{C}^r$) by $\mathcal{O}^r$ (respectively, $\mathcal{M}^r$) and $\mathcal{O}^{r\times r}$ (respectively, $\mathcal{M}^{r\times r}$) will denote the $r\times r$-matrix analogues. When convenient, elements in $\mathbb{C}^r$(respectively, $\mathcal{O}^r$) will be considered as row vectors (respectively, row vector functions). The regular elements in $\mathcal{O}^{r\times r}$ (respectively, $\mathcal{M}^{r\times r}$) will be denoted by $\mathcal{GL}(r,\mathcal{O})$ (respectively, $\mathcal{GL}(r,\mathcal{M})$). We let $\mathcal{GL}(r,\mathcal{O})$ act on $\mathcal{GL}(r,\mathcal{M})$ on the left. An element Θ_p in $(\mathcal{GL}(r,\mathcal{M})/\mathcal{GL}(r,\mathcal{O}))_p$ is called a (rank r) *local matrix divisor* at p. A *matrix divisor* is a section Θ of $\mathcal{GL}(r,\mathcal{M})/\mathcal{GL}(r,\mathcal{O})$. See, [13] for a slightly different definition of matrix divisor. The set of values $p \in M$ (necessarily finite) where $\Theta_p \neq I$ will be called the support of Θ. The value Θ_p of a matrix divisor is a set of germs at p of the form $[HF]_p$, where F is a fixed regular $r \times r$-meromorphic matrix function defined in a neighborhood of p and H varies over invertible $r \times r$-analytic matrix functions defined in a neighborhood of p. Two divisors Θ and $\widetilde{\Theta}$ are said to be *linearly equivalent* in case there is a globally defined regular $r \times r$-meromorphic matrix K such that $\widetilde{\Theta} = \Theta K$. If F is a regular $r\times r$-meromorphic matrix function defined in a neighborhood of p, then we define the *null-pole subspace* associated with F at p as

$$\mathcal{O}_p^r[F]_p = \{[\mathbf{f}]_p : \mathbf{f} = \mathbf{g}F,\ \mathbf{g} \text{ is (a row vector) } \mathbb{C}^r\text{-valued and analytic at } p\}. \quad (1)$$

This defines stalks of a locally free sheaf which is dual to the standard bundle determined by Θ. Obviously, if $[F]_p$ and $[\widetilde{F}]_p$ belong to the divisor Θ_p, then $\mathcal{O}_p^r[F]_p = \mathcal{O}_p^r[\widetilde{F}]_p$. Conversely, this last equality implies $[F]_p$ and $[\widetilde{F}]_p$ belong to the same divisor at p. As a consequence, if Θ_p is a matrix divisor at p, it determines a *null-pole subspace* $\mathcal{S}(\Theta_p)$ given by (1) in $\mathcal{M}_p^r$.

It develops that local matrix divisors admit a more concrete description in terms of matrix pole-zero data. In fact, there is a natural correspondence between local matrix divisors and similarity orbits of local null-pole triples. We briefly present the concept of local null-pole triple. Suppose F is an $r \times r$-meromorphic matrix function defined in a neighborhood of $z = 0$ in the complex plane. The pole-zero structure of F at $z = 0$ can be encoded in a *minimal right null-pole triple*

$$\Upsilon = ((B_\zeta, A_\zeta), (A_\pi, C_\pi), S), \quad (2)$$

where (B_ζ, A_ζ) is a minimal right zero pair of F, (A_π, C_π) is a minimal left pole pair of F and the $n_\pi \times n_\zeta$-null-pole coupling matrix S satisfies

$$A_\pi S - S A_\zeta = C_\pi B_\zeta. \quad (3)$$

To say that (B_ζ, A_ζ) is a minimal right zero pair of F means that A_ζ is an $n_\zeta \times n_\zeta$-nilpotent matrix, B_ζ is an $r \times n_\zeta$-matrix with

$$\bigcap_{j=0}^{n_\zeta} \ker(B_\zeta A_\zeta^j) = \{\mathbf{0}\} \quad (4)$$

and there is an $n_\zeta \times n$-matrix $\widetilde{C}$ such that

$$F^{-1}(z) - B_\zeta(zI - A_\zeta)^{-1}\widetilde{C}$$

is analytic at zero. Note that if (B_ζ, A_ζ) is a left zero pair for F, then for any invertible matrix U

$$(\, B_\zeta U, U^{-1}A_\zeta U) \tag{5}$$

is also a left zero pair for F. Moreover, any left zero pair for F will have the form (5) for some invertible matrix U. In a dual manner, to say that (A_π, C_π) is a right pole pair of F means that A_π is an $n_\pi \times n_\pi$-nilpotent matrix, C_π is an $n_\pi \times r$-matrix with

$$\sum_{j=0}^{n_\pi} \operatorname{im}(A_\pi^j C_\pi) = \mathbb{C}^{n_\pi} \tag{6}$$

and there is an $r \times n_\pi$-matrix $\widetilde{B}$ such that

$$F(z) - \widetilde{B}(zI - A_\pi)^{-1}C_\pi$$

is analytic at zero. If (A_π, C_π) is a right-pole pair of F, then for any invertible V

$$(V^{-1}A_\pi V, V^{-1}C_\pi) \tag{7}$$

is a right pole pair of F. Any right pole pair for F will have the form (7) for some invertible matrix V.

The fact that S acts as a coupling operator means the following: Given an r-dimensional row vector function $\mathbf{h} = \mathbf{h}(z)$ analytic at zero, then one can write

$$(\mathbf{h}F)(z) = \mathbf{x}(zI - A_\pi)^{-1}C_\pi + \mathbf{k}(z) \tag{8}$$

where $\mathbf{x}$ is an n_π-dimensional row vector and $\mathbf{k} = \mathbf{k}(z)$ is analytic at zero. Moreover, every n_π-dimensional row vector $\mathbf{x}$ occurs in such a decomposition for an appropriate choice of $\mathbf{h}$. The coupling operator S satisfies

$$\mathbf{x}S = \operatorname{res}_{z=0}\left[\mathbf{k}(z)B_\zeta(zI - A_\zeta)^{-1}\right] \tag{9}$$

where res denotes the residue. By combining (1), (8) and (9) we see that the null-pole subspace associated with F at $z = 0$ has the explicit description

$$\begin{aligned}\mathcal{O}_0^r[F]_0 = \{&\mathbf{x}(zI - A_\pi)^{-1}C_\pi + \mathbf{k}(z) :\\ &\mathbf{x} \in \mathbb{C}^{n_\pi}, \mathbf{k} \in \mathcal{O}_0^r\ ,\ \mathbf{x}S = \operatorname{res}_{z=0}\left[\mathbf{k}(z)B_\zeta(zI - A_\zeta)^{-1}\right]\},\end{aligned} \tag{10}$$

where $((B_\zeta, A_\zeta), (A_\pi, C_\pi), S)$ is a null-pole triple for F at $z = 0$. (For a self-contained complete proof of this statement, see Theorem 12.3.1 of [4].)

It is possible to construct a canonical null-pole triple for a given $r \times r$-matrix function F meromorphic at $z = 0$. The details of this construction can be found in [4]. Following the usual convention, in case no matrix entry of F (respectively, of F^{-1}) has a pole at $z = 0$, a null-pole triple for F will be written simply as a zero pair (respectively, a pole pair).

A triple $\Upsilon = ((B_\zeta, A_\zeta), (A_\pi, C_\pi), S)$ consisting of a pair of matrices (A_π, C_π) (of sizes $n_\pi \times n_\pi$ and $n_\pi \times r$) satisfying (6), a pair (B_ζ, A_ζ) (of sizes $r \times n_\zeta$ and $n_\zeta \times n_\zeta$) satisfying (4) and with S satisfying (3) will be called a rank r *admissible triple.* Given a rank r admissible triple Υ there is an $r \times r$-matrix function F meromorphic at $z = 0$ such that Υ is the null-pole triple of F at $z = 0$.

If U and V are invertible matrices of appropriate size, then the triple

$$\widetilde{\Upsilon} = ((\, B_\zeta U, U^{-1} A_\zeta U), (V^{-1} A_\pi V, V^{-1} C_\pi), V^{-1} S U) \tag{11}$$

is also a null-pole triple for F . One says the null-pole triples Υ and $\widetilde{\Upsilon}$ are similar. If Υ is an admissible triple, then the collection $\mathcal{S}(\Upsilon)$ of triples of the form (11), where U and V vary over invertible matrices of the appropriate size, will be called the similarity orbit of Υ. An important result from [6] establishes a one-to-one correspondence between the similarity orbits of admissible rank r triples and rank r local matrix divisors at $z = 0$ in $\mathbb{C}$. This correspondence is a consequence of the following result: Let F_1 and F_2 be regular $r \times r$-meromorphic matrix functions defined in a neighborhood of $z = 0$. The matrix functions F_1 and F_2 are associated with similar null-pole triples if and only if for some invertible analytic $r \times r$-matrix function H, $F_2 = HF_1$ in a neighborhood of $z = 0$. Thus if $\mathcal{S}(\Upsilon)$ is the similarity orbit of a rank r admissible triple, there is a unique rank r local matrix divisor Θ_0 at $z = 0$ associated with $\mathcal{S}(\Upsilon)$. This divisor Θ_0 consists of the set of germs at $z = 0$ of regular $r \times r$-meromorphic matrix functions F such that every element in $\mathcal{S}(\Upsilon)$ is a null-pole triple of F.

At a point p on a Riemann surface, it is possible to specify a local matrix divisor Θ_p using the similarity orbit $\mathcal{S}(\Upsilon)$ of an admissible triple together with local coordinates (s, V), where s maps the neighborhood V of p into $\mathbb{C}$ with $s(p) = 0$. This matrix divisor consists of the collection of germs of regular $r \times r$-meromorphic matrix functions L that have the form $L(q) = F(s(q))$ in a neighborhood of p, where F admits $\mathcal{S}(\Upsilon)$ as a set of null-pole triples at $s = 0$. A less concrete but coordinate free approach to the null-pole triple can be given as follows. The value of the matrix divisor Θ_p at a point $p \in M$ defines the null-pole subspace $\mathcal{S}(\Theta_p)$ of the space $\mathcal{M}_p^r$ of r-dimensional meromorphic row vector germs. Introduce the pole space $\mathcal{P}_p = [\mathcal{S}(\Theta_p) + \mathcal{O}_p^r]/\mathcal{O}_p^r$ and the null space $\mathcal{N}_p = \mathcal{O}_p^r/[\mathcal{O}_p^r \bigcap \mathcal{S}(\Theta_p)]$. The spaces $\mathcal{P}_p$ and $\mathcal{N}_p$ are finite dimensional. The triple $(\mathcal{N}_p, \mathcal{P}_p, T)$, where $T : \mathcal{N}_p \to \mathcal{P}_p$ is a linear transformation can be used as an alternative to the null-pole triple introduced above. See, e.g., [16],[5].

2. Interpolation problems

The concrete prescription of null-pole triple data in interpolation problems will be given in fixed local coordinates at the interpolation nodes. Let $z_1, \ldots, z_K$ be fixed points on the Riemann surface M and (s_j, V_j) be local coordinates at z_j with $s_j(z_j) = 0$, $j = 1, \ldots, K$. Let

$$\Upsilon_j = ((B_{\zeta_j}, A_{\zeta_j}), (A_{\pi_j}, C_{\pi_j}), S_j) \; j = 1, \ldots, K$$

be rank r admissible triples. We do allow the possibility that Υ_j consists only of a zero or pole pair. The collection

$$\mathcal{D} : \{(s_1, z_1, \Upsilon_1), \ldots, (s_K, z_K, \Upsilon_K)\} \tag{12}$$

will be referred to as an admissible rank r interpolation data set.

First Basic Interpolation Problem: Given the admissible rank r interpolation data set (12), determine whether there exists a regular meromorphic function G on M such that in a neighborhood of z_j

$$G(p) = F_j(s_j(p)),$$

where F_j is a regular $r \times r$-meromorphic matrix function at $s = 0$ having Υ_j as a null-pole triple for $j = 1, \ldots, K$ and such that at other points of M, the matrix function G is a non-singular analytic matrix function.

A solution of the *First Basic Interpolation Problem* will be presented below.

In order to discuss multiple valued solutions of our interpolation problems, it is natural to work in the environment of the universal cover. To this end let $\rho : \widetilde{M} \to M$ be "the" universal cover of M. For convenience whenever local coordinates (s, V) are chosen at a point $p \in M$ it will be assumed that $\rho^{-1}(V)$ is a disjoint collection of neighborhoods of points in $\rho^{-1}(\{p\})$ and, therefore, $s \circ \rho$ provides local coordinates at points in $\rho^{-1}(\{p\})$. The data

$$\rho^*\mathcal{D} : \{(s_1 \circ \rho, \rho^{-1}(\{z_1\}), \Upsilon_1), \ldots, (s_K \circ \rho, \rho^{-1}(\{z_K\}), \Upsilon_K)\} \tag{13}$$

will be called admissible rank r interpolation data on $\widetilde{M}$.

Second Basic Interpolation Problem: Given the admissible rank r interpolation data set (13) on $\widetilde{M}$ determine whether there exists a regular $r \times r$-meromorphic matrix function G with constant matrix factor of automorphy on $\widetilde{M}$ such that in a neighborhood of a point in $\rho^{-1}(\{p_j\})$

$$G(p) = F_j(s_j(\rho(p)))$$

where F_j is a regular $r \times r$-meromorphic matrix function at $s = 0$ having Υ_j as a null-pole triple for $j = 1, \ldots, K$ and such that at other points of $\widetilde{M}$, the matrix function G is a non-singular analytic matrix function.

A solution G to the *Second Basic Interpolation Problem* will be called a *flat solution.*

Note that these interpolation problems only depend on the similarity orbits $\mathcal{S}(\Upsilon_j)$ of the admissible triples Υ_j, $j = 1, \ldots, K$ and the data $\mathcal{D}$ given in (12) can be taken in the form

$$\mathcal{S}D : \{(s_1, z_1, \mathcal{S}(\Upsilon_1)), \ldots, (s_K, z_K, \mathcal{S}(\Upsilon_K))\}. \tag{14}$$

Indeed, it is important to recognize that the data $\mathcal{S}D$ given in (14) determines a unique matrix divisor $\Theta_{\mathcal{S}D}$ on M and, conversely, once coordinates are fixed at points in its support a matrix divisor Θ on M determines a unique set of data $\mathcal{S}D_\Theta$ of the form (14).

It is necessary to assemble interpolation data as follows: Let $z_1, \ldots, z_{N_0}$ be a list of the points z_j, where a zero pair appears in some Υ_j and $w_1, \ldots, w_{N_\infty}$ a list of the points z_j where a pole pair appears in some Υ_j. One will have $z_i = w_j$ for a pair (i,j) whenever there is a coupling matrix at $z_i = w_j$. In the sequel, we will frequently assume the data set has been split into three cases corresponding to "zero only", "pole only" or "pole-zero coupling." The points $z_1, \ldots, z_{N_0^0}$ will denote the interpolation points $z_1, \ldots, z_{N_0}$ where the data consists only of a zero pair and $z_{N_0^0+1}, \cdots, z_{N_0^0}$ the interpolation points $z_1, \ldots, z_{N_0}$ where there is a nontrivial coupling matrix; $w_1, \ldots, w_{N_\infty^0}$ will denote the interpolation points $w_1, \ldots, w_{N_\infty}$ where the data consists only of a pole pair and $w_{N_\infty^0+1}, \cdots, v_{N_\infty}$ will be a list of the interpolation points where there is a nontrivial coupling matrix. Obviously, $N_c = N_0 - N_0^0 = N_\infty - N_\infty^0$ and it can be assumed that $z_{N_0^0+j} = w_{N_\infty^0+j}$, $j = 1, \cdots, N_c$. Local coordinates at z_j will be denoted by t_j, $j = 1, \ldots, N_0$ and at w_i, by s_i, $i = 1, \ldots, N_\infty$. In addition, whenever $w_i = z_j$ we will take the local parameters to coincide: $s_i = t_j$. In the sequel, we will sometimes drop the subscripts and write $s = s(p)$ for the fixed coordinates at a node. To avoid confusion about which index is used for the associated coupling matrix, we write S_{ij} for the coupling matrix associated with points $w_i = z_j$.

3. Vector bundles and interpolation data

A rank r matrix divisor or a collection of admissible rank r interpolation data corresponds in a natural way to a rank r complex vector bundle over the Riemann surface M. We briefly describe this correspondence. First, suppose that Θ is a rank r matrix divisor and $\{V_\alpha\}_{\alpha \in A}$ is an open cover of M with the property that there is a regular $r \times r$-meromorphic matrix function L_α on V_α such that $[L_\alpha]_p$ belongs to the value of Θ at $p \in M$. The invertible holomorphic 1-cocycle $\{\Phi_{\alpha\beta}\}_{(\alpha,\beta)\in A\times A}$ given by

$$\Phi_{\alpha\beta}(p) = L_\alpha(p) L_\beta^{-1}(p), \; p \in V_\alpha \cap V_\beta \tag{15}$$

defines a rank r vector bundle E_Θ over M. Indeed, by this construction Θ corresponds to a well-defined class e_Θ of holomorphically equivalent vector bundles over M. Further, $e_\Theta = e_{\widetilde{\Theta}}$ if and only if the divisors Θ and $\widetilde{\Theta}$ are linearly equivalent. On the other hand, given admissible rank r interpolation data $\mathcal{SD}$ as in (14) one can find local solutions L_α to the interpolation problem with this data on domains V_α, where $\{V_\alpha\}_{\alpha \in A}$ covers M. The cocycle (15) defines a rank r vector bundle $E_{\mathcal{SD}}$ on M. The corresponding equivalence class of bundles will be denoted $e_{\mathcal{SD}}$. Using the above notations one has

$$e_\Theta = e_{\mathcal{SD}_\Theta} \text{ and } e_{\mathcal{SD}} = e_{\Theta_{\mathcal{SD}}}.$$

It is easy to see that the *First Basic Interpolation Problem* with data $\mathcal{D}$ has a solution if and only if the bundle $E_{\mathcal{SD}}$ is holomorphically equivalent to the trivial bundle. The *Second Basic Interpolation Problem* has a solution if and only if the bundle $E_{\mathcal{SD}}$ is holomorphically equivalent to a flat bundle.

The degree of a vector bundle V is by definition the degree of the associated determinant line bundle det V (i.e., the line bundle with transition functions

$\{\det \Phi_{\alpha\beta}\}$ where $\{\Phi_{\alpha\beta}\}$ are the transition functions for V). The degree of a line bundle in turn is the number of zeros minus the number of poles of any holomorphic section. One can show that the degree of a bundle of the form E_{SD} is the integer

$$d = d_{SD} = \sum_{j=1}^{K}(n_{\pi_j} - n_{\zeta_j}),$$

where A_{π_j} is of size $n_{\pi_j} \times n_{\pi_j}$ and A_{ζ_j} is of size $n_{\zeta_j} \times n_{\zeta_j}$, $j = 1, \ldots, K$. This follows from the connection between the null-pole triple Υ and the local Smith-McMillan form for an associated interpolant L_α (see Theorem 3.1.2 of [4]). A flat bundle necessarily has degree zero.

In the sequel it will be important to consider line bundles of degree $g-1$ which have no holomorphic sections; such bundles are characterized explicitly by the fact that their image under the Abel-Jacobi map (appropriately translated) does not lie on the divisor of the classical Riemann theta function. If we fix a base point p_0 in M, these line bundles correspond to divisors λ of the form

$$\lambda = p_1 + \cdots + p_g - p_0$$

where $\mu = p_1 + \cdots + p_g$ is a non-special divisor in the g-fold symmetric product $M^{(g)}$. The notation $\mathbb{L}_\lambda$ will be used for the line bundle corresponding to the divisor λ. When λ is as above, the condition $h^0(\mathbb{L}_\lambda) = 0$ means that there is no nonconstant meromorphic function with poles only in μ and vanishing at p_0. Any degree $g-1$ line bundle $\mathbb{L}$ satisfying $h^0(\mathbb{L}) = 0$ will be called a *non-special line bundle.* The significance of such line bundles can be seen in the following result:

Proposition 3. *Let E be a complex vector bundle of degree zero on the closed Riemann surface M of genus g. A sufficient condition that for E to be flat is the existence of a non-special line bundle $\mathbb{L}$ of degree $g-1$ such that $h^0(\mathbb{L} \otimes E) = 0$. In the case where $g = 1$, this condition is also necessary.*

Proof. Assume $E = E_1 \oplus \cdots \oplus E_J$ is a decomposition of E into indecomposable bundles E_i of rank r_i, $i = 1, \ldots, J$. Then $h^0(\mathbb{L} \otimes E_i) = 0$, $i = 1, \ldots, J$. By the Riemann-Roch Theorem

$$-h^1(\mathbb{L} \otimes E_i) = r_i(g-1) + \deg E_i + r_i(1-g) = \deg E_i, \ i = 1, \ldots, J.$$

Since $\deg E_1 + \cdots + \deg E_J = 0$, we conclude $\deg E_1 = \cdots = \deg E_J = 0$. The classical result of Weil (see, e.g., [10])implies that the bundle E is flat. In the case where $g = 1$, the result of Atiyah described earlier gives the representation of E in the form

$$E = \mathbb{L}_1 \otimes F_{h_1} \oplus \cdots \oplus \mathbb{L}_s \otimes F_{h_s}$$

where the line bundles $\mathbb{L}_1, \ldots, \mathbb{L}_s$ have degree zero. It is easy to see that there are line bundles $\mathbb{L}$ of degree zero such that $h^0(\mathbb{L} \otimes \mathbb{L}_i) = 0$, $i = 1, \ldots, s$. With such a choice of $\mathbb{L}$, $h^0(\mathbb{L} \otimes E) = 0$. This completes the proof.

Remark 1. It is not hard to construct examples where $g > 1$ and where the converse of the result in this last proposition doesn't hold. Let $\mathbb{L}_\xi, \mathbb{L}_\eta$ be line bundles on a compact Riemann surface of genus $g \geq 4$ that satisfy

$$0 < \deg \mathbb{L}_\xi = -\deg \mathbb{L}_\eta : h^1(\mathbb{L}_\xi \mathbb{L}_\eta^{-1}) \neq 0.$$

For example, $\mathbb{L}_\xi$ could be the line bundle corresponding to the divisor $\xi = p$ consisting of a single point p and $\mathbb{L}_\eta$ the line bundle corresponding to the divisor $\eta = -p_1 - p_2 + z_1$, for distinct points p_1, p_2, z_1 distinct from p. It follows from the Riemann-Roch Theorem that $h^1(\mathbb{L}_\xi \mathbb{L}_\eta^{-1}) \neq 0$. Using a non-zero element σ from $H^1(\mathbb{L}_\xi \mathbb{L}_\eta^{-1})$ one constructs an indecomposable rank 2 vector bundle E using the transition matrices

$$\Phi_{\alpha\beta} = \begin{bmatrix} \xi_{\alpha\beta} & \sigma_{\alpha\beta} \\ 0 & \eta_{\alpha\beta} \end{bmatrix},$$

where $\{\xi_{\alpha\beta}\}$ and $\{\eta_{\alpha\beta}\}$ are transition functions for the bundles $\mathbb{L}_\xi$ and $\mathbb{L}_\eta$ relative to a suitable cover of M. Let $\mathbb{L}_\lambda$ be a line bundle of degree $g-1$. Then $\mathbb{L}_\lambda \otimes E$ has a "triangular" form with 1,1-entry $\mathbb{L}_{\lambda+\xi}$. Since this line bundle has sections, we conclude that $h^0(\mathbb{L}_\lambda \otimes E) \neq 0$, for any line bundle $\mathbb{L}_\lambda$ of degree $g-1$.

Remark 2. In case $g > 2$, there are examples of semi-stable degree zero bundles E satisfying $h^0(\mathbb{L} \otimes E) > 0$, for every non-special line bundle $\mathbb{L}$ of degree $g-1$. Recall that a bundle E is called *semi-stable* in case

$$\mu_F \equiv \frac{\deg F}{rank F} \leq \mu_E = \frac{\deg E}{rank E}$$

for all subbundles $F \subset E$ (see, [12], [14]). Every semi-stable bundle of degree zero is flat. This again follows from the aforementioned result of Weil. Indeed, if E is semi-stable of degree zero, then $\deg E_i \leq 0$ for each summand in the decomposition $E = E_1 \oplus \cdots \oplus E_J$, where $E_1, \ldots, E_J$ are indecomposable. Since $\sum_{i=0}^J \deg E_i = \deg E = 0$, we have $\deg E_i = 0$, $i = 1, \ldots, J$, and, consequently, E is flat. An example, of a semi-stable bundle E of degree zero such that $h^0(\mathbb{L} \otimes E) > 0$ for all degree $g-1$ line bundles $\mathbb{L}$ can be found in [11].

Remark 3. In the case M has genus one, every flat bundle E is semi-stable. This follows from the representation $E = \mathbb{L}_1 \otimes F_{h_1} \oplus \cdots \oplus \mathbb{L}_s \otimes F_{h_s}$ of Atiyah and the fact that direct sums of semi-stable bundles with the same slope are semi-stable (see, e.g. [14]).

4. The flat case

Let w be a fixed point of M and (s, V) be fixed local coordinates at w, where as usual we assume $s(w) = 0$. Then with λ satisfying $h^0(\mathbb{L}_\lambda) = 0$, for any $k \geq 1$ an integer, there is a unique meromorphic function f_{kw}^λ whose divisor satisfies $(f_{kw}^\lambda) + \lambda + kw \geq 0$ and such that in the coordinate s, this function f_{kw}^λ has principal Laurent part at $s = 0$ equal to $\frac{1}{s^k(p)}$. To see this, note that the dimension of the space of meromorphic functions f whose divisor satisfies $(f) + \lambda + kw \geq 0$, which is equal to the dimension $h^0(\mathbb{L}_{\lambda+kw})$ of the space of holomorphic sections of the line bundle $\mathbb{L}_{\lambda+kw}$ corresponding to the divisor $\lambda + kw$, equals k. To see this note first from the Riemann-Roch theorem $h^0(\mathbb{L}_{\lambda+kw}) \geq \deg(\lambda + kw) - (g-1) = k$. On the other hand, it can be easily verified from the assumption $h^0(\mathbb{L}_\lambda) = 0$,that $h^0(\mathbb{L}_{\lambda+kw}) \leq k$. Thus there exists a one-dimensional space of meromorphic functions f with $(f) + \lambda + kw \geq 0$ with exactly a kth-order pole at w; by normalizing the principal part, we obtain a uniquely defined function f^λ_{kw}.

Suppose that A is the $n \times n$ Jordan cell

$$\begin{bmatrix} 0 & 1 & \cdots & 0 \\ 0 & 0 & \ddots & \vdots \\ \vdots & \cdots & \cdots & 1 \\ 0 & \cdots & 0 & 0 \end{bmatrix}.$$

We introduce the $n \times n$ matrix function

$$f^\lambda_{w,A} = \begin{bmatrix} f^\lambda_w & f^\lambda_{2w} & \cdots & f^\lambda_{nw} \\ 0 & f^\lambda_w & \ddots & \vdots \\ \vdots & \ddots & \ddots & f^\lambda_{2w} \\ 0 & \cdots & 0 & f^\lambda_w \end{bmatrix}.$$

We extend this definition to an arbitrary nilpotent matrix by following the conventions

$$f^\lambda_{w,SAS^{-1}} = Sf^\lambda_{w,A}S^{-1} \text{ and } f^\lambda_{w,A_1\oplus A_2} = f^\lambda_{w,A_1} \oplus f^\lambda_{w,A_2},$$

where, S is an invertible matrix. Then for any nilpotent matrix N, the difference

$$f^\lambda_{w,N}(p) - (s(p)I - N)^{-1}$$

is analytic at $p = w$. Note this last identity reflects the fact that the local (left) pole pair of $f^\lambda_{w,N}$ at w in the coordinate s has the form $(C_\pi, A_\pi) = (I_r, N)$, where N is $r \times r$.

We will provide a concrete description of the space of holomorphic sections of the bundle

$$\mathbb{L}_\lambda \otimes E^*_{\mathcal{D}}$$

where $\mathbb{L}_\lambda$ is a non-special degree $g-1$ line bundle and $E^*_{\mathcal{D}}$ is the dual of the bundle $E_{\mathcal{D}}$. The divisor λ determining the line bundle $\mathbb{L}_\lambda$ will be assumed to have the form $\lambda = p_1 + \cdots + p_g - p_0$, where the points $p_0, p_1, \ldots, p_g$ are distinct from the interpolation nodes and $\mu = p_1 + \cdots + p_g$ is a non-special divisor.

Let $\{\Phi_{\alpha\beta}\}$ be a holomorphic cocycle determining a vector bundle E and $\Phi_{\alpha\beta} = L_\alpha L_\beta^{-1}$ a trivialization of E by a family $\{L_\alpha\}_{\alpha\in A}$ of regular $r \times r$-meromorphic matrix functions relative to the open cover $\{V_\alpha\}_{\alpha\in\mathcal{A}}$. We introduce the collection of vector-valued meromorphic functions of the form

$$L^\vee(E) = \{\mathbf{h} \in \mathcal{M}^r(M) : [\mathbf{h}]_p \in \mathcal{O}^r_p[L_\alpha]_p,\ p \in V_\alpha\}.$$

Note that the condition $[\mathbf{h}]_p = [\mathbf{h}_\alpha]_p[L_\alpha]_p$, where $\mathbf{h}_\alpha$ is holomorphic "transposes" to the condition $(L_\alpha^t)^{-1}L_\beta^t\mathbf{h}_\beta^t = \mathbf{h}_\alpha^t$, ("$t$" denotes the transpose operation) on the intersection $V_\alpha \bigcap V_\beta$. Thus the space $L^\vee(E)$ is naturally isomorphic to the space $\Gamma(E^*)$ of holomorphic sections of the dual bundle E^* determined by the cocycle $\{(\Phi_{\alpha\beta}^t)^{-1}\}$.

In order to separate the role of λ, we will assume that $\mathbb{L}_\lambda$ and $E_\mathcal{D}$ are trivialized separately relative to the open cover $\{V_\alpha\}_{\alpha\in\mathcal{A}}$. That is, it will be assumed that $k_\alpha, \alpha \in \mathcal{A}$ is a family of scalar meromorphic functions which interpolates the divisor $\lambda = p_1 + \cdots + p_g - p_0$ and that $L_\alpha, \alpha \in \mathcal{A}$ is a family of $r \times r$ meromorphic matrix functions which locally interpolates the divisor $\mathcal{D} : \{(s_1, z_1, \Upsilon_1), \ldots, (s_K, z_K, \Upsilon_K)\}$. The notation L_α^λ will be used for the matrix functions $L_\alpha^\lambda = k_\alpha^{-1}L_\alpha, \alpha \in \mathcal{A}$, and $E_\mathcal{D}^\lambda$ for the corresponding vector bundle $\mathbb{L}_{-\lambda} \otimes E_\mathcal{D}$.

We first claim that the elements in $L^\vee(E_\mathcal{D}^\lambda)$ necessarily have the form

$$\mathbf{h} = \sum_{i=1}^{N_\infty} \mathbf{u}_j f_{w_i,A_{\pi_i}}^\lambda C_{\pi_i}, \tag{16}$$

where $\mathbf{u}_i$ is an n_{π_i}-dimensional row vector for $i = 1, \ldots, N_\infty$. One sees this last claim as follows: Let $\mathbf{h}$ be an element of $L^\vee(E_\mathcal{D}^\lambda)$. Then for each point w_i, we know that $[\mathbf{h}]_{w_i}$ is in the null-pole subspace $\mathcal{O}_{w_i}^r[L_\alpha^\lambda]_{w_i}$. From the formula (10) we see that there is a (necessarily unique) vector $\mathbf{u}_i$ such that $\mathbf{h}(s^{-1}(z))$ and $\mathbf{u}_i(zI - A_{\pi_i})^{-1}C_{\pi_i}$ have the same principal part at $z = 0$ (here s denotes local coordinates at $p = w_i$). This implies $\mathbf{h} - \mathbf{u}_i f_{w_i,A_{\pi_i}}^\lambda C_{\pi_i}$ has an analytic continuation to $p = w_i$. If we set $\mathbf{g} = \sum_{i=1}^{N_\infty} \mathbf{u}_i f_{w_i,A_{\pi_i}}^\lambda C_{\pi_i}$, then it follows that $\mathbf{h} - \mathbf{g}$ has analytic continuation through each of the points $w_1, \ldots, w_{N_\infty}$. From the construction, it is now easy to check that each scalar component $h_k - g_k$ $(k = 1, \ldots, r)$ of the vector function $\mathbf{h} - \mathbf{g}$ has divisor $(h_k - g_k)$ satisfying

$$(h_k - g_k) + \lambda \geq 0. \tag{17}$$

Since $h^0(\mathbb{L}_\lambda) = 0$, we conclude that $\mathbf{h} = \mathbf{g}$ and, therefore, $\mathbf{h}$ has the claimed form.

The following lemmas will be used to complete the description of $L^\vee(E_\mathcal{D}^\lambda)$.

Lemma 4. *Suppose $w_i \neq z_j$, $j = 1, \ldots, N_0$. If $\mathbf{h} \in \mathcal{M}^r$, then $[\mathbf{h}]_{w_i} \in \mathcal{O}_{w_i}^r[L_\alpha^\lambda]_{w_i}$ if and only if*

$$[\mathbf{h}]_{w_i} - [\mathbf{u}_i f_{w_i,A_{\pi_i}}^\lambda C_{\pi_i}]_{w_i}, \tag{18}$$

has analytic continuation at $p = w_i$ for some row vector $\mathbf{u}_i \in \mathbb{C}^{n_{\pi_i}}$.

Proof. This result follows immediately from (10) and the fact that $f_{w_i,A_{\pi_i}}^\lambda(p) - (s_i(p)I - A_{\pi_i})^{-1}$ is analytic at w_i.

Before stating the next lemma we introduce the notations

$$\Gamma_{ij}^\lambda = -\operatorname{res}_{z=0}[f_{w_i,A_{\pi_i}}^\lambda(t_j^{-1}(z))C_{\pi_i}B_{\zeta j}(z - A_{\zeta_j})^{-1}],$$

for all i, j with $z_j \neq w_i$,

$$\Gamma_{ij}^{\lambda} = S_{ij} - res_{z=0}[\{f_{w_i,A_{\pi_i}}^{\lambda}(s_i^{-1}(z)) - (z - A_{\pi_i})^{-1}\}C_{\pi_i}B_{\zeta_j}(z - A_{\zeta j})^{-1}],$$

for i, j with $z_j = w_i$, and

$$\Gamma^{\lambda} = \left[\Gamma_{ij}^{\lambda}\right]_{N_0 \times N_{\infty}}. \tag{19}$$

Lemma 5. *If* $\mathbf{h}$ *has the form (16), then for* $1 \leq j \leq N_0$, *the row vector function* $[\mathbf{h}]_{z_j}$ *is in* $\mathcal{O}_{z_j}^r[L_\alpha^\lambda]_{z_j}$ *if and only if*

$$\sum_{i=0}^{N_{\infty}} \mathbf{u}_i \Gamma_{ij}^{\lambda} = \mathbf{0}.$$

Proof. The arguments for the cases $1 \leq j \leq N_0^0$ and $N_0^0 + 1 \leq j \leq N_0$ are similar. We will only present the details for the latter case. Suppose j satisfies $N_0^0 + 1 \leq j \leq N_0$. Without loss of generality assume that $w_1 = z_j$. Write $\mathbf{h}$ in the form

$$\begin{aligned}
\mathbf{h}(s_1^{-1}(z)) &= \mathbf{u}_1 f_{w_1,A_{\pi_1}}^{\lambda}(s_1^{-1}(z))C_{\pi_1} + \sum_{i=2}^{N_{\infty}} \mathbf{u}_i f_{w_i,A_{\pi_i}}^{\lambda}(s_1^{-1}(z))C_{\pi_i} \\
&= \mathbf{u}_1(z - A_{\pi_1})^{-1}C_{\pi_1} + \mathbf{u}_1[f_{w_1,A_{\pi_1}}^{\lambda}(s_1^{-1}(z)) - (z - A_{\pi_1})^{-1}]C_{\pi_1} \\
&\quad + \sum_{i=2}^{N_{\infty}} \mathbf{u}_i f_{w_i,A_{\pi_i}}^{\lambda}(s_1^{-1}(z))C_{\pi_i} \\
&= \mathbf{u}_1(z - A_{\pi_1})^{-1}C_{\pi_1} + \mathbf{k}(z),
\end{aligned}$$

where $\mathbf{k}$ is analytic at $z = 0$. Using the description of $\mathcal{O}_{z_i}^r[L_\alpha^\lambda]_{z_i}$ in (10) we learn that $\mathbf{h}$ lies in this subspace if and only if

$$\begin{aligned}
\mathbf{u}_1 S_{1j} &= \text{res}_{z=0}(\mathbf{k}(z)B_{\zeta_j}(z - A_{\zeta_j})^{-1}) \\
&= \mathbf{u}_1 \,\text{res}_{z=0}[\{f_{w_1,A_{\pi_1}}^{\lambda}(s_1^{-1}(z)) - (z - A_{\pi_1})^{-1}\}C_{\pi_1}B_{\zeta_j}(z - A_{\zeta_j})^{-1}] \\
&\quad + \sum_{i=2}^{N_{\infty}} \mathbf{u}_i \,\text{res}_{z=0}[f_{w_i,A_{\pi_i}}^{\lambda}(s_1^{-1}(z))C_{\pi_i}B_{\zeta_j}(z - A_{\zeta_j})^{-1}] \\
&= \mathbf{u}_1[S_{1j} - \Gamma_{i1}^{\lambda}] - \sum_{i=2}^{N_{\infty}} \mathbf{u}_i \Gamma_{ij}^{\lambda}.
\end{aligned}$$

The result follows.

Theorem 6. *Let* $\mathbb{L}_\lambda$ *be a line bundle of degree* $g - 1$ *determined by the divisor* $p_1 + \cdots + p_g - p_0$, *where* $\mu = p_1 + \cdots + p_g$ *is a non-special divisor and let the pole-zero interpolation data be given by* $\mathcal{D} : \{(s_1, z_1, \Upsilon_1), \ldots, (s_K, z_K, \Upsilon_K)\}$. *The vector space of sections of the vector bundle* $\mathbb{L}_\lambda \otimes E_{\mathcal{D}}^*$ *is isomorphic to the collection of row vector meromorphic functions of the form* $\mathbf{h} = \sum_{i=1}^{N_{\infty}} \mathbf{u}_i f_{w_i,A_{\pi_i}}^{\lambda} C_{\pi_i}$, *where the row vector* $\mathbf{u} = [\mathbf{u}_1, \ldots, \mathbf{u}_{N_{\infty}}]$ *satisfies* $\mathbf{u}\Gamma^{\lambda} = \mathbf{0}$. *In particular,* $h^0(\mathbb{L}_\lambda \otimes E_{\mathcal{D}}^*)$ *equals the dimension of the left-kernel of* Γ^{λ}.

Proof. The above lemmas imply that the space $L^\vee(E_{\mathcal{D}}^\lambda)$ is isomorphic to the collection of row vector meromorphic functions of the form $\mathbf{h} = \sum_{i=1}^{N_\infty} \mathbf{u}_i f_{w_i,A_{\pi_i}}^\lambda C_{\pi_i}$, where the row vector $\mathbf{u} = [\mathbf{u}_1, \ldots, \mathbf{u}_{N_\infty}]$ satisfies $\mathbf{u}\Gamma^\lambda = \mathbf{0}$. The result now follows from the isomorphism between $L^\vee(E_{\mathcal{D}}^\lambda)$ and $\Gamma(\mathbb{L}_\lambda \otimes E_{\mathcal{D}}^*)$ which was described above. This ends the proof.

The following result represents a partial solution to the *Second Basic Interpolation Problem* and is an immediate corollary of the last theorem and Proposition 3:

Corollary 7. *Let $\mathcal{D} : \{(s_1, z_1, \Upsilon_1), \ldots, (s_K, z_K, \Upsilon_K)\}$ be given pole-zero interpolation data of degree zero on a closed Riemann surface M with genus $g \geq 1$. If there exists a divisor $\lambda = p_1 + \cdots + p_g - p_0$ of degree $g-1$ with $h^0(\mathbb{L}_\lambda) = 0$ such that the matrix Γ^λ is invertible, then there is an $r \times r$ meromorphic function F on the universal cover $\rho : \widetilde{M} \to M$ with flat factor of automorphy interpolating the pole-zero data $\rho^*\mathcal{D}$. In the case, where $g = 1$, such a matrix function F exists if and only if Γ^λ is invertible for some non-zero degree $g-1$ divisor λ with $h^0(\mathbb{L}_\lambda) = 0$.*

5. The single valued case

In this section we solve the *First Basic Interpolation Problem* which was introduced above. We begin with the remark that a necessary condition for the existence of a solution is that the bundle $E_{\mathcal{D}}$ be trivial and, consequently, for every degree $g-1$ divisor λ with $h^0(\mathbb{L}_\lambda) = 0$, we have $h^0(\mathbb{L}_\lambda \otimes E_{\mathcal{D}}^*) = 0$. Thus for all such divisors λ, the matrix Γ^λ is invertible.

In order to simplify the discussion, we will assume the Riemann surface is presented as a fundamental domain R_0 on the universal cover $\rho : \widetilde{M} \to M$ and that the interpolation points $z_1, \ldots, z_K$ as well as points in the divisors $\lambda = p_1 + \cdots + p_g - p_0$ are in R_0. The functions $\underset{\sim}{f_{kw}^\lambda}$ will be assumed to be functions in global coordinates on the universal cover $\widetilde{M}$ which can be assumed to be $\mathbb{C}$ or the unit disc $\mathbb{D}$. The pole-zero interpolation data will be taken in the form

$$\mathcal{D} : \{(z_1, \Upsilon_1), \ldots, (z_K, \Upsilon_K)\},$$

where we suppress writing the coordinates $s_j(u) = u - z_j$ at the points z_j, $j = 1, \ldots, K$.

Fix a divisor $\lambda = p_1 + \cdots + p_g - p_0$ satisfying $h^0(\mathbb{L}_\lambda) = 0$, with $p_0, \ldots, p_g$ distinct from $z_1, \ldots, z_K$. If there exists a solution F of the first basic interpolation problem, it can be assumed to satisfy $F(p_0) = I_r$ and, therefore, has the form

$$F = I_r + \sum_{i=1}^{K} U_i f_{w_i,A_{\pi_i}}^\lambda C_{\pi_i}, \tag{20}$$

where U_i are $r \times n_{\pi_i}$-matrices. This last remark uses the fact that $h^0(\mathbb{L}_\lambda) = 0$. The rows of F are obviously in $\mathcal{O}_{z_i}^r[F]_{z_i}$ for $i = 1, \ldots, K$ and as in the proof of Theorem 4, one sees that

$$[U_1, \ldots, U_K]\Gamma^\lambda = [B_{\zeta_1}, \ldots, B_{\zeta_K}] \tag{21}$$

or, equivalently, the matrices $U_1, \ldots, U_K$, are given by

$$[U_1, \ldots, U_K] = [B_{\zeta_1}, \ldots, B_{\zeta_K}](\Gamma^\lambda)^{-1}.$$

The following proposition is an immediate consequence of the above discussion.

Proposition 8. *Let $\mathcal{D} : \{(z_1, \Upsilon_1), \ldots, (z_K, \Upsilon_K)\}$ be given pole-zero interpolation data and $\lambda = p_1 + \cdots + p_g - p_0$ a divisor satisfying $h^0(\mathbb{L}_\lambda) = 0$, with $p_0, \ldots, p_g$ distinct from $z_1, \ldots, z_K$. The matrix Γ^λ is invertible if and only if there exists an $r \times r$-meromorphic matrix function satisfying:*

- *F and F^{-1} are analytic off $\{z_1, \ldots, z_K, p_0, \ldots, p_g\}$ with $F(p_0) = I_r$.*
- *F interpolates the divisor $\mathcal{D}$ at the points $z_1, \ldots, z_K$.*
- *The entries of F have at most simple poles at $p_1, \ldots, p_g$.*

Further, when the matrix Γ^λ is invertible, the unique F satisfying these last conditions is given by (20) with the matrices $U_1, \ldots, U_K$ given by (21).

In order that the matrix function F described in the preceding proposition interpolate only the data $\mathcal{D}$ one must ensure that the residues of F at $p_1, \ldots, p_g$ be zero. This involves additional linear conditions on the matrices $U_1, \ldots, U_K$ which we now describe. Introduce the notation

$$R_{ij} = \mathrm{res}_{u=p_j}[f^\lambda_{w_i, A_{\pi_i}}(u) C_{\pi_i}], \ i = 1, \ldots, K; j = 1, \ldots, g.$$

The matrix function F given by (20) has the property that $\mathrm{res}_{u=p_j}[F(u)]$, $j = 1, \ldots, g$ is the zero matrix if and only if

$$\begin{bmatrix} U_1 & \cdots & U_K \end{bmatrix} \begin{bmatrix} R_{11} & \cdots & R_{1g} \\ \vdots & & \vdots \\ R_{K1} & \cdots & R_{Kg} \end{bmatrix} = UR = B(\Gamma^\lambda)^{-1} R = 0,$$

where we are using the notations

$$R = \begin{bmatrix} R_{11} & \cdots & R_{1g} \\ \vdots & & \vdots \\ R_{K1} & \cdots & R_{Kg} \end{bmatrix}, \ U = \begin{bmatrix} U_1 & \cdots & U_K \end{bmatrix} \text{ and } B = \begin{bmatrix} B_{\zeta_1} & \cdots & B_{\zeta_K} \end{bmatrix}. \tag{22}$$

The following theorem is the main result of this section. It represents a generalization of the genus one result in [4] and also generalizes results in [2] and [8] to the case where the poles and zeros have multiplicity.

Theorem 9. *Let $\mathcal{D} : \{(z_1, \Upsilon_1), \ldots, (z_K, \Upsilon_K)\}$ be a given divisor on M. There is an $r \times r$-meromorphic matrix function F interpolating $\mathcal{D}$ if and only if for some degree $g-1$ divisor λ with $h^0(\mathbb{L}_\lambda) = 0$ the matrix Γ^λ given in (19) is invertible and $B(\Gamma^\lambda)^{-1} R = 0$, where the matrices R and B are given by (22). In this case the unique solution F of the interpolation problem satisfying $F(p_0) = I_r$ is given by (20).*

6. The case of genus 1

As mentioned in the introduction, in the case of simple data, the matrix Γ^λ has a nice form when M is of genus 1 and is realized in the form $M = \mathbb{C}/(\mathbb{Z} + \tau\mathbb{Z})$, where $\operatorname{Im}\tau > 0$. In this section, we will give more details on this representation of Γ^λ when M has genus 1.

The divisor data $\mathcal{D}$ will be collected as follows. Let

$$z_1, \ldots, z_N; \zeta_1, \ldots, \zeta_K; w_1, \ldots, w_N$$

be distinct points in the complex plane lying in a fundamental domain

$$R_0 = \{u = x + i\tau y : 0 \leq x, y < 1\})$$

for $\mathbb{C}/(\mathbb{Z} + \tau\mathbb{Z})$. We will write the data $\mathcal{D}$ in the form

$$\mathcal{D} : (z_1, \underline{b}_1), \ldots, (z_N, \underline{b}_N); (w_1, \mathbf{c}_1), \ldots, (w_N, \mathbf{c}_N);$$
$$(\zeta_1, (\underline{\beta}_1, \gamma_1, S_1)), \ldots, (\zeta_k, (\underline{\beta}_K, \gamma_K, S_K))$$

where the data $(z_1, \underline{b}_1)$, $\ldots$, $(z_N, \underline{b}_N)$ consists of only simple right zero data, $(w_1, \mathbf{c}_1)$, $\ldots$, $(w_N, \mathbf{c}_N)$ consists of only simple left pole data and $(\zeta_1$, $(\underline{b}_{N+1}$, $\mathbf{c}_{N+1}$, $S_1))$, $\ldots$, $(\zeta_1$, $(\underline{b}_{N+K}$, $\mathbf{c}_{N+K}$, $S_K))$ consists of data at points $\zeta_1, \ldots, \zeta_K$, where we have pole zero coupling numbers $S_1, \ldots, S_K$. In this notation, we suppress writing the 1×1 zero matrices A_{ζ_i} and A_{π_j}. Moreover, $\mathbf{c}_j \underline{b}_j = 0$, $j = N+1, \ldots, N+K$.

The $r \times r$-meromorphic function $F = F(u)$ on $\mathbb{C}$ will solve the interpolation problem with data $\mathcal{D}$ in case:

- The matrix function F is holomorphic and invertible off
$$\{z_1, \ldots, z_N, w_1, \ldots, w_N, \zeta_1, \ldots, \zeta_K\} + \mathbb{Z} + \tau\mathbb{Z}.$$
- The only poles of entries of the matrix function F are at most simple poles at points in $\{w_1, \ldots, w_N, \zeta_1, \ldots, \zeta_K\} + \mathbb{Z} + \tau\mathbb{Z}$.
- The matrix function F is holomorphic at $z_i + \mathbb{Z} + \tau\mathbb{Z}$ and $\underline{b}_i$ spans $\ker[F]$ at these points.
- The matrix function F^{-1} is holomorphic at $w_j + \mathbb{Z} + \tau\mathbb{Z}$ and $\mathbf{c}_j$ spans the left kernel of F^{-1} at these points.
- At $u = \zeta_i + \mathbb{Z} + \tau\mathbb{Z}$, the singular subspace $\mathcal{O}_u^r[F]_u$ has the description

$$\mathcal{O}_u^r[F]_u = \{\frac{\mu \mathbf{c}_{N+i}}{z-u} + \mathbf{k}(z) : \mu \in \mathbb{C}, \mathbf{k} \in \mathcal{O}_u^r \text{ such that } \mu S_i = \operatorname{res}_{z=u}\left[\frac{\mathbf{k}(z)\underline{b}_{N+i}}{z-u}\right]\}.$$

Given distinct p_0, p_1 in the fundamental domain R_0 the divisor $\lambda = p_1 - p_0$ satisfies $h^0(\mathbb{L}_\lambda) = 0$. Indeed, by fixing p_0 and varying p_1, the divisors $\lambda = p_1 - p_0$ realize every degree zero divisor λ with $h^0(\mathbb{L}_\lambda) = 0$. The functions f_w^λ have a very explicit form in terms of the function

$$\theta_*(u) = \sum_{n \in \mathbb{Z}} \exp\left\{2\pi i[\frac{1}{2}(n+\frac{1}{2})\tau(n+\frac{1}{2}) + (n+\frac{1}{2})(u+\frac{1}{2})]\right\}.$$

Note that the function θ_* has the automorphic behavior

$$\theta_*(u+m+n\tau) = \exp\{\pi i(m-n-n^2\tau)\}\exp\{-2\pi inu\}\theta_*(u),\ m,n \in \mathbb{Z}.$$

If w in R_0 is distinct from the points p_0, p_1, for any constant C_w^λ the function

$$f_w^\lambda(u) = C_w^\lambda \frac{\theta_*(u-p_0)\theta_*(u+p_0-p_1-w)}{\theta_*(u-w)\theta_*(u-p_1)}$$

is single valued with divisor $(f_w^\lambda) = p_0 + q - w - p_1$, where $q = p_1 + w - p_0$. The choice

$$C_w^\lambda = \frac{\theta_*'(w)\theta_*(w-p_1)}{\theta_*(w-p_0)\theta_*(p_0-p_1)}$$

normalizes f_w^λ so that it has the requisite principal part at $u = w$. The matrix Γ^λ has the form

$$\Gamma^\lambda = \begin{bmatrix} \Gamma_{11}^\lambda & \Gamma_{12}^\lambda \\ \Gamma_{21}^\lambda & \Gamma_{22}^\lambda \end{bmatrix}$$

where

$$\Gamma_{11}^\lambda = \begin{bmatrix} -f_{w_1}^\lambda(z_1)\mathbf{c}_1\underline{b}_1 & \cdots & -f_{w_1}^\lambda(z_N)\mathbf{c}_1\underline{b}_N \\ \vdots & & \vdots \\ -f_{w_N}^\lambda(z_1)\mathbf{c}_N\underline{b}_1 & \cdots & -f_{w_N}^\lambda(z_N)\mathbf{c}_N\underline{b}_N \end{bmatrix}$$

$$\Gamma_{12}^\lambda = \begin{bmatrix} -f_{w_1}^\lambda(\zeta_1)\mathbf{c}_1\underline{b}_{N+1} & \cdots & -f_{w_1}^\lambda(\zeta_K)\mathbf{c}_N\underline{b}_{N+K} \\ \vdots & & \vdots \\ -f_{w_N}^\lambda(\zeta_1)\mathbf{c}_N\underline{b}_{N+1} & \cdots & -f_{w_N}^\lambda(\zeta_K)\mathbf{c}_N\underline{b}_{N+K} \end{bmatrix}$$

$$\Gamma_{21}^\lambda = \begin{bmatrix} -f_{\zeta_1}^\lambda(z_1)\mathbf{c}_{N+1}\underline{b}_1 & \cdots & -f_{\zeta_1}^\lambda(z_N)\mathbf{c}_{N+1}\underline{b}_N \\ \vdots & & \vdots \\ -f_{\zeta_K}^\lambda(z_1)\mathbf{c}_{N+K}\underline{b}_1 & \cdots & -f_{\zeta_K}^\lambda(z_N)\mathbf{c}_{N+K}\underline{b}_N \end{bmatrix}$$

$$\Gamma_{22}^\lambda = \begin{bmatrix} S_1 & S_2 - r_{\zeta_1}^\lambda(\zeta_2)\mathbf{c}_{N+1}\underline{b}_{N+2} & \cdots & S_K - r_{\zeta_1}^\lambda(\zeta_K)\mathbf{c}_{N+1}\underline{b}_{N+K} \\ S_1 - r_{\zeta_2}^\lambda(\zeta_1)\mathbf{c}_{N+2}\underline{b}_{N+1} & S_2 & \cdots & S_K - r_{\zeta_2}^\lambda(\zeta_K)\mathbf{c}_{N+2}\underline{b}_{N+K} \\ \vdots & \vdots & & \vdots \\ S_1 - r_{\zeta_K}^\lambda(\zeta_1)\mathbf{c}_{N+K}\underline{b}_{N+1} & S_2 - r_{\zeta_K}^\lambda(\zeta_2)\mathbf{c}_{N+K}\underline{b}_{N+2} & \cdots & S_K \end{bmatrix}$$

with $r_{\zeta_i}^\lambda(u) = f_{\zeta_i}^\lambda(u) - \frac{1}{u-\zeta_i}$, $i = 1, \ldots, K$.

It follows immediately from Corollary 7 that the interpolation problem corresponding to the data $\mathcal{D}$ has a flat solution if and only if Γ^λ is invertible for some non-zero λ. In this case the zeros of $\det(\Gamma^\lambda)$ in $R_0\backslash\{0\}$ correspond precisely to the non-trivial divisors $\lambda_1, \dots, \lambda_s$ providing the decomposition

$$E_{\mathcal{D}} = \mathbb{L}_{\lambda_1} \otimes F_{h_1} \oplus \cdots \oplus \mathbb{L}_{\lambda_s} \otimes F_{h_s}$$

of Atiyah which was used earlier (notice that each Atiyah bundle F_h is equivalent to its dual). Moreover, the dimension of the kernel of Γ^λ counts the number of summands in this Atiyah decomposition where $\mathbb{L}_\lambda$ appears as a factor. In particular, the bundle $E_{\mathcal{D}}$ is equivalent to a direct sum of non-trivial line bundles if and only if $\det(\Gamma^\lambda)$ has r zeros (counting multiplicity) in $R_0\backslash\{0\}$. At the other extreme, $E_{\mathcal{D}}$ is equivalent to a direct sum of Atiyah bundles F_h if and only if $\det(\Gamma^\lambda)$ doesn't vanish on $R_0\backslash\{0\}$.

In the case where there is no pole-zero coupling ($K = 0$), the $N \times N$-matrix Γ^λ has the simpler form Γ^λ_{11}, with ij-entry

$$\frac{\theta_*'(w_i)\theta_*(w_i - p_1)}{\theta_*(w_i - p_0)\theta_*(\lambda)} \frac{\theta_*(w_i - z_j + \lambda)\mathbf{c}_i \underline{b}_j}{\theta_*(w_i - z_j)} \frac{\theta_*(z_j - p_0)}{\theta_*(z_j - p_1)}.$$

Clearly, the zeros of $\det \Gamma^\lambda$ coincide with the zeros of the determinant of the matrix function

$$\Gamma^\lambda_0 = \left[\frac{\theta_*(w_i - z_j + \lambda)\mathbf{c}_i \underline{b}_j}{\theta_*(w_i - z_j)}\right]_{N\times N}.$$

These last remarks complete the proof of Theorem 1.

References

[1] M.F. Atiyah, Vector bundles over an elliptic curve, *Proc. London Math. Soc.* **7** (1957), 414–452.

[2] Ball, Joseph A.; Clancey, Kevin F., Interpolation with meromorphic matrix functions. *Proc. Amer. Math. Soc.* 121 (1994), no. 2, 491–496.

[3] Ball, Joseph A.; Clancey, Kevin F.; Vinnikov, Victor, Linear equivalence of matrix divisors on Riemann surfaces, in preparation.

[4] Ball, Joseph A.; Gohberg, Israel; Rodman, Leiba, *Interpolation of Rational Matrix Functions. Operator Theory: Advances and Applications*, 45. Birkhäuser Verlag, Basel, 1990.

[5] J.A. Ball, M. Rakowski and B. Wyman, Coupling operators, Wedderburn-Forney spaces, and generalized inverses, *Linear Algebra and its Applications*, 203–204 (1994), 111–138.

[6] J.A. Ball and A.C.M. Ran, Local inverse spectral problems for rational matrix functions, *Integral Equations and Operator Theory* **10** (1987), 349–415.

[7] Ball, Joseph A.; Vinnikov, Victor, Zero-pole interpolation for meromorphic matrix functions on an algebraic curve and transfer functions of 2*D* systems. *Acta Appl. Math.* 45 (1996), no. 3, 239–316.

[8] Ball, Joseph A.; Vinnikov, Victor, Zero-pole interpolation for matrix meromorphic functions on a compact Riemann surface and a matrix Fay trisecant identity. *Amer. J. Math.* 121 (1999), no. 4, 841–888.

[9] I. Gohberg, M.A. Kaashoek, L. Lerer and L. Rodman,Minimal divisors of rational matrix functions with prescribed zero and pole structure, in *Topics in Operator Theory, Systems and Networks* (Ed. H. Dym and I. Gohberg), *Operator Theory: Advances and Applications*, 12, Birkhäuser Verlag, Basel-Berlin-Boston, 1984, pp. 241–275.

[10] R.C. Gunning, *Lectures on Vector Bundles over Riemann Surfaces*, Princeton University Press, Princeton, 1967.

[11] Raynaud, Michel, Sections des fibrés vectoriels sur une courbe. (French) [Sections of vector bundles over a curve.] *Bull. Soc. Math. France* 110 (1982), no. 1, 103–125.

[12] C.S. Seshadri, *Fibrés Vectoriels sur les Courbes Algébriques*, Astérisque #96, Société Mathématique de France, Paris, 1982.

[13] Tjurin, A.J.; Classification of n-dimensional vector bundles over an algebraic curve of arbitrary genus (Russian). Izv. Akad. Nauk SSSR Ser. Mat. 30 (1966), 1353–1366.

[14] Tu, Loring; Semistable bundles over an elliptic curve. *Advances in Mathematics* 98 (1993), 1–26.

[15] A. Weil; Généralization des fonctions abéliennes, *J. math. pures appl.* **17** (1938), 47–87.

[16] Wyman, Bostwick F.; Sain, Michael K.; Conte, Giuseppe; Perdon, Anna-Maria; On the zeros and poles of a transfer function. *Linear Algebra and its Applications*, 122/123/124 (1989), 123–144.

Joseph A. Ball
Department of Mathematics
Virginia Tech
Blacksburg, Virginia 24061, USA
e-mail: `ball@math.vt.edu`

Kevin F. Clancey
Department of Mathematics
University of Georgia
Athens, Georgia 30602, USA
Current address:
Department of Mathematics
University of Louisville
Louisville, Kentucky 40292, USA
e-mail: `k.clancey@louisville.edu`

Victor Vinnikov
Department of Mathematics
Ben Gurion University of the Negev
POB 653, 84105 Beer-Sheva, Israel
e-mail: `vinnikov@cs.bgu.ac.il`

Operator Theory:
Advances and Applications, Vol. 134, 157–185

On Realizations of Rational Matrix Functions of Several Complex Variables

M.F. Bessmertnyĭ

0. Introduction

Holomorphic functions of one complex variable with a non-negative imaginary part, and the related classes of J-contractive matrix functions (contractive in a space with an indefinite metric, defined in a standard way by a hermitian matrix J such that $J^2 = I$) were actively studied by many mathematicians in the past years. A series of problems of mathematical analysis and its applications lead to the necessity of studying analogous classes of holomorphic functions of several complex variables. Holomorphic functions with a non-negative imaginary part in a tubular domain over a cone and in the polydisk were studied by V.S. Vladimirov (see [Vla2], [Vla5]–[Vla8]) and by Vladimirov and Drozhzhinov [VlaDr]. In bounded strictly star-shaped domains, and in particular in the classical symmetric domains, they were studied by L.A. Aizenberg and Sh.A. Dautov [AD]. W. Rudin (see [Rud]) gives a "parametrical" representation of scalar rational inner functions in the polydisk $\mathbb{D}^n$. The parameter is an arbitrary polynomial non-vanishing in $\mathbb{D}^n$.

For the additive class of holomorphic functions mapping the right half-plane into itself, there is a well-known decomposition in terms of a sum of simple fractions and a representation as a continued fraction. The elementary functions of the representation belong to the same class. J-contractive functions which are J-unitary on the imaginary axis admit a multiplicative decomposition [Pot1], [EfPo]. Darlington's theorem (see [Me]; [EfPo]: Chapter V) allows to represent any positive rational function $f(\lambda)$ (i.e. a function mapping the right half-plane into itself) as a linear fractional transformation of a positive constant. The coefficient matrix of the transformation is J-contractive in the right half-plane and J-unitary on the imaginary axis. The possibility to represent a rational function non-negative on the imaginary axis as a sum of squares of rational functions analytic in the right half-plane is essential here. The various representations described above play an important role in various interpolation problems, applications to physics, etc.

This paper is a translation, prepared by D. Alpay and V. Katsnelson, of a part from the introduction and of the first chapter from the author's Ph. D. thesis entitled "Functions of Several Variables in the Theory of Finite Linear Structures", Kharkov, 1982. The manuscript, entitled "Realization of Rational Functions of Several Variables by Means of a Long Resolvent" was deposited at VINITI (86 pages, submitted by the Kharkov University, July 8, 1981, No. 3352-81).

The situation becomes complicated when one goes to the case of functions of several variables. Although the sum of positive functions is still a positive function in this case (invariance of the class under addition), it is nevertheless impossible to obtain an additive representation for an arbitrary (positive) function. This is related to the fact that among polynomials of several variables one can find irreducible polynomial of arbitrarily large degree, and, therefore, the "simplest" components (in the representation of a function as a sum of fractions) may have arbitrarily large degrees. To trace down positivity here is therefore impossible. Using Artin's solution (see [Lan]) of Hilbert's 17th problem on the representation of a non-negative rational function of several variables as a sum of squares of rational functions to obtain an analogue of Darlington's representation of positive functions of one variable is difficult for two reasons: Artin's theorem says nothing on the location of the singularities of the functions in the decomposition, and, moreover, the proof of the existence of the representation is not constructive.

Recently, a number of authors studied actively the use of functions of several complex variables in the theory of mutidimensional systems. See [Vla2], [Vla3], [Vla9]: Chapter III. The Institute of Electrical and Electronics Engineers (IEEE) devoted an issue [Mult1] to "multidimensional systems".

A convenient model for multidimensional systems consists of an electrical circuit, the characteristic matrix of which is a function of the impedances $z_1, \dots, z_n$ of the elements of the circuit, and not of the frequency λ as customary in the analytic theory of electrical circuits [Kar], [SeRe], [EfPo]. Along with the problem of analysing such circuits (that is, obtaining a theoretical functional characterization of the classes of matrix-valued functions which are characteristic matrices of the circuits) arises the inverse problem of reconstructing the circuit from its characteristic matrix (the synthesis problem). A fundamental difficulty to solve the inverse problem lies in the necessity to represent the circuit functions in a way convenient for these aims. In the study of multidimensional circuits and their generalization, there appears a class of rational positive functions of several variables, analogous to the class of positive functions of one complex variable.

Koga (see [Kog]) made an attempt to prove that every positive rational matrix function of several variables is the characteristic matrix of a multidimensional circuit. But, as was pointed out by Bose [Bos1], the proof relies on the erroneous statement that a positive polynomial of several variables, the degree of which with respect to each variable (the others being fixed) is equal to two, can be represented as a sum of squares of polynomials.

In the approach of R. Kalman (R. Kalman, P. Falb, M. Arbib [KFA]), the model of a system consists in finding a realization of a rational function $W(\lambda)$ of one variable in the form of a resolvent

$$W(\lambda) = H(\lambda I - F)^{-1}G,$$

where H, F, G are constant matrices.

It turns out that a representation of this form admits a generalization to the case of rational matrix-valued functions of several variables. Moreover, this generalization is closely related to the theory of multidimensional linear systems (and in particular to multidimensional circuits). In connection to this we consider the following construction, which we will meet often in the sequel.

We denote by $\mathcal{H}$ a finite dimensional vector space over the complex field; $\mathbb{C}^n = \mathbb{C} \times \cdots \times \mathbb{C}$ denotes the direct product of n copies of the complex plane $\mathbb{C}$; $z = (z_1, \ldots, z_n)$ denotes a point of $\mathbb{C}^n$. In the sequel, $'$ indicates the operation of transposition and $\overline{}$ indicates the passage to the complex conjugate elements: if A is a matrix with complex entries, then A' is the transposed matrix, and $\overline{A}$ is the matrix with the complex conjugate entries. $A^* = \overline{(A')} = (\overline{A})'$ is the matrix hermitian conjugate to the matrix A. If $z = (z_1, \ldots, z_n) \in \mathbb{C}^n$, then $\overline{z} \stackrel{\text{def}}{=} (\overline{z}_1, \ldots, \overline{z}_n)$.

Let $A_0, A_1, \ldots, A_n$ be *square* matrices with entries that are complex numbers. We consider a linear matrix bundle $A(z) = A_0 + z_1 A_1 + \cdots + z_n A_n$. This bundle is said to be *non-singular* if the matrix $(A_0 + z_1^{(0)} A_1 + \cdots + z_n^{(0)} A_n)$ is invertible for at least one point $z_0 \in \mathbb{C}^n$.

If the linear bundle is non-singular, then the matrix

$$A(z)^{-1} = (A_0 + z_1 A_1 + \cdots + z_n A_n)^{-1}$$

exists for all z, with a possible exception of some "thin" subset of $\mathbb{C}^n$. (This thin set is a zero-set of a non-zero polynomial.)

Definition 0.1. The matrix function $A(z)^{-1}$ is said to be the long resolvent of the linear matrix bundle $A(z) = A_0 + z_1 A_1 + \cdots + z_n A_n$.

Let, moreover, $\mathcal{E}$ be a subspace of $\mathcal{H}$, and let $\pi = \pi^2 = \pi^*$ be the orthogonal projection from $\mathcal{H}$ onto $\mathcal{E}$. We form

$$f(z) = \left[\pi \left(A_0 + z_1 A_1 + \cdots + z_n A_n\right)^{-1} \pi^*\right]^{[-1]} \qquad (R)$$

(where $C^{[-1]}$ denotes the inverse of C on the C-invariant subspace $\mathcal{E}$). The matrix function $f(z)$ is clearly rational. It is symmetric if all the matrices A_k are symmetric. It is real if all the matrices A_k are real.

We note that (R) can be rewritten in a somewhat different way. Indeed, let

$$\pi = \begin{bmatrix} I_k & 0 \\ 0 & 0 \end{bmatrix}.$$

We consider the block decomposition of the matrix function $A(z)$ as

$$A(z) = \begin{bmatrix} A_{11}(z) & A_{12}(z) \\ A_{21}(z) & A_{22}(z) \end{bmatrix},$$

where $A_{11}(z)$ is $\mathbb{C}^{k \times k}$-valued. Using the formula to compute the inverse of a block matrix (if $A_{22}(z)^{-1}$ exists), we obtain

$$f(z) = A_{11}(z) - A_{12}(z) A_{22}(z)^{-1} A_{21}(z). \qquad (R_0)$$

Unlike (R), the expression (R_0) makes sense for a matrix function satisfying the condition $\det f(z) \equiv 0$.

Let us agree to call a function $g(z)$ *nonsingular* if it holds that $\det g(z) \not\equiv 0$. In case of nonsingular matrix functions, the representations (R) and (R_0) are clearly equivalent.

Definition 0.2. *Both representations (R) and (R_0) are said to be a representation of $f(z)$ in the form of a long resolvent.*

In the sequel we will often need the following subsets of $\mathbb{C}^n$:

$$\begin{aligned}
\mathcal{D}_R^+ &= \{z : z \in \mathbb{C}^n, \operatorname{Re} z_1 > 0, \dots, \operatorname{Re} z_n > 0\}\\
\mathcal{D}_R^- &= \{z : z \in \mathbb{C}^n, \operatorname{Re} z_1 < 0, \dots, \operatorname{Re} z_n < 0\}\\
\mathcal{D}_J^+ &= \{z : z \in \mathbb{C}^n, \operatorname{Im} z_1 > 0, \dots, \operatorname{Im} z_n > 0\}\\
\mathcal{D}_J^- &= \{z : z \in \mathbb{C}^n, \operatorname{Im} z_1 < 0, \dots, \operatorname{Im} z_n < 0\}.
\end{aligned}$$

Definition 0.3. A rational matrix functions $f(z) = f(z_1, \dots, z_n)$ of the complex variables $z_1, \dots, z_n$ is said to be **positive** if the following *positivity conditions* hold:

$$\begin{aligned}
f(z) + f(z)^* \geq 0 \quad &\text{for} \quad z \in \mathcal{D}_R^+,\\
f(z) + f(z)^* \leq 0 \quad &\text{for} \quad z \in \mathcal{D}_R^-,\\
i\,(f(z)^* - f(z)) \geq 0 \quad &\text{for} \quad z \in \mathcal{D}_J^+,\\
i\,(f(z)^* - f(z)) \leq 0 \quad &\text{for} \quad z \in \mathcal{D}_J^-.
\end{aligned}$$

We will denote by $\mathcal{P}$ the class of positive matrix functions.

The aims of this paper are:

1. The study of the possibility of representing rational matrix functions of several variables by means of the long resolvent of a linear bundle of constant matrices and the clarification of the connections between the properties of the functions and the properties of the constant matrices in the representation.
2. The study of the class of positive real matrix functions of several variables.

In Section 1 we prove the existence of a representation of a rational (square) matrix-valued function of several variables by means of the long resolvent of a linear bundle of matrices. In Section 2 we study the properties of positive matrix functions of several variables. Section 3 is devoted to the study of the properties of functions which can be represented by means of the long resolvent of a bundle of positive definite matrices.

1. Realization of rational functions by means of a long resolvent

1. The main aim of this section is to prove the following assertion:

Theorem 1.1. (Main theorem.) *Every rational $\mathbb{C}^{k\times k}$-valued matrix function $f(z_1, \dots, z_n)$ of the variables $z_1, \dots z_n$ can be represented in the form*

$$f(z) = A_{11}(z) - A_{12}(z)A_{22}(z)^{-1}A_{21}(z) \tag{1.1}$$

where $A = (A_{ij}(z))_{i,j=1,2}$ *is the block decomposition of the matrix function* $A(z)$,

$$A(z) = \begin{bmatrix} A_{11}(z) & A_{12}(z) \\ A_{21}(z) & A_{22}(z) \end{bmatrix} = A_0 + z_1 A_1 + \cdots + z_n A_n.$$

into blocks (so that $A_{11}(z)$ *is* $\mathbb{C}^{k\times k}$*-valued and* $A_{22}(z)$ *is square non-singular) and where* $A_0, A_1, \ldots, A_n$ *are square matrices with complex entries.*
Moreover:

a) *if* $f(z) = \overline{f(\bar{z})}$, *one can choose the matrices* A_j $(j = 0, 1, \ldots, n)$ *to be real.*

b) *if* $f(z) = f(z)'$, *one can choose the matrices* A_j $(j = 0, 1, \ldots, n)$ *to be symmetric.*

c) *if* $f(\lambda z) = \lambda f(z)$, *then* $A_0 = 0$.

2. We make the following observation. In the introduction it was noticed that, in the case of a nonsingular $f(z)$, the representation (1.1) is equivalent to the representation

$$f(z) = (\pi A(z)^{-1}\pi^*)^{-1} \tag{1.2}$$

where

$$\pi = \begin{bmatrix} I_k & 0 \\ 0 & 0 \end{bmatrix}.$$

Then, it follows from the trivial equality

$$A(z)A(z)^{-1}\pi^* f(z) = \begin{bmatrix} f(z) & 0 \\ 0 & 0 \end{bmatrix}$$

that there exists a matrix function $\Phi(z)$ such that the identity

$$A(z)\begin{bmatrix} I_k \\ \Phi(z) \end{bmatrix} = \begin{bmatrix} f(z) \\ 0 \end{bmatrix} \tag{1.3}$$

holds. Conversely, if (1.3) holds and if $A(z)$ is nonsingular, then (1.2) is valid. Therefore, to prove the theorem in the case of a nonsingular $f(z)$, it is enough to build a linear matrix bundle $A(z)$ satisfying condition (1.3).

In the following lemma, the bundle $A(z)$ is as in (1.3), but possibly singular. Moreover we consider right away the case of an arbitrary (i.e. possibly singular) function $f(z)$.

Lemma 1.1. *Assume that the matrix functions*

$$A(z) = \begin{bmatrix} a(z) & b(z) \\ c(z) & d(z) \end{bmatrix} \quad \text{and} \quad \Psi(z) = \begin{bmatrix} I_k \\ \Phi(z) \end{bmatrix}$$

where $a(z)$ *is* $\mathbb{C}^{k\times k}$*-valued and where* $d(z) \not\equiv 0$, *satisfy the conditions*

$$A(z)\Psi(z) = \begin{bmatrix} f(z) \\ 0 \end{bmatrix}, \qquad \Psi(z)' A(z) = (f(z)\ 0). \tag{1.4}$$

Then there exists a submatrix

$$A_1(z) = \begin{bmatrix} a(z) & b_1(z) \\ c_1(z) & d_1(z) \end{bmatrix},$$

of $A(z)$ which is symmetric when $A(z)$ is symmetric, and such that:

a) $\det d_1(z) \not\equiv 0$,

b) $f(z) = a(z) - b_1(z)d_1(z)^{-1}c_1(z)$.

Proof. If $\det d(z) \not\equiv 0$, the statement is trivial. Let $\det d(z) \equiv 0$ and let $r = \max \operatorname{rank} d(z)$. We have $r \geq 1$ since $d(z) \not\equiv 0$. Without loss of generality, we may assume that the first r lines and the first r columns of the matrix $d(z)$ are linearly independent. This can always be achieved by a permutation of the lines and of the columns of the matrix $A(z)$, and, when $A(z)$ is symmetric, by a permutation of the lines and of the corresponding columns. Then, if $d(z)$ is divided into blocks

$$d(z) = \begin{bmatrix} d_1(z) & d_{12}(z) \\ d_{21}(z) & d_{22}(z) \end{bmatrix} \tag{1.5}$$

so that $d_1(z)$ is $\mathbb{C}^{r\times r}$-valued, then:

1) $\det d_1(z) \not\equiv 0$;

2) There exist rational $r \times r$ matrix functions $s_j(z)$ $(j = 1, 2)$ such that

$$\begin{aligned} (s_1(z)\ \ I)\begin{bmatrix} d_1(z) & d_{12}(z) \\ d_{21}(z) & d_{22}(z) \end{bmatrix} &= (0\ \ 0), \\ \begin{bmatrix} d_1(z) & d_{12}(z) \\ d_{21}(z) & d_{22}(z) \end{bmatrix}\begin{bmatrix} s_2(z) \\ I \end{bmatrix} &= \begin{bmatrix} 0 \\ 0 \end{bmatrix}. \end{aligned}$$

We decompose according to (1.5) the matrices $A(z)$ and $\Psi(z)$

$$A(z) = \begin{bmatrix} a(z) & b_1(z) & b_2(z) \\ c_1(z) & d_1(z) & d_{12}(z) \\ c_2(z) & d_{21}(z) & d_{22}(z) \end{bmatrix}, \qquad \Psi(z) = \begin{bmatrix} I_k \\ \Phi_2(z) \\ \Phi_3(z) \end{bmatrix},$$

and consider the matrix function

$$A_1(z) = S_1(z)A(z)S_2(z),$$

where

$$S_1(z) = \begin{bmatrix} I_k & 0 & 0 \\ 0 & I_r & 0 \\ 0 & s_1(z) & I \end{bmatrix}, \qquad S_2(z) = \begin{bmatrix} I_k & 0 & 0 \\ 0 & I_r & s_2(z) \\ 0 & 0 & I \end{bmatrix}.$$

Then,

$$A_1(z) = \begin{bmatrix} a(z) & b_1(z) & \tilde{b}_2(z) \\ c_1(z) & d_1(z) & 0 \\ \tilde{c}_2(z) & 0 & 0 \end{bmatrix}$$

and accordingly, (1.4) can be written in the form

$$\begin{aligned} \Psi'(z)S_1(z)^{-1}A_1(z) &= (f(z)\ ,\ 0)\,, \\ A_1(z)S_2(z)^{-1}\Psi(z) &= \begin{bmatrix} f(z) \\ 0 \end{bmatrix}. \end{aligned}$$

It follows from these equations that $\tilde{b}_2(z) = 0$ and $\tilde{c}_2(z) = 0$, and therefore,

$$\begin{bmatrix} a(z) & b_1(z) \\ c_1(z) & d_1(z) \end{bmatrix} \begin{bmatrix} I_k \\ \Phi_0(z) \end{bmatrix} = \begin{bmatrix} f(z) \\ 0 \end{bmatrix} \tag{1.6}$$

where

$$\Phi_0(z) = \Phi_2(z) - s_2(z)\Phi_3(z).$$

Eliminating $\Phi_0(z)$ from (1.6) we obtain

$$f(z) = a(z) - b_1(z)d_1(z)^{-1}c_1(z),$$

as wanted. □

3. The proof of the main theorem will be carried out in a number of steps. First we build in Lemma 1.3 a linear matrix bundle $A(w)$ for the elementary scalar functions $g(w) = g(w_1, \dots, w_{2m+1})$ of the form

$$g(w) = \frac{w_{m+1} \cdots w_{2m+1}}{w_1 \cdots w_m}.$$

Making use of Lemma 1.3, we obtain a matrix bundle $A(z)$ for scalar rational functions which satisfy the condition

$$f(\lambda z_1, \dots, \lambda z_n) = \lambda f(z_1, \dots z_n) = \lambda \overline{f(\overline{z_1}, \dots, \overline{z_n})}.$$

After that, we build the matrices A_j $(j = 0, 1, \dots n)$ of the representation (1.1) for arbitrary matrix functions $f(z)$ satisfying the conditions of the main theorem.

4. Before turning to the proof we fix the notation and consider an auxiliary lemma, Lemma 1.2. Let α be a multiindex such that

$$\alpha = (\alpha^{(1)}, \dots, \alpha^{(d)}),$$

where the components $\alpha^{(j)}$ $(j = 1, \dots, d)$ are non-negative integers. The number d is called the *dimension of the multiindex* α $(d = \dim \alpha)$.

In the set of all multiindices, we introduce the addition of two multiindices of same dimension as componentwise addition. Moreover, we order the set of multiindices by the relation

$$\alpha \succ 0 \overset{\text{def.}}{\Longleftrightarrow} \alpha^{(1)} \geq 0, \dots, \alpha^{(d)} \geq 0.$$

We will not be interested in arbitrary multiindices, but only in those for which

1. $\dim \alpha = 2m + 1$,
2. All the components of α are 0 or 1.
3. $\langle \alpha, \alpha \rangle = m\,, \quad (\langle \alpha, \beta \rangle \overset{\text{def.}}{=} \alpha^{(1)}\beta^{(1)} + \dots + \alpha^{(2m+1)}\beta^{(2m+1)})$.

We will denote by $\mathcal{U}$ the set of multiindices satisfying the conditions 1), 2) and 3).

Clearly, the multiindex

$$\delta = (1, \dots 1 \,|\, 0, \dots 0) \tag{1.7}$$

(with m times 1 and $m + 1$ times 0) belongs to $\mathcal{U}$. We will write the other multiindices of $\mathcal{U}$ in the form

$$\alpha = (\alpha^{(1)}, \, \dots, \, \alpha^{(m)} \,|\, \alpha^{(m+1)}, \, \dots, \, \alpha^{(2m+1)}),$$

dividing their components with a vertical bar into two groups: m components on the left of the line and $m+1$ on the right. The set $\mathcal{U}$ can be divided into equivalence classes using the element δ defined in (1.7). By definition,

$$\alpha \sim \beta \Longleftrightarrow \langle \alpha, \delta \rangle = \langle \beta, \delta \rangle.$$

We will denote by $\mathcal{U}_k$ the equivalence class for which $\langle \alpha, \delta \rangle = k$.

The following proposition is at the basis of our subsequent investigations:

Lemma 1.2. *Let $\gamma = (1, \dots 1 \,|\, 1, 1, \dots 1)$ and let α_k be fixed multiindieces in the equivalence classes $\mathcal{U}_k$ with respect to the element $\delta = (1, \dots, 1 \,|\, 0, 0, \dots 0)$. Then the inequality*

$$(\gamma - \alpha_k) - \alpha \succ 0$$

has exactly $m+1$ solutions in $\mathcal{U}$, with moreover:

a) $(m-k)$ solutions belonging to the class $\mathcal{U}_{m-k-1}$.

b) $(k+1)$ solutions belonging to the class $\mathcal{U}_{m-k}$.

Proof. By definition, $\langle \alpha_k, \delta \rangle = k$, $\langle \alpha_k, \alpha_k \rangle = m$. Therefore the multiindex $(\gamma - \alpha_k)$ has $(m-k)$ components equal to 1 on the left of the vertical line, and $(k+1)$ components equal to 1 on the right of the vertical line. Since for any $\alpha \in \mathcal{U}$ the condition $\langle \alpha, \alpha \rangle = m$ holds, the inequality

$$(\gamma - \alpha_k) - \alpha \succ 0$$

will hold when the multiindex $(\gamma - \alpha_k) - \alpha$ has exactly one component equal to 1 and the others equal to 0. Thus every solution α of the inequality $(\gamma - \alpha_k) - \alpha \succ 0$ can be obtained by replacing in $(\gamma - \alpha_k)$ one component equal to 1 by 0. Replacing one component equal to 1 by 0 in the multiindex $(\gamma - \alpha_k)$ can be done in $(m-k)$ places to the left of the vertical bar, and for these we obtain a solution in the class $\mathcal{U}_{m-k-1}$, and in $(k+1)$ places to the right of the bar, and these lead to solutions in $\mathcal{U}_{m-k}$. The lemma is proved. □

5. If $w = (w_1, \dots, w_{2m+1}) \in \mathbb{C}^{2m+1}$ and

$$\alpha = (\alpha^{(1)}, \dots, \alpha^{(m)} \,|\, \alpha^{(m+1)}, \dots, \alpha^{(2m+1)})$$

is a multiindex in $\mathcal{U}$, we denote by w^α the monomial

$$w^\alpha = w_1^{\alpha_1} \cdots w_{2m+1}^{\alpha_{2m+1}}.$$

Lemma 1.3. *For every natural number m there exists a matrix bundle of real symmetric matrices A_j, independent of w_k $(k = 1, \dots 2m+1)$,*

$$A(w) = w_1 A_1 + \cdots + w_{2m+1} A_{2m+1}$$

such that

$$\begin{bmatrix} w^\beta \\ 0 \\ \vdots \\ 0 \end{bmatrix} = A(w) \begin{bmatrix} w^{\alpha_1} \\ w^{\alpha_2} \\ \vdots \\ w^{\alpha_M} \end{bmatrix} \tag{1.8}$$

where $\beta = (0,\dots,0\,|\,1,1,\dots,1)$, $\alpha_1 = (1,\dots,1\,|\,0,0,\dots,0)$, $M = \#\mathcal{U}$ *(the cardinality of the set* $\mathcal{U}$: $M = \frac{(2m+1)!}{m!\,(m+1)!}$*)*, $\{\alpha_1,\dots,\alpha_M\}$ *is the set of all multiindices in* $\mathcal{U}$ *with some ordering.*

Proof. We set

$$A(w) = \left[c_{k\nu} w^{\gamma-\alpha_k-\alpha_\nu}\right]_{k,\nu=1}^{M}$$

where $\gamma = (1,\dots 1\,|\,1,1,\dots,1)$ and

$$\begin{array}{lll} c_{k\nu} = 0, & \text{if} & \gamma-\alpha_k-\alpha_\nu \not\succ 0, \\ c_{k\nu} = \varphi_m(k_0,\nu_0) & \text{if} & \gamma-\alpha_k-\alpha_\nu \succ 0, \quad \alpha_k \in \mathcal{U}_{k_0} \quad \text{and} \quad \alpha_\nu \in \mathcal{U}_{\nu_0}, \end{array} \tag{1.9}$$

where

$$\varphi_m(k,\nu) = (-1)^{m+k+\nu}\frac{k!\nu!(m-k)!(m-\nu)!}{m!(m+1)!}. \tag{1.10}$$

Since the inequality $\gamma-\alpha_k-\alpha_\nu \succ 0$ holds only if the multiindex $\gamma-\alpha_k-\alpha_\nu$ has exactly one component equal to 1 and the others equal to 0, the matrix function $A(w)$ is linear:

$$A(w) = w_1A_1 + \cdots + w_{2m+1}A_{2m+1}.$$

By construction, $c_{k\nu}w^{\gamma-\alpha_k-\alpha_\nu} = c_{\nu k}w^{\gamma-\alpha_\nu-\alpha_k}$ and the coefficients $c_{k\nu}$ are real. Therefore, the matrices A_j are symmetric and real.

We compute the components $a_k(w)$ of the column vector

$$A(w)\cdot\begin{bmatrix} w^{\alpha_1} \\ \vdots \\ w^{\alpha_M}\end{bmatrix} = \begin{bmatrix} a_1(w) \\ \vdots \\ a_M(w)\end{bmatrix}:$$

$$a_k(w) = \sum_{1\le\nu\le M} c_{k\nu}\cdot w^{\gamma-\alpha_k-\alpha_\nu}\cdot w^{\alpha_\nu} = w^{\gamma-\alpha_k}\cdot\sum_{1\le\nu\le M} c_{k\nu} \tag{1.11}$$

Let $\alpha_k \in \mathcal{U}_{k_0}$. Then, by Lemma 1.2, the inequality $\gamma-\alpha_k-\alpha_\nu \succ 0$ has exactly $(m-k_0)$ solutions in the class $\mathcal{U}_{m-k_0-1}$ and (k_0+1) solutions in the class $\mathcal{U}_{m-k_0}$. Therefore in the k-th line of the matrix $A(w)$ only those coefficients $c_{k\nu}$ can differ from zero for which the inequality $\gamma-\alpha_k-\alpha_\nu \succ 0$ holds and moreover either $\alpha_\nu \in \mathcal{U}_{m-k_0-1}$ or $\alpha_\nu \in \mathcal{U}_{m-k_0}$.

In accordance with this, the non-zero coefficients $c_{k\nu}$ in (1.11) may be divided into two groups:

1. $(m-k_0)$ coefficients $c_{k\nu}$, equal to $\varphi_m(k_0, m-k_0-1)$;
2. (k_0+1) coefficients $c_{k\nu}$, equal to $\varphi_m(k_0, m-k_0)$.

Then,

$$\begin{aligned} a_k(w) &= w^{\gamma-\alpha_k}\cdot\sum_{1\le\nu\le M} c_{k\nu} \\ &= w^{\gamma-\alpha_k}\big\{(k_0+1)\,\varphi_m(k_0,\,m-k_0) + (m-k_0)\,\varphi_m(k_0,\,m-k_0-1)\big\}. \end{aligned}$$

It follows from this and from (1.10) that

1) If $k_0 \neq m$, then $a_k(w) = 0$;
2) If $k_0 = m$, then the class $\mathcal{U}_m$ consists of one element, namely δ (which is defined in (1.7)). This case corresponds to $k = 1$ in (1.11), i.e. $\alpha_1 = \delta$. Then, $\gamma - \alpha_1 = \beta$, and

$$a_1(w) = w^{\gamma-\alpha_1}(m+1)\,\varphi_m(m\,,\,0)) = w^{\beta},$$

as was required. □

Corollary 1.1. *Given two arbitrary monomials z^{β_0} and z^{α_0} of n variables and of degrees respectively $m+1$ and m, there exists a matrix bundle*

$$A(z) = z_1 A_1 + \cdots + z_n A_n \qquad (A_j = A_j' = \overline{A_j}),$$

such that

$$A(z)\begin{bmatrix} z^{\alpha_0} \\ z^{\alpha_1} \\ \vdots \\ z^{\alpha_N} \end{bmatrix} = \begin{bmatrix} z^{\beta_0} \\ 0 \\ \vdots \\ 0 \end{bmatrix},$$

where $\{z^{\alpha_j}\}$ $(j = 0, 1, \ldots N)$ is the set of all different monomials of degree m in the variables $z_1, \ldots, z_n$.

Proof. For the proof, it is enough to make a substitution of each w_j $(j = 1, \ldots, 2m+1)$ by some z_i $(i = 1, \ldots, n)$ in such a way that w^{β} is transformed into z^{β_0} and w^{α_1} is transformed into z^{α_0}. This is possible because the monomials w^{α_1} and w^{β} do not contain identical variables w_j: $w^{\alpha_1} = w_1 \cdots w_m$, $w^{\beta} = w_{m+1} \cdots w_{2m+1}$. Since after substituting w by z there can appear identical monomials in the resulting column vector

$$\begin{bmatrix} z^{\alpha_0} \\ z^{\alpha_{k_1}} \\ \vdots \\ z^{\alpha_{k_{M-1}}} \end{bmatrix},$$

the matrix $A(z)$ should be "reduced" by a transformation $T'A(z)T$, where the rectangular matrix T is such that

$$\begin{bmatrix} z^{\alpha_0} \\ z^{\alpha_{k_1}} \\ \vdots \\ z^{\alpha_{k_{M-1}}} \end{bmatrix} = T \cdot \begin{bmatrix} z^{\alpha_0} \\ z^{\alpha_1} \\ \vdots \\ z^{\alpha_N} \end{bmatrix},$$

where all the monomials z^{α_j} in the column vector in the right-hand side are different and this column vector contains all the monomials of degree m. □

Lemma 1.4. *Let $f(z) = P(z)/Q(z)$ be a rational scalar function in the variables $z_1, \ldots, z_n$, satisfying the condition*

$$f(\lambda z_1, \ldots, \lambda z_n) = \lambda f(z_1, \ldots, z_n), \quad \forall \lambda \in \mathbb{C}.$$

Then there exists a matrix bundle

$$A(z) = z_1 A_1 + \cdots + z_n A_n, \qquad (A_j' = A_j)$$

such that

$$A(z) \begin{bmatrix} Q(z) \\ z^{\alpha_2} \\ \vdots \\ z^{\alpha_N} \end{bmatrix} = \begin{bmatrix} P(z) \\ 0 \\ \vdots \\ 0 \end{bmatrix},$$

where the z^{α_j} $(j = 2, \dots N)$ are monomials in the variables $z_1, \dots, z_n$ of the same degree as the polynomial $Q(z)$.
The matrices A_j are symmetric[1] *$(A_j = A_j')$. If, moreover, $\overline{f(\overline{z})} = f(z)$, then the matrices A_j may be chosen real $(\overline{A_j} = A_j)$.*

Proof. Since $f(\lambda z) = \lambda f(z)$, $P(\lambda z) = \lambda^{m+1} P(z)$ and $Q(z) = \lambda^m Q(z)$, i.e. $P(z)$ and $Q(z)$ are homogeneous polynomials of degrees $m+1$ and m respectively. Therefore,

$$P(z) = a_1 z^{\beta_1} + \cdots + a_\nu z^{\beta_\nu},$$

where the z^{β_j} $(j = 1, \dots, \nu)$ are monomials of degree $m+1$ of n variables and

$$Q(z) = b_1 z^{\alpha_1} + \cdots + b_\mu z^{\alpha_\mu},$$

where the z^{α_j} $(j = 1, \dots, \mu)$ are monomials of degree m of n variables. Moreover we can assume that $\beta_1 \neq 0$.

Consider one of the monomials z^{β_j} appearing in the numerator of $f(z)$. Making use of the results of Lemma 1.3, we build matrix functions

$$A_{jk}(z) = z_1 A_1^{(j,k)} + \cdots + z_n A_n^{(j,k)},$$

such that

$$k\text{-th line} \quad \begin{bmatrix} 0 \\ \vdots \\ 0 \\ z^{\beta_j} \\ 0 \\ \vdots \\ 0 \end{bmatrix} = A_{jk}(z) \begin{bmatrix} z^{\alpha_1} \\ \vdots \\ z^{\alpha_{k-1}} \\ z^{\alpha_k} \\ z^{\alpha_{k+1}} \\ \vdots \\ z^{\alpha_N} \end{bmatrix}, \qquad (k = 1, \dots \mu).$$

[1] A scalar function is always symmetric.

Multiplying the k-th equality by the coefficient b_k of the monomial z^{α_k} of the denominator and summing up, we obtain

$$\begin{bmatrix} b_1 z^{\beta_j} \\ \vdots \\ b_\mu z^{\beta_j} \\ 0 \\ \vdots \\ 0 \end{bmatrix} = A_j(z) \begin{bmatrix} z^{\alpha_1} \\ \vdots \\ z^{\alpha_\mu} \\ z^{\alpha_{\mu+1}} \\ \vdots \\ z^{\alpha_N} \end{bmatrix} \tag{1.12}$$

where

$$A_j(z) = b_1 A_{j1}(z) + \cdots + b_\mu A_{j\mu}(z)\,, \quad (j = 1, \ldots, \nu)\,.$$

Let

$$T = \begin{bmatrix} b_1 & b_2 & \cdots & b_\mu & 0 & \cdots & 0 \\ 0 & 1 & \cdots & 0 & 0 & \cdots & 0 \\ \vdots & \vdots & \ddots & \vdots & \cdots & \ddots & \vdots \\ 0 & 0 & \cdots & 1 & 0 & \cdots & 0 \\ 0 & 0 & \cdots & 0 & 1 & \cdots & 0 \\ \vdots & \vdots & \ddots & \vdots & \vdots & \ddots & \vdots \\ 0 & 0 & \cdots & 0 & 0 & \cdots & 1 \end{bmatrix}, \quad (b_1 \neq 0)\,.$$

Then

$$T'^{-1} \begin{bmatrix} b_1 P(z) \\ \vdots \\ b_\mu P(z) \\ 0 \\ \vdots \\ 0 \end{bmatrix} = T'^{-1} \widetilde{A}(z) T^{-1} T \begin{bmatrix} z^{\alpha_1} \\ \vdots \\ z^{\alpha_\mu} \\ z^{\alpha_{\mu+1}} \\ \vdots \\ z^{\alpha_N} \end{bmatrix},$$

where $\widetilde{A}(z) = a_1 A_1(z) + \cdots + a_\nu A_\nu(z)$, or, after computations,

$$\begin{bmatrix} P(z) \\ 0 \\ \vdots \\ 0 \end{bmatrix} = A(z) \begin{bmatrix} Q(z) \\ z^{\alpha_2} \\ \vdots \\ z^{\alpha_N} \end{bmatrix}, \tag{1.13}$$

where

$$A(z) = T'^{-1} \widetilde{A}(z) T^{-1},$$

as required.

Remark 1.1. We mention that in the proof of the main theorem we will need the fact that for any set of rational functions of the form $f_k(z) = P_k(z)/Q(z)$, satisfying the properties of the theorem with one and the same denominator $Q(z)$, the relation (1.13) can be written in the form

$$\begin{bmatrix} f_k(z) \\ 0 \\ \vdots \\ 0 \end{bmatrix} = A_k(z) \begin{bmatrix} 1 \\ \varphi_2(z) \\ \vdots \\ \varphi_N(z) \end{bmatrix},$$

where $\varphi_j(z) = z^{\alpha_j}/Q(z)$, $(j = 2, \dots, N)$ are the same for all the functions $f_k(z)$.

6. We turn now to the proof of the main theorem. First of all, if a matrix function $f(z)$ satisfies the conditions of the theorem, then the matrix function of $(n+1)$ variables

$$f_0(z_0, z_1, \dots, z_n) = z_0 f(z_1/z_0, \dots, z_n/z_0)$$

also satisfies the conditions of the theorem and is moreover a homogeneous matrix function of degree of homogeneity 1. Since

$$f(z_1, \dots, z_n) = f_0(1, z_1, \dots, z_n),$$

it is enough to prove the theorem for homogeneous matrix functions.

Let

$$f(z) = \{f_{ij}(z)\}_{i,j=1}^{k}$$

be a homogeneous matrix function of degree of homogeneity 1, satisfying the conditions of the theorem. Without loss of generality, we may assume that the matrix functions $f_{ij}(z)$ have the same denominator. Then, making use of Lemma 1.4 we construct matrix bundles $A_{ij}(z)$ such that

$$\begin{bmatrix} a_{11}^{(ij)}(z) & a_{12}^{(ij)}(z) \\ a_{12}^{(ij)\prime}(z) & a_{22}^{(ij)}(z) \end{bmatrix} \begin{bmatrix} 1 \\ \varphi(z) \end{bmatrix} = \begin{bmatrix} f_{ij}(z) \\ 0 \end{bmatrix} \tag{1.14}$$

where $\varphi(z)$ is a vector-valued function independent of $i, j = 1, \dots k$.

We consider the matrix functions

$$\widehat{A}(z) = \left[\begin{array}{ccc|ccc} a_{11}^{(11)} & \cdots & a_{11}^{(1k)} & a_{12}^{(11)} & \cdots & a_{12}^{(1k)} \\ \vdots & \ddots & \vdots & \vdots & \ddots & \vdots \\ a_{11}^{(k1)} & \cdots & a_{11}^{(kk)} & a_{12}^{(k1)} & \cdots & a_{12}^{(kk)} \\ \hline a_{12}^{(11)\prime} & \cdots & a_{12}^{(1k)\prime} & a_{22}^{(11)} & \cdots & a_{22}^{(1k)} \\ \vdots & \ddots & \vdots & \vdots & \ddots & \vdots \\ a_{12}^{(k1)\prime} & \cdots & a_{12}^{(kk)\prime} & a_{22}^{(k1)} & \cdots & a_{22}^{(kk)} \end{array}\right];$$

$$\Psi(z) = \begin{bmatrix} 1 & \cdots & 0 \\ \vdots & \ddots & \ddots \\ 0 & \cdots & 1 \\ -- & -- & -- \\ \varphi(z) & \cdots & 0 \\ \vdots & \ddots & \ddots \\ 0 & \cdots & \varphi(z) \end{bmatrix}.$$

By construction, the matrix bundle $\widehat{A}(z)$ inherits the properties of the matrix function $f(z)$:

a) if $f(z)' = f(z)$, then $\widehat{A}(z)' = \widehat{A}(z)$;

b) if $\overline{f(\overline{z})} = f(z)$, then $\overline{\widehat{A}(\overline{z})} = \widehat{A}(z)$.

Since the matrices in (1.14) are symmetric, we obtain from there

$$\widehat{A}(z)\Psi(z) = \begin{bmatrix} f(z) \\ 0 \end{bmatrix} \quad ; \quad \Psi(z)'\widehat{A}(z) = (f(z) \ , \ 0)\,. \tag{1.15}$$

If $\widehat{A}_{22}(z) \not\equiv 0$, then, by Lemma 1.1, $\widehat{A}(z)$ has a submatrix

$$A(z) = \begin{bmatrix} A_{11}(z) & A_{12}(z) \\ A_{21}(z) & A_{22}(z) \end{bmatrix} \tag{1.16}$$

such that $\det A_{22}(z) \not\equiv 0$ and

$$f(z) = A_{11}(z) - A_{12}(z)A_{22}(z)^{-1}A_{21}(z),$$

as required. Moreover, $A(z)$ is symmetric if $\widehat{A}(z)$ is symmetric. If $\widehat{A}_{22}(z) \equiv 0$, then it follows from (1.15) that also $\widehat{A}_{12}(z) \equiv 0$ and $\widehat{A}_{21}(z) \equiv 0$. In this case the function $f(z)$ is linear:

$$f(z) = A_{11}(z)\,.$$

This equality can be considered as the degenerate case of the representation (1.16) in which the dimension of the matrix $A_{22}(z)$ is equal to zero: $A_{22}(z)$ is the 0×0 matrix. □

2. Positive real matrix functions

In this section we obtain a number of theorems which describe the properties of rational positive real matrix functions of several variables. We remind that a rational matrix-function $f(z)$ is said to be positive real if it is holomorphic in the domains[2] $\mathcal{D}_R^+$, $\mathcal{D}_R^-$, $\mathcal{D}_J^+$, $\mathcal{D}_J^-$ and satisfies the conditions

1.	$f(z)+f(z)^*$	≥ 0	for	$z\in\mathcal{D}_R^+$,
2.	$f(z)+f(z)^*$	≤ 0	for	$z\in\mathcal{D}_R^-$,
3.	$i\,(f(z)^*-f(z))$	≥ 0	for	$z\in\mathcal{D}_J^+$,
4.	$i\,(f(z)^*-f(z))$	≤ 0	for	$z\in\mathcal{D}_J^-$,
5.	$\overline{f(\overline{z})}=f(z)$			

1. The following theorem shows us that for rational matrix functions the analyticity is the consequence of the inequalities.

Theorem 2.1. *If a rational matrix function $f(z)$ satisfies the inequality*

$$f(z)+f^*(z)\geq 0, \qquad z\in\mathcal{D}_R^+,$$

then $f(z)$ is holomorphic in the domain $\mathcal{D}_R^+$.

Proof. Let $z_0 = (z_1^0, \dots, z_n^0)$ by an arbitrary fixed point from $\mathcal{D}_R^+$. Then the functions

$$f_1(\lambda) = f(\lambda, z_2^0, \dots, z_n^0);$$

$$\dots\dots\dots\dots\dots\dots\dots\dots\dots$$

$$f_n(\lambda) = f(z_1^0, \dots, z_0^{n-1}, \lambda)$$

are rational and satisfy the condition

$$f_j(\lambda)+f_j^*(\lambda)\geq 0 \qquad \text{for} \qquad \operatorname{Re}\lambda>0.$$

From here it follows that $f_j(\lambda)$ are holomorphic for $\operatorname{Re}\lambda>0$. Therefore, $f(z)$ is holomorphic with respect to each variable z_j separately for $z=(z_1, \dots\ z_n)$ in $\mathcal{D}_R^+$. According to Hartogs' Theorem (see, for example, [Vla1], I.4.2 or [Ful], Theorem 1.6), $f(z)$ is jointly holomorphic for $z=(z_1, \dots\ z_n)$ in $\mathcal{D}_R^+$. □

Remark 2.1. The analyticity of $f(z)$ in the domains $\mathcal{D}_R^-$, $\mathcal{D}_J^+$, $\mathcal{D}_J^-$ can be proved analogously.

Theorem 2.2. *Every rational positive real matrix functions $f(z)$ is homogeneous of degree one:*

$$f(\lambda z_1, \dots, \lambda z_n)=\lambda f(z_1, \dots, z_n).$$

Proof. Let $f(z)\in\mathcal{P}$ and $x_j>0$ $(j=1,2,\dots,n)$. Let us consider the slice-function of the variable λ:

$$\varphi_x(\lambda)=f(\lambda x_1, \dots, \lambda x_n).$$

From the definition of the class $\mathcal{P}$:

$$\begin{aligned}
\varphi_x(\lambda)+\varphi_x^*(\lambda) &\geq 0 \quad \text{for} \quad \operatorname{Re}\lambda>0;\\
\varphi_x(\lambda)+\varphi_x^*(\lambda) &\leq 0 \quad \text{for} \quad \operatorname{Re}\lambda<0;\\
i(\varphi_x^*(\lambda)-\varphi_x(\lambda)) &\geq 0 \quad \text{for} \quad \operatorname{Im}\lambda>0;\\
i(\varphi_x^*(\lambda)-\varphi_x(\lambda)) &\leq 0 \quad \text{for} \quad \operatorname{Im}\lambda<0.
\end{aligned}$$

---

[2]The definition of the domains $\mathcal{D}_R^+$, $\mathcal{D}_R^-$, $\mathcal{D}_J^+$, $\mathcal{D}_J^-$ was given in the introduction.

From here it follows that $\varphi_x(\lambda)$ is of the form

$$\varphi_x(\lambda) = \lambda A(x)\,.$$

Setting $\lambda = 1$, we obtain

$$A(x) = f(x_1, \ldots, x_n).$$

Thus,

$$f(\lambda x_1, \ldots, \lambda x_n) = \lambda f(x_1, \ldots, x_n)$$

for real $x_1, \ldots, x_n$. Since the set

$$\{z_1, \ldots, z_n : x_1 > 0, \ldots, x_n > 0, y_1 = 0, \ldots, y_n = 0\}$$

is the uniqueness set for the class of rational functions,

$$f(\lambda z_1, \ldots, \lambda z_n) \equiv \lambda f(z_1, \ldots, z_n)\,.$$

□

Using Theorems 2.1 and 2.2, we can give an equivalent definition of matrix-functions of the class $\mathcal{P}$:

Theorem 2.3. *For a rational matrix-function $f(z)$ to be positive real it is necessary and sufficient that the following three conditions are satisfied:*

$$1)\quad f(z) + f^*(z) \geq 0 \quad \textit{for} \quad z \in \mathcal{D}_R^+;$$
$$2)\quad f(\lambda z) = \lambda f(z) \quad \textit{for every} \quad \lambda \in \mathbb{C}, z \in \mathbb{C}^n;$$
$$3)\quad f(z) = f'(z) = \overline{f(\overline{z})}\,.$$

Proof. The necessity is the consequence of Theorems 2.1 and 2.2. To prove the sufficiency, we remark that if $z \in \mathcal{D}_R^+$ then $-z \in \mathcal{D}_R^-$. Therefore if $z \in \mathcal{D}_R^-$, then

$$f(z) + f^*(z) = -[f(-z) + f^*(-z)] \leq 0\,.$$

If $z \in \mathcal{D}_J^+$, then $(-iz) \in \mathcal{D}_R^+$. Thus,

$$i[f^*(z) - f(z)] = f(-iz) + f^*(-iz) \geq 0\,.$$

For the domain $\mathcal{D}_J^-$ the situation is analogous. □

Let us remark that the condition

$$f(z) + f^*(z) \geq 0 \quad \text{for} \quad z \in \mathcal{D}_R^+$$

can be weakened slightly. Namely, the following result holds:

Theorem 2.4. *Let $f(z)$ be a rational matrix function. Let us assume that for every $j = 1, 2, \ldots, n$ and for every real $\tau_1, \tau_2, \ldots, \tau_n$, the functions*

$$f_j = f(\,i\tau_1, \ldots, i\tau_{j-1}, z_j\,, i\tau_{j+1}\,, \ldots, i\tau_n\,)$$

satisfy the condition

$$f_j + f_j^* \geq 0 \quad \textit{for} \quad \operatorname{Re} z_j > 0\,.$$

Then the inequality

$$f(z) + f^*(z) \geq 0 \quad \textit{for} \quad z \in \mathcal{D}_R^+$$

holds.

Proof. For a vector ξ that does not depend on z, let us consider the function $f_\xi(z) = \xi^* f(z)\xi$. It is clear that

$$2\operatorname{Re} f_\xi(z) = \xi^*[f(z) + f^*(z)]\xi\,.$$

Therefore if the inequality

$$\operatorname{Re} f_\xi\,(z) \geq 0 \quad \text{for} \quad z \in \mathcal{D}_R^+$$

holds for every ξ, then the inequality

$$f(z) + f^*(z) \geq 0 \quad \text{for} \quad z \in \mathcal{D}_R^+$$

holds as well. Therefore, it is enough to consider scalar functions only. Let a scalar function $f(z)$ satisfy the assumptions of the theorem. We consider the rational function

$$u(\zeta_1, \dots, \zeta_n) = \frac{f\left(\frac{1+\zeta_1}{1-\zeta_1}, \dots, \frac{1+\zeta_n}{1-\zeta_n}\right) - 1}{f\left(\frac{1+\zeta_1}{1-\zeta_1}, \dots, \frac{1+\zeta_n}{1-\zeta_n}\right) + 1}. \tag{2.1}$$

It is clear, that for every $j = 1, 2, \dots, n$ and for every $t_1 \in \mathbb{T}, \dots, t_{j-1} \in \mathbb{T}, t_{j+1} \in \mathbb{T}, \dots, t_n \in \mathbb{T}$, the function $u(t_1, \dots, t_{j-1}, \zeta_j, t_{j+1}, \dots, t_n)$ is holomorphic for $|\zeta_j| < 1$ and satisfies the inequality

$$|u(t_1, \dots, t_{j-1}, \zeta_j, t_{j+1}, \dots, t_n)| < 1 \quad \text{for} \quad |\zeta_j| < 1\,. \tag{2.2}$$

Let us show that the function $u(\zeta_1, \dots, \zeta_n)$ is holomorphic in the polydisk $\mathbb{D}^n = \{\zeta : |\zeta_1| < 1, \dots, |\zeta_n| < 1\}$ and satisfies the inequality

$$|u(\zeta_1, \dots, \zeta_n)| < 1 \quad \text{for} \quad \zeta \in \mathbb{D}^n\,.$$

To prove this, we consider the Fourier coefficients $\hat{u}(k_1, \dots, , k_n)$ of the function $u(t)$ considered on the torus $\mathbb{T}^n = \{(t_1, \dots, t_n) : |t_1| = 1, \dots, |t_n| = 1\}$:

$$\hat{u}(k_1, \dots, , k_n) = \int\limits_{\mathbb{T}^n} \cdots \int u(t_1, \dots, t_n) t_1^{-k_1} \cdots t_n^{-k_n}\, m(dt_1) \cdots m(dt_n)$$

($m(dt)$ is the one-dimensional normalized Lebesgue measure). The function u is contractive on $\mathbb{T}^n$: $|u(t)| \leq 1$ for $t \in \mathbb{T}^n$. Therefore, its Fourier coefficients $\hat{u}(k_1, \dots, , k_n)$ exist. If $k_j < 0$ at least for one $j = 1, \dots, n$, then $\hat{u}(k_1, \dots, , k_n) = 0$. Indeed, for definiteness, let $k_1 < 0$. Then

$$\hat{u}(k_1, \dots, , k_n) = \int\limits_{\mathbb{T}^{n-1}} \cdots \int t_2^{-k_2} \cdots t_n^{-k_n}\, m(dt_2) \cdots m(dt_n)$$
$$\times \int\limits_{\mathbb{T}} u(t_1, \dots, t_n) t_1^{-k_1} m(dt_1)\,,$$

By condition (2.2), the inner integral vanishes. Therefore, the Fourier coefficients $\hat{u}(k_1, \dots, , k_n)$ determine the function

$$g(\zeta_1, \dots, \zeta_n) \stackrel{\text{def}}{=} \sum_{\forall k} \hat{u}(k_1, \dots, k_n)\, \zeta_1^{k_1} \cdots \zeta_n^{k_n}$$

which is holomorphic in $\mathbb{D}^n$. On the other hand, denoting $\zeta_j = r_j t_j$ $(r_j \geq 0,\ t_j \in \mathbb{T})$, we obtain

$$g(\zeta_1, \dots, \zeta_n) = \sum_k r_1^{k_1} \cdots r_n^{k_n} \cdot \hat{u}(k_1, \dots, k_n)\, t_1^{k_1} \cdots t_n^{k_n}.$$

Therefore, $g(\zeta_1, \dots, \zeta_n)$ is the convolution of the function $u(t_1, \dots, t_n)$ and of the Poisson kernel

$$P(r,t) = \sum_k r_1^{|k_1|} \cdots r_n^{|k_n|}\, t_1^{k_1} \cdots t_n^{k_n}.$$

Since $|u(t_1, \dots, t_n)| \leq 1$ on $\mathbb{T}^n$ and

$$\int \cdots \int_{\mathbb{T}^n} P(r_1, \dots, r_n; t_1, \dots, t_n)\, m(dt_1) \dots\, m(dt_n) = 1,$$

we have

$$|g(\zeta_1, \dots, \zeta_n)| \leq 1 \quad \text{for} \quad \zeta \in \mathbb{D}^n.$$

Since the Poisson kernel is an approximate identity,

$$\lim_{r \to 1-0} g(rt_1, \dots, rt_n) = u(t_1, \dots, t_n)$$

in every point $t = (t_1, \dots, t_n) \in \mathbb{T}^n$ where the function u is continuous. By the uniqueness theorem,

$$u(\zeta) \equiv g(\zeta) \text{ for } \zeta = (\zeta_1, \dots, \zeta_n) \in \mathbb{D}^n.$$

Thus,

$$|u(\zeta_1, \dots, \zeta_n)| \leq 1 \quad \text{for} \quad \zeta \in \mathbb{D}^n.$$

Returning to $f(z_1, \dots, z_n)$ by means of the transformation that is inverse to the transformation (2.1) we obtain the statement of the theorem. □

Theorem 2.5. *If $f(z)$ is a positive real matrix-function, then*

$$\frac{\partial f}{\partial z_j}(x_1, \dots, x_n) \geq 0$$

for every $j = 1, \dots, n$ and for every real $x_1, \dots, x_n$.

Proof. Let $g(z) = \xi^* f(z) \xi$ where ξ is an arbitrary column vector with complex entries. Then $g(z)$ is a scalar rational function satisfying the conditions

$$\operatorname{Im} g(z) \geq 0 \qquad \text{for} \quad \operatorname{Im} z_j > 0;$$

$$\operatorname{Im} g(z) = 0 \qquad \text{for} \quad \operatorname{Im} z_j = 0.$$

Denoting the real and the imaginary part of $g(z)$ by $u(z)$ and $v(z)$ respectively and using the Cauchy-Riemann equations, we obtain

$$\frac{\partial g}{\partial z_j}(x_1, \dots, x_n) = \frac{\partial u}{\partial x_j} = \frac{\partial v}{\partial y_j} = \frac{\partial\,[\operatorname{Im} g]}{\partial y_j} \geq 0,$$

where $z_j = x_j + iy_j$. Since ξ is an arbitrary vector,

$$\frac{\partial f}{\partial z_j}(x_1, \ldots, x_n) \geq 0 \quad \text{for} \quad x_j \in \mathbb{R} \ (j = 1, \ldots, n).$$

□

To prove the following property, we need

Theorem 2.6. *Let $n > 1$, $P(z)$ and $Q(z)$ be coprime polynomials of n variables $z_1, \ldots, z_n$ such that $P(0) = 0$, $Q(0) = 0$. Let Ω be an arbitrary neighborhood of the origin in $\mathbb{C}^n$ and $f(z) = \frac{P(z)}{Q(z)}$. Then for an arbitrary complex number α, there exists a point z_0 in Ω such that*

$$Q(z_0) \neq 0 \quad \text{and} \quad f(z_0) = \alpha .$$

A proof can be found in [Rud]: Theorem 1.3.2.

Theorem 2.7. *Let $f(z) = \frac{P(z)}{Q(z)}$ be a scalar rational positive real function where $P(z)$ and $Q(z)$ are coprime polynomials. Then the polynomials $P(z)$ and $Q(z)$ do not vanish in the domains $\mathcal{D}_R^+$, $\mathcal{D}_R^-$, $\mathcal{D}_J^+$, $\mathcal{D}_J^-$.*

Proof. For definiteness, consider the domain $\mathcal{D}_R^+$. Let $z_0 \in \mathcal{D}_R^+$ and $Q(z_0) = 0$. Since f is holomorphic at the point z_0, $P(z_0) = 0$. Let $\Omega \in \mathcal{D}_R^+$ be a neighborhood of the point z_0. Take α : Re$\alpha < 0$. According to Theorem 2.6, we find $z_1 \in \Omega$ such that $Q(z_1) \neq 0$ and $f(z_1) = \alpha$. Since $\operatorname{Re} f(z_1) \geq 0$, this is impossible. To prove that $P(z)$ does not vanish in $\mathcal{D}_R^+$, we consider the function $f^{-1}(z)$: $\operatorname{Re} f^{-1}(z) \geq 0$ for $z \in \mathcal{D}_R^+$. □

Theorem 2.8. *Let $f(z) = \frac{P(z)}{Q(z)}$ be a scalar rational positive real matrix function where $P(z)$ and $Q(z)$ are coprime polynomials. Then for every $j = 1, \ldots, n$, the functions*

$$f_1(z) = \frac{Q(z)}{\frac{\partial Q}{\partial z_j}(z)} \qquad \text{and} \qquad f_2(z) = \frac{P(z)}{\frac{\partial P}{\partial z_j}(z)}$$

are positive real.

Proof. (For definiteness, we take $j = 1$.) First of all we remark that

$$f_1(\lambda z) = \lambda f_1(z) \quad \text{and} \quad f_2(\lambda z) = \lambda f_2(z) .$$

Let us show that $\operatorname{Re} f_1(z) \geq 0$ for $z \in \mathcal{D}_R^+$. This condition is equivalent to the condition

$$\operatorname{Re} f_1^{-1}(z) \geq 0 \quad \text{for} \quad z_j \in \mathcal{D}_R^+ .$$

Let us fix $z_0' = (z_2^0, \ldots, z_n^0)$; $\operatorname{Re} z_j^0 > 0$, $j = 2, \ldots, n$. Since $Q(z)$ is a denominator of a rational positive function, $Q(z_1, z_2^0, \ldots, z_n^0)$ does not vanish in $\operatorname{Re} z_1 > 0$. Therefore

$$Q(z_1, z_0') = \alpha_0 \prod_{1 \leq j \leq m} [z_1 - \beta_j(z_0')]^{k_j} ,$$

where $\operatorname{Re}\beta_j(z_0') \leq 0$. Hence

$$f_1^{-1}(z) = \sum_{1\leq j\leq m} \frac{k_j}{z_1 - \beta_j(z_0')}$$

and

$$\operatorname{Re} f_1^{-1}(z) = \sum_{1\leq j\leq m} \frac{k_j\left[z_1 + \overline{z}_1 - \left(\beta_j(z_0') + \overline{\beta_j(z_0')}\right)\right]}{\left|z_1 - \beta_j(z_0')\right|^2} \geq 0 \quad \text{for} \quad \operatorname{Re} z_1 \geq 0\,.$$

For the function f_2, the reasoning is analogous. □

2. In the sequel, if $f(z) = \left\{f_{ij}(z)\right\}_{i,j=1}^{k}$ is a rational matrix-function and $q(z)$ is a common denominator of the entries $f_{ij}(z)$, we represent $f(z)$ in the form

$$f(z) = \frac{P(z)}{q(z)}$$

where

$$P(z) = \{p_{ij}(z)\}_{i,j=1}^{k}$$

is a polynomial matrix function and $q(z)$ is a polynomial. Clearly, in order for a matrix function $f(z)$ to be positive real, it is necessary and sufficient that for any constant vector ξ of appropriate size, the scalar function

$$g(z) = \xi f(z)\xi^*$$

is positive real.

Theorem 2.9. *Let $f(z) = P(z)/q(z)$ be a rational positive real matrix function, with $P(z)$ a matrix polynomial and $q(z)$ a polynomial. Then, for every $j = 1,\dots n$, the matrix function*

$$f_j(z) = \frac{\frac{\partial P}{\partial z_j}(z)}{\frac{\partial q}{\partial z_j}(z)}$$

is positive real.

Proof. Let ξ be an arbitrary vector. Then, the scalar function of $(n+1)$ variables

$$z_{n+1} + \frac{\xi P(z)\xi^*}{q(z)} = \frac{z_{n+1}q(z) + \xi P(z)\xi^*}{q(z)}$$

is positive real.

In view of Theorems 2.7 and 2.8, the polynomial

$$z_{n+1}\frac{\partial q}{\partial z_j}(z) + \xi\frac{\partial P}{\partial z_j}(z)\xi^*$$

does not vanish for $z \in \mathcal{D}_R^+$, $\operatorname{Re} z_{n+1} > 0$. Since the equation (with respect to z_{n+1})

$$z_{n+1}\,\frac{\partial q}{\partial z_j}(z) + \xi\,\frac{\partial P}{\partial z_j}(z)\,\xi^* = 0$$

has a unique solution

$$z_{n+1}^0 = -\frac{\xi \dfrac{\partial P}{\partial z_j} \xi^*}{\dfrac{\partial q}{\partial z_j}},$$

we conclude that $\operatorname{Re} z_{n+1}^0 \leq 0$. Thus $\operatorname{Re} \xi \dfrac{\dfrac{\partial P}{\partial z_j}}{\dfrac{\partial q}{\partial z_j}} \xi^* \geq 0 \quad \text{for} \quad z \in \mathcal{D}_R^+ \cdot$ □

Theorem 2.10. *Let $f(z)$ be a positive $m \times m$ matrix function and*

$$k = \sup_{z \in \mathbb{C}^n} \operatorname{rank} f(z).$$

Then there exists a constant (i.e. not depending on z) unitary matrix U (i.e. $UU^* = I_m$) such that

$$f(z) = U \begin{bmatrix} f_1(z) & 0 \\ 0 & 0 \end{bmatrix} U^*,$$

where $f_1(z)$ is a non-singular positive $k \times k$ matrix-function.

Proof. Since $\operatorname{rank} f(z)$ takes integer values, there exists a point

$$x_0 = (x_1^0, \ldots, x_n^0) \qquad (x_j > 0)$$

such that $\operatorname{rank} f(x_0) = k$.[3]

To prove the theorem it is enough to prove that if $f(z_0)\xi_0 = 0$ for some vector ξ_0, then $f(z)\xi_0 \equiv 0$ for all $z \in \mathbb{C}^n$. Since $f(z)$ is positive, the function

$$S(\zeta_1, \ldots, \zeta_n) = \left[f\left(\frac{1+\zeta_1}{1-\zeta_1}, \ldots, \frac{1+\zeta_n}{1-\zeta_n}\right) - I_m\right]\left[f\left(\frac{1+\zeta_1}{1-\zeta_1}, \ldots, \frac{1+\zeta_n}{1-\zeta_n}\right) + I_m\right]^{-1}. \quad (2.3)$$

is well defined for $\zeta = (\zeta_1, \ldots, \zeta_n)$ from the unit polydisk $\mathbb{D}^n$. From the identity

$$I_m - S(z)^* S(z) = 2f + I_m)^{*-1}(f + f^*)(f + I_m)^{-1}$$

we conclude that

$$S(\zeta)^* S(\zeta) \leq I_m \quad \text{for} \quad \zeta \in \mathbb{D}^n,$$

and, moreover, $S(\zeta)$ is a matrix function analytic in the polydisk $\mathbb{D}^n$.

We note that that $f\xi = 0$ for some vector ξ if and only if $S\xi = -\xi$. This is easily seen from (2.3) and from its equivalent form $S = (f + I_m)^{-1}(f - I_m)$.

[3] *Translators' Note:* Of course there is no a priori reason why the maximal rank k should be achieved at a point all of whose coordinates are real and positive. However it follows from the reasoning below that given any point x_0 all of whose coordinates are real and positive, with $\operatorname{rank} f(x_0) = k_0$, there exists a required factorization with $f_1(z)$ a non-singular positive $k_0 \times k_0$ matrix-function. Hence necessarily $k_0 = k$.

Let now $x_0 = (x_1^0, \dots, x_n^0)$ $(x_j^0 > 0, \forall j)$, such that $f(x_0)\xi = 0$. Then, for some inner point ζ_0 of the polydisk $\mathbb{D}^n$,

$$\eta(\zeta_0) = S(\zeta_0)\xi = -\xi \tag{2.4}$$

where $\eta(\zeta) \overset{\text{def}}{=} S(\zeta)\xi$ is a vector function holomorphic in $\mathbb{D}^n$.

Since $\eta(\zeta)^*\eta(\zeta) \leq \xi^*\xi$ for $\zeta \in \mathbb{D}^n$ and $\eta(\zeta_0)^*\eta(\zeta_0) = \xi^*\xi$, $\eta(\zeta)$ is a constant vector function: $\eta(\zeta) \equiv -\xi_0$. Therefore

$$S(\zeta)\xi_0 \equiv -\xi_0 \quad \text{and} \quad f(z)\,\xi_0 \equiv 0$$

Making a diagonalization of $f(x_0)$ we obtain the assertion of the theorem. □

3. Properties of functions representable as the resolvent of a bundle of non-negative matrices

Before turning to the examination of the properties of functions which admit a representation as the resolvent of a bundle of non-negative definite (positive semidefinite) matrices, we prove the following simple lemma:

Lemma 3.1. *Let the matrix function*

$$g(z) = \begin{bmatrix} a(z) & b(z) \\ b'(z) & c(z) \end{bmatrix}$$

be positive real, and $\det c(z) \not\equiv 0$. *Then the matrix function*

$$f(z) = a(z) - b(z)c(z)^{-1}b'(z)$$

is positive real.

Proof. Let the dimensions of the block matrix $a(z)$ be $k \times k$. If $\det g(z) \neq 0$, then clearly,

$$f(z) = \left(\pi_k g^{-1}(z)\pi_k^*\right)^{[-1]},$$

where

$$\pi_k = \begin{bmatrix} I_k & 0 \\ 0 & 0 \end{bmatrix}.$$

The claim of the lemma follows then from the equality

$$f(z) \pm f(z)^* = f(z)\pi_k g^{-1}(z)(g(z) \pm g(z)^*)g^{-1}(z)^*\pi_k^* f(z)^*. \tag{3.1}$$

Assume $\det g(z) \equiv 0$ and suprank $g(z) = r + (n-k)$. It follows from Theorem 2.10 that there exists a constant orthogonal matrix

$$U = \begin{bmatrix} U_{11} & U_{12} \\ U_{21} & U_{22} \end{bmatrix}$$

such that

1.
$$g(z) = \begin{bmatrix} U_{11} & U_{12} & 0 \\ U_{21} & U_{22} & 0 \\ 0 & 0 & I \end{bmatrix} \begin{bmatrix} a_1(z) & 0 & b_1(z) \\ 0 & 0 & 0 \\ b_1'(z) & 0 & c(z) \end{bmatrix} \begin{bmatrix} U_{11}' & U_{21}' & 0 \\ U_{12}' & U_{22}' & 0 \\ 0 & 0 & I \end{bmatrix}$$

2.
$$g_1(z) = \begin{bmatrix} a_1(z) & b_1(z) \\ b_1'(z) & c(z) \end{bmatrix} \in \mathcal{P};$$

3. $\det g_1(z) \not\equiv 0$.

Then,
$$\begin{aligned} f(z) &= a(z) - b(z)c(z)^{-1}b'(z) \\ &= U \begin{bmatrix} a_1(z) - b_1(z)c(z)^{-1}b_1'(z) & 0 \\ 0 & 0 \end{bmatrix} U' \\ &= U \begin{bmatrix} \left(\pi_r g_1(z)^{-1}\pi_r^*\right)^{[-1]} & 0 \\ 0 & 0 \end{bmatrix} U'. \end{aligned}$$

Writing for $f(z)$ a formula analogous to (3.1) we obtain the claim of the lemma. □

2. Let $f(z)$ be a $\mathbb{C}^{k\times k}$-valued matrix function which can be represented in the form $f(z) = A_{11}(z) - A_{12}(z)A_{22}(z)^{-1}A_{12}(z)'$, where
$$\begin{bmatrix} A_{11}(z) & A_{12}(z) \\ A_{12}'(z) & A_{22}(z) \end{bmatrix} = z_1A_1 + \cdots + z_nA_n = A(z),$$
and the matrices A_j $(j = 1, \dots n)$ are real non-negative definite.

Clearly, $A(z) \in \mathcal{P}$ and, in accordance with Lemma 3.1, the matrix function $f(z)$ is positive real. By Theorem 2.10, it is enough to consider the nonsingular case.

Everywhere in what follows we consider nonsingular matrix functions which can be represented in the form
$$f(z) = \left[\pi A(z)^{-1}\pi^*\right]^{[-1]}, \tag{3.2}$$
where
$$A(z) = z_1A_1 + \cdots + z_nA_n$$
is a bundle of non-negative definite real matrices.

3. Let $f(z) = P(z)/q(z)$ be a rational positive real matrix function, where $P(z)$ and $q(z)$ are respectively a matrix polynomial and a polynomial which are relatively prime. Then, by Theorem 2.5, the matrix polynomial
$$\mathcal{F}(z) = q(z)\frac{\partial P}{\partial z_j}(z) - P(z)\frac{\partial q}{\partial z_j}(z) = q(z)^2\frac{\partial f}{\partial z_j}(z)$$
is non-negative definite for real $z_k = x_k$ $(k = 1, \dots, n)$.

The following theorem shows that if $f(z)$ is moreover of the form (3.2), then $\mathcal{F}(z)$ admits a representation as a sum of "squares" of rational matrix functions, analytic in the sets $\mathcal{D}_R^+$, $\mathcal{D}_R^-$, $\mathcal{D}_J^+$ and $\mathcal{D}_J^-$. The problem of representing a non-negative polynomial function of several variables as a sum of squares of rational functions (Hilbert's 17th problem) was solved by Artin, see [Lan]. It is interesting that the following theorem allows to construct examples of polynomials of several variables which admit a representation as sums of squares of rational functions analytic in certain domains.

Theorem 3.1. *Set $f(z) = P(z)/q(z)$, where $P(z)$ and $q(z)$ are respectively a matrix polynomial and a polynomial which are relatively prime. Assume that f can be represented as*

$$f(z) = \left(\pi(z_1 A_1 + \cdots + z_n A_n)^{-1}\right)^{[-1]},$$

where the A_j $(j = 1, \ldots, n)$ are real non-negative definite matrices.

Then, the matrix function

$$\mathcal{F}(z) = q(z)\frac{\partial P}{\partial z_j}(z) - P(z)\frac{\partial q}{\partial z_j}(z)$$

admits a representation as a sum of squares of rational functions $S_k(z)$ analytic in the domains $\mathcal{D}_R^+$, $\mathcal{D}_R^-$, $\mathcal{D}_J^+$ and $\mathcal{D}_J^-$:

$$\mathcal{F}(z) = \sum_1^N S_k(z) S_k(z)'.$$

Proof. Since $A_k \geq 0$ $(k = 1, \ldots, n)$, $A(z)^{-1}$ and $f(z)$ are holomorphic in the domains $\mathcal{D}_R$, $\mathcal{D}_-$, $\mathcal{D}_J^+$ and $\mathcal{D}_J^-$. Then,

$$\frac{\partial f}{\partial z_j}(z) = f(z)\pi A(z)^{-1} A_j A(z)^{-1} \pi^* f(z).$$

It is clear that the matrix function $f(z)\pi A(z)^{-1}$ is holomorphic in the required domains. Moreover, since the A_j are real non-negative definite, then

$$A_j = T_j(\Lambda_{j1} + \cdots + \Lambda_{jn})T_j',$$

where the Λ_{jk} $(k = 1, \ldots n)$ are diagonal matrices of rank one, the entries of which are 0 or 1. Therefore, setting

$$S_k(z) = q(z) f(z) \pi A(z)^{-1} \Lambda_{jk},$$

we obtain

$$\mathcal{F}(z) = q(z)^2 \frac{\partial f}{\partial z_j}(z) = \sum_{k=1}^n S_k(z) S_k(z)',$$

as required. □

4. In conclusion of this section we consider yet another representation for matrix functions of the form (3.2). This representation is the analogue of the representation of Hefer[4] for functions holomorphic in a domain of holomorphy $\mathcal{G}$ by means

[4] This theorem is presented in the paper [Hef]. In a footnote to the paper, Behnke and Stein state that the author died in 1941 and that the paper represents a part of his 1940 Munster dissertation. They also mention that papers by Oka [Jap. J. Math. 17, 523–531 (1941)] and Cartan [Ann. Sci.

of functions $p_j(z,\zeta)$ holomorphic in the domain $\mathcal{G} \times \mathcal{G}$

$$f(z) - f(\zeta) = \sum_{j=1}^{n} (z_j - \zeta_j) p_j(z,\zeta).$$

Theorem 3.2. *Let a matrix-function $f(z)$ be represented in the form*

$$f(z) = \left[\pi\,(z_1 A_1 + \ldots + z_n A_n)^{-1}\,\pi^*\right]^{[-1]},$$

where A_j are non-negative real matrices.

Then $f(z)$ is representable in the form

$$f(z) = \sum_{1 \le j \le n} z_j \Phi_j(z,\, \zeta),$$

where $\Phi_j(z,\, \zeta)$ are matrix-functions which are holomorphic in the domains $\mathcal{D}_R^+ \times \mathcal{D}_R^+$, $\mathcal{D}_R^- \times \mathcal{D}_R^-$, $\mathcal{D}_J^+ \times \mathcal{D}_J^+$, $\mathcal{D}_J^- \times \mathcal{D}_J^-$ and satisfy the following condition: for every set of points $\{z_k\}_{1 \le k \le N}$ the inequality

$$\begin{bmatrix} \Phi_j(z_1,\, \overline{z}_1) & \ldots & \Phi_j(z_1,\, \overline{z}_N) \\ \vdots & \ddots & \vdots \\ \Phi_j(z_N,\, \overline{z}_1) & \ldots & \Phi_j(z_N,\, \overline{z}_N) \end{bmatrix} \ge 0 \tag{3.3}$$

holds.

Remark 3.1. Except the case $n = 2$, the functions $\Phi_j(z,\, \zeta)$ are determined non-uniquely from $f(z)$. In the case $n = 2$,

$$\Phi_1(z,\, \zeta) = \frac{\zeta_2 f(z) - z_2 f(\zeta)}{z_1 \zeta_2 - z_2 \zeta_1};$$

$$\Phi_2(z,\, \zeta) = \frac{\zeta_1 f(z) - z_1 f(\zeta)}{z_1 \zeta_2 - z_2 \zeta_1}.$$

In the case of two variables, the inequalities (3.3) are a consequence of the Schwarz-Pick inequality for functions with positive real part in the right half-plane. Therefore the inequalities (3.3) can be considered as a generalization of the Schwarz-Pick inequality to functions in the class $\mathcal{P}$ representable as the resolvent of a bundle of non-negative matrices.

Proof. From the representation

$$f(z) = \left[\pi\,(z_1 A_1 + \ldots + z_n A_n)^{-1}\,\pi^*\right]^{[-1]},$$

Ecole Norm. Sup. (3) 61, 149–197 (1944)] which have appeared since then contain Hefer's results proved by different methods. The formulation of Hefer's Theorem can also be found in [Fu1]: Theorem 22.1, its proof in [Fu2]: Theorem 7.3.

it follows that

$$f(z) \pm f(\zeta) = \left[\pi A^{-1}(z)\pi^*\right]^{[-1]} \pm \left[\pi A^{-1}(\zeta)\pi^*\right]^{[-1]}$$
$$= f(z)\pi A^{-1}(z)\left[A(z) \pm A(\zeta)\right]A^{-1}\zeta)\pi^* f(\zeta)\,.$$

Denoting $f(z)\pi A^{-1}(z)$ by $\varphi(z)$, we obtain

$$f(z) \pm f(\zeta) = \sum_{1\le j\le n} (z_j \pm \zeta_j)\varphi(z)A_j\varphi'(\zeta)\,,$$

or

$$f(z) \pm f(\zeta) = \sum_{1\le j\le n} (z_j \pm \zeta_j)\Phi_j(z,\,\zeta)$$

where

$$\Phi_j(z,\zeta) = \varphi(z)A_j\varphi'(\zeta)\,.$$

Since $A_k \ge 0$, the matrix functions Φ_j are holomorphic in the domains $\mathcal{D}_R^+ \times \mathcal{D}_R^+$, $\mathcal{D}_R^- \times \mathcal{D}_R^-$, $\mathcal{D}_J^+ \times \mathcal{D}_J^+$, $\mathcal{D}_J^- \times \mathcal{D}_J^-$. The inequality (3.3) is equivalent to the inequality

$$[\Phi(z_1),\,\ldots\,,\,\Phi(z_N)]\begin{bmatrix} A_j & \cdots & A_j \\ \vdots & \ddots & \vdots \\ A_j & \cdots & A_j \end{bmatrix}\begin{bmatrix} \Phi(z_1) \\ \vdots \\ \Phi(z_N) \end{bmatrix} \ge 0\cdot$$

Since $A_j \ge 0$, the latter inequality holds. □

References

[AD] Aizenberg, L.A. and Sh.A. Dautov: *Golomorfnye funktsii mnogikh peremennykh s neotritsatel'noĭ deĭstvitel'noĭ chast'yu. Sledy golomorfnykh i plyurigarmonicheskikh funktsiĭ na granitse Shilova.* (Russian). Matem. Sbornik, (Novaya. Ser.) **99(141)**, pp. 342–355 (1976). English transl.: Holomorphic functions of several complex variables with non-negative real part. Traces of holomorphic and pluriharmonic functions on the Silov boundary. (English) Math. USSR, Sb. **28** (1976), 301–313.

[Bos1] Bose, N.K.: *New techniques and results in multidimensional problems.* Journal of the Franklin Inst., **301**, no. 1/2 (Jan.–Feb. 1976), Special issue: *Recent trends in systems theory*, pp. 83–101.

[Bos2] Bose, N.K.: *Problems and progress in multidimensional system theory.* In [Mult1], pp. 824–840.

[EfPo] Efimov, A.V. and V.P. Potapov: *J-rastyagivayushchie matritsy-funktsii i ikh rol' v analiticheskoĭ teorii elektricheskikh tsepeĭ* (Russian). Uspekhi Matematicheskikh Nauk, **28**:1 (1973), pp. 65–130. English transl: *J-expanding matrix functions and their role in the analytical theory of electrical circuits.* Russ. Math. Surveys, **28**:1 (1973), pp. 69–140.

[Fu1] FUKS, B.A.: *Vvedenie v Teoriyu Analiticheskikh Funktsiĭ Mnogikh Kompleksnykh Peremennykh* (Russian). Fizmztgiz, Moscow 1962, 419 p. English transl.: *Introduction to the Theory of Analytic Functions of Several Complex Variables.* (Series: Translations of mathematical monographs, vol. 8). American Mathematical Society, Providence, R. I., 1963, x+388 p.

[Fu2] FUKS, B.A.: *Spetsial'nye Glavy Teorii Analiticheskikh Funktsiĭ Mnogikh Kompleksnykh peremennykh* (Russian). Fizmztgiz, Moscow 1963, 427 p. English transl.: *Special chapters in the theory of analytic functions of several complex variables.* (Series: Translations of mathematical monographs, vol. **14**). American Mathematical Society, Providence, R. I.,1965, vi+357 p.

[Hef] HEFER, H.: *Zur Funktionentheorie mehrerer Veränderlichen. Über die Zerlegung analytischer Funktionen und die Weilische Integraldarstellung.* Math. Ann., **122** (1950), pp. 276–280.

[Kar] KARNI, SH.: *Network theory; analysis and synthesis.* (Allyn and Bacon series in electrical engineering). Allyn and Bacon, Boston 1966, xi+483 p. Russian transl.: *Teoriya tsepeĭ: Analiz i Sintez.*, Izdat. "Svyaz", Moscow 1968, 368 p.

[KFA] KALMAN, R., P. FALB and M. ARBIB: *Topics in Mathematical Systems Theory*, McGraw-Hill, New York, 1969. Russian Transl.: *Očerki po Matematičeskoĭ Teorii Sistem*, Mir, Moskwa, 1971.

[Kog] KOGA, T.: *Synthesis of finite passive n-ports with prescribed positive real matrices of several variables.* IEEE Trans. Circuit Theory **CT 15**, no.1 (March 1968), pp. 2–23.

[Lan] LANG, S.: *Algebra.* Addison-Wesley Publishing Co., Inc., Reading, Mass. 1965 xvii+508 pp.

[Liv] LIVŠIČ, M.S.(=LIVSHITS, M.S.): *Operatory, Kolebaniya, Volny. Otkrytye Sistemy* (Russian). English transl. (edited by R. HERDEN.): *Operators, oscillations, waves (open systems).* (Translations of Mathematical Monographs, Vol. **34**.) American Mathematical Society, Providence, R.I., 1973. vi+274 pp.

[Me] MELAMUD, E.YA.: *Ob odnom obobshchenii teoremy Darlingtona.* [A certain generalization of Darlington's theorem](Russian). Izvestiya Akad. Nauk Armyanskoĭ SSR, ser. Matem., **7**:3 (1972), pp. 183–195, 226.

[Mult1] Proceedings of the IEEE, **65**:7 (June 1977). *Special issue on multidimensional system.* (BOSE, N.K.-ed.), pp. 819–980.

[Mult2] *Multidimensional Systems Theory.* (*Progress, Directions and Open Problems in Multidimensional Systems*). D. Reidel Publishing Company, Dordrecht·Boston·Lancaster 1985.[5]

[Pot1] POTAPOV, V.P.: *Mul'tiplikativnaya struktura J-rastyagivayushchikh matritsfunktsiĭ* (Russian). Trudy Moskovskogo Matematicheskogo Obshchestba, **4** (1955), pp. 125–236. English transl.: *The multiplicative structure of J-contractive matrix functions.* Amer. Math. Soc., Transl., (Ser. **2**), **15** (1960), pp. 131–243.

[Pot2] POTAPOV, V.P.: *Drobno-lineĭnye preobrazovaniya matrits* (Russian). In: Issledovaniya po Teorii Operatorov i ikh Prilozheniyam. [Studies in the Theory of Operators and their Applications], (MARCHENKO, V.A. and V.YA. GOLODETS

[5] This bibliography item was added in the translation.

– editors.) Naukova Dumka, Kiev 1979, pp. 75–97, 177. English transl.: *Linear fractional transformations of matrices.* Amer. Math. Soc. Translations (Ser. **2**), vol.**138** (1988), (Seven papers translated from the Russian. Edited by Ben Silver), American Mathematical Society, Providence, RI, 1988, viii+77 pp., pp. 21–35.

[Rud] Rudin, W.: *Function theory in polydiscs.* W.A. Benjamin, New York · Amsterdam 1969, vii+188 pp. Russian transl.: *Teoriya Funktsiĭ v polikruge.* Mir, Moscow 1974.

[SeRe] Seshu, S. and M.B. Reed: *Linear Graphs and Electrical Networks.* (Addison-Wesley series in the engineering sciences. Electrical and control systems). Addison-Wesley, Reading, MA, 1961. 315 p. Russian transl.: *Lineĭnye grafy i elektricheskie tsepi.* Izdat. "Vysshaya shkola", Moskow 1971, 448 p.

[Vla1] Vladimirov, V.S.: *Metody teorii funktsiĭ mnogikh kompleksnykh peremennykh.* (Russian). [Methods in the theory of functions of several complex variables] With a Foreword by N.N. Bogoljubov. Izdat. "Nauka", Moscow 1964, 411 pp. English transl: *Methods of the theory of functions of many complex variables.* Translation edited by Leon Ehrenpreis. The M.I.T. Press, Cambridge, MA·London 1966 xii+353 pp. French transl.: *Les fonctions de plusieurs variables complexes et leur application à la théorie quantique des champs.* (Travaux et Recherches Mathématiques, No. **14**). Dunod, Paris 1967 xv+338 pp.

[Vla2] Vladimirov, V.S.: *Golomorfnye funktsii mnogikh peremennykh s neotritsatel'noĭ deĭstvitel'noĭ chast'yu v trubchatoĭ oblasti nad konusom.* (Russian). Matem. Sbornik, **79** (1969), pp. 128–152. English transl.: *Holomorphic functions with non-negative imaginary part in a tubular domain over a cone.* Math. USSR, Sbornik, **8**, (1969), 125–146.

[Vla3] Vladimirov, V.S.: *Lineĭnye passivnye sistemy* (Russian). Teoret. i Mat. Fiz. **1** (1969), pp. 67–94. English transl.: *Linear passive systems.* Theoret. and Math. Phys. (Consultants Bureau), **1** (1969), pp. 51–72.

[Vla4] Vladimirov, V.S.: *Mnogomernye lineĭnye passivnye sistemy* [Multidimensional passive linear systems] (Russian). In: *Mekhanika Sploshnoĭ Sredy i Rodstvennye Problemy Analiza.* [Continuum mechanics and related problems of analysis] (Russian). (On the occasion of the eightieth birthday of Academician N.I. Musheliŝvili), pp. 121–134. Izdat. "Nauka", Moscow, 1972.

[Vla5] Vladimirov, V.S.: *Golomorfnye funktsii s polozhitel'noĭ mnimoĭ chast'yu v trube budushchego* (Russian). Matem. Sbornuk (Nov. Ser.), **93** (**135**) (1974), pp. 3–17. English transl.: *Holomorphic functions with positive imaginary part in the future tube.* Math. USSR, Sbornik **22** (1974), pp. 1–16.

[Vla6] Vladimirov, V.S.: *Golomorfnye funktsii s polozhitel'noĭ mnimoĭ chast'yu v trube budushchego. II* (Russian). Matem. Sbornuk (Nov. Ser.), **94** (**136**), (1974), pp. 3–17. English transl.: *Holomorphic functions with positive imaginary part in the future tube. II.* Math. USSR, Sbornik **22** (1974), pp. 1–16.

[Vla7] Vladimirov, V.S.: *Golomorfnye funktsii s polozhitel'noĭ mnimoĭ chast'yu v trube budushchego. III* (Russian). Matem. Sbornuk (Nov. Ser.), **98** (**140**), (1975), pp. 292–297. English transl.: *Holomorphic functions with positive imaginary part in the future tube. III.* Math. USSR, Sbornik **27** (1975), pp. 263–268.

[Vla8] VLADIMIROV, V.S.: *Golomorfnye funktsii s polozhitel'noĭ mnimoĭ chast'yu v trube budushchego. IV* (Russian). Matem. Sbornuk (Nov. Ser.), **104** (**146**), (1977), pp. 341–370. English transl.: *Holomorphic functions with positive imaginary part in the future tube. IV.* Math. USSR, Sbornik **33** (1977), pp. 301–325.

[Vla9] VLADIMIROV, V.S.: *Obobshennye Funktsii v Matematicheskoĭ fizike.* (Second edition). (Russian). Nauka, Moscow 1979. English transl.: *Generalized functions in mathematical physics.* Moscow: Mir Publishers. 362 p. French transl.: *Distributions en physique mathématique. 280 pp.* Moscou, Editions Mir 1979. Italian transl.: *Le distribuzioni nella fisica matematica.* 320 pp. Mosca: Edizioni Mir 1981.

[VlaDr] VLADIMIROV, V.S. and YU.N. DROZHZHINOV (=JU.N. DROŽŽINOV): *Golomorfnye funktsii v polikruge s neotritsatel'noĭ mnimoĭ chast'yu.* (Russian). Matem. zametki, **15** (1974), pp. 55–61. English transl.: *Holomorphic functions in a polycircle with non-negative imaginary part.* Math. Notes, **15**, (1974), 31–34.

M.F. Bessmertnyĭ
Svolody Square, 4
Department of Mathematics
Faculty of Physics
Kharkov National University
61077, Kharkov, Ukraina

Operator Theory:
Advances and Applications, Vol. 134, 187–208

The Poincaré-Hardy Inequality on the Complement of a Cantor Set

Cristian S. Calude and Boris Pavlov

Dedicated to Harry Dym, scientist and person of highest excellence.

Abstract. The Poincaré-Hardy inequality on the complement of the Cantor set E

$$\int \frac{|u|^2}{\text{dist}^2(x,E)}\,dm \le 4\mathcal{K}^2 \cdot \int |\nabla u|^2 dm$$

holds for every $u \in W_2^1(\mathcal{R}_3)$. Corresponding higher-order inequalities will be also derived. We use a special self-similar tiling and a natural metric on the space of trajectories generated by a Mauldin-Williams graph which is homeomorphic with the space of tiles endowed with the Euclidean distance. A crude estimation of the constant $\mathcal{K}$ is 2,100. Three applications will be briefly discussed. In the second one, the constant $\frac{1}{2}\mathcal{K}^{-1} \approx 0.0002$ plays the role of an estimate for the dimensionless Planck constant in the corresponding uncertainty principle.

1. Introduction

In Classical Analysis the Poincaré-Hardy inequality (see, for example, Hardy, Littlewood, Polya [10] or [8]; for a recent overview see Davies [6]) is one of most popular tools for comparing the generalized smoothness of a given function and its square integrability with a singular weight-function. The inequality is also used in Quantum Mechanics for deriving the uncertainty principle, Schiff [22], and in Mathematical Hydrodynamics for proving the existence and uniqueness of solutions of Navier-Stokes equations, Ladyzhenskaja [15]. Combined with the Garding inequality [9] it proves a surprisingly sharp instrument of qualitative spectral analysis of differential operators [2]; it even appears as a central point of the proof of semi-boundedness of solvable few-body Hamiltonians in Quantum Scattering [19]. A version of the Poincaré-Hardy inequality on the complement of a uniformly δ-regular set was derived in [1] in connection with the question on the uniqueness of the solution of the Dirichlet problem for second order elliptic equations in a domain with a uniformly δ-regular boundary. The uniform δ-regularity is equivalent to the existence of the corresponding superharmonic strong barrier function (see

Theorem 2 in [1]) and is invariant under conformal transformations of the space (an equivalent of uniform perfectness), [21, 12].

For our study of Dirichlet forms in Hilbert spaces of square integrable functions with singular weights we need Poincaré-Hardy inequalities in multidimensional spaces on complements of perfect zero-measure sets (actually fractals) with *explicit estimates of corresponding constants.* An exact description (in terms of capacities) of all "Hardy weights" d^{-1} for which the Poincaré-Hardy inequality can be written in the form

$$\int_\Omega |u|^2 \frac{dm}{[d]^2} \leq 4\mathcal{K}^2 \cdot \int_\Omega |\nabla u|^2 dm$$

is given by Maz'ja [17] (in Chapter 2). Methods of straightforward estimation of capacities for Cantor sets in geometric terms were suggested by Carleson in [5] (Chapter 4).

In the present note we derive the simplest version of the Poincaré-Hardy inequality and corresponding higher-order inequalities on the complement of a Cantor set in $\mathcal{R}_3$. Our approach is "non-capacitory"; it is based on combinatorial properties of a special self-similar tiling of the domain. We have chosen the Cantor set because of its simplicity and usefulness (Cantor sets are highly useful mathematical models for physical phenomena which include, for example, the distribution of galaxies in the universe and the fractal structure of the rings of Saturn, see Pickover [20], or [11]). We reduce the proof of the Poincaré-Hardy inequality to the estimation of a discretized integral which appears from the analysis of an analog of the strong barrier function, see Theorem 3.2 below. This estimation is based on the generating Mauldin-Williams graph of the Cantor set together with a proper measure constructed on all cylinder sets of trajectories produced by the generating finite automaton, see Calude [4].[1] The above measure leads to a metric space homeomorphic with the space of tiles endowed with the Euclidean distance.

This paper describes a simple case study of a connection between the analysis of smooth functions on the complement of a uniformly δ-regular set (or just a zero-measure perfect set), on one hand, and Symbolic Dynamics (see, for example, Schuster [23], Lind and Marcus [16]), on the other hand. Although the phenomenon studied is analytically trivial, still the characteristic features of a possible general construction can be already seen here:

- a special self-similar tiling of a neighborhood of a singular set, parameterized by trajectories generated by some Mauldin-Williams graph which defines the automorphisms of the set.
- a homeomorphism between an Euclidean metric structure on the tiling and the metric space of trajectories.

It is obvious that the above structures contain more information on the underlying set than just the uniform δ-regularity, so they may be used for a more precise

[1] Trajectories may be represented as paths on a binary Bruhat-Tits tree, [3].

estimation of the constant in the Poincaré-Hardy inequality (or even for deriving new versions of it).

In what follows we will also compute an estimation of the constant $\mathcal{K}^2$ appearing in the Poincaré-Hardy inequality. Our constant is certainly not the best; sharper estimates need more accurate operations with integrals on tiles.

2. Prerequisites

We denote by Σ the binary alphabet $\{0,1\}$ and by Σ^* the set of all non-empty binary strings, i.e., $\Sigma^* = \{0,1,00,01,10,11,000,\ldots\}$. If $a = a_1a_2\ldots a_n$ is a string of n digits, then its length is denoted by $|a| = n$. By Σ^l we denote the set of strings of length l. The *concatenation* of two strings a, c is denoted by ac. A string a is a prefix of a string b (we write $a \subset b$) in case $b = ac$, for some $c \in \Sigma^*$. The *negation* of a string $a \in \{0,1\}$ is denoted by $\bar{a} = a - 1$, so that $\bar{0} = 1$, $\bar{1} = 0$. For $a, d \in \Sigma^*$ we denote by $a \cap d$ the maximum common prefix of the strings a, d. Clearly, $|a \cap b| \leq \min\{|a|,|b|\}$, and $|a \cap d| = |a|$ if and only if $a \subset d$. Let Σ^ω be the set of all infinite binary sequences. In analogy with the case of strings, if σ and τ are two distinct sequences, then $\sigma \cap \tau$ denotes the maximum common prefix of σ and τ; of course, $\sigma\cap\tau$ is a string. If σ and τ are two distinct sequences in Σ^ω and r is a real number in the unit interval $(0,1)$, then $\delta_r(\sigma,\tau) = r^{|\sigma\cap\tau|}$ is an *ultrametric* on Σ^ω. The space $(\Sigma^\omega, \delta_r)$ is complete, compact and separable. For different r, s in $(0,1)$, the spaces $(\Sigma^\omega, \delta_r)$ and $(\Sigma^\omega, \delta_s)$ are homeomorphic. For more information see Edgar [7].

A *middle third Cantor set* is constructed by removing successive open middle thirds from a sequence of closed intervals. In the traditional construction, the one we are going to use in this paper, we are starting from the interval $\Delta = [0,1]$ (the pre-Cantor set of zero order) from which we remove the "middle third" $(1/3, 2/3)$ on the first step, leaving the union of closed intervals $\Delta_0 = [0, \frac{1}{3}]$ and $\Delta_1 = [\frac{2}{3}, 1]$. The set $\Delta_0 \cup \Delta_1$ is called the *pre-Cantor set* of the first order. The endpoints of the closed intervals constitute its *skeleton*. In the second step we remove the middle thirds $(1/9, 2/9)$ and $(7/9, 8/9)$ respectively from the intervals Δ_0, Δ_1, and thus obtain the closed intervals

$$\Delta_{00} = [0, \frac{1}{3^2}],\ \Delta_{01} = [\frac{2}{3^2}, \frac{1}{3}],\ \Delta_{10} = [\frac{2}{3}, \frac{7}{3^2}],\ \Delta_{11} = [\frac{8}{3^2}, 1],$$

which constitute the pre-Cantor set of the second order, and so on. For example, the skeleton of Δ is $\mathcal{E}_0 = \{0,1\}$, the skeleton of $\Delta_0 \cup \Delta_1$ is $\mathcal{E}_1 = \{0, \frac{1}{3}, \frac{2}{3}, 1\}$. This procedure continues indefinitely. The Cantor set E is defined as the intersection of the countable sequence of pre-Cantor sets E_a formed by all closed intervals enumerated by all binary strings a length $|a| = l$:

$$E = \bigcap_{l=0}^{\infty} E_l,\ E_l = \bigcup_{|a|=l} \Delta_a.$$

The endpoints of intervals constituting the pre-Cantor set E_l of order l, form the corresponding skeleton $\mathcal{E}_l$ and are enumerated by all binary strings length $l+1$, that is, two strings $a0, a1$ correspond to each interval Δ_a. The first steps of this construction are pictured in Figure 1. The Cantor set is compact, perfect and has length zero.

FIGURE 1. Cantor set

A convenient way to work with the Cantor set is to consider the Mauldin-Williams graph (see Edgar [7]; equivalently, we could use a non-deterministic automaton as in [14]) in Figure 2, the contracting ratio list $(r_0, r_1) = (1/3/, 1/3)$, and the functions $f_0, f_1 : [0,1] \to [0,1]$ defined by $f_0(x) = x/3, f_1(x) = (x+2)/3$. Let $r_\Lambda = 1$ (Λ is the empty string), $r_{\alpha i} = r_\alpha \cdot r_i$, for $\alpha \in \Sigma^*, i \in \Sigma$ and define $\delta(\sigma,\tau) = r_{\sigma \cap \tau}$. It is seen that δ is an ultrametric and, in fact, $\delta(\sigma,\tau) = 3^{-|\sigma\cap\tau|} = \rho_{1/3}(\sigma,\tau)$. According to Theorem 4.2.3 in [7] there exists a unique continuous function $h : \Sigma^\omega \to [0,1]$ satisfying the following two conditions:

1. $h(i\sigma) = f_i(h(\sigma))$, for all $i \in \Sigma, \sigma \in \Sigma^\omega$,
2. $h(\Sigma^\omega) = E$.

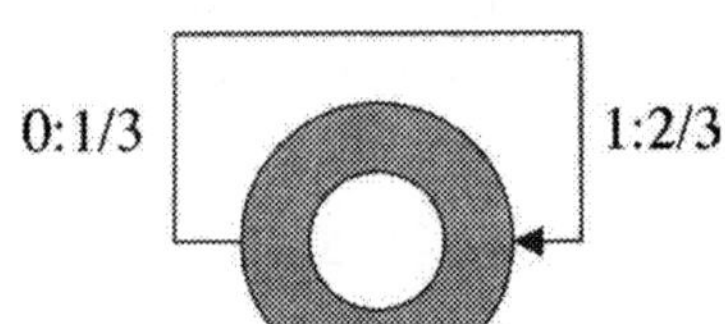

FIGURE 2. Mauldin-Williams graph for the Cantor set

The function h can be defined inductively by the following equations:

$$h(0\sigma) = \frac{h(\sigma)}{3}, \ h(1\sigma) = \frac{h(\sigma)+2}{3}, \tag{1}$$

for all $\sigma \in \Sigma^\omega$, see Edgar [7]. For example, $h(0101010101\ldots) = 1/4$ because of the equality $h(0101010101\ldots) = \frac{1}{9}(2 + h(0101010101\ldots))$.

The map h has a "bounded distortion" with respect to the ultrametric $\delta_{1/3}$, that is, for every $\sigma, \tau \in \Sigma^\omega$,

$$\frac{1}{3}\delta_{1/3}(\sigma,\tau) \leq | \ h(\sigma) - h(\tau) \ | \leq \ \delta_{1/3}(\sigma,\tau). \tag{2}$$

Re-phrased, the ultrametric $\delta_{1/3}(\sigma, \tau)$ on the set of binary sequences is equivalent to Euclidean distance between $h(\sigma), h(\tau)$. Note that h does not have the above property with respect to any other ultrametric δ_r with $r \neq 1/3$.

We consider now the middle-third Cantor set situated on x_1-axis in R_3 and define a *special tiling* of a neighborhood of the Cantor set by extending the map h to the elliptic body Ω with foci 0, 1

$$\Omega = \{x : |x| + |x-1| \leq 5/3\}, \ \text{diam}\ \Omega = 5/3.$$

We shall see below that the sum of all tiles enumerated by these sequences gives an elliptic body Ω_a, diam $\Omega_a = 5{\cdot}3^{-|a|-1}$, and the metric space of trajectories is homeomorphic to the space of tiles endowed with the Euclidean distance, see Lemma 4.1.

We denote by $W_2^1(\mathcal{R}_3)$ the Sobolev space of all square-integrable functions on $\mathcal{R}_3$ which have square-integrable derivatives of the first order. This is a complete Hilbert space endowed with the dot-product

$$\langle u \cdot v\rangle_{W_2^1(\mathcal{R}_3)} = \int_{\mathcal{R}_3} \left(\langle \overline{\nabla u} \nabla u\rangle + \bar{u}v\right) dx^3,$$

and the corresponding norm

$$|u|_{W_2^1(\mathcal{R}_3)} = \sqrt{\langle u \cdot u\rangle_{W_2^1(\mathcal{R}_3)}}\,.$$

For more details about Sobolev classes which will be used below see [17].

We denote by dist the Euclidean distance. A set E is bounded in case $\sup_{x\in E}$ dist$(x, 0) < \infty$. A *fractal* in $\mathcal{R}_3$ is a self-similar perfect set E, that is, a closed, zero-measure set with no isolated points. We assume that E possesses a self-similar tiling (see below the construction of a tiling for the Cantor set).

Note that the function $d_E(x) = \text{dist}(x, E)$ is generally Lip$_1$-continuous:

$$|d_E(x) - d_E(y)| \leq C(K) \cdot |x - y|,$$

on each compact subset K of the complement E' of E in $\mathcal{R}_3$.

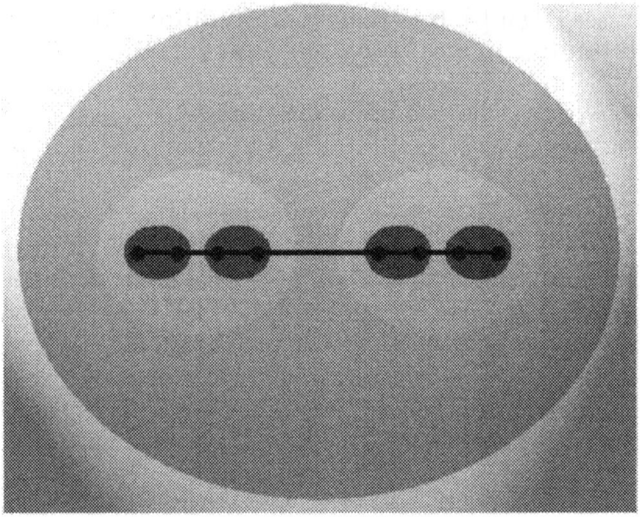

FIGURE 3. Cantor tiling

3. Poisson construction

The Lebesgue measure $\mu(\delta)$ of the δ-neighborhood $E_\delta = \{x \,|\, \mathrm{dist}(x, E) < \delta\}$ of E in $\mathcal{R}_3$ is a "sufficiently smooth" function of δ and can be generally estimated, for small δ, as

$$\mu(\delta) = \int_{E_\delta} dm \leq C(\alpha)\delta^{3-\alpha},$$

with finite non-negative $\alpha \leq 3$ and some positive $C(\alpha)$. The lower bound α_E of values of the parameter α for which this estimate holds is called the *Minkowski dimension* $\dim_E = \alpha_E$ of the set E, see for instance Edgar [7] and the literature quoted there. The Minkowski dimension of sets in $\mathcal{R}_n$ may be defined in a similar way; it does not depend on the dimension n of the space $\mathcal{R}_n$ and may take any non-negative value less than n. In the most interesting cases the Minkowski dimension coincides with the *Hausdorff dimension* [7]. In particular, the Minkowski dimension of the above Cantor set $E \in \mathcal{R}_3$ is equal to $\frac{\log 2}{\log 3}$.

Remark 3.1. The Minkowski dimension α_E defines the order p of admissible singularity of the integrand in convergent integrals on the complement of a compact set E in $\mathcal{R}_n$. For instance, the singular integral

$$\int_{d_E(x)<1} \frac{1}{d_E^p(x)} dm = p \int_0^1 \frac{\mu(\delta)}{\delta^{1+p}}\, d\mu(\delta) + \mu(1),$$

is convergent if $p < n - \alpha_E$. In particular, the integral on the complement of the Cantor set in a large ball in $\mathcal{R}_3$ is convergent if $p < 2 < 3 - \frac{\log 2}{\log 3}$.

The following general statement serves as a base for our calculations in the next section.

Theorem 3.2. *For every function $u \in W_2^1(\mathcal{R}_3)$ and every bounded perfect set $E \in \mathcal{R}_3$ with $\dim_E < 1$ which fulfills the condition*

$$\mathcal{K}_E = \sup_{x \in E} \frac{d_x}{4\pi} \int \frac{1}{|x-s|^2} \frac{dm(s)}{d_E^2(s)} < \infty, \tag{3}$$

the Poincaré-Hardy inequality holds with the constant $\mathcal{K}_E^2$:

$$\int \frac{u^2}{d_E^2(x)}\, dm \leq 4\, \mathcal{K}_E^2 \cdot \int |\bigtriangledown u|^2\, dm. \tag{4}$$

Proof. It is sufficient to obtain the inequality (4) for any smooth function u with a compact support in the complement E' of E in $\mathcal{R}_3$.

If the Minkowski dimension α_E is less than 1, then the function $d_E^{-2}(x) = \mathrm{dist}^{-2}(x, E)$ being Lip_1-continuous on any compact in $E' = \mathcal{R}_3 \setminus E$ is integrable on any bounded neighborhood E_δ. Indeed, following the above general remark, we may rewrite the integral $\int_{E_1} d_E^{-2}(x)\, dm$ as $\int_0^1 \delta^{-2}\, d\mu(\delta)$ and then reduce it via integration by parts for any $\alpha \in (\alpha_E, 1)$ to the following form:

$$\lim_{\delta \to 0} \left(2 \int_\delta^1 s^{-3} \mu(s)\, ds + \mu(1) - \frac{\mu(\delta)}{\delta^2} \right)$$
$$\leq \lim_{\delta \to 0} \left(\mu(1) + C(\alpha) \frac{1}{1-\alpha} (1 - \delta^{1-\alpha}) \right) < \infty.$$

Hence the function d_E^{-2} is integrable on any bounded domain in $\mathcal{R}_3$. Then we consider a Poisson equation

$$-\bigtriangleup v + \kappa^2 u = \frac{1}{d_E^2},\ \kappa^2 > 0, \tag{5}$$

and represent its generalized solution via the corresponding Green function

$$v(x) = \int_{\mathcal{R}_3} \frac{e^{-\kappa|x-s|}}{4\pi|x-s|} \frac{dm(s)}{d_E^2(s)}, \tag{6}$$

on any compact subdomain of the complement E' of E in $\mathcal{R}_3$. The generalized solution (6) is twice continuously differentiable, on any compact $K \in \mathcal{R}_3 \backslash E = E'$, $v \in C^{2+\beta}(K)$, $\beta > 0$, which permits the integration by parts for any smooth real function u with a compact support $K_u \in K$ in E':

$$\int \frac{u^2}{d_E^2(x)}\, dm = -\int u^2 \bigtriangleup v\, dm + \kappa^2 \int u^2 v\, dm$$
$$= 2 \int u < \bigtriangledown u, \bigtriangledown v > dm + \kappa^2 \int u^2 v\, dm,$$

so, the following estimate holds true for any positive κ:

$$\int \frac{u^2}{d_E^2(x)}\, dm \le \left(\int \frac{u^2}{d_E^2(x)}\, dm\right)^{1/2} \cdot \left(4 \cdot \int |\nabla u|^2 \left(|\nabla v| d_E(x)\right)^2 dm\right)^{1/2} + \kappa^2 \cdot \int u^2 v\, dm. \quad (7)$$

We can estimate ∇v as

$$|\nabla v(x)| = \left| \int (1+\kappa|x-s|) \frac{e^{-\kappa|x-s|}}{4\pi|x-s|^2} \frac{x-s}{|x-s|} \frac{dm(s)}{d_E^2(s)} \right|$$
$$\le \int \frac{e^{-\kappa|x-s|}}{4\pi|x-s|^2} \frac{dm(s)}{d_E^2(s)} + \kappa \cdot \int \frac{e^{-\kappa|x-s|}}{4\pi|x-s|} \frac{dm(s)}{d_E^2(s)}. \quad (8)$$

Together with (7), for fixed u, (8) gives:

$$\int \frac{u^2}{d_E^2(x)}\, dm \le \lim_{\kappa\to 0} \left\{ \left(4 \cdot \int \frac{u^2}{d_E^2(x)}\, dm\right)^{1/2} \right.$$
$$\cdot \left(\int |\nabla u(x)|^2 d_E(x) \cdot \left[\int \frac{e^{-\kappa|x-s|}}{4\pi|x-s|^2} \frac{dm(s)}{d_E^2(s)} + \kappa \cdot \int \frac{e^{-\kappa|x-s|}}{4\pi|x-s|} \frac{dm(s)}{d_E^2(s)} \right]^2 dm(x) \right)^{1/2}$$
$$\left. + \kappa^2 \cdot \int u^2 v\, dm \right\}. \quad (9)$$

which implies, after passing to the limit $\kappa \to 0$, the inequality:

$$\int \frac{u^2}{d_E^2(x)}\, dm \le 4 \left[\sup_x \frac{d_x}{4\pi} \int \frac{1}{|x-s|^2} \frac{dm(s)}{d_E^2(s)} \right]^2 \cdot \int |\nabla u(x)|^2\, dm(x). \quad (10)$$

The final result can be obtained now via taking the closure of (10) in the Sobolev space $W_2^1(\mathcal{R}_3)$. □

In Section 5 we shall derive a crude estimate for the constant $\mathcal{K}^2$ for the Cantor set E. Our estimate is not optimal; however, our analysis of the discretized integral representing $\mathcal{K}^2$ shows that the main part of this constant appears from an estimate of some infinite sum over a special tiling. This tiling appears from the extension $\mathcal{H}$ of the parameterizing map h of the Cantor set onto some neighbourhood in $\mathcal{R}_3$ (see the construction in the next section).

4. A special tiling

Consider the Cantor set E on x-axis in $\mathcal{R}_3$ and denote by e_1 the unit vector looking at the positive direction of the x-axis. Consider a tiling of the whole space $\mathcal{R}_3$ formed by the complement $\mathcal{R}_3 \setminus \Omega$ of the rotation-symmetric elliptic body Ω

bordered by the ellipsoid Ω with foci in $0, e_1$, that is on the skeleton of zero-order pre-Cantor set $\Delta = [0, 1]$ on the x-axis:

$$\Omega = \left\{ x : |x| + |x - e_1| \leq \frac{5}{3} \right\}.$$

Next we consider the map $\mathcal{H} : \Sigma \times \Omega \to \Omega$ defined for each $x \in \Omega$ as a splitting of one point x into two images:[2]

$$\mathcal{H}(0, x) = \frac{x}{3}, \; \mathcal{H}(1, x) = \frac{2e_1 + x}{3}.$$

The function $\mathcal{H}$ can be extended in a natural way to a function, also denoted by $\mathcal{H}$, from $\Sigma^* \times \Omega$ into Ω by

$$\mathcal{H}(ia, x) = \mathcal{H}(i, \mathcal{H}(a, x)),$$

for all $i \in \Sigma, a \in \Sigma^*$ and $x \in \Omega$. Clearly, $\mathcal{H}(ab, x) = \mathcal{H}(a, \mathcal{H}(b, x))$, for all $a, b \in \Sigma^*$ and $x \in \Omega$.

The image $\mathcal{H}(\Sigma \times \Omega)$ consists of two components — two similar elliptic bodies $\Omega_0 = \mathcal{H}(0, \Omega) = \frac{1}{3}\Omega$, $\Omega_1 = \mathcal{H}(1, \Omega) = (\frac{2e_1}{3} + \frac{1}{3}\Omega)$,

$$\Omega_0 = \left\{ x : |x| + |x - \frac{1}{3}e_1| \leq \frac{5}{3^2} \right\}, \quad \Omega_1 = \left\{ x : |x - \frac{2}{3}e_1| + |x - e_1| \leq \frac{5}{3^2} \right\},$$

with foci at the *skeleton* $\mathcal{E}_1$ of the first-order pre-Cantor set $E_1 = \Delta_0 \cup \Delta_1$, $\mathcal{E}_1 = \{r_{00}, r_{01}, r_{10}, r_{11}\}$:

$$r_{00} = 0, \; r_{01} = \frac{1}{3}, \; r_{10} = \frac{2}{3}, \; r_{11} = 1,$$

and the basic tile ω is formed as a complement $\Omega \setminus \mathcal{H}(\Sigma \times \Omega) = \Omega \setminus (\Omega_0 \cup \Omega_1)$.

On the next step we form two tiles ω_0, ω_1 of the first order which are similar to ω and are defined respectively as the complement

$$\omega_0 = \mathcal{H}(0, \Omega) \setminus (\mathcal{H}(00, \Omega) \cup \mathcal{H}(01, \Omega)) = \Omega_0 \setminus (\Omega_{00} \cup \Omega_{01})$$

of the bodies

$$\Omega_{00} = \left\{ x : |x - 0| + |x - \frac{1}{9}e_1| \leq \frac{5}{3^3} \right\},$$

$$\Omega_{01} = \left\{ x : |x - \frac{2}{9}e_1| + |x - \frac{1}{3}e_1| \leq \frac{5}{3^3} \right\}$$

in Ω_0 and the complement

$$\omega_1 = \mathcal{H}(1, \Omega) \setminus (\mathcal{H}(10, \Omega) \cup \mathcal{H}(11, \Omega)) = \Omega_1 \setminus (\Omega_{10} \cup \Omega_{11})$$

of the bodies

$$\Omega_{10} = \left\{ x : |x - \frac{6}{9}e_1| + |x - \frac{7}{9}e_1| \leq \frac{5}{3^3} \right\},$$

$$\Omega_{11} = \left\{ x : |x - \frac{8}{9}e_1| + |x - e_1| \leq \frac{5}{3^3} \right\}$$

[2]Note that $\mathcal{H}$ is the extension of h defined by (1).

in Ω_1. The foci of ellipsoids bordering Ω_{00}, Ω_{01}, Ω_{10}, Ω_{11} form the skeleton $\mathcal{E}_2$ of the second-order pre-Cantor set $E_2 = \Delta_{00} \cup \Delta_{01} \cup \Delta_{10} \cup \Delta_{11}$ and are enumerated by all binary strings of length 3: $r_{00} = r_{000} = 0$, $r_{00} + \frac{1}{3^2} = r_{001} = \frac{1}{3^2}$, $r_{01} = r_{010} = \frac{2}{3^2}$, $r_{01} + \frac{2}{3^2} = r_{011} = \frac{3}{3^2} = \frac{1}{3}$, $r_{10} = r_{100} = \frac{2}{3}$, $r_{10} + \frac{1}{3^2} = r_{101} = \frac{7}{3^2}$, $r_{10} + \frac{2}{3^2} = r_{110} = \frac{8}{3^2}$, $r_{10} + \frac{3}{3^2} = r_{111} = 1$.

The construction of the following tiles can be described by induction. On each step l, $|a| = l - 1$, we begin from the result of the previous step — the set of 2^{l-1} non-intersecting elliptic bodies Ω_a bordered by the ellipsoids

$$\Omega_a = \left\{ x : |x - r_{a0}| + |x - r_{a1}| \leq \frac{5}{3^l} \right\},$$

with foci at the skeleton $\mathcal{E}_l$ of the pre-Cantor set $E_l = \cup_{|b|=l} \Delta_b$ enumerated by all possible binary strings $b = a0, a1$ of length l. Then we continue the construction by forming the tiles ω_a as complements $\omega_a = \Omega_a \setminus (\Omega_{a0} \cup \Omega_{a1})$ in Ω_a of the elliptic bodies, respectively bordered by the ellipsoids

$$\Omega_{a0} = \left\{ x : |x - r_{a00}| + |x - r_{a01}| = \frac{5}{3^{l+1}} \right\},$$

and

$$\Omega_{a1} = \left\{ x : |x - r_{a10}| + |x - r_{a11}| = \frac{5}{3^{l+1}} \right\},$$

and so on. Hence, for every $a \in \Sigma^*$, $\mathcal{H}(a, \Omega) = \Omega_a$ and

$$\omega_a = \mathcal{H}(a, \Omega) \setminus (\mathcal{H}(a0, \Omega) \cup \mathcal{H}(a1, \Omega)) = \Omega_a \setminus (\Omega_{a0} \cup \Omega_{a1}).$$

The following Lemma 4.1 will be used to derive bilateral estimates for the coefficient $\mathcal{K}^2$ in (4) in terms of the constructed tiling. We enumerate the tiles by binary strings b.[3]

Lemma 4.1. *The sets* $\{\omega, \omega_1, \omega_2, \omega_c\}$, *enumerated by all possible binary strings* c, $|c| \geq 0$ *form a tiling for the elliptic body* Ω *with the following properties:*

1. *The distance* $d_E(x)$ *from the point* $x \in \omega_c$ *to the Cantor set* E *may be bilaterally estimated by the distance* $d_{|c|}(x)$ *from* x *to the skeleton* $\mathcal{E}_{|c|}$ *of the pre-Cantor set* $E_{|c|}$. *In particular, the ratio* $d_{|c|}(x)/d_E(x)$ *takes the minimal and maximal values on the border* $\partial\Omega_c$, $\partial\Omega_{c0}$, $\partial\Omega_{c1}$ *of the tile* ω_c *and*

$$1 \leq \frac{d_{|c|}(x)}{d_E(x)} \leq 4, \; x \in \partial\Omega_{c0} \cup \partial\Omega_{c1},$$

$$1 \leq \frac{d_{|c|}(x)}{d_E(x)} \leq \frac{5}{\sqrt{17}}, \; x \in \partial\Omega_c. \tag{11}$$

The Euclidean volume of the tile ω_c *is equal to* $10^3\pi 3^{-3|c|+7}$ *and the distance from the Cantor set* E *to* $x \in \omega_c$ *can be bilaterally estimated as*

$$3^{-|c|-2} \leq d_E(x)|_{x\in\omega_c} \leq \frac{\sqrt{17}}{2} \cdot 3^{-|c|-1}.$$

[3] Recall that $a \cap b$ is the maximal common prefix of the strings a, b.

2. *The distance between the points $x_a \in \omega_a$ and $x_b \in \omega_c$ may be estimated from above as:*

$$|x_a - x_c| \leq 5 \cdot 3^{-|a \cap c|-1}. \tag{12}$$

If the tiles ω_a, ω_c do not contact each other (that is, do not have a common piece of the boundary), then the distance between the points $x_a \in \omega_a$ and $x_c \in \omega_c$ may be estimated from below as

$$|x_a - x_c| \geq 3^{-|a \cap c|-2}. \tag{13}$$

Proof. The above constructed tiling is self-similar, hence the estimate (11), if derived for the basic tile ω and the tiles ω_0, ω_1 of the first order, remains true, under proper scaling, for the whole tiling. Note, for instance, that the ratio $d_1(x)/d_E(x)$ takes the minimal and maximal values on the boundary of the tile ω and can be estimated as

$$1 \leq \frac{d_1(x)}{d_E(x)} \leq 4, \ x \in \partial\Omega_0 \cup \partial\Omega_1,$$

and

$$1 \leq \frac{d_1(x)}{d_E(x)} \leq \frac{5}{17}, \ x \in \partial\Omega. \tag{14}$$

Similarly, the last part of the first statement follows from the estimate

$$\frac{1}{9} \leq d_E(x)|_{x \in \omega} \leq \frac{\sqrt{17}}{6}.$$

To prove the last part of the second statement we notice that from the condition $U_a \cap \omega_c = \emptyset$ follows that either $\omega_a \in \omega_{d0}$, $\omega_c \in \omega_{d1}$, for some string d of length k, or vice versa. This implies the announced inequality:

$$|x_a - x_c| \geq \text{dist}\,(\omega_{d0}, \omega_{d1}) = 3^{-|k|-2}. \qquad \square$$

Notice, that for every string a, the map $\mathcal{H}(a, \cdot)$ acts transitively on the constructed tiling, transferring each l-generation of tiles $\cup_{|a|=l} \omega_a$ into the following $l+1$-generation $\cup_{|a|=l+1} \omega_a$. The same function maps the l-generation of elliptic bodies $\mathcal{H}(\Sigma^l, \Omega) = \cup_{|a|=l} \Omega_a$ into itself. One can easily see that the Cantor set E is an invariant set of the map $\mathcal{H}$. The restriction $\mathcal{H}$ onto $[0,1]$ coincides with h defined by (1); see the corresponding property of the parameterizing map $h : [0,1] \to [0,1]$). This is the exact meaning of the statement at the end of the previous section, that the special tiling is formed by the extension $\mathcal{H}$ of the parameterizing map h onto Ω. The transitive action of the map $\mathcal{H}$ on the tiling permits to represent the generating map $\mathcal{H}(a, \cdot)$ as an analog of the unilateral shift on the orthogonal sum of Hilbert spaces $\oplus \sum_a \mathcal{L}_2(\omega_a)$.

Considering the basic mappings

$$\mathcal{H}_0 = \mathcal{H}(0, \cdot) : x \to \frac{x}{3},$$

$$\mathcal{H}_1 = \mathcal{H}(1, \cdot) : x \to \frac{2}{3} e_1 + \frac{x}{3},,$$

we form the strings $\mathcal{H}(b,\cdot) = \mathcal{H}_b = \mathcal{H}_{b_l}\mathcal{H}_{b_{l-1}}\ldots\mathcal{H}_{b_1}$ enumerated by the binary strings $b = b_l b_{l-1}\ldots b_1$. These strings form a *Cantor scaling* of the above tiling, mapping the basic tile ω onto the corresponding tile $\omega_b = \mathcal{H}_b\,\omega$ such that a given point r_0 is transferred into $r_b = \mathcal{H}_b\, r_0$. This scaling will be used in the construction of the sequence of test-functions in Theorem 7.1.

5. Estimates for the discretized integral

We begin this section with a few preliminary results. Together with a given tile ω_b we consider a triple of its closest neighbours: its *mother* ω_a of ω_b such that $b = a0$ or $b = a1$ and two *daughters* ω_{b0}, ω_{b1} which form together with ω_b the *cut* U_b of the corresponding Bruhat-Tits tree at the level b: $\omega_b \cap \omega_a \cap \omega_{b0} \cap \omega_{b1} = U_b$. We consider its complement $\Omega\backslash U_b$ in Ω which is represented by joining of all remaining tiles

$$\Omega \setminus U_b = \bigcup_c \omega_c,\ c \neq a, b, b0, b1. \tag{15}$$

First note that *for $x \in \omega_1$, the integral over $U_1 = \omega \cup \omega_1 \cup \omega_{10} \cup \omega_{11}$,*

$$\mathcal{J}_1(x) = \frac{1}{4\pi}\int_{U_1} \frac{1}{|x-s|^2}\frac{dm(s)}{d_E^2(x)},$$

is a uniformly continuous function of $x \in \omega_1$, and there exist two absolute constants A_1, B_1 such that

$$A_1 \le \frac{d_E(x)}{4\pi}\int_{U_1} \frac{1}{|x-s|^2}\frac{dm(s)}{d_E^2(s)} \le B_1. \tag{16}$$

An obviously crude but still reasonable numerical estimate is:

$$A_1 = \frac{1}{3},\ B_1 = 150. \tag{17}$$

Indeed, due to the first statement in Lemma 4.1, the distance $d_E(s)$ from the set E on the cut U_1 can be estimated by the distance $d_2(s)$ from the skeleton $\mathcal{E}_2$:

$$\frac{1}{d_2^2(s)} \le \frac{1}{d_E^2(s)} \le 16\cdot\frac{1}{d_2^2(s)}.$$

Hence,

$$\begin{aligned}\frac{d_E(x)}{4\pi}\int_{U_1} \frac{16}{|x-s|^2}\frac{dm(s)}{d_E^2(s)} &\le \frac{d_E(x)}{4\pi}\int_{U_1} \frac{16}{|x-s|^2}\frac{dm(s)}{d_2^2(s)}\\ &\le \frac{4d_E(x)}{\pi}\int_{U_1}\frac{1}{|x-s|^2}\left[\sum_{s_{ik}\in\mathcal{E}_2}\frac{1}{|s-s_{ik}|^2}\right]dm(s).\end{aligned}$$

Using the following estimate for the standard integral

$$\frac{1}{3} \le \frac{1}{4\pi}\int_{\mathcal{R}_3}\frac{1}{|x-1|\,|x|}\,dm \le \frac{7}{3}, \tag{18}$$

we obtain, after the change of variables, the estimation from above:

$$4\frac{d_E(x)}{\pi}\int_{U_1}\frac{1}{|x-s|^2}\left[\sum_{s_{ik}\in\mathcal{E}_2}\frac{1}{|s-s_{ik}|^2}\right]dm(s)\le 150\cdot\frac{d_E(x)}{d_2(x)}\le 150. \tag{19}$$

The estimate from below may be obtained as follows:

$$\begin{aligned}\frac{d_E(x)}{4\pi}\int_{U_1}\frac{1}{|x-s|^2}\frac{dm(s)}{d_E^2(s)} &\ge \frac{d_E(x)}{4\pi}\int_{U_1}\frac{1}{|x-s|^2}\frac{dm}{d_2^2(s)}\\ &\ge \max_{s_{ik}\in\mathcal{E}_2}\frac{d_E(x)}{4\pi}\int_{U_1}\frac{1}{|x-s|^2}\frac{dm}{|s-s_{ik}|^2}\\ &\ge \frac{1}{3}\cdot\frac{d_E(x)}{d_2(s)}\\ &\ge \frac{1}{3}.\end{aligned}$$

It follows from the self-similarity of the tiling that the same estimate holds for the corresponding integral over any cut U_c, for every string c and $x\in\omega_c$:

$$A_1\le\frac{d_E(x)}{4\pi}\int_{U_c}\frac{1}{|x-s|^2}\frac{dm(s)}{d_E^2(s)}\le B_1. \tag{20}$$

The integral $C_\omega=\frac{1}{4\pi}\int_\omega\frac{dm(s)}{d_E^2(s)}$ *can be estimated as* $3^{-2}\le C_\omega\le 16$. *Consequently, due to the self-similarity, all integrals* $\frac{1}{4\pi}\int_{\omega_c}\frac{dm(s)}{d_E^2(s)}$ *can be estimated uniformly:*

$$3^{-|s|-2}\le\frac{1}{4\pi}\int_{\omega_c}\frac{dm(s)}{d_E^2(s)}\le 16\cdot 3^{-|c|}. \tag{21}$$

Indeed, due to the first statement in Lemma 4.1 the integral

$$\frac{1}{4\pi}\int_\omega\frac{dm(s)}{d_E^2(s)}$$

may be estimated as

$$\frac{25}{17}\cdot\frac{1}{4\pi}\int_\omega\frac{dm(s)}{d_2^2(s)}\le 4\cdot\frac{25}{17}\int_0^{1/3}\int_0^{4/6}\frac{\rho d\rho dh}{h^2+\rho^2}\le 6\cdot\int_0^{1/3}\ln\frac{h^2+(2/3)^2}{h^2}\,dh\le 16.$$

An estimate of the integral from below may be derived from the estimate of d_E in above Lemma 4.1

Next note that

$$\sup_{x\in\mathcal{R}_3\setminus(\Omega_0\cup\Omega_1)}\frac{d_E\,(x)}{4\pi}\int\frac{1}{|x-s|^2}\frac{dm(s)}{d_E^2(s)}$$

can be estimated from above by the sum

$$4\cdot\frac{d_E}{\pi}\int_{\mathcal{R}_3\setminus(\cup_{ik}\Omega_{ik})}\frac{1}{|x-s|^2}\frac{dm(s)}{d_2^2(s)}+\frac{d_E\,(x)}{4\pi(d_E\,(x)-2\,.\,3^{-3})^2}\int_{(\bigcup_{ik}\Omega_{ik})}\frac{1}{d_E^2}\,dm.$$

Due to (18) we have

$$\frac{448}{3} \cdot \frac{d_E(x)}{d_2(x)} < 150 \cdot \frac{d_E(x)}{d_2(x)} < 150,$$

in view of (21) and self-similarity,

$$\frac{3^{-3} \cdot d_E(x)}{(d_E(x) - 2 \cdot 3^{-3})^2} \cdot \frac{8}{3} \leq 200,$$

hence *the integral* $\frac{d_E\,(x)}{4\pi} \int \frac{1}{|x-s|^2} \frac{dm(s)}{d_E^2(s)}$, *for* $x \in \mathcal{R}_3 \backslash (\Omega_0 \cup \Omega_1)$, *is estimated from above by* 350 .

We obtain further the *dominating* estimate for the integral

$$\frac{d_E\,(x)}{4\pi} \int \frac{1}{|x-s|^2} \frac{dm(s)}{d_E^2(s)}$$

for $x \in \omega_a$, $|a| \geq 1$.

Lemma 5.1. *The integral coefficient*

$$\frac{d_E\,(x)}{4\pi} \int \frac{1}{|x-s|^2} \frac{dm(s)}{d_E^2(s)}$$

can be discretized and estimated for $x \in \omega_a$ *as follows:*

$$\frac{d_E(x)}{4\pi} \int_{\mathcal{R}_3} \frac{1}{|x-s|^2} \frac{dm(s)}{d_E^2(s)} = \frac{d_E(x)}{4\pi} \int_{\mathcal{R}_3 \backslash \Omega_0 \cup \Omega_1} \frac{1}{|x-s|^2} \frac{dm(s)}{d_E^2(s)}$$

$$+ \frac{d_E(x)}{\pi} \int_{U_a} \frac{1}{|x-s|^2} \frac{dm(s)}{d_E^2(s)} + \sum_b \frac{d_E(x)}{\pi} \int_{\omega_b,\, |b| \geq 1, b \cap U_a = \emptyset} \frac{1}{|x-s|^2} \frac{dm(s)}{d_E^2(s)}$$

$$\leq 300 + 900 \cdot \sum_{\omega_b,\, |b| \geq 1, b \cap U_a = \emptyset} 3^{2 \cdot |a \cap b| - |ab|}.$$

Proof. The proof of the first statement is based on (18):

$$\frac{d_E(x)}{4\pi} \int_{\mathcal{R}_3 \backslash (\Omega_0 \cup \Omega_1} \frac{1}{|x-s|^2} \frac{dm(s)}{d_E^2(s)} \leq \frac{112}{3} \cdot \frac{d_E(x)}{d_1(x)} \leq 150.$$

In view of (17) and (20) we get:

$$\sup_{\mathbf{x} \in U_{\omega_1}} \frac{d_E(x)}{4\pi} \int_{U_{\omega_1}} \frac{1}{|x-s|^2} \frac{dm(s)}{d_E^2(s)} \leq 150.$$

To get an upper bound for the third term we use the estimate (21) for integrals

$$\frac{d_E(x)}{4\pi} \int_{\omega_b,\, b \neq a} \frac{1}{|x-s|^2} \frac{dm(s)}{d_E^2(s)}, \; x \in \omega_a$$

and the second statement of Lemma 4.1:

$$\frac{d_E(x)}{4\pi} \int_{\omega_b,\, |b| \geq 1, b \cap U_a = \emptyset} \frac{1}{|x-s|^2} \frac{dm(s)}{d_E^2(s)} \leq 900 \cdot \sum_{\omega_b,\, |b| \geq 1, b \cap U_a = \emptyset} 3^{2 \cdot |a \cap b| - |ab|}. \quad \square$$

The next statement, of algebraic nature, completes the estimation of the integral representing the constant $\mathcal{K}$.

Lemma 5.2. *The following inequality holds true for every string $b \in \Sigma^*$:*

$$\sum_{a\in S} 3^{2\cdot|a\cap b|-|ab|} \leq 4. \tag{22}$$

Proof. Assume that $|b| = m$ and $b = b_1 b_2 \dots b_m$. First, decompose the series in the left-hand side of (22) into two disjoint series:

$$\begin{aligned}\sum_{a\in S} 3^{2\cdot|a\cap b|-|ab|} &= \sum_{k=1}^{\infty}\sum_{|a|=k} 3^{2\cdot|a\cap b|-|ab|}\\ &= \sum_{k=1}^{m-1}\sum_{|a|=k} 3^{2\cdot|a\cap b|-|ab|} + \sum_{k=m}^{\infty}\sum_{|a|=k} 3^{2\cdot|a\cap b|-|ab|}.\end{aligned}$$

A typical string $a = a_1 a_2 \dots a_k$ of length $k \leq m-1$ will be of one of the following two forms: $b_1 b_2 \dots b_i \bar{b}_{i+1} \dots a_k$ (for some $0 \leq i \leq k-1$) or $b_1 b_2 \dots b_{k-1} b_k$. We have 2^{k-i-1} different strings of the first form and exactly one string of the last form. Similarly, a typical string $a = a_1 a_2 \dots a_k$ of length $k \geq m$ will be of one of the following two forms: $b_1 b_2 \dots b_i \bar{b}_{i+1} \dots a_m \dots a_k$ (for some $0 \leq i \leq m-1$) or $b_1 b_2 \dots b_{m-1} b_m \dots a_{m+1} a_k$. We have 2^{k-i-1} different strings of the first form and 2^{k-m} strings of the last form. An elementary computation, based on the above combinatorial analysis, justifies the following two inequalities which combine to prove (22):

$$\begin{aligned}\sum_{k=1}^{m-1}\sum_{k=1}^{m-1}\sum_{|a|=k} 3^{2\cdot|a\cap b|-|ab|} &= 3^{-m-k}\left(\sum_{i=0}^{k-1} 3^{2i}\cdot 2^{k-i-1} + 3^{2k}\right)\\ &= \frac{1}{7\cdot 3^m}\cdot\sum_{k=1}^{m-1}\left(\frac{2}{3}\right)^k\cdot\left(\left(\frac{9}{2}\right)^m - 1\right) + \frac{1}{2}\left(1-\frac{1}{3^{m-1}}\right) \leq \frac{4}{7},\end{aligned}$$

and

$$\begin{aligned}\sum_{k=m}^{\infty}\sum_{|a|=k} 3^{2\cdot|a\cap b|-|ab|} &= \sum_{k=m}^{\infty} 3^{-m-k}\left(\sum_{i=0}^{m-1} 3^{2i}\cdot 2^{k-i-1} + 3^{2m}\cdot 2^{k-m}\right)\\ &= \frac{1}{7\cdot 3^{m-1}}\cdot\left(\left(\frac{9}{2}\right)^m - 1\right)\left(\frac{2}{3}\right)^m + 3 \quad\leq\quad \frac{24}{7}. \qquad\square\end{aligned}$$

For a string $b \in \Sigma^*$ of length greater than one let b' be the prefix of b of length $|b| - 1$. We can use now the inequality (22) to deduce the following upper bound:

$$\sum_{\omega_b,\,|b|\geq 1, b\cap U_a=\emptyset} 3^{2\cdot|a\cap b|-|ab|} \leq \sum_{a\in S\setminus\{b,b0,b1,b'\}} 3^{2\cdot|a\cap b|-|ab|} \leq 2,$$

which leads directly to

Theorem 5.3. *The Poincaré-Hardy inequality on the complement of the Cantor set E*

$$\int \frac{|u|^2}{dist^2(x,E)}\, dm \leq 1764 \cdot 10^4 \cdot \int |\nabla\, u|^2 dm$$

holds for every $u \in W_2^1(\mathcal{R}_3)$.

6. Higher-order Poincaré-Hardy inequalities

In this section we obtain higher-order Poincaré-Hardy inequalities by iterating the inequality (4). From now on assume that $\Omega = \mathcal{R}_3 \setminus E$.

Lemma 6.1. *The function* $d_E^{-p}(x)$, $p \geq 1$, *is a super-harmonic function in a weak sense: for each smooth non-negative function* φ *with a compact support in* Ω,

$$\int_\Omega \triangle\varphi(x)\; d_E^{-p}(x) dm \geq 0. \tag{23}$$

Proof. Consider a smooth non-negative spherically-symmetric averaging kernel $\eta(x) = \eta(|x|)$ of zero order with compact support in the unit ball B_1:

$$\int_{B_1} \eta(x) dx = 1.$$

Each continuous function u may be approximated by the corresponding averaged functions

$$u_\delta(x) = \frac{1}{\delta^3} \int \eta\left(\frac{|x-s|}{\delta}\right) u(s) ds.$$

The Laplacian of u can be obtained as a spherical derivative

$$\triangle u(x) = \lim_{\delta\to 0} \frac{1}{\delta^2}\left(u_\delta(x) - u(x)\right). \tag{24}$$

The integral operator

$$\triangle_\delta u = \frac{1}{\delta^2}\left(u_\delta(x) - u(x)\right)$$

in the right side of (24) is bounded and symmetric on the class of all continuous functions with a compact support in the sub-domain

$$\Omega_{\sqrt{\delta}} = \left\{x \in \Omega :\, d_E(x) > \sqrt{\delta}\right\}.$$

Using the second Weierstrass Theorem for any $x \in \Omega$ we find a point $\xi \in E$ such that $d_E^{-p}(x) = |x-\xi|^{-p}$. The function $|x-\xi|^{-p}$, $p \geq 1$ is obviously super-harmonic, $\triangle|x-\xi|^{-p} \geq 0$, hence for δ small enough

$$\triangle_\delta |x-\xi|^{-p} \geq 0,$$

if $x \in \Omega_{\sqrt{\delta}}$. On the other hand,

$$0 \leq \triangle_\delta |x-\xi|^{-p} \;\; \leq \;\; \frac{1}{\delta^2}\left(\int_\Omega |s-\xi|^{-p} dm(s) - |x-\xi|^{-p}\right)$$

$$\begin{aligned} &\leq \frac{1}{\delta^2}\left[\frac{1}{\delta^3}\int_\Omega \max_{\zeta\in E}|s-\zeta|^{-p}\eta\left(\frac{|x-s|}{\delta}\right)dm(s)-|x-\xi|^{-p}\right] \\ &= \triangle_\delta d_E^{-p}(x). \end{aligned}$$

This implies, due to the symmetry of $\triangle_\delta$,

$$\int \varphi(x)\,\triangle_\delta\, d_E^{-p}(x)dm = \int d_E^{-p}(x)\,\triangle_\delta\,\varphi(x)dm \geq 0, \tag{25}$$

for any non-negative smooth function φ with a compact support in $\Omega_{\sqrt{\delta}}$, $\delta > 0$. Passing to the limit $\delta \to 0$ we obtain the announced inequality (23). □

Theorem 6.2. *For every smooth function u with compact support vanishing near E and $l \geq 1$, the following Poincaré-Hardy inequality is true:*

$$\int_\Omega \frac{|u|^2(x)}{d_E^{2l}(x)}dm \leq (4\,\mathcal{K}^2)^l\cdot\int_\Omega \left|\nabla^l u(x)\right|^2 dm. \tag{26}$$

Proof. We derive the inequality (26) for real functions and then extend it to complex functions using quadratic forms. Consider the averaged function, the smoothened inverse distance, $\left(d_E^{-1}\right)_\delta(x)$. It is infinitely-differentiable in $\Omega_{\sqrt{\delta}}$ and can be estimated as follows:

$$\frac{1-\sqrt{\delta}}{1+\sqrt{\delta}} \leq \left(d_E^{-1}\right)_\delta d_E \leq \frac{1+\sqrt{\delta}}{1-\sqrt{\delta}}.$$

Then for each smooth real function with a compact support in Ω_δ

$$\int_\Omega \frac{u^2(x)}{d_E^{2l}(x)}\,dm \leq \left(\frac{1+\sqrt{\delta}}{1-\sqrt{\delta}}\right)^{2l-2}\cdot\int_\Omega u^2(x)\frac{1}{d_E^2(x)}\left[\left(\frac{1}{d_E(x)}\right)_\delta\right]^{2l-2}(x)\,dm,$$

where the last integral, due to the inequality (4), does not exceed

$$4\,\mathcal{K}^2\cdot\int_\Omega\left[\nabla\left(u(x)\left[\left(\frac{1}{d_E}\right)_\delta\right]^{l-1}\right)\right]^2(x)\,dm.$$

Integrating by parts we get

$$\begin{aligned} \int_\Omega\left[\nabla\left(u(x)\left[\left(\frac{1}{d_E}\right)_\delta\right]^{l-1}\right)\right]^2(x)\,dm &= \int_\Omega[\nabla u(x)]^2\left[\left(\frac{1}{d_E}\right)_\delta\right]^{2l-2}(x)\,dm \\ &+\int_\Omega\nabla\left[u^2(x)\left[\left(\frac{1}{d_E}\right)_\delta\right]^{l-1}(x)\,\nabla\left(\frac{1}{d_E}\right)_\delta^{l-1}(x)\right]dm \\ &-\int_\Omega u^2(x)\,\triangle\left[\left(\frac{1}{d_E}\right)_\delta\right]^{2l-2}(x)\,dm. \end{aligned} \tag{27}$$

The last integral in the right side can be written, via integration by parts, as

$$\int_\Omega \triangle u^2(x)\left[\left(\frac{1}{d_E}\right)_\delta\right]^{2l-2}(x)\,dm.$$

Its limit when $\delta \to 0$

$$\int_\Omega \Delta u^2(x) \left(\frac{1}{d_E}\right)^{2l-2} (x)\, dm$$

exists and is positive due to Lemma 6.1. The second integral in (6) vanishes due to the Stokes Theorem, hence

$$\begin{aligned} \int_\Omega \frac{u^2(x)}{d_E^{2n}(x)} dx &\leq \lim_{\delta\to 0} \left(\frac{1+\sqrt{\delta}}{1-\sqrt{\delta}}\right)^{2l-2} \cdot \int_\Omega u^2 \frac{1}{d_E^2} \left[\left(\frac{1}{d_E}\right)_\delta\right]^{2l-2} dm \\ &\leq 4\mathcal{K}^2 \cdot \int_\Omega (\nabla u(x))^2 \left(\frac{1}{d_E}\right)^{2l-2} (x)\, dm. \end{aligned}$$

Then continuing by induction we obtain after l steps the inequality (26). □

Corollary 6.3. *The statement of Theorem 6.2 remains true, if we consider the closure in the Sobolev norm W_2^l in the corresponding Sobolev class $W_2^l(\Omega^0)$ of functions vanishing on E.*

7. Three applications

The inequality (4) can be used to derive several useful facts. In what follows we will present three such applications.

A. *Consider a real measurable function q, locally bounded on each compact in the complement E' of the fractal E. Then the Dirichlet form $\int_{\mathcal{R}_3} \left(|\nabla u|^2 + q(x)|u|^2\right) dm$ is closed in $W_2^1(\mathcal{R}_3)$ if q satisfies the following additional condition:*

$$\lim_{d_E(x)\to 0} |q(x)|\, d_E^2(x) = 0. \tag{28}$$

To prove the above statement we have to check that the inequality (4) implies the strong subordination of the quadratic form of the potential $\int_{\mathcal{R}_3} q(x)|u|^2\, dm$ to the Dirichlet form $\int_{\mathcal{R}_3} |\nabla u|^2\, dm$ (see Reed and Simon[24]). Indeed, for any $\varepsilon > 0$ we can choose a positive constant C such that

$$|q(x)| \leq C + \frac{\varepsilon}{\mathcal{K}_E d_E^2(x)},$$

which implies the strong subordination:

$$\int_{\mathcal{R}_3} |q(x)||u|^2\, dm \leq \varepsilon \cdot \int_{\mathcal{R}_3} |\nabla u|^2\, dm + C \cdot \int_{\mathcal{R}_3} |u|^2\, dm.$$

B. The constant $\mathcal{K}$ plays the role of an estimate for the dimensionless Planck constant in the corresponding uncertainty principle. To verify this statement, let us consider the self-adjoint operator (unbounded in $\mathcal{L}_2(\mathcal{R}_3)$) of multiplication by the function $\varepsilon(x)d_E(x)$, where the factor $\varepsilon(x) = \pm 1$ is chosen such that for a given smooth function u with a compact support in E' the mean value of the "balanced distance" with respect to some unitary-valued sign-factor $\varepsilon(x)$, $\varepsilon(x)d_E(x)$ to the singular set E is equal to zero: $\int_{\mathcal{R}_3} \varepsilon(x) d_E(x) |u|^2 \, dm = 0$.

We assume that the mean value of momentum is also zero: $\int_{\mathcal{R}_3} \nabla u \, \bar{u} \, dm = 0$.

Under the above hypotheses we may estimate from below the product of the mean quadratic errors of the balanced distance and the mean quadratic error of the momentum (the Dirichlet integral) as

$$\begin{aligned}\frac{1}{2\mathcal{K}} \cdot \int_{\mathcal{R}_3} |u|^2 \, dm &\leq \left(\int_{\mathcal{R}_3} |d_E(x) u|^2 \, dm \right)^{1/2} \cdot \left(\frac{1}{4\mathcal{K}^2} \cdot \int_{\mathcal{R}_3} \frac{|u|^2}{d_E^2(x)} \, dm \right)^{1/2} \\ &\leq \left(\int_{\mathcal{R}_3} |d_E(x)|^2 \, |u|^2 \, dm \right)^{1/2} \cdot \left(\int_{\mathcal{R}_3} |\nabla u|^2 \, dm \right)^{1/2},\end{aligned}$$

to obtain an analog of the classical dimensionless Heisenberg's uncertainty relation:

$$\frac{1}{2\mathcal{K}} \cdot \int_{\mathcal{R}_3} |u|^2 \, dm \leq \left(\int_{\mathcal{R}_3} |x|^2 |u|^2 \, dm \right)^{1/2} \cdot \left(\int_{\mathcal{R}_3} |\nabla u|^2 \, dm \right)^{1/2}.$$

The constant $\frac{1}{2}\mathcal{K}^{-1} \approx 0.0002$ plays the role of an estimate for analog of the classical "dimensionless Planck constant" 1/2. It defines the attainable precision of simultaneous measurements of deviation of the coordinate of the quantum particle from the singular set and the deviation of its momentum from zero.

C. The inequality (26) can be applied to the spectral problem for Schrödinger equation with potentials singular on fractal sets or to polar equations with singular densities. Both applications are based on the following embedding result:

Theorem 7.1. *The unit ball B_1 in $W_2^l(\Omega^0)$, $2l > 3$, is compact in the space of weighted square-integrable functions $L_2(\rho, \Omega)$ with non-negative locally bounded measurable weight ρ if and only if the weight fulfills the following condition on the tiling Ω_a covering the neighborhood E_{δ_0} of the singular set E:*

$$\lim_{|b| \to \infty} 3^{(-2l+3)|b|} \int_{\Omega_b} \rho(x) \, dm = 0. \tag{29}$$

Proof. We use the embedding of $W_2^l(\Omega^0)$ into the class of continuous functions, see [2, 17]. If the condition (29) is not fulfilled, then there exists a sequence of tiles Ω_b, $|b| = n \geq n_0$ such that $3^{(-2l+3)|b|} \int_{\Omega_b} \rho(x) dx \geq \varepsilon_0 > 0$. We can assume that this sequence begins with Ω_{00} and is enumerated by binary strings ending with 0. With the first tile of the sequence Ω_{00} we associate the smooth function η_{00} which is equal to 1 on the tile Ω_{00} and equal to zero outside of the corresponding cut $U_{00} = \Omega_0 \cap \Omega_{00} \cap \Omega_{000} \cap \Omega_{001}$. We can also assume that triples of different tiles of the considered sequence are disjoint. Then we can construct an orthogonal and almost normalized in $L_2(\rho, \Omega)$ self-similar sequence of smooth functions η_b obtained by scaling $\eta_b = (\mathcal{H}_b)\, \eta_0$ from the first one η_0:

$$\eta_b(x) = 3^{-|b| \frac{2l-3}{2}} \eta \left((\mathcal{H}_b)^{-1} x \right).$$

The corresponding sequence of W_2^l norms is bounded:

$$|\eta_b|_{W_2^l} = |\eta_0|_{W_2^l}.$$

On the other hand, the sequence is orthogonal in $L_2(\rho, \Omega)$ and almost normalized, since

$$\int |\eta_b|^2 \, \rho \, dx \geq 3^{-|b|(2l-3)} \int_{\Omega_0} \rho(x) dx \geq \varepsilon_0 > 0.$$

Hence it is not compact, though bounded in $W_2^l(\Omega^0)$. This implies the necessity of the condition (29).

To prove that the condition is sufficient we use the fact that embedding the Sobolev class $W_2^l(\mathcal{R}_3)$ into both $L_2(\rho, \Omega)$ and the space of continuous functions on any bounded domain in $\mathcal{R}_3$ with smooth boundary is compact. Hence to finish the proof we have to derive the uniform estimate of *tails* of elements in the unit ball B_1 of $W_2^l(\Omega^0)$ in $L_2(\rho, \Omega)$-norm:

$$\sum_{|b|>M} \int_{\Omega_b} u^2 \rho \, dm \leq \varepsilon, \qquad \text{if } u \in B_1, \, M \geq M_\varepsilon.$$

Due to the embedding result cited above there exist two absolute constants C_1', C_l' such that for any $\xi \in \Omega_b$:

$$u^2(x) \leq u^2(\xi) + 3^{-(2l-3)|b|} C_1' \cdot \int_{\Omega_b} \frac{|\nabla u|^2}{d_E^{2l-2}} \, dm + 3^{-|b|(2l-3)} C_l' \cdot \int_{\Omega_b} |\nabla^l u|^2 \, dm. \quad (30)$$

Now we multiply both parts of (30) by ρ, integrate once more over Ω_b and choose ξ such that

$$u^2(\xi) = \frac{\int_{\Omega_b} u^2 d_E^{-2l} \, dm}{\int_{\Omega_b} d_E^{-2l} \, dm}.$$

It is clear that

$$\left(\int_{\Omega_b} d_E^{-2l} dx \right)^{-1} = 3^{-(2l-3)|b|} C_0' = C_0 \cdot d_{E,b}^{2l-3},$$

for some absolute constant C_0' (here d_E is bilaterally estimated by the distance $d_{E,b} = 3^{-|b|}$ between Ω_b and E). Then, after proper re-normalization of constants $C_1 = \frac{C_0}{C_0'} C_1'$, $C_l = \frac{C_0}{C_0'} C_l'$ and summation over all tiles $|b| > M$, we obtain the uniform estimate for the $L_2(\rho, \Omega)$-norm of the tail of the function u from the unit ball of $W_2^l(\Omega^0)$:

$$\int_{\Omega_b} u^2 \rho \, dm \leq d_E^{2l-3} \int_{\Omega_b} \rho \, dm$$

$$\cdot \left(C_0 \cdot \int_{\Omega_b} \frac{u^2}{d_E^{2l}} \, dm + C_1 \cdot \int_{\Omega_b} \frac{|\nabla u|^2}{d_E^{2l-2}} \, dm + C_l \cdot \int_{\Omega_b} |\nabla^l u|^2 \, dm \right)$$

Summarizing the whole history and using Theorem 7.1 we get

$$\begin{aligned}\sum_{|b|>M}\int_{\Omega_b} u^2\rho\,dm &\le \sup_{|b|>M} d_{E,b}^{2l-3}\int_{\Omega_b}\rho\,dm\\ &\le \left(C_0\,(4\mathcal{K}^2)^l + C_1\,(4\mathcal{K}^2)^{l-1} + C_l\right)\cdot\int_{\mathcal{R}_3} |\nabla^l u|^2\,dm,\end{aligned}$$

which proves formula (29). □

Corollary 7.2. *The spectrum of the operator $\rho^{-1}(-\Delta)^l$ defined by the quadratic form $\int_{\Omega^0}|\nabla^l u|^2\,dm$ in $W_2^l(\Omega^0)\cap L_2(\rho,\Omega)$ is discrete if and only if the condition* (29) *is fulfilled.*

Other applications of Theorem 7.1 can be derived using the general technique of quadratic forms developed in [2].

Acknowledgment. The authors are grateful to Dr. V. Oleinik for bibliographic suggestions and discussions regarding properties of uniformly perfect sets and Carleson measures. Fruitful discussions with Professors V. Maz'ja and M. Solomjak are also acknowledged. We thank the anonymous referees for constructive criticism.

References

[1] A. Ancona. On strong barriers and an inequality of Hardy for domains in $\mathcal{R}_n$, *J. London Math. Soc.* (2) **34** (1986), 274–290.

[2] M.S. Birman. On spectrum of boundary value problems, *Mat. Sb. (N.S.)* **55**, (1961), 125–174. (in Russian)

[3] F. Bruhat and J. Tits. Groupes réductifs sur un corps local, *Inst. Hautes Études Sci. Publ. Math.* **41** (1972), 5–251. (in French)

[4] C. Calude. *Information and Randomness — An Algorithmic Perspective*, Springer-Verlag, Berlin, (1994).

[5] L. Carleson. *Selected Problems on Exceptional Sets*, Van-Nostrand, Princeton (1967).

[6] E.B. Davies. A review of Hardy inequalities, in *The Maz'ya Anniversary Collection*, Vol.2 (Rostock, 1998), 55–67, *Oper. Theory Adv. Appl.*, 110, Birkhäuser, Basel, (1999).

[7] Gerald A. Edgar. *Measure, Topology, and Fractal Geometry*, Springer-Verlag, New York (1990).

[8] W.N. Everitt (ed.). *Inequalities: Fifty Years on from Hardy, Littlewood and Polya*, Proceedings of the International Conference held at the University of Birmingham, Birmingham, July 13–17, 1987, Lecture Notes in Pure and Applied Mathematics, 129, Marcel Dekker, Inc., New York, (1991).

[9] L. Garding. Dirichlet problem for linear elliptic differential equations, *Math. Scand.* **1** (1953), 55–72.

[10] G.H. Hardy, J.E. Littlewood, J.E. Polya. *Inequalities*, Cambridge Univ. Press, Cambridge, (1934).

[11] D.J.L. Herrmann, T. Janssen. On spectral properties of Harper-like models, *J. Math. Physics*, **40** 3 (1999), 1197–1214.

[12] P. Järvi, M. Vuorinen. Uniformly perfect sets and quasiregular mappings, *J. London Math. Soc.* (2) **54** (1966), 515–529.

[13] P. Järvi, M. Vuorinen. Self-similar Cantor sets and quasiregular mappings. *J. Reine Angew. Math.* **424** (1992), 31–45.

[14] P. Kurka. Simplicity criteria for dynamical systems, in *Analysis of Dynamical and Cognitive Systems (Stockholm, 1993)*, 189–225, Lecture Notes in Comput. Sci., 888, Springer-Verlag, Berlin, (1995).

[15] O.A. Ladyzhenskaja. *Problems in the Dynamics of Viscous Incompressible Flow*, Gordon & Breach, New York, (1963).

[16] D. Lind, B. Marcus. *An Introduction to Symbolic Dynamics and Coding*, Cambridge University Press, Cambridge, (1995).

[17] V.M. Maz'ja. *Sobolev Spaces*, Springer-Verlag, Berlin, (1985).

[18] V.L. Oleinik. Carleson measures and uniformly perfect sets, *Zap. Nauchn. Sem. S.-Peterburg. Otdel. Mat. Inst. Steklov. (POMI)* **255** (1998), *Issled. po Linein. Oper. i Teor. Funkts.* **26**, 92–103, 251 (in Russian).

[19] B. Pavlov. Boundary conditions on thin manifolds and the semi-boundedness of the three-particle Schrödinger operator with pointwise potential, *Math. USSR Sbornik*, **64**, 1 (1989), 161–175.

[20] C.A. Pickover. *Mazes for the Mind: Computers and the Unexpected,* St. Martin's Press, New York, (1992).

[21] Ch. Pommerenke. Uniformly perfect sets and the Poincaré metric, *Arch. Math.* **32** (1979), 192–199.

[22] L.I. Schiff. *Quantum Mechanics*, 3rd edition McGraw-Hill, New York, (1968).

[23] H.G. Schuster. *Deterministic Chaos*, Physik-Verlag, Weiheim, (1984).

[24] M. Reed, B. Simon. *Methods of Modern Mathematical Physics*, Vol. 3, Academic Press, 3rd edition, New York, (1987).

Cristian S. Calude
Department of Computer Science
University of Auckland, Private Bag 92019
Auckland, New Zealand
e-mail: `cristian@cs.auckland.ac.nz`

Boris Pavlov
Department of Mathematics
University of Auckland, Private Bag 92019
Auckland, New Zealand
e-mail: `pavlov@math.auckland.ac.nz.`

Operator Theory:
Advances and Applications, Vol. 134, 209–224

Finite Section Method for Linear Ordinary Differential Equations on the Full Line

I. Gohberg, M.A. Kaashoek, and F. van Schagen

Dedicated to Harry Dym with admiration and friendship, on the occasion of his sixtieth birthday

Abstract. Sufficient conditions are given in order that the solution of a linear ordinary differential equation on the full line is obtained as the limit of solutions of corresponding equations on finite intervals with boundary conditions or on half lines with initial conditions. Both the time-variant and the time-invariant case are considered, and in the latter case the sufficient conditions are also shown to be necessary. Included are applications to integral equations with semi-separable kernels.

1. Introduction

In this paper we study solutions of linear ordinary differential equations as limits of solutions of corresponding equations on smaller intervals. We focus on full line equations which are viewed as limits of equations either on finite intervals with boundary conditions or on half lines with initial conditions. Our results extend those of [7], where half line equations were studied as limits of finite interval equations.

To describe the problems treated here in more detail, we start with approximations by finite interval equations. Consider the equation

$$\dot{x}(t) - A(t)x(t) = f(t), \quad -\infty < t < \infty, \tag{1}$$

and a corresponding equation on the finite interval of the form

$$\begin{cases} \dot{x}(t) - A(t)x(t) = f(t), & \sigma \le t \le \tau, \\ M(\sigma)x(\sigma) + N(\tau)x(\tau) = 0. \end{cases} \tag{2}$$

Here $A(t)$ is a locally integrable $n \times n$ matrix function on $\mathbb{R}$, and $M(\sigma)$ and $N(\tau)$ are $n \times n$ matrices such that $\operatorname{rank} M(\sigma) + \operatorname{rank} N(\tau) = n$. Throughout this paper we shall impose the following conditions on the function f, the coefficient $A(t)$ and the boundary conditions in (2). We assume that the right-hand side f in (1) is in $L_p^n(-\infty, \infty)$, where $1 \le p < \infty$ is fixed, and we require the solution to be in $L_p^n(-\infty, \infty)$. In order that the equation (1) has a unique solution in $L_p^n(-\infty, \infty)$

for each right-hand side f in $L_p^n(-\infty,\infty)$, we also assume that there exists an *exponential dichotomy* P of

$$\dot{x}(t) - A(t)x(t) = 0, \qquad -\infty < t < \infty. \tag{3}$$

The latter means that P is a projection and there exist positive real constants M and α such that

$$\begin{aligned} \|U(t)PU(s)^{-1}\| &\leq M\mathrm{e}^{-\alpha(t-s)}, \quad s \leq t, \\ \|U(t)(I-P)U(s)^{-1}\| &\leq M\mathrm{e}^{-\alpha(s-t)}, \quad t \leq s, \end{aligned} \tag{4}$$

where $U(t)$ is the fundamental matrix of (3). Furthermore, we shall restrict ourselves to the case when for τ and $-\sigma$ sufficiently large ($\tau \geq \tau_\circ$ and $\sigma \leq \sigma_\circ$, say) the equation (2) has a unique solution for each right-hand side f in $L_p^n(-\infty,\infty)$. The latter happens (see [5]) if and only if

$$\det\big(M(\sigma)U(\sigma) + N(\tau)U(\tau)\big) \neq 0, \quad \tau \geq \tau_\circ, \quad \sigma \leq \sigma_\circ.$$

Given $f \in L_p^n(-\infty,\infty)$, our aim is to approximate the unique solution x of (1) by the solution $x_{\sigma,\tau}$ of (2) for $\tau, -\sigma \to \infty$. More precisely, we consider the problem of finding conditions guaranteeing that for $\tau, -\sigma \to \infty$ the solution $x_{\sigma,\tau}$ of the equation (2) converges in L_p to the solution x of (1), where convergence of $x_{\sigma,\tau}$ to x in L_p means that

$$\lim_{\tau,-\sigma\to\infty} \int_\sigma^\tau \|x_{\sigma,\tau}(t) - x(t)\|^p dt = 0.$$

The first main result is that for each f in $L_p^n(-\infty,\infty)$ the unique solution x of (1) in $L_p^n(-\infty,\infty)$ is obtained as the limit in L_p of the unique solution $x_{\sigma,\tau}$ ($\tau \geq \tau_\circ, \sigma \leq \sigma_\circ$) of equation (2) whenever the boundary conditions $M(\sigma)$ and $N(\tau)$ satisfy

$$\sup_{\tau\geq\tau_\circ,\sigma\leq\sigma_\circ} \left\| \begin{pmatrix} U(\sigma) \\ U(\tau) \end{pmatrix} \big(M(\sigma)U(\sigma) + N(\tau)U(\tau)\big)^{-1} \begin{pmatrix} M(\sigma) & N(\tau) \end{pmatrix} \right\| < \infty.$$

In this case we say that for the equation (1) the *finite section method with respect to the boundary value matrices* $\{M(\sigma)\}$ *and* $\{N(\tau)\}$ *converges in* L_p.

The result mentioned in the previous paragraph is specified further for the time-invariant case when $A(t)$ does not depend on t. In this case we also assume that $M(s)$ and $N(s)$ do not depend on s. We show (see Theorem 3.1) that for this time-invariant case the sufficient condition for convergence of the finite section method mentioned above is also necessary.

The results are applied to integral equations

$$\phi(t) + \int_{-\infty}^{\infty} k(t,s)\phi(s)ds = f(t), \quad -\infty < t < \infty, \tag{5}$$

with a semi-separable kernel given by

$$k(t,s) = \begin{cases} C(t)U(t)PU(s)^{-1}B(s), & -\infty < s < t < \infty, \\ -C(t)U(t)(I-P)U(s)^{-1}B(s), & -\infty < t < s < \infty. \end{cases}$$

Here, as usual (cf., [4], Section III.3), the finite section method means that the solutions of (5) are approximated by solutions of the equation

$$\phi(t) + \int_\sigma^\tau k(t,s)\phi(s)ds = f(t), \quad \sigma \le t \le \tau.$$

Analogous results are also obtained for approximations of (1) by half line equations, both in the time-variant and the time-invariant case.

2. A finite section method for differential equations with dichotomy

Throughout this section $U(t)$ is the fundamental matrix of the differential equation (3), i.e., $U(t)$ is absolutely continuous on finite intervals, $U(0)$ is the $n \times n$ identity matrix and $(\frac{d}{dt})U(t) = A(t)U(t)$ a.e. on $-\infty < t < \infty$. Let P be a projection of $\mathbb{C}^n$. If P is an exponential dichotomy of $U(t)$, i.e., the inequalities (4) hold true, then P is also called an exponential dichotomy for the equation (3). A dichotomy P (assuming that it exists) is unique. This is the contents of the next proposition (see [3], pp. 16, 17).

Proposition 2.1. *The exponential dichotomy of* (3) *is unique. In fact, if P is an exponential dichotomy of* (3), *and $U(t)$ is the fundamental matrix of* (3), *then for any p with $1 \le p < \infty$*

$$\operatorname{Im} P = \{x \in \mathbb{C}^n \mid U(t)x \in L_p^n[0,\infty)\}, \quad \operatorname{Ker} P = \{x \in \mathbb{C}^n \mid U(t)x \in L_p^n(-\infty,0]\}.$$

In the time-invariant case, i.e., when $A(t) = A$ for each $t \in \mathbb{R}$, the equation (3) has an exponential dichotomy if and only if A has no eigenvalues on the imaginary axis, and in that case the exponential dichotomy P is the spectral projection of A corresponding to eigenvalues in the left half plane.

Fix $f \in L_p^n(-\infty,\infty)$ $(1 \le p < \infty)$, and consider the equation

$$\dot{x}(t) - A(t)x(t) = f(t), \quad -\infty < t < \infty, \tag{6}$$

and a corresponding equation on the finite interval

$$\begin{cases} \dot{x}(t) - A(t)x(t) = f(t), & \sigma \le t \le \tau, \\ M(\sigma)x(\sigma) + N(\tau)x(\tau) = 0. \end{cases} \tag{7}$$

Recall that throughout this paper we assume that $A(t)$ is a locally integrable $n \times n$ matrix function on $\mathbb{R}$, that P is an exponential dichotomy for the equation $\dot{x}(t) = A(t)x(t)$, and that $M(\sigma)$ and $N(\tau)$ are $n \times n$ matrices such that rank $M(\sigma)+$ rank $N(\tau) = n$ for each $\sigma < \tau$. Also, $1 \le p < \infty$ is fixed.

Theorem 2.2. *Assume that $M(\sigma)U(\sigma) + N(\tau)U(\tau)$ is invertible for each $\tau \ge \tau_\circ$ and $\sigma \le \sigma_\circ$. If, in addition, the sequence of projections $P_{\sigma,\tau}$,*

$$P_{\sigma,\tau} = \begin{pmatrix} U(\sigma) \\ U(\tau) \end{pmatrix} \left(M(\sigma)U(\sigma) + N(\tau)U(\tau)\right)^{-1} \begin{pmatrix} M(\sigma) & N(\tau) \end{pmatrix}, \tag{8}$$

is uniformly bounded in norm for $\sigma \to -\infty$ *and* $\tau \to \infty$, *then for* (6) *the finite section method relative to the boundary value matrices* $\{M(\sigma)\}$ *and* $\{N(\tau)\}$ *converges in* L_p.

Before we prove this theorem, let us note that one can always choose the boundary value matrices $M(\sigma)$ and $N(\tau)$ in such a way that the projections $P_{\sigma,\tau}$ given by (8) are uniformly bounded in norm for $\sigma \to -\infty$ and $\tau \to \infty$. Indeed, the latter happens if one chooses $M(\sigma) = (I-P)U(\sigma)^{-1}$ and $N(\tau) = PU(\tau)^{-1}$.

We precede the proof of Theorem 2.2 with two lemmas providing integral representations of the solution of (6) and (7), respectively.

Lemma 2.3. *Assume that* $U(t)$ *has an exponential dichotomy* P *on* $(-\infty,\infty)$. *Then the unique solution of* (6) *is given by*

$$x(t) = \int_{-\infty}^{\infty} \gamma(t,s)f(s)ds, \quad -\infty < t < \infty, \tag{9}$$

where

$$\gamma(t,s) = \begin{cases} U(t)PU(s)^{-1}, & s < t \\ -U(t)(I-P)U(s)^{-1}, & t < s. \end{cases}$$

Furthermore, the linear operator $T : L_p^n(-\infty,\infty) \to L_p^n(-\infty,\infty)$ *defined by*

$$(Tf)(t) = \int_{-\infty}^{\infty} \gamma(t,s)f(s)ds \tag{10}$$

is bounded.

Proof. From the estimates (4) it follows that $x(t)$ is well defined and that the operator T is bounded. Differentiating the right-hand side of (9) yields that $x(t)$ is indeed the solution of (6). □

Lemma 2.4. *Assume that* $M(\sigma)U(\sigma) + N(\tau)U(\tau)$ *is invertible. Then the solution of* (7) *is unique and given by*

$$x_{\sigma,\tau}(t) = \int_{\sigma}^{\tau} \gamma_{\sigma,\tau}(t,s)f(s)ds, \quad \sigma \le t \le \tau, \tag{11}$$

where

$$\gamma_{\sigma,\tau}(t,s) = \begin{cases} U(t)L(\sigma,\tau)U(s)^{-1}, & \sigma \le s < t \le \tau \\ -U(t)\big(I - L(\sigma,\tau)\big)U(s)^{-1}, & \sigma \le t < s \le \tau, \end{cases}$$

and the projection $L(\sigma,\tau)$ *is given by*

$$L(\sigma,\tau) = \big(M(\sigma)U(\sigma) + N(\tau)U(\tau)\big)^{-1} M(\sigma)U(\sigma).$$

Proof. The uniqueness of the solution follows from the well-posedness of the boundary conditions. By differentiating, and by computing the right-hand side of (11) for $t = \sigma$ and $t = \tau$ one checks that $x(t)$ is a solution of (7). □

Proof of Theorem 2.2. For each $\tau > \sigma$ let $E_{\sigma,\tau}$ be the canonical embedding of $L_p^n[\sigma,\tau]$ in $L_p^n(-\infty,\infty)$ and $R_{\sigma,\tau}$ be the restriction operator from $L_p^n(-\infty,\infty)$ to $L_p^n[\sigma,\tau]$. Note that both operators $E_{\sigma,\tau}$ and $R_{\sigma,\tau}$ are of norm one. Since

$$E_{\sigma,\tau}R_{\sigma,\tau}f \to f$$

for $\sigma \to -\infty$ and $\tau \to \infty$, the fact that the operator T given by (10) and the operators $R_{\sigma,\tau}$ are bounded implies that

$$\|R_{\sigma,\tau}Tf - R_{\sigma,\tau}TE_{\sigma,\tau}R_{\sigma,\tau}f\| \to 0$$

for $\sigma \to -\infty$ and $\tau \to \infty$. Therefore, it is sufficient to show that for an arbitrary fixed $f \in L_p$ one has

$$\lim_{\sigma\to-\infty,\tau\to\infty} \int_\sigma^\tau (\gamma_{\sigma,\tau}(t,s) - \gamma(t,s))f(s)ds = 0,$$

where the limit is a limit in the norm in L_p. Now

$$\begin{aligned}\gamma_{\sigma,\tau}(t,s) - \gamma(t,s) = U(t)\begin{pmatrix} P & I-P \end{pmatrix}\begin{pmatrix} U(\sigma)^{-1} & 0 \\ 0 & U(\tau)^{-1} \end{pmatrix} \\ \cdot P_{\sigma,\tau}\begin{pmatrix} U(\sigma) & 0 \\ 0 & U(\tau) \end{pmatrix}\begin{pmatrix} I-P \\ -P \end{pmatrix} U(s)^{-1}.\end{aligned}$$

Since P is an exponential dichotomy, for each $v \in \mathbb{C}^{2n}$ the function

$$U(t)\begin{pmatrix} P & I-P \end{pmatrix}\begin{pmatrix} U(\sigma)^{-1} & 0 \\ 0 & U(\tau)^{-1} \end{pmatrix} v, \quad \sigma \le t \le \tau,$$

belongs to $L_p^n[\sigma,\tau]$, and as a function of σ and τ its norm is uniformly bounded for $\sigma \to -\infty$ and $\tau \to \infty$. By assumption $P_{\sigma,\tau}$ is uniformly bounded.

It remains to show that

$$\lim_{-\sigma,\tau\to\infty} \int_\sigma^\tau \begin{pmatrix} U(\sigma) & 0 \\ 0 & U(\tau) \end{pmatrix}\begin{pmatrix} I-P \\ -P \end{pmatrix} U(s)^{-1}f(s)ds = 0.$$

For the first block component it follows from standard results on convolutions that

$$\begin{aligned}\|\int_\sigma^\tau U(\sigma)(I-P)U(s)^{-1}f(s)ds\| &\le \int_\sigma^\tau M\mathrm{e}^{-\alpha(s-\sigma)}\|f(s)\|ds \\ = \int_0^{\tau-\sigma} M\mathrm{e}^{-\alpha t}\|f(t+\sigma)\|dt &\le \int_0^\infty M\mathrm{e}^{-\alpha t}\|f(t+\sigma)\|dt \to 0 \quad (\sigma \to -\infty).\end{aligned}$$

Similarly, we have for the second block component:

$$\begin{aligned}\|\int_\sigma^\tau U(\tau)PU(s)^{-1}f(s)ds\| &\le \int_\sigma^\tau M\mathrm{e}^{-\alpha(\tau-s)}\|f(s)\|ds \\ = \int_0^{\tau-\sigma} M\mathrm{e}^{-\alpha t}\|f(\tau-t)\|dt &\le \int_0^\infty M\mathrm{e}^{-\alpha t}\|f(\tau-t)\|dt \to 0 \quad (\tau \to \infty),\end{aligned}$$

which completes the proof. □

3. The time-invariant case

In this section we consider the differential equation with constant coefficients

$$\dot{x}(t) - Ax(t) = f(t), \quad -\infty < t < \infty, \tag{12}$$

and a corresponding equation on the finite interval

$$\begin{cases} \dot{x}(t) - Ax(t) = f(t), & \sigma \leq t \leq \tau, \\ Qx(\sigma) + (I - Q)x(\tau) = 0. \end{cases} \tag{13}$$

Here Q is a projection, and the right-hand side $f \in L_p^n(-\infty, \infty)$ $(1 \leq p < \infty)$. Throughout this section we assume that A has no eigenvalue on the imaginary axis, and that P is the spectral projection of A with respect to the left half plane. The next result gives a necessary and sufficient condition in terms of P and Q for the convergence of the finite section method.

Theorem 3.1. *Let* rank Q = rank P. *The finite section method for* (12) *relative to the boundary value matrices Q and $I - Q$ converges in L_p if and only if $(I - P)(I - Q) + PQ$ is invertible.*

Notice that the invertibility of $(I - P)(I - Q) + PQ$ implies rank Q = rank P.

Proof. To prove the theorem we shall use a finite section method result from [2] for the integral operators that appear as input-output operators of input-output systems. For this purpose, we embed both (12) and (13) in linear input-output systems, as follows:

$$\begin{cases} \dot{x}(t) - Ax(t) = f(t), & -\infty < t < \infty, \\ g(t) = -Cx(t) + f(t), & -\infty < t < \infty, \end{cases} \tag{14}$$

and

$$\begin{cases} \dot{x}(t) - Ax(t) = f(t), & \sigma \leq t \leq \tau, \\ g(t) = -Cx(t) + f(t), & \sigma \leq t \leq \tau, \\ Qx(\sigma) + (I - Q)x(\tau) = 0. \end{cases} \tag{15}$$

The matrix C will be an invertible matrix and will be specified later on. Given f, one can compute x easily from g. Therefore we concentrate on the question when the output $g_{\sigma,\tau}$ of (15) converges for $\sigma \to -\infty$ and $\tau \to \infty$ to the output g of (14). We would like to view g and $g_{\sigma,\tau}$ as solutions of integral equations. To arrive at such a point of view, we first invert the systems (15) and (14). So we get

$$\begin{cases} \dot{x}(t) - (A + C)x(t) = g(t), & -\infty < t < \infty, \\ f(t) = Cx(t) + g(t), & -\infty < t < \infty, \end{cases} \tag{16}$$

and

$$\begin{cases} \dot{x}(t) - (A + C)x(t) = g(t), & \sigma \leq t \leq \tau, \\ f(t) = Cx(t) + g(t), & \sigma \leq t \leq \tau, \\ Qx(\sigma) + (I - Q)x(\tau) = 0. \end{cases} \tag{17}$$

Next we specify the matrix C. We choose a matrix $A_\circ^\times$ such that Q is the spectral projection of $A_\circ^\times$ with respect to the left half plane. This can be done as follows. Since rank $Q =$ rank P, we know that the projections Q and P are similar. Let S be the similarity, i.e., S is such that $Q = SPS^{-1}$. Choose $A_\circ^\times = SAS^{-1}$. Then Q is indeed the spectral projection of $A_\circ^\times$ with respect to the left half plane. Now consider $A_\circ^\times - A$. If 0 is an eigenvalue, then there is a number $\epsilon > 0$ such that Q is the spectral projection of $\epsilon I + A_\circ^\times$ with respect to the left half plane and 0 is not an eigenvalue of $\epsilon I + A_\circ^\times - A$. Now choose $C = A_\circ^\times - A$, if 0 is not an eigenvalue of $A_\circ^\times - A$, and $C = \epsilon I + A_\circ^\times - A$, if 0 is an eigenvalue of $A_\circ^\times - A$. Thus C is invertible and Q is the spectral projection with respect to the left half plane of $A^\times = A + C$.

Let us compute $x(t)$ and $f(t)$ from (16). We find

$$x(t) = \int_{-\infty}^{\infty} \gamma^\times(t,s)g(s)ds, \quad -\infty < t < \infty, \tag{18}$$

and

$$f(t) = g(t) + C\int_{-\infty}^{\infty} \gamma^\times(t,s)g(s)ds, \quad -\infty < t < \infty. \tag{19}$$

Here

$$\gamma^\times(t,s) = \begin{cases} Q\mathrm{e}^{(t-s)A^\times}, & t \le s, \\ (I-Q)\mathrm{e}^{-(t-s)A^\times}, & s \le t. \end{cases} \tag{20}$$

Next, compute $x_{\sigma,\tau}(t)$ and $f(t)$ from (17). One obtains with $\gamma^\times$ given by (20) that

$$x_{\sigma,\tau}(t) = \int_{\sigma}^{\tau} \gamma^\times(t,s)g(s)ds, \quad \sigma \le t \le \tau, \tag{21}$$

and

$$f(t) = g(t) + C\int_{\sigma}^{\tau} \gamma^\times(t,s)g(s)ds, \quad \sigma \le t \le \tau. \tag{22}$$

Consider (19) as an integral equation in the unknown function g. Then the equation (22) is a finite section of (19). Now use the results of [2], Section 5 to see that the solution $g_{\sigma,\tau}$ of (22) converges in L_p to the solution g of (19) for each right-hand side f if and only if $(I-Q)(I-P)+QP$ is invertible. Since $g(t) = Cx(t)+f(t)$ and C is invertible, we obtain that the function $x_{\sigma,\tau}$ from (21) converges in L_p to the function x from (18) for each f if and only if $(I-Q)(I-P)+QP$ is invertible. Finally note that $(I-Q)(I-P)+QP$ is invertible if and only if $(I-P)(I-Q)+PQ$ is invertible. □

Let us remark that the above method of proof works equally well for the half line case when solutions on the half line $[\sigma,\infty)$ are approximated by solutions on a finite interval $[\sigma,\tau]$ for a fixed σ and τ going to infinity. Also convergence of solutions on a half line to solutions on the full line can be treated in this way. This is based on the fact that for the time invariant case the corresponding results on integral equations are available, see [2], Section 5.

The next proposition shows that the necessary and sufficient condition for convergence of the finite section method in Theorem 3.1 implies the necessary condition of Theorem 2.2 when specified for the time invariant case.

Proposition 3.2. *If $(I-P)(I-Q)+PQ$ is invertible, then there exist $\sigma_\circ$ and $\tau_\circ$ such that $Q\mathrm{e}^{\sigma A}+(I-Q)\mathrm{e}^{\tau A}$ is invertible for $\sigma<\sigma_\circ$ and $\tau>\tau_\circ$, and such that*

$$P_{\sigma,\tau}=\begin{pmatrix}\mathrm{e}^{\sigma A}\\ \mathrm{e}^{\tau A}\end{pmatrix}\left(Q\mathrm{e}^{\sigma A}+(I-Q)\mathrm{e}^{\tau A}\right)^{-1}\begin{pmatrix}Q & I-Q\end{pmatrix} \tag{23}$$

is uniformly bounded in norm for $\sigma\to-\infty$ and $\tau\to\infty$.

Proof. First notice that $P_{\sigma,\tau}$ is bounded if and only if $P_{\sigma,\tau}\begin{pmatrix}I\\ I\end{pmatrix}$ is bounded. Hence it is sufficient to show that $\left(Q\mathrm{e}^{(\sigma-\tau)A}+(I-Q)\right)^{-1}$ and $\left(Q+(I-Q)\mathrm{e}^{(\tau-\sigma)A}\right)^{-1}$ exist and are uniformly bounded for $\sigma\to-\infty$ and $\tau\to\infty$. Put $\rho=\tau-\sigma$. We will show that $\left(Q\mathrm{e}^{-\rho A}+(I-Q)\right)^{-1}$ and $\left(Q+(I-Q)\mathrm{e}^{\rho A}\right)^{-1}$ exist and are uniformly bounded in norm for ρ sufficiently large (in fact, we shall prove that the limit exists for $\rho\to\infty$). The condition $(I-P)(I-Q)+PQ$ is invertible implies that $\mathbb{C}^n=\operatorname{Im}Q\oplus\operatorname{Ker}P$ and $\mathbb{C}^n=\operatorname{Im}P\oplus\operatorname{Ker}Q$. In what follows P, Q, and A are represented as 2×2 block matrices with respect to the direct sum decomposition $\mathbb{C}^n=\operatorname{Im}Q\oplus\operatorname{Ker}P$. For P and Q this yields

$$P=\begin{pmatrix}I & 0\\ R & 0\end{pmatrix},\quad Q=\begin{pmatrix}I & S\\ 0 & 0\end{pmatrix},$$

with $I+SR$ invertible because $\mathbb{C}^n=\operatorname{Ker}Q\oplus\operatorname{Im}P$. For A we get

$$A=\begin{pmatrix}A_{11} & 0\\ RA_{11}-A_{22}R & A_{22}\end{pmatrix},$$

with the eigenvalues of A_{11} and A_{22} all in the left and right half plane, respectively. It follows that

$$\mathrm{e}^{tA}=\begin{pmatrix}\mathrm{e}^{tA_{11}} & 0\\ R\mathrm{e}^{tA_{11}}-\mathrm{e}^{tA_{22}}R & \mathrm{e}^{tA_{22}}\end{pmatrix}.$$

Thus

$$\begin{aligned}Q+(I-Q)\mathrm{e}^{\rho A} &= \begin{pmatrix}I & S\\ 0 & 0\end{pmatrix}+\begin{pmatrix}-SR\mathrm{e}^{\rho A_{11}}+S\mathrm{e}^{\rho A_{22}}R & -S\mathrm{e}^{\rho A_{22}}\\ R\mathrm{e}^{\rho A_{11}}-\mathrm{e}^{\rho A_{22}}R & \mathrm{e}^{\rho A_{22}}\end{pmatrix}\\ &= \begin{pmatrix}I & S(\mathrm{e}^{-\rho A_{22}}-I)\\ 0 & I\end{pmatrix}\begin{pmatrix}h(\rho) & 0\\ R\mathrm{e}^{\rho A_{11}}-\mathrm{e}^{\rho A_{22}}R & \mathrm{e}^{\rho A_{22}}\end{pmatrix},\end{aligned}$$

with $h(\rho)=I+SR-S\mathrm{e}^{-\rho A_{22}}R\mathrm{e}^{\rho A_{11}}$. Because $\mathrm{e}^{-\rho A_{22}}\to0$, $\mathrm{e}^{\rho A_{11}}\to0$ there exists a number $\rho_\circ$ such that $h(\rho)$ is invertible for $\rho>\rho_\circ$, and hence

$$\begin{aligned}&\left(Q+(I-Q)\mathrm{e}^{\rho A}\right)^{-1}\\ &=\begin{pmatrix}h(\rho)^{-1} & 0\\ -\mathrm{e}^{-\rho A_{22}}(R\mathrm{e}^{\rho A_{11}}-\mathrm{e}^{\rho A_{22}}R)h(\rho)^{-1} & \mathrm{e}^{-\rho A_{22}}\end{pmatrix}\begin{pmatrix}I & -S(\mathrm{e}^{-\rho A_{22}}-I)\\ 0 & I\end{pmatrix}\end{aligned}$$

$$= \begin{pmatrix} h(\rho)^{-1} & -h(\rho)^{-1}S(\mathrm{e}^{-\rho A_{22}} - I) \\ (R - \mathrm{e}^{-\rho A_{22}} R \mathrm{e}^{\rho A_{11}})h(\rho)^{-1} & (R - \mathrm{e}^{-\rho A_{22}} R \mathrm{e}^{\rho A_{11}})h(\rho)^{-1}S(\mathrm{e}^{-\rho A_{22}} - I) + \mathrm{e}^{-\rho A_{22}} \end{pmatrix}$$

Now since $\mathrm{e}^{-\rho A_{22}} \to 0$, $\mathrm{e}^{\rho A_{11}} \to 0$, it follows that $h(\rho)^{-1} \to (I+SR)^{-1}$ for $\rho \to \infty$. Thus for $\rho \to \infty$ we see

$$\left(Q + (I - Q)\mathrm{e}^{\rho A}\right)^{-1} \to \begin{pmatrix} (I+SR)^{-1} & (I+SR)^{-1}S \\ R(I+SR)^{-1} & R(I+SR)^{-1}S \end{pmatrix}$$

$$= \begin{pmatrix} I \\ R \end{pmatrix} (I+SR)^{-1} \begin{pmatrix} I & S \end{pmatrix}. \quad (24)$$

This shows that $\left(Q + (I - Q)\mathrm{e}^{\rho A}\right)^{-1}$ is uniformly bounded in norm for $\rho \to \infty$. (The right-hand side in (24) is the projection of $\mathbb{C}^n$ along $\operatorname{Ker} Q$ onto $\operatorname{Im} P$.)

Next

$$Q\mathrm{e}^{-\rho A} + (I - Q) = \begin{pmatrix} (I+SR)\mathrm{e}^{-\rho A_{11}} - S\mathrm{e}^{-\rho A_{22}}R & -S + S\mathrm{e}^{-\rho A_{22}} \\ 0 & I \end{pmatrix}.$$

Inverting gives

$$\left(Q\mathrm{e}^{-\rho A} + (I - Q)\right)^{-1} = \begin{pmatrix} \mathrm{e}^{\rho A_{11}}h(\rho)^{-1} & -\mathrm{e}^{\rho A_{11}}h(\rho)^{-1}\left(-S + S\mathrm{e}^{-\rho A_{22}}\right) \\ 0 & I \end{pmatrix},$$

which converges to

$$\begin{pmatrix} 0 & 0 \\ 0 & I \end{pmatrix}$$

for $\rho \to \infty$. (Note that this operator matrix represents the projection of $\mathbb{C}^n$ along $\operatorname{Im} Q$ onto $\operatorname{Ker} P$.) It follows that $\left(Q\mathrm{e}^{-\rho A} + (I - Q)\right)^{-1}$ is uniformly bounded in norm for $\rho \to \infty$. □

By combining Proposition 3.2 with Theorem 2.2 we see that the invertibility of $(I - P)(I - Q) + PQ$ implies that for (12) the finite section method relative to the boundary value matrices Q and $I - Q$ converges in L_p. In other words, for the time-invariant case Proposition 3.2 provides an alternative proof for the sufficiency part of Theorem 3.1.

4. From interval to full line via half line

We consider the equation

$$\begin{cases} \dot{x}(t) - A(t)x(t) = f(t), & \sigma \le t \le \infty, \\ M(\sigma)x(\sigma) = 0. \end{cases} \quad (25)$$

We assume that the fundamental matrix $U(t)$ of $\dot{x}(t) = A(t)x(t)$ has an exponential dichotomy L on $[\sigma, \infty)$, i.e., L is a projection such that there exist positive numbers

M and α such that

$$\begin{aligned} \|U(t)LU(s)^{-1}\| &\le M e^{-\alpha(t-s)}, \quad \sigma \le s \le t, \\ \|U(t)(I-L)U(s)^{-1}\| &\le M e^{-\alpha(s-t)}, \quad \sigma \le t \le s. \end{aligned} \tag{26}$$

Recall that an exponential dichotomy on $[\sigma,\infty)$ is not unique. In fact only the image of an exponential dichotomy on a half line is fixed. For (25) to have a unique solution in $L_p^n[\sigma,\infty)$ for each right-hand side $f \in L_p^n[\sigma,\infty)$, it is known to be necessary that $\operatorname{rank} M(\sigma) = \operatorname{rank} L$ and $M(\sigma)U(\sigma) + (I-L)$ is invertible. The next result gives the solution of (25).

Lemma 4.1. *Let L be an exponential dichotomy of the fundamental matrix $U(t)$ of $\dot{x}(t) = A(t)x(t)$ on $[\sigma,\infty)$, and let the boundary value operators $M(\sigma)$ in (25) be such that* $\operatorname{rank} M(\sigma) = \operatorname{rank} L$ *and $M(\sigma)U(\sigma) + (I-L)$ is invertible. Then the unique solution of* (25) *is*

$$x(t) = \int_\sigma^\infty \gamma_\sigma(t,s) f(s)ds, \quad \sigma \le t < \infty, \tag{27}$$

where

$$\gamma_\sigma(t,s) = \begin{cases} U(t)L(\sigma)U(s)^{-1}, & \sigma \le s < t < \infty, \\ -U(t)(I-L(\sigma))U(s)^{-1}, & \sigma \le t < s < \infty, \end{cases}$$

with

$$L(\sigma) = \big(M(\sigma)U(\sigma) + (I-L)\big)^{-1} M(\sigma)U(\sigma). \tag{28}$$

The linear operator $T_\sigma : L_p^n(\sigma,\infty) \to L_p^n(\sigma,\infty)$ defined by

$$(T_\sigma f)(t) = \int_\sigma^\infty \gamma_\sigma(t,s) f(s)ds$$

is bounded.

Proof. From (26) it follows that the right-hand side of (27) is well defined, and that T_σ is a bounded linear operator. By differentiation and substitution of $t=\sigma$ in (27) one verifies that $x(t)$ is a solution of (25) indeed. □

Notice that for $L(\sigma)$ given by (28) one has

$$I - L(\sigma) = \big(M(\sigma)U(\sigma) + (I-L)\big)^{-1}(I-L),$$

and hence $\operatorname{Ker}(I-L(\sigma)) = \operatorname{Ker}(I-L)$ and $\operatorname{Im} L(\sigma) = \operatorname{Im} L$, which tells us that the projection $L(\sigma)$ is an exponential dichotomy for $\dot{x}(t) = A(t)x(t)$ on $[\sigma,\infty)$. Moreover, $\operatorname{Ker} L(\sigma) = \operatorname{Ker} M(\sigma)U(\sigma)$. So the initial value condition in (25) can also be written as $L(\sigma)U(\sigma)^{-1}x(\sigma) = 0$.

From now on we assume that $\operatorname{rank} M(\sigma) = \operatorname{rank} L$ and that $M(\sigma)U(\sigma) + (I-L)$ is invertible. We recall that the solution of the equation

$$\begin{cases} \dot{x}(t) - A(t)x(t) = f(t), & \sigma \le t \le \tau, \\ M(\sigma)x(\sigma) + N(\tau)x(\tau) = 0, \end{cases} \tag{29}$$

on the finite interval is given in Lemma 2.4. Our assumption that $\operatorname{rank} M(\sigma) + \operatorname{rank} N(\tau) = n$ implies that $\operatorname{rank} N(\tau) = \operatorname{rank}(I-L)$ and that $L(\sigma,\tau)$, as given in Lemma 2.4, is a projection, which is called the canonical boundary projection (cf., [5]).

Theorem 4.2. *Assume that* $\operatorname{rank} M(\sigma) = \operatorname{rank} L$, *that* $M(\sigma)U(\sigma) + (I - L)$ *is invertible, and that* $M(\sigma)U(\sigma)+N(\tau)U(\tau)$ *is invertible for* $\tau \geq \tau_\circ$. *If, in addition,*

$$\sup_{\tau \geq \tau_\circ} \|U(\tau)\big(M(\sigma)U(\sigma) + N(\tau)U(\tau)\big)^{-1}N(\tau)\| < \infty,$$

then for $\tau \to \infty$ *the solution* $x_{\sigma,\tau}$ *of* (29) *converges to the solution* x_σ *of* (25) *in* L_p.

Proof. Like in the full line case, cf., the proof of Theorem 2.2, it is sufficient to show that

$$\lim_{\tau\to\infty} \| \int_\sigma^\tau (\gamma_\sigma(t,s) - \gamma_{\sigma,\tau}(t,s))f(s)ds\|_{L_p^n[\sigma,\tau]} = 0.$$

Because $\operatorname{Ker} L(\sigma,\tau) = \operatorname{Ker} L(\sigma)$, we have $L(\sigma) = L(\sigma)L(\sigma,\tau)$ and $L(\sigma,\tau) = L(\sigma,\tau)L(\sigma)$. It follows that

$$\begin{aligned}\gamma_\sigma(t,s) - \gamma_{\sigma,\tau}(t,s) &= U(t)(L(\sigma) - L(\sigma,\tau))U(s)^{-1}\\ &= -U(t)(I - L(\sigma))U(\tau)^{-1}U(\tau)L(\sigma,\tau)U(\tau)^{-1}U(\tau)L(\sigma)U(s)^{-1}.\end{aligned}$$

First we show that

$$\lim_{\tau\to\infty} \int_\sigma^\tau \|U(\tau)L(\sigma)U(s)^{-1}\|\|f(s)\|ds = 0. \tag{30}$$

Since $L(\sigma)$ is an exponential dichotomy for $U(t)$, we have

$$\|U(\tau)L(\sigma)U(s)^{-1}\| \leq M\mathrm{e}^{-\alpha(\tau-s)}, \quad \sigma \leq s \leq \tau.$$

Thus, in order to prove (30), it suffices to show that

$$\lim_{\tau\to\infty} \int_\sigma^\tau M\mathrm{e}^{-\alpha(\tau-s)}\|f(s)\|ds = 0.$$

Notice that

$$\begin{aligned}\int_\sigma^\tau M\mathrm{e}^{-\alpha(\tau-s)}\|f(s)\|ds &= \int_0^{\tau-\sigma} M\mathrm{e}^{-\alpha t}\|f(\tau-t)\|dt\\ &\leq \int_0^\infty M\mathrm{e}^{-\alpha t}\|f(\tau-t)\|dt \to 0 \quad (\tau\to\infty).\end{aligned}$$

which follows from a well-known property of convolutions. Secondly, use again that $L(\sigma)$ is an exponential dichotomy to see that

$$\begin{aligned}\int_\sigma^\tau \|U(t)(I - L(\sigma))U(\tau)^{-1}\|^p dt &\leq \int_\sigma^\tau M^p\mathrm{e}^{-\alpha(\tau-t)p}dt\\ = M^p\int_0^{\tau-\sigma} \mathrm{e}^{-\alpha ps}ds &\leq M^p\int_0^\infty \mathrm{e}^{-\alpha ps}ds.\end{aligned}$$

Finally, since

$$U(\tau)L(\sigma,\tau)U(\tau)^{-1} = U(\tau)(M(\sigma)U(\sigma) + N(\tau)U(\tau))^{-1}N(\tau)$$

is bounded by assumption, we conclude that

$$\begin{aligned}\lim_{\tau\to\infty} \| \int_\sigma^\tau (\gamma_\sigma(t,s) - \gamma_{\sigma,\tau}(t,s))f(s)ds\|_{L_p^n[\sigma,\tau]} \\ \leq \lim_{\tau\to\infty} M^p \int_0^\infty e^{-\alpha ps} ds \| U(\tau)(M(\sigma)U(\sigma) \\ + N(\tau)U(\tau))^{-1} N(\tau)\| \int_\sigma^\tau M e^{-\alpha(\tau-s)} \|f(s)\| ds = 0,\end{aligned}$$

as desired. □

A further examination of the above proof yields the following result.

Corollary 4.3. *Assume that* rank $M(\sigma)$ = rank L, *that* $M(\sigma)U(\sigma) + (I - L)$ *is invertible, and that* $M(\sigma)U(\sigma) + N(\tau)U(\tau)$ *is invertible for* $\tau \geq \tau_\circ$ *and* $\sigma \leq \sigma_\circ$. *If, in addition,*

$$\sup_{\sigma\leq\sigma_\circ, \tau\geq\tau_\circ} \|U(\tau)\big(M(\sigma)U(\sigma) + N(\tau)U(\tau)\big)^{-1} N(\tau)\| < \infty,$$

then for $\tau \to \infty$ *the solution* $x_{\sigma,\tau}$ *of* (29) *converges to the solution* x_σ *of* (25) *in* L_p *uniformly in* σ.

Next we consider the convergence from half line equation (25) to full line equation (6). Recall that P is the dichotomy for (6). We suppose that the boundary matrix $M(\sigma)$ of (25) has the following properties: rank $M(\sigma)$ = rank P and $M(\sigma)U(\sigma) + I - P$ is invertible for $\sigma < \sigma_\circ$.

Theorem 4.4. *Assume that* rank $M(\sigma)$ = rank P *and that* $M(\sigma)U(\sigma) + I - P$ *is invertible for* $\sigma < \sigma_\circ$. *If* $U(\sigma)\big(M(\sigma)U(\sigma)+I-P\big)^{-1}M(\sigma)$ *is bounded for* $\sigma \to -\infty$, *then for* $\sigma \to -\infty$ *the solution* x_σ *of* (25) *converges in* L_p *to the solution* x *of* (6).

Proof. As in the proof of Theorem 2.2, it is sufficient to show that in L_p-sense

$$\lim_{\sigma\to-\infty} \int_\sigma^\infty \gamma(t,s) - \gamma_\sigma(t,s)f(s)ds = 0.$$

Put $L(\sigma) = \big(M(\sigma)U(\sigma) + I - P)\big)^{-1} M(\sigma)U(\sigma)$. Then $L(\sigma)$ is a projection with $\operatorname{Im} L(\sigma) = \operatorname{Im} P$. Thus $L(\sigma)$ is an exponential dichotomy for the half line equation on $[\sigma, \infty)$. Write

$$\begin{aligned}\gamma(t,s) - \gamma_\sigma(t,s) &= U(t)(P - L(\sigma))U(s)^{-1} \\ &= -U(t)PU(\sigma)^{-1}U(\sigma)L(\sigma)U(\sigma)^{-1}U(\sigma)(I-P)U(s)^{-1}.\end{aligned}$$

We consider three factors. First $\{U(t)PU(\sigma)^{-1}v : v \in \mathbb{R}^n \text{ and } \|v\| < M\}$ is a collection of functions which is uniformly bounded in L_p for $\sigma \to -\infty$. Secondly, the factor $U(\sigma)L(\sigma)U(\sigma)^{-1}$ is bounded for $\sigma \to -\infty$ by assumption. Finally

$$\lim_{\sigma\to-\infty} \int_\sigma^\infty U(\sigma)(I-P)U(s)^{-1}f(s)ds = 0,$$

because $\|U(\sigma)(I-P)U(s)^{-1}\| \leq Me^{-\alpha(s-\sigma)}$. By combining these three results the theorem follows. □

From Theorem 4.4 and Corollary 4.3 we obtain the following result. As before P is the dichotomy for (6). Notice that the sufficient conditions are different from those in Theorem 2.2.

Theorem 4.5. *Assume that* $\operatorname{rank} M(\sigma) = \operatorname{rank} P$, *that* $M(\sigma)U(\sigma) + I - P$ *is invertible for* $\sigma \leq \sigma_\circ$ *and that* $M(\sigma)U(\sigma) + N(\tau)U(\tau)$ *is invertible for* $\tau \geq \tau_\circ$ *and* $\sigma \leq \sigma_\circ$. *If, in addition,*

$$\sup_{\sigma\leq\sigma_\circ,\tau\geq\tau_\circ} \|U(\tau)\big(M(\sigma)U(\sigma) + N(\tau)U(\tau)\big)^{-1}N(\tau)\| < \infty,$$

$$\sup_{\sigma\leq\sigma_\circ} \|U(\sigma)\big(M(\sigma)U(\sigma) + I - P\big)^{-1}M(\sigma)\| < \infty,$$

then for (6) *the finite section method relative to the boundary value matrices* $\{M(\sigma)\}$ *and* $\{N(\tau)\}$ *converges in* L_p.

Proof. Consider the full line equation (6) with a fixed right-hand side f. We will prove that the solution $x_{\sigma,\tau}$ of (29) converges in L_p to the solution x of (6). Let $\epsilon > 0$. According to Theorem 4.4 there exists a number σ_1 such that $\|x_\sigma - x\| < \frac{\epsilon}{2}$ for all $\sigma < \sigma_1$, where x_σ is the solution of (25) and x is the solution of (6). Next it follows from Corollary 4.3 that there exists a number τ_1 such that $\|x_{\sigma,\tau} - x_\sigma\| < \frac{\epsilon}{2}$ for each $\tau > \tau_1$ and all $\sigma < \sigma_\circ$, where $x_{\sigma,\tau}$ is the solution of (29). Thus for $\tau > \tau_1$ and $\sigma < \sigma_1$ we have $\|x_{\sigma,\tau} - x\| < \epsilon$, which shows that for $\sigma \to -\infty$ and $\tau \to \infty$ the solution of (29) converges to the solution of (7). □

5. Application to integral equations

In this section we apply the finite section method for differential equations with dichotomy to integral equations with semi-separable kernels. Throughout this section the projection P is an exponential dichotomy for the differential equation $\dot{x}(t) = A(t)x(t)$ on $-\infty < t < \infty$, and $U(t)$ is the fundamental matrix of this differential equation. We consider the integral equation

$$\phi(t) + \int_{-\infty}^{\infty} k(t,s)\phi(s)ds = f(t), \quad -\infty < t < \infty, \tag{31}$$

where

$$k(t,s) = \begin{cases} C(t)U(t)PU(s)^{-1}B(s), & -\infty < s < t < \infty, \\ -C(t)U(t)(I-P)U(s)^{-1}B(s), & -\infty < t < s < \infty. \end{cases} \tag{32}$$

Here $B(t)$ is an $n \times m$ matrix and $C(t)$ is an $m \times n$ matrix, and the entries of both matrices are bounded measurable functions on the full line. We require both the right-hand side and the solution of (31) to be functions in $L_p^m(-\infty,\infty)$. Our aim is to get the solution of (31) as a limit for $\sigma \to -\infty$ and $\tau \to \infty$ of the solution of the corresponding equation on the interval $[\sigma, \tau]$.

In our analysis the matrix function $A^\times(t) = A(t) - B(t)C(t)$ will play an important role. Since the entries of both $B(t)$ and $C(t)$ are assumed to be bounded measurable functions on $(-\infty, \infty)$, the matrix function $A^\times(t)$ is again locally integrable on $-\infty < t < \infty$, and hence the differential equation

$$\dot{x}(t) - A^\times(t)x(t) = 0, \quad -\infty < t < \infty, \tag{33}$$

has a well-defined fundamental matrix, which we shall denote by $U^\times(t)$.

We shall say that *the finite section method for the integral equation* (31) converges if there exist numbers $\sigma_\circ$ and $\tau_\circ$ such that for every $f \in L_p^m(-\infty, \infty)$ and each $\sigma \leq \sigma_\circ$ and $\tau \geq \tau_\circ$ the integral equation

$$\phi(t) + \int_0^\tau k(t,s)\phi(s)ds = f(t), \quad \sigma \leq t \leq \tau, \tag{34}$$

has a unique solution $\phi_{\sigma,\tau} \in L_p^m[\sigma, \tau]$, which converges in L_p for $-\sigma, \tau \to \infty$ to the solution ϕ of (31).

The next theorem is the counterpart of Theorem 2.2 for integral operators.

Theorem 5.1. *Let $k(t,s)$ be given by* (32). *Put $A^\times(t) = A(t) - B(t)C(t)$, and assume that the fundamental matrix $U^\times(t)$ of* (33) *has an exponential dichotomy $P^\times$. Assume that there exist numbers $\sigma_\circ$ and $\tau_\circ$ such that the matrix function $PU(\sigma)^{-1}U^\times(\sigma) + (I-P)U(\tau)^{-1}U^\times(\tau)$ is invertible for $\sigma < \sigma_\circ$ and $\tau > \tau_\circ$. If the 2×2 matrix function*

$$\begin{pmatrix} U^\times(\sigma) \\ U^\times(\tau) \end{pmatrix} \left(PU(\sigma)^{-1}U^\times(\sigma) + (I-P)U(\tau)^{-1}U^\times(\tau)\right)^{-1} \cdot \begin{pmatrix} PU(\sigma)^{-1} & (I-P)U(\tau)^{-1} \end{pmatrix} \tag{35}$$

is uniformly bounded for $\sigma < \sigma_\circ$ and $\tau > \tau_\circ$, then the finite section method for (31) *converges.*

Proof. According to [6], Theorem I.2.3, a function $\phi \in L_p^m(-\infty, \infty)$ is a solution of (31) if and only if there exists a (unique) function $\rho \in L_p^n(-\infty, \infty)$ such that with input $u = \phi$ the system

$$\begin{cases} \dot{\rho}(t) &= A(t)\rho(t) + B(t)u(t), \quad -\infty < t < \infty, \\ y(t) &= C(t)\rho(t) + u(t), \quad -\infty < t < \infty, \end{cases} \tag{36}$$

has output $y = f$. Hence to solve (31) one inverts the system (36), i.e., one passes to the inverse system:

$$\begin{cases} \dot{\rho}(t) &= A^\times(t)\rho(t) + B(t)y(t), \quad -\infty < t < \infty, \\ u(t) &= -C(t)\rho(t) + y(t), \quad -\infty < t < \infty, \end{cases} \tag{37}$$

For the latter system we know that if the input $y = f$, then the output $u = \phi$. Since we assumed that the fundamental matrix $U^\times(t)$ has an exponential dichotomy $P^\times$ the system (37) has a unique solution $\rho \in L_p^n(-\infty, \infty)$.

Next we consider the equation (34). According to [5], Theorem 2.1, a function $\phi_{\sigma,\tau} \in L_p^m[\sigma,\tau]$ is a solution of (34) if and only if there exists a (unique) function $\rho_{\sigma,\tau} \in L_p^n[\sigma,\tau]$ such that with input $u = \phi_{\sigma,\tau}$ the system

$$\begin{cases} \dot\rho(t) = A(t)\rho(t) + B(t)u(t), & \sigma \le t \le \tau, \\ y(t) = C(t)\rho(t) + u(t), & \sigma \le t \le \tau, \\ PU(\sigma)^{-1}\rho(\sigma) + (I-P)U(\tau)^{-1}\rho(\tau) = 0 \end{cases} \tag{38}$$

has output $y = f$. In order to solve (34) one inverts the system (38) to get

$$\begin{cases} \dot\rho(t) = A^\times(t)\rho(t) + B(t)y(t), & \sigma \le t \le \tau, \\ u(t) = -C(t)\rho(t) + y(t), & \sigma \le t \le \tau, \\ PU(\sigma)^{-1}\rho(\sigma) + (I-P)U(\tau)^{-1}\rho(\tau) = 0 \end{cases} \tag{39}$$

which has the output $u = \phi$ if the input $y = f$. Since $PU(\sigma)^{-1}U^\times(\sigma) + (I - P)U(\tau)^{-1}U^\times(\tau)$ is assumed to be invertible, it follows that the system (39) is uniquely solvable.

Consider the differential equations

$$\dot\rho(t) = A^\times(t)\rho(t) + B(t)f(t), \quad \sigma \le t < \infty,$$

and

$$\begin{cases} \dot\rho(t) = A^\times(t)\rho(t) + B(t)f(t), & \sigma \le t \le \tau, \\ PU(\sigma)^{-1}\rho(\sigma) + (I-P)U(\tau)^{-1}\rho(\tau) = 0. \end{cases}$$

One sees that the condition (35) implies that we may apply Theorem 2.2 to conclude that indeed $\rho_{\sigma,\tau}$ converges in L_p to ρ for $-\sigma, \tau \to \infty$. Now remark that

$$\phi_{\sigma,\tau}(t) = C(t)\rho_{\sigma,\tau}(t) + B(t)f(t), \quad \phi(t) = C(t)\rho(t) + B(t)f(t),$$

and conclude that $\phi_{\sigma,\tau}$ converges in L_p to ϕ for $-\sigma, \tau \to \infty$. □

There is another way to prove Theorem 5.1. To see this, let us consider the operator $K_{\sigma,\tau}$ on $L_p^m[\sigma,\tau]$ defined by

$$(K_{\sigma,\tau}\phi)(t) = \phi(t) + \int_\sigma^\tau k(t,s)\phi(s)ds,$$

where $k(t,s)$ is given by (32). Condition (35) and the argument used in the proof of Theorem 2.2 show that the operators $K_{\sigma,\tau}^{-1}$ are uniformly bounded in the operator norm. Hence by the general theory of the projection method (see [4], Theorem II.2.1, and also [1], Theorem 4.4) it follows that the finite section method converges.

References

[1] A. Böttcher, Infinite Matrices and Projection Methods, *in* "Lectures on Operator Theory and Its Applications", Amer. Math. Soc., Providence (RI), 1996.

[2] H. Bart, I. Gohberg, M.A. Kaashoek, Convolution equations and linear systems, *Integral Equations and Operator Theory*, **5** (1982), 283–340.

[3] W.A. Coppel, "Dichotomies in Stability Theory", Lecture Notes in Mathematics, Springer-Verlag, Berlin, Heidelberg, New York, 1978.

[4] I. Gohberg, I.A. Feldman, "Convolution Equations and Projection Methods for Their Solution", Amer. Math. Soc. Transl. of Math. Monographs 41, Providence (RI), 1974 (Russian Original: Nauka, Moscow 1971).

[5] I. Gohberg, M.A. Kaashoek, Time varying systems with boundary conditions and integral operators, I. The transfer operator and its properties, *Integral Equations and Operator Theory,* **7** (1984), 325–391.

[6] I. Gohberg, M.A. Kaashoek, F. van Schagen, Non-compact integral operators with semi-separable kernels and their discrete analogues: inversion and Fredholm properties, *Integral Equations and Operator Theory,* **7** (1984), 642–703.

[7] I. Gohberg, M.A. Kaashoek, F. van Schagen, Finite section method for linear ordinary differential equations, *Journal of Differential Equations,* to appear.

I. Gohberg
School of Mathematical Sciences
Raymond and Beverly Sackler Faculty of Exact Sciences
Tel Aviv University
Ramat Aviv 69978, Israel

M.A. Kaashoek
Division of Mathematics and Computer Science
Vrije Universiteit
De Boelelaan 1081a
1081 HV Amsterdam
The Netherlands

F. van Schagen
Division of Mathematics and Computer Science
Vrije Universiteit
De Boelelaan 1081a
1081 HV Amsterdam
The Netherlands

Operator Theory:
Advances and Applications, Vol. 134, 225–241

On the Spectral Radius of Multi-Matrix Functions

Daniel Hershkowitz

Abstract. Several problems that deal with certain spectral properties of multi-matrix functions are discussed:
(i) Denote by $\rho(A)$ the spectral radius of a nonnegative square matrix A. Known characterizations of all multi-variable functions $f : \mathbb{R}_+^m \longrightarrow \mathbb{R}_+$ such that the Hadamard function $f(A_1, \ldots, A_m)$ satisfies

$$\rho(f(A_1, \ldots, A_m)) \le f(\rho(A_1), \ldots, \rho(A_m)), \quad \forall A_1, \ldots, A_m \in \mathbb{R}_+^{nn}, \quad \forall n \in \mathbb{N},$$

or

$$\rho(f(A_1, \ldots, A_m)) \ge f(\rho(A_1), \ldots, \rho(A_m)), \quad \forall A_1, \ldots, A_m \in \mathbb{R}_+^{nn}, \quad \forall n \in \mathbb{N},$$

are reviewed. The study is then extended to the investigation of functions that satisfy the above conditions for just *some* n.
(ii) For a nonnegative square matrix A denote by $\sigma(A)$ the minimal real eigenvalue of its comparison matrix $M(A) = 2\mathrm{diag}(a_{ii}) - A$. Denote by HP_n the set of all n-by-n nonnegative H-matrices, i.e. the nonnegative matrices A for which $\sigma(A) \ge 0$. The relations between Hadamard functions that preserve HP_n and functions that satisfy the conditions above are reviewed.
(iii) Known results on the behavior of the spectral radius of products of certain one cycle matrices as a function of the lengths of the cycles are reviewed.

1. Introduction

In this paper we discuss several problems that deal with certain spectral properties of multi-matrix functions.

Let A be a nonnegative (entrywise) $n \times n$ matrix and let $\rho(A)$ be the spectral radius of A, that is, the largest absolute value of an eigenvalue of A. It is well known from the Perron-Frobenius spectral theory for nonnegative matrices that $\rho(A)$ itself is eigenvalue of A, e.g. [8, 1, 9]. Furthermore, if a nonnegative $n \times n$ matrix B satisfies $b_{ij} \ge a_{ij}$ for all i and j then $\rho(B) \ge \rho(A)$. In that respect, it was proven in [5] that if we increase every positive element of A by 1 and if $\rho(A) > 0$ then the spectral radius of the matrix increases by at least 1. This result could be stated as follows. Define the function $\mathrm{sgn}(x)$ on the real numbers by

$$\mathrm{sgn}(x) = \begin{cases} 1, & x > 0 \\ 0, & x = 0 \\ -1, & x < 0 \end{cases}.$$

For a nonnegative $n \times n$ matrix A consider the Hadamard function $\operatorname{sgn}(A)$, that is, the $n \times n$ matrix whose elements are

$$(\operatorname{sgn}(A))_{ij} = \operatorname{sgn}(a_{ij}), \qquad i,j = 1,\ldots,n.$$

Then we have

$$\rho(A + \operatorname{sgn}(A)) \geq \rho(A) + \operatorname{sgn}(\rho(A)). \tag{1}$$

Another interesting result was proven in [10] and in [6]. Let $A_1,\ldots,A_m$ be nonnegative $n \times n$ matrices and let $\alpha_1,\ldots,\alpha_m$ be positive numbers such that $\alpha_1 + \ldots + \alpha_m \geq 1$. Define the nonnegative $n \times n$ matrix C by

$$c_{ij} = (A_1)_{ij}^{\alpha_1} \cdot \ldots \cdot (A_m)_{ij}^{\alpha_m}, \qquad i,j = 1,\ldots,n.$$

Then

$$\rho(C) \leq \rho(A_1)^{\alpha_1} \cdot \ldots \cdot \rho(A_m)^{\alpha_m}.$$

This result too can be restated in terms similar to those used in the previous problem. Let $f(x_1,\ldots,x_m) = x_1^{\alpha_1} \cdot \ldots \cdot x_m^{\alpha_m}$ be a function from $\mathbb{R}_+^m$ to $\mathbb{R}_+$. For nonnegative $n \times n$ matrices $A_1,\ldots,A_m$ consider the Hadamard function $f(A_1,\ldots,A_m)$ defined by

$$(f(A_1,\ldots,A_m))_{ij} = f((A_1)_{ij},\ldots,(A_m)_{ij}).$$

Then we have $\rho(f(A_1,\ldots,A_m)) \leq f(\rho(A_1),\ldots,\rho(A_m))$.

These two problems raise the natural question of characterizing all multi-variable functions $f : \mathbb{R}_+^m \longrightarrow \mathbb{R}_+$ such that the Hadamard function $f(A_1,\ldots,A_m)$ satisfies

$$\rho(f(A_1,\ldots,A_m)) \leq f(\rho(A_1),\ldots,\rho(A_m)), \quad \forall A_1,\ldots,A_m \in \mathbb{R}_+^{nn}, \quad \forall n \in \mathbb{N}, \tag{2}$$

or

$$\rho(f(A_1,\ldots,A_m)) \geq f(\rho(A_1),\ldots,\rho(A_m)), \quad \forall A_1,\ldots,A_m \in \mathbb{R}_+^{nn}, \quad \forall n \in \mathbb{N}, \tag{3}$$

where $\mathbb{R}_+^{nn}$ denotes the set of all $n \times n$ nonnegative matrices and $\mathbb{N}$ denotes the set of all positive integers. This question is investigated in [3]. The results of that paper are discussed in Sections 2 and 3. Furthermore, in these sections we extend the study of [3] by discussing functions that do not necessarily satisfy (2) or (3) but they do satisfy the weaker condition

$$\rho(f(A_1,\ldots,A_m)) \leq f(\rho(A_1),\ldots,\rho(A_m)), \quad \forall A_1,\ldots,A_m \in \mathbb{R}_+^{nn}, \tag{4}$$

or

$$\rho(f(A_1,\ldots,A_m)) \geq f(\rho(A_1),\ldots,\rho(A_m)), \quad \forall A_1,\ldots,A_m \in \mathbb{R}_+^{nn}, \tag{5}$$

for *some* positive integer n.

For a nonnegative square matrix A we denote by $\sigma(A)$ the minimal real eigenvalue of its *comparison matrix* $M(A) = 2\operatorname{diag}(a_{ii}) - A$. We denote by HP_n the set of all n-by-n nonnegative H-matrices, i.e. the nonnegative matrices A for which $\sigma(A) \geq 0$. These are the nonnegative matrices whose comparison matrices are M-*matrices*.

Let m be a positive integer, let $f : \mathbb{R}_+^m \longrightarrow \mathbb{R}_+$, and let K be a set of nonnegative matrices of same size. We say that the Hadamard function f *preserves* K if $A_1, \ldots, A_m \in K$ implies $f(A_1, \ldots, A_m) \in K$. The functions that preserve HP_n are characterized in [2]. In particular, it is proven there that these functions satisfy (2) as well. It turns out that these are exactly the functions f of the form $f(x_1, \ldots, x_m) = cx_1^{\alpha_1} \ldots x_m^{\alpha_m}$ where $\alpha_i \geq 0$, $c \geq 0$ and $\Sigma_{i=1}^m \alpha_i \geq 1$. (The latter condition is required whenever $n > 2$.) These functions are also characterized by the inequality

$$\sigma(f(A_1, \ldots, A_k)) \geq f(\sigma(A_1), \ldots, \sigma(A_k)),$$

for all $A_i \in HP_n$ for some n, $n > 2$. These results are discussed in the conclusion of Section 3.

Another interesting question comes up in connection with investigations of iterative methods for solving a linear system $x = Bx + c$, where B is a nonnegative square matrix and $\rho(B) < 1$. In such a case, the basic iteration $x^{(i+1)} = Bx^{(i)} + c$ converges to the solution $x^* = (I - B)^{-1}c$ of the system. Splitting the work to calculate $Bx + c$ between several parallel processors, operating independently one of each other in an asynchronous manner, and where the assignment of subtasks and storage for the current iterate is done by a central processor, leads to an iterative procedure whose convergence rate depends on the spectral radius of a product of $n \times n$ matrices $A_{d_1,c}, \ldots, A_{d_p,c}$, where the matrix $A_{d_k,c}$ is defined by

$$(A_{d_k,c})_{ij} = \begin{cases} 1, & i = j + 1 \\ c, & (i,j) = (1, d_k) \\ 0, & \text{otherwise} \end{cases} .$$

Here c is a positive scalar, $c < 1$, and d_k is a positive integer, $d_k \leq n$. The behavior of the spectral radius of that product of matrices as a function of the sequence $(d_1, \ldots, d_p)$ is investigated in [4]. In Section 4 we discuss some of the results of [4]. While in the previous sections the functions under discussion are Hadamard functions of matrices, in Section 5 we discuss "regular" multi-variable matrix functions.

Most of this article is a survey, reviewing results of [2, 3, 4]. Nevertheless, in several places we have slightly changed the exposition from the original papers, and also added some examples to illustrate the assertions. In Sections 2 and 3 we have also included a discussion with original results of the functions satisfying (4) and (5). The paper is based on an invited talk given at the Applied Linear Algebra Workshop dedicated to Ludwig Elsner on the occasion of his 60th birthday, held in Bielefeld, Germany, on January 21–23, 1999, and an invited talk given at the Workshop on Operator Theory and its Applications dedicated to Harry Dym on the occasion of his 60th birthday, held in Rehovot, Israel, on March 7–12, 1999.

2. The inequality $\rho(f(A_1,\ldots,A_m)) \le f(\rho(A_1),\ldots,\rho(A_m))$

In this section we review results of [3] characterizing Hadamard functions that satisfy (2). We also extend the study to obtain results on functions satisfying the weaker condition (4). We start with functions of one variable.

Theorem 2.1. [3, Theorem 2.1] *Let $f : \mathbb{R}_+ \longrightarrow \mathbb{R}_+$. The following are equivalent:*
(i) *We have*

$$\rho(f(A)) \le f(\rho(A)), \qquad \forall A \in \mathbb{R}_+^{nn}, \quad \forall n \in \mathbb{N}.$$

(ii) *We have*

$$\begin{cases} f(x) + f(y) \le f(x+y) \\ \sqrt{f(x)f(y)} \le f(\sqrt{xy}) \end{cases}, \qquad \forall x, y \in \mathbb{R}_+.$$

A similar result holds for continuous multi-variable functions. Here, for two matrices A and B of same size we denote by $A \circ B$ the Hadamard product of A and B, that is, the matrix (of same size) whose elements are the products of the corresponding elements of A and B.

Theorem 2.2. [3, Theorem 2.1] *Let m be a positive integer and let $f : \mathbb{R}_+^m \longrightarrow \mathbb{R}_+$ be a continuous function. The following are equivalent:*
(i) *We have*

$$\rho(f(A_1,\ldots,A_m)) \le f(\rho(A_1),\ldots,\rho(A_m)), \qquad \forall A_1,\ldots,A_m \in \mathbb{R}_+^{nn}, \quad \forall n \in \mathbb{N}.$$

(ii) *We have*

$$\begin{cases} f(\mathbf{x}) + f(\mathbf{y}) \le f(\mathbf{x}+\mathbf{y}) \\ \sqrt{f(\mathbf{x})f(\mathbf{y})} \le f(\sqrt{\mathbf{x} \circ \mathbf{y}}) \end{cases}, \qquad \forall \mathbf{x}, \mathbf{y} \in \mathbb{R}_+^m.$$

If the function f is not continuous then Condition (ii) in Theorem 2.2 should be modified, as follows.

Theorem 2.3. [3, Theorem 2.1] *Let m be a positive integer and let f be a function $f : \mathbb{R}_+^m \longrightarrow \mathbb{R}_+$. The following are equivalent:*
(i) *We have*

$$\rho(f(A_1,\ldots,A_m)) \le f(\rho(A_1),\ldots,\rho(A_m)), \qquad \forall A_1,\ldots,A_m \in \mathbb{R}_+^{nn}, \quad \forall n \in \mathbb{N}.$$

(ii) *We have*

$$\begin{cases} f(\mathbf{x}) + f(\mathbf{y}) \le f(\mathbf{x}+\mathbf{y}), & \forall \mathbf{x}, \mathbf{y} \in \mathbb{R}_+^m \\ \left(\prod_{k=1}^{s} f(\mathbf{x}_k)\right)^{1/s} \le f((\mathbf{x}_1 \circ \ldots \circ \mathbf{x}_s)^{1/s}), & \forall \mathbf{x}_1,\ldots \mathbf{x}_s \in \mathbb{R}_+^m, \quad s = 2,3,\ldots \end{cases}$$

The following result follows from the previous theorems and corresponds to combinations of functions satisfying (2).

Theorem 2.4. [3, Proposition 2.5] *Let m and p be positive integers, let $f, g : \mathbb{R}_+^m \longrightarrow \mathbb{R}_+$, let $h : \mathbb{R}_+ \longrightarrow \mathbb{R}_+$, and let $q_1, \ldots, q_m : \mathbb{R}_+^p \longrightarrow \mathbb{R}_+$, all satisfy (2). Then so do*

(i) $h(f(\mathbf{x}))$, $\mathbf{x} \in \mathbb{R}_+^m$.
(ii) $f(q_1(\mathbf{x}), \ldots, q_m(\mathbf{x}))$, $\mathbf{x} \in \mathbb{R}_+^p$.
(iii) $\min\{f(\mathbf{x}), g(\mathbf{x})\}$, $\mathbf{x} \in \mathbb{R}_+^m$.
(iv) $f(c\mathbf{x})$, $\mathbf{x} \in \mathbb{R}_+^m$, $c > 0$.
(v) $cf(\mathbf{x})$, $\mathbf{x} \in \mathbb{R}_+^m$, $c > 0$.
(vi) $f^\alpha(\mathbf{x})g^\beta(\mathbf{x})$, $\mathbf{x} \in \mathbb{R}_+^m$, *whenever* $\alpha, \beta > 0$, $\alpha + \beta \geq 1$.

As an immediate corollary of Theorems 2.1 and 2.4, one can obtain the result of [10] and [6] mentioned in the introduction.

Corollary 2.5. *Let $\alpha_1, \ldots, \alpha_m$ be positive numbers such that $\alpha_1 + \ldots + \alpha_m \geq 1$. Then the function $f(x_1, \ldots, x_m) = x_1^{\alpha_1} \cdot \ldots \cdot x_m^{\alpha_m}$ satisfies (2).*

The functions that satisfy (2) do not form a convex cone, as is demonstrated by the following example.

Example 2.6. By Corollary 2.5, the functions $f(x) = x$ and $g(x) = x^2$ satisfy (2). Nevertheless, the function $h(x) = f(x) + g(x)$ does not satisfy (2). For example, for the matrix

$$A = \begin{bmatrix} 0 & 1 \\ 2 & 0 \end{bmatrix}$$

we have

$$\rho(h(A)) = \rho\left(\begin{bmatrix} 0 & 1 \\ 2 & 0 \end{bmatrix} + \begin{bmatrix} 0 & 1 \\ 4 & 0 \end{bmatrix}\right) = 3.4641 > h(\rho(A)) = h(1.4142) = 3.4141.$$

In [3] the authors characterize all functions f that satisfy (2). It would be interesting to study functions that satisfy the weaker condition (4) for *some* positive integer n. Clearly, every function f satisfies

$$\rho(f(A_1, \ldots, A_m)) = f(\rho(A_1), \ldots, \rho(A_m)), \quad \forall A_1, \ldots, A_m \in \mathbb{R}_+^{11}.$$

In order to address larger n's we prove

Lemma 2.7. *Let m and n be positive integers, $n \geq 2$, and let f be a function $f : \mathbb{R}_+^m \longrightarrow \mathbb{R}_+$. If f satisfies (4) then $f(\mathbf{0}) = 0$.*

Proof. If we choose $A_1, \ldots, A_m$ to be the zero $n \times n$ matrix then $f(A_1, \ldots, A_m)$ is an $n \times n$ matrix all of whose are entries are equal to $f(\mathbf{0})$. Therefore, it is a rank 1 matrix, and since $f(\mathbf{0}) \geq 0$ it now follows that

$$n\,f(\mathbf{0}) = \rho(f(A_1, \ldots, A_m)) \leq f(\rho(A_1), \ldots, \rho(A_m)) = f(\mathbf{0}).$$

Since $f(\mathbf{0}) \geq 0$, it follows that $f(\mathbf{0}) = 0$. □

We remark that the assertion of Lemma 2.7 is proven in the proof of Theorem 2.1 in [3]. We provided a proof here for the sake of completeness since in [3] it is proven that if (4) is satisfied for $n = 2$ then the conclusion follows, while here we state it for any n, $n \geq 2$.

Corollary 2.8. *Let m and n be positive integers, $n \geq 2$, and let f be a function $f : \mathbb{R}_+^m \longrightarrow \mathbb{R}_+$. If f satisfies (4) then we have*

$$\rho(f(A_1,\dots,A_m)) \leq f(\rho(A_1),\dots,\rho(A_m)), \quad \forall A_1,\dots,A_m \in \mathbb{R}_+^{kk}, \quad k=1,\dots,n.$$

Proof. Let $1 \leq k < n$, and let $A_1,\dots,A_m \in \mathbb{R}_+^{kk}$. We append zero rows and columns to $A_1,\dots,A_m$ to obtain $n \times n$ matrices $B_1,\dots,B_m$. In view of Lemma 2.7 we have

$$\begin{aligned}\rho(f(A_1,\dots,A_m)) &= \rho(f(B_1,\dots,B_m)) \\ &\leq f(\rho(B_1),\dots,\rho(B_m)) = f(\rho(A_1),\dots,\rho(A_m)),\end{aligned}$$

proving our assertion. □

We use Lemma 2.7 and Corollary 2.8 to obtain the following necessary condition for the inequality (4) to hold.

Theorem 2.9. *Let m and n be positive integers, $n \geq 2$, and let f be a function $f : \mathbb{R}_+^m \longrightarrow \mathbb{R}_+$. If f satisfies (4) then we have*

$$\begin{cases} f(\mathbf{x}) + f(\mathbf{y}) \leq f(\mathbf{x}+\mathbf{y}), & \forall \mathbf{x},\mathbf{y} \in \mathbb{R}_+^m \\ \left(\prod_{k=1}^s f(\mathbf{x}_k)\right)^{1/s} \leq f((\mathbf{x}_1 \circ \dots \circ \mathbf{x}_s)^{1/s}), & \forall \mathbf{x}_1,\dots\mathbf{x}_s \in \mathbb{R}_+^m, \quad s=2,\dots,n. \end{cases} \tag{6}$$

Proof. The first inequality of (6) follows exactly as in the proof of Theorem 2.1 in [3, p. 112], since the proof in [3] uses only 2×2 matrices, and by Corollary 2.8 the function f satisfies (4) for 2×2 matrices. To prove the second inequality of (6) note that, in view of Corollary 2.8, it is enough to prove that that inequality holds just for $s = n$. The proof of this assertion actually exists in the proof of Theorem 2.1 of [3]. Nevertheless, since it is not stated there explicitly, we provide it here for the sake of completeness.

Let $\mathbf{x}_k = (x_{ik},\dots,x_{mk})$, $k \in \{1,\dots,n\}$. Define the matrices

$$A_k = \begin{bmatrix} 0 & x_{k1} & \cdots & 0 \\ \vdots & \vdots & \ddots & \vdots \\ 0 & 0 & \cdots & x_{k,n-1} \\ x_{kn} & 0 & \cdots & 0 \end{bmatrix}, \qquad k=1,\dots,m.$$

Observe that we have $\rho(A_k) = \left(\prod_{j=1}^n x_{kj}\right)^{1/n}$, $k=1,\dots,m$. Therefore, we have

$$f(\rho(A_1),\dots,\rho(A_m)) = f((\mathbf{x}_1 \circ \dots \circ \mathbf{x}_s)^{1/n}). \tag{7}$$

By Lemma 2.7 we have

$$f(A_1,\dots,A_m) = \begin{bmatrix} 0 & f(\mathbf{x}_1) & \cdots & 0 \\ \vdots & \vdots & \ddots & \vdots \\ 0 & 0 & \cdots & f(\mathbf{x}_{n-1}) \\ f(\mathbf{x}_n) & 0 & \cdots & 0 \end{bmatrix},$$

and hence

$$\rho(f(A_1,\ldots,A_m)) = \left(\prod_{k=1}^{n} f(\mathbf{x}_k)\right)^{1/n}. \tag{8}$$

Our claim now follows from (7), (8) and the fact that f satisfies (4). □

It now follows that we can characterize continuous functions f satisfying (4) for *some* n, $n \geq 2$, by strengthening Theorem 2.2 to say the following

Theorem 2.10. *Let m be a positive integer and let $f : \mathbb{R}_+^m \longrightarrow \mathbb{R}_+$ be a continuous function. The following are equivalent:*
(i) *We have*

$$\rho(f(A_1,\ldots,A_m)) \leq f(\rho(A_1),\ldots,\rho(A_m)), \qquad \forall A_1,\ldots,A_m \in \mathbb{R}_+^{nn}, \quad \forall n \in \mathbb{N}.$$

(ii) *We have*

$$\rho(f(A_1,\ldots,A_m)) \leq f(\rho(A_1),\ldots,\rho(A_m)), \qquad \forall A_1,\ldots,A_m \in \mathbb{R}_+^{nn},$$

for some positive integer n, $n \geq 2$.
(iii) *We have*

$$\begin{cases} f(\mathbf{x}) + f(\mathbf{y}) \leq f(\mathbf{x}+\mathbf{y}) \\ \sqrt{f(\mathbf{x})f(\mathbf{y})} \leq f(\sqrt{\mathbf{x}\circ\mathbf{y}}) \end{cases}, \qquad \forall \mathbf{x},\mathbf{y} \in \mathbb{R}_+^m.$$

(iv) *We have*

$$\begin{cases} f(\mathbf{x}) + f(\mathbf{y}) \leq f(\mathbf{x}+\mathbf{y}), & \forall \mathbf{x},\mathbf{y} \in \mathbb{R}_+^m \\ \left(\prod_{k=1}^{s} f(\mathbf{x}_k)\right)^{1/s} \leq f((\mathbf{x}_1 \circ \ldots \circ \mathbf{x}_s)^{1/s}), & \forall \mathbf{x}_1,\ldots \mathbf{x}_s \in \mathbb{R}_+^m, \quad s = 2,3,\ldots. \end{cases}$$

Proof. (i) $\Longrightarrow$ (ii) is trivial.
(ii) $\Longrightarrow$ (iii) follows from Theorem 2.9.
(iii) $\Longrightarrow$ (i) is in Theorem 2.2.
(i) $\Longleftrightarrow$ (iv) is in Theorem 2.3. □

Theorem 2.10 does not hold in general for non-continuous functions f. It is shown in [3] that the function

$$f = \begin{cases} xy, & xy > 1 \\ 1, & xy = 1, \quad x = e^{k/2^m}, \text{ where } k \text{ and } m \text{ are positive integers,} \\ 0, & \text{otherwise} \end{cases}$$

satisfies Condition (iii) of Theorem 2.10 but not Conditions (i) and (iv). Also, it does not satisfy Condition (ii) whenever $n \geq 3$.

We have no counter-example that shows that the converse of Theorem 2.9 does not hold. Therefore, we leave it as the following open problem.

Open Problem 2.11. Let m and n be positive integers, $n \geq 2$, and let f be a function $f : \mathbb{R}_+^m \longrightarrow \mathbb{R}_+$ satisfying

$$\begin{cases} f(\mathbf{x}) + f(\mathbf{y}) \leq f(\mathbf{x}+\mathbf{y}), & \forall \mathbf{x}, \mathbf{y} \in \mathbb{R}_+^m \\ \left(\prod_{k=1}^s f(\mathbf{x}_k)\right)^{1/s} \leq f((\mathbf{x}_1 \circ \ldots \circ \mathbf{x}_s)^{1/s}), & \forall \mathbf{x}_1, \ldots \mathbf{x}_s \in \mathbb{R}_+^m, \quad s = 2, \ldots, n. \end{cases}$$

Does it follow that

$$\rho(f(A_1, \ldots, A_m)) \leq f(\rho(A_1), \ldots, \rho(A_m)), \quad \forall A_1, \ldots, A_m \in \mathbb{R}_+^{nn}?$$

3. The inequality $\rho(f(A_1, \ldots, A_m)) \geq f(\rho(A_1), \ldots, \rho(A_m))$

The characterization of functions satisfying (3) is more complicated and is not quite complete. Nevertheless, paper [3] contains an extensive study of the subject. In this section we review some of the corresponding results of [3]. We also extend the study to obtain results on functions satisfying the weaker condition (5).

It is shown in [3] that the behavior of multi-variable functions with respect to the inequality (3) strongly relates to the reductions of these functions. We therefore use the following notation, which is a slight generalization of the notation used in [3].

Notation 3.1. (i) Let m be a positive integer and let α be a subset of $\{1, \ldots, m\}$. For a vector $\mathbf{x} = (x_1, \ldots, x_m) \in \mathbb{R}_+^m$ we denote by $\mathbf{x}_\alpha$ the vector defined by

$$(\mathbf{x}_\alpha)_k = \begin{cases} x_k, & k \in \alpha \\ 0, & \text{otherwise.} \end{cases}$$

(ii) For a function $f : \mathbb{R}_+^m \longrightarrow \mathbb{R}_+$ we denote by f_α the function from R_+^m to $\mathbb{R}_+$ defined by $f_\alpha(\mathbf{x}) = f(\mathbf{x}_\alpha)$. Note that f can be regarded also as a function from $R_+^{|\alpha|}$ to $\mathbb{R}_+$.
(iii) Let k be an integer, $1 \leq k \leq m$. We denote by f_k the function $f_{\{k\}}$, considering it as a function from R_+ to itself.

The following theorem strengthens Proposition 3.2 of [3] in two directions. First, it addresses the inequality (5) rather than the stronger inequality (3). Second, it gives a condition on f in terms of f_α and f_{α^c} for any subset α of $\{1, \ldots, n\}$ rather than the functions f_k only. In our proof we very much imitate the proof of Proposition 3.2 of [3].

Theorem 3.2. *Let m be a positive integer, let $f : \mathbb{R}_+^m \longrightarrow \mathbb{R}_+$ be a function satisfying $f(\mathbf{0}) = 0$, let α be a subset of $\{1, \ldots, m\}$, and let n be a positive integer, $n \geq 2$. The following are equivalent:*
(i) *We have*

$$\rho(f(A_1, \ldots, A_m)) \geq f(\rho(A_1), \ldots, \rho(A_m)), \qquad \forall A_1, \ldots, A_m \in \mathbb{R}_+^{nn}. \tag{9}$$

(ii) *We have*

$$f(\mathbf{x}) = \max\{f_\alpha(\mathbf{x}), f_{\alpha^c}(\mathbf{x})\}, \quad \forall \mathbf{x} \in \mathbb{R}_+^m \tag{10}$$

and

$$\begin{cases} \rho(f_\alpha(A_1,\dots,A_m)) \ge f_\alpha(\rho(A_1),\dots,\rho(A_m)), & \forall A_1,\dots,A_m \in \mathbb{R}_+^{nn} \\ \rho(f_{\alpha^c}(A_1,\dots,A_m)) \ge f_{\alpha^c}(\rho(A_1),\dots,\rho(A_m)), & \forall A_1,\dots,A_m \in \mathbb{R}_+^{nn}. \end{cases} \tag{11}$$

Proof. (i) $\Longrightarrow$ (ii). It is easy to check that (9) implies (11). We now prove that (10) holds. For a vector $\mathbf{x} = (k_1,\dots,x_m) \in \mathbb{R}_+^m$ we define $n \times n$ matrices $A_1,\dots,A_m$ by

$$(A_k)_{ij} = \begin{cases} x_k, & (i,j) = (1,2) \\ x_k, & (i,j) = (2,1),\ k \in \alpha, \\ 0, & \text{otherwise} \end{cases} \qquad k = 1,\dots,m.$$

Note that

$$\rho(A_k) = \begin{cases} x_k, & k \in \alpha, \\ 0, & \text{otherwise} \end{cases} \qquad k = 1,\dots,m. \tag{12}$$

Also

$$(f(A_1,\dots,A_m))_{ij} = \begin{cases} f(\mathbf{x}), & (i,j) = (1,2) \\ f_\alpha(\mathbf{x}), & (i,j) = (2,1) \\ 0, & \text{otherwise} \end{cases},$$

and hence

$$\rho(f(A_1,\dots,A_m)) = \sqrt{f(\mathbf{x})f_\alpha(\mathbf{x})}. \tag{13}$$

It now follows from (9), (12) and (13) that $\sqrt{f(\mathbf{x})f_\alpha(\mathbf{x})} \ge f_\alpha(\mathbf{x})$, implying that

$$f(\mathbf{x}) \ge f_\alpha(\mathbf{x}). \tag{14}$$

Similarly, we prove that

$$f(\mathbf{x}) \ge f_{\alpha^c}(\mathbf{x}). \tag{15}$$

Now, define $n \times n$ matrices $A_1,\dots,A_m$ by

$$(A_k)_{ij} = \begin{cases} x_k, & (i,j) = (1,1),\ k \in \alpha \\ x_k, & (i,j) = (2,2),\ k \in \alpha^c, \\ 0, & \text{otherwise} \end{cases} \qquad k = 1,\dots,m.$$

Note that

$$\rho(A_k) = x_k, \quad k = 1,\dots,m. \tag{16}$$

Also

$$(f(A_1,\dots,A_m))_{ij} = \begin{cases} f_\alpha(\mathbf{x}), & (i,j) = (1,1) \\ f_{\alpha^c}(\mathbf{x}), & (i,j) = (2,2), \\ 0, & \text{otherwise} \end{cases}$$

and hence

$$\rho(f(A_1,\dots,A_m)) = \max\{f_\alpha(\mathbf{x}), f_{\alpha^c}(\mathbf{x})\}. \tag{17}$$

It now follows from (9), (16) and (17) that

$$\max\{f_\alpha(\mathbf{x}), f_{\alpha^c}(\mathbf{x})\} \ge f(\mathbf{x}). \tag{18}$$

The inequalities (14), (15) and (18) give (10).

(ii) $\Longrightarrow$ (i). Let $A_1,\dots,A_m \in \mathbb{R}_+^{nn}$. By (10) we have

$$f(\rho(A_1),\dots,\rho(A_m)) \le \begin{cases} f_\alpha(\rho(A_1),\dots,\rho(A_m)) \\ f_{\alpha^c}(\rho(A_1),\dots,\rho(A_m)) \end{cases}. \tag{19}$$

By (11) we have

$$\begin{cases} f_\alpha(\rho(A_1),\ldots,\rho(A_m)) \le \rho(f_\alpha(A_1,\ldots,A_m)) \\ f_{\alpha^c}(\rho(A_1),\ldots,\rho(A_m)) \le \rho(f_{\alpha^c}(A_1,\ldots,A_m)) \end{cases}. \tag{20}$$

It follows from (10) that

$$f(A_1,\ldots,A_m) \ge \begin{cases} f_\alpha(A_1,\ldots,A_m) \\ f_{\alpha^c}(A_1,\ldots,A_m) \end{cases}.$$

Thus, by the Perron-Frobenius spectral theory for nonnegative matrices we have

$$\rho(f(A_1,\ldots,A_m)) \ge \begin{cases} \rho(f_\alpha(A_1,\ldots,A_m)) \\ \rho(f_{\alpha^c}(A_1,\ldots,A_m)) \end{cases}. \tag{21}$$

Inequalities (19), (20) and (21) give (9). □

A repeated application of Theorem 3.2 to f_α and f_{α^c} yields the following.

Theorem 3.3. *Let m be a positive integer, let $f : \mathbb{R}_+^m \longrightarrow \mathbb{R}_+$ be a function satisfying $f(\mathbf{0}) = 0$, and let n be a positive integer, $n \ge 2$. The following are equivalent:*
(i) *We have*

$$\rho(f(A_1,\ldots,A_m)) \ge f(\rho(A_1),\ldots,\rho(A_m)), \qquad \forall A_1,\ldots,A_m \in \mathbb{R}_+^{nn}.$$

(ii) *We have*

$$f(\mathbf{x}) = \max_{k=1,\ldots,m} f_k(x_k)$$

and

$$\rho(f_k(A)) \ge f_k(\rho(A)), \qquad \forall A \in \mathbb{R}_+^{nn}.$$

As a corollary we now obtain Proposition 3.2 of [3].

Corollary 3.4. *Let m be a positive integer and let $f : \mathbb{R}_+^m \longrightarrow \mathbb{R}_+$ be a function satisfying $f(\mathbf{0}) = 0$. The following are equivalent:*
(i) *We have*

$$\rho(f(A_1,\ldots,A_m)) \ge f(\rho(A_1),\ldots,\rho(A_m)), \qquad \forall A_1,\ldots,A_m \in \mathbb{R}_+^{nn}, \quad \forall n \in \mathbb{N}.$$

(ii) *We have*

$$f(\mathbf{x}) = \max_{k=1,\ldots,m} f_k(x_k)$$

and

$$\rho(f_k(A)) \ge f_k(\rho(A)), \qquad \forall A \in \mathbb{R}_+^{nn}, \quad \forall n \in \mathbb{N}.$$

We remark that the condition $f(\mathbf{0}) = 0$ is not necessary for a function to satisfy (5) or not even (3), as is demonstrated by the following (original) example.

Example 3.5. Let $f(x) = x + 1$ and let A be an $n \times n$ matrix. Denote by J the $n \times n$ matrix all of whose elements are 1, and note that $f(A) = A + J$. Since the matrix $A + J$ is entrywise greater than or equal to the matrix $A + \text{sgn}(A)$, it follows from the Perron-Frobenius spectral theory for nonnegative matrices that $\rho(A + J) \ge \rho(A + \text{sgn}(A))$. It now follows by (1) that

$$\rho(f(A)) = \rho(A + J) \ge \rho(A + \text{sgn}(A)) \ge \rho(A) + \text{sgn}(\rho(A)). \tag{22}$$

If $\rho(A) > 0$ then it follows from (22) that $\rho(f(A)) \geq \rho(A) + 1$. If $\rho(A) = 0$ then, since $A + J \geq J$, it follows from the Perron-Frobenius theory that $\rho(A + J) \geq \rho(J) = n \geq 1 = \rho(A)+1$. So, in either case we have $\rho(f(A)) \geq \rho(A)+1 = f(\rho(A))$, and so although $f(0) \neq 0$, the function f satisfies (3).

In view of Theorem 3.3 and Corollary 3.4, it is essential to study one variable functions that satisfy (3) or (5). The following is a technical improvement of Proposition 3.3 in [3]. It particular, it refers also to functions satisfying the weaker condition (5). The claim involves a set of functions that is larger than the set of continuous functions. Recall that a function is said to be *bounded above at a point* if there exist a neighborhood of the point such that the function is bounded above in this neighborhood.

Theorem 3.6. *Let $f : \mathbb{R}_+ \longrightarrow \mathbb{R}_+$ be a function that is bounded above at some point $\alpha \in (0, \infty)$ and such that $f(0) = 0$. The following are equivalent:*
(i) *We have*

$$\rho(f(A)) \geq f(\rho(A)), \qquad \forall A \in \mathbb{R}_+^{nn}, \quad \forall n \in \mathbb{N}.$$

(ii) *We have*

$$\rho(f(A)) \geq f(\rho(A)), \qquad \forall A \in \mathbb{R}_+^{nn}$$

for some integer n, $n \geq 2$.
(iii) *We have*

$$\rho(f(A)) \geq f(\rho(A)), \qquad \forall A \in \mathbb{R}_+^{22}.$$

(iv) *We have*

$$\begin{cases} f(x) + f(y) \geq f(x+y) \\ \sqrt{f(x)f(y)} \geq f(\sqrt{xy}) \end{cases}, \qquad \forall x, y \in \mathbb{R}_+.$$

Proof. (i)$\Longrightarrow$(ii) is trivial.
(ii)$\Longrightarrow$(iii). Let $A \in \mathbb{R}_+^{22}$. We append zero rows and columns to A to obtain an $n \times n$ matrix B. Since $f(0) = 0$, it follows from (ii) that

$$\rho(f(A)) = \rho(f(B)) \geq f(\rho(B)) = f(\rho(A)).$$

(iii)$\Longrightarrow$(iv) is actually proven in the proof of the implication (i)$\Longrightarrow$(iii) in Proposition 3.3 in [3], since only 2×2 matrices are used there.
(iv)$\Longrightarrow$(i) is proven in [3]. □

Remark 3.7. The condition $f(0) = 0$ in Theorem 3.6 is needed only to prove the implications (ii)$\Longrightarrow$(iii)$\Longrightarrow$(iv), while the implications (iv)$\Longrightarrow$(i)$\Longrightarrow$(ii) hold also without that condition.

As a corollary of Theorems 3.3 and 3.6 we now obtain the following improvement of Theorem 3.1 of [3].

Theorem 3.8. *Let m be a positive integer and let $f : \mathbb{R}_+^m \longrightarrow \mathbb{R}_+$ be a function that is bounded above at some point $\alpha \in \operatorname{int} \mathbb{R}_+^m$ and such that $f(\mathbf{0}) = 0$. The following*

are equivalent:
(i) *We have*

$$\rho(f(A_1,\ldots,A_m)) \geq f(\rho(A_1),\ldots,\rho(A_m)), \qquad \forall A_1,\ldots,A_m \in \mathbb{R}_+^{nn}, \quad \forall n \in \mathbb{N}.$$

(ii) *We have*

$$\rho(f(A_1,\ldots,A_m)) \geq f(\rho(A_1),\ldots,\rho(A_m)), \qquad \forall A_1,\ldots,A_m \in \mathbb{R}_+^{nn},$$

for some n, $n \geq 2$.
(iii) *We have*

$$\rho(f(A_1,\ldots,A_m)) \geq f(\rho(A_1),\ldots,\rho(A_m)), \qquad \forall A_1,\ldots,A_m \in \mathbb{R}_+^{22}.$$

(iv) *We have*

$$f(x_1,\ldots,x_m) = \max_{k=1,\ldots,m} f_k(x_k)$$

and

$$\begin{cases} f_k(x)+f_k(y) \geq f_k(x+y) \\ \sqrt{f_k(x)f_k(y)} \geq f_k(\sqrt{xy}) \end{cases}, \qquad \forall x,y \in \mathbb{R}_+, \quad k=1,\ldots,m.$$

Since the condition $f(\mathbf{0}) = 0$ is not necessary for a matrix to satisfy (3), it follows that Theorem 3.8 does not provide a characterization not even for all *continuous* functions f that satisfy (3). Other two sets of functions that satisfy (3) are given by the following theorem.

Theorem 3.9. [3, Theorem 3.8] *Let m be a positive integer and let $f : \mathbb{R}_+^m \longrightarrow \mathbb{R}_+$. Then*
(i) *if f is componentwise decreasing then f satisfies* (3).
(ii) *if there exists a positive scalar c such that $c \leq f(\mathbf{x}) \leq 2c$ for all $\mathbf{x} \in \mathbb{R}_+^m$ then f satisfies* (3).

The following result on combinations of functions satisfying (3) easily follows from inequality (3).

Theorem 3.10. [3, Proposition 3.4] *Let m and p be positive integers, let $f, g : \mathbb{R}_+^m \longrightarrow \mathbb{R}_+$, let $h : \mathbb{R}_+ \longrightarrow \mathbb{R}_+$, and let $q_1,\ldots,q_m : \mathbb{R}_+^p \longrightarrow \mathbb{R}_+$, all satisfy* (3). *Then so do*

(i) $h(f(\mathbf{x}))$, $\mathbf{x} \in \mathbb{R}_+^m$, *whenever h is increasing.*
(ii) $f(q_1(\mathbf{x}),\ldots,q_m(\mathbf{x}))$, $\mathbf{x} \in \mathbb{R}_+^p$, *whenever f is componentwise increasing.*
(iii) $\max\{f(\mathbf{x}), g(\mathbf{x})\}$, $\mathbf{x} \in \mathbb{R}_+^m$.
(iv) $f(c\mathbf{x})$, $\mathbf{x} \in \mathbb{R}_+^m$, $c > 0$.
(v) $cf(\mathbf{x})$, $\mathbf{x} \in \mathbb{R}_+^m$, $c > 0$.

Another result on combinations is the following.

Theorem 3.11. [3, Proposition 3.7] *Let $f, g : \mathbb{R}_+ \longrightarrow \mathbb{R}_+$ be continuous functions on $(0,\infty)$. If both f and g satisfy* (3) *then so does $f+g$.*

The inequality (1), proven in [5], is an immediate corollary of Theorems 3.6 and 3.11.

We remark that the results brought here do not cover all functions that satisfy (3). It is observed in [3] that, by Theorem 3.10, the function $f(x) = \max\{x, \frac{1}{1+x}\}$ satisfies (3) although it does not belong to any of the sets discussed above.

We conclude this section by noting that the study of functions satisfying (2) is related to the study of functions that preserve the set HP_n. In particular, it is shown in [2] that if for some positive integers m and n a function f, $f : \mathbb{R}_+^m \longrightarrow \mathbb{R}_+$ preserves HP_n then

$$\rho(f(A_1, \ldots, A_m)) \leq f(\rho(A_1), \ldots, \rho(A_m)). \quad \forall A_1, \ldots, A_m \in \mathbb{R}_+^{nn}.$$

It is also shown there that for the set $\tilde{H}P_n$ of all nonnegative H-matrices with *constant diagonal* we have

Theorem 3.12. [2, Theorem 3.1] *Let $f : \mathbb{R}_+^m \longrightarrow \mathbb{R}_+$ be a continuous function. The following are equivalent.*

(i) *f preserves $\tilde{H}P_n$ for some n, $n \geq 3$.*
(ii) *f preserves $\tilde{H}P_n$ for all n, $n = 2, 3, \ldots$.*
(iii) *f satisfies* (2).

4. Spectral radii of product of iteration matrices

While in the previous sections we discussed the spectral radius of Hadamard functions of matrices, in this section we discuss "regular" multi-variable functions of matrices. More specifically, we review results of [4] on the behavior of the spectral radius of that product of iteration matrices $A_{d_1,c}, \ldots, A_{d_p,c}$ as a function of the sequence $\delta = (d_1, \ldots, d_p)$.

Definition 4.1. (i) We denote by $\Delta(\delta)$ the arc-weighted digraph with vertex set $\{1, \ldots, p\}$, and with an arc from i to j with weight d_i whenever $j - i \equiv d_i (\mathrm{mod}\, p)$. Such an arc is denoted by $i_{d_i}j$. The weight d_i of the arc is also called the length of that arc.
(ii) Let γ be a cycle in $\Delta(\delta)$. We denote by $\mu(\gamma)$ the cycle mean of γ, that is, the average of weights of arcs in γ. We denote by $\mu(\Delta(\delta))$ the maximal cycle mean of $\Delta(\delta)$, that is, the maximal $\mu(\gamma)$ where γ is a cycle in $\Delta(\delta)$.
(iii) A cycle γ in $\Delta(\delta)$ is said to be *maximal* if $\mu(\gamma) = \mu(\Delta(\delta))$. A cycle γ in $\Delta(\delta)$ is said to be *minimal* if $\mu(\gamma) \leq \mu(\gamma')$ for every cycle γ' in $\Delta(\delta)$.

Theorem 4.2. [4, Theorem 3.5] *The spectral radius $\rho(\delta, c)$ of the product $A_{d_1,c} \cdot \ldots \cdot A_{d_p,c}$ is equal to $c^{p/\mu(\Delta(\delta))}$.*

In view of Theorem 4.2, in order to study $\rho(\delta, c)$ it is enough to study $\mu(\Delta(\delta))$. Indeed, in [4] the authors study maximal cycle means of digraphs $\Delta(\delta)$. Since $0 < c < 1$, it also follows from Theorem 4.2 that the bigger $\mu(\Delta(\delta))$ is the bigger $\rho(\delta, c)$ is. This observtion motivates the following definition.

Definition 4.3. We denote by $\delta \geq \delta'$ the case where $d_i \geq d_i'$, $i = 1, \ldots, p$. The sequence δ is said to be *downward optimal* if $\mu(\Delta(\delta)) \geq \mu(\Delta(\delta'))$ whenever $\delta \geq \delta'$. The sequence δ is said to be *upward optimal* if $\mu(\Delta(\delta)) \leq \mu(\Delta(\delta'))$ whenever $\delta \leq \delta'$.

In order to find conditions for sequences to be downward or upward optimal we define

Definition 4.4. We denote by $\tilde{\Delta}(\delta)$ the digraph whose vertices are the positive integers, and with an arc from i to j whenever $j - i = d_i$. Such an arc (i, j) is said to be of *length* d_i. The total length of the arcs in a path $\tilde{\gamma}$ in $\tilde{\Delta}(\delta)$ is said to be the *length* of $\tilde{\gamma}$.

Remark 4.5. There is a correspondence between an arc $i_{d_i}j$ in $\Delta(\delta)$ and all arcs (k, l) in $\tilde{\Delta}(\delta)$ such that $i \equiv (k-1)(\mathrm{mod}\, p)+1$. Therefore, a path in $\tilde{\Delta}(\delta)$ corresponds to a unique path in $\Delta(\delta)$, but a path in $\Delta(\delta)$ corresponds to infinitely many paths in $\tilde{\Delta}(\delta)$ (with different starting points). Also, a path in $\tilde{\Delta}(\delta)$ whose length is divisible by p corresponds to a unique cycle in $\Delta(\delta)$, and a cycle in $\Delta(\delta)$ corresponds to infinitely many paths in $\tilde{\Delta}(\delta)$, where the length of each is equal to the total weight of the arcs in γ.

Theorem 4.6. [4, Theorem 4.15] *If there exists a cycle γ in $\Delta(\delta)$ and a positive integer k such that for every k consecutive arcs of γ, a path in $\tilde{\Delta}(\delta)$ corresponding to those arcs does not lie in the interior of the union of any k arcs of $\tilde{\Delta}(\delta)$, then γ is a maximal cycle and δ is a downward optimal sequence.*

Corollary 4.7. [4, Corollary 4.16] *If there exists a positive integer k such that for every k consecutive arcs in $\Delta(\delta)$, a path in $\tilde{\Delta}(\delta)$ corresponding to those arcs does not lie in the interior of the union of any k arcs of $\Delta(\delta)$, then every cycle in $\Delta(\delta)$ is a maximal cycle, and δ is a downward optimal sequence.*

It is shown in [4], by means of an example, that the condition, proven in Theorem 4.6 and Corollary 4.7 to be sufficient for a sequence δ to be downward optimal, is not necessary. However, this condition is also necessary in the case of sequences of two elements. In this case, the graphs $\Delta(d_1, d_2)$ and $\Delta(d_2, d_1)$ are the same up to relabeling of the vertices, and therefore it may be assumed, without loss of generality, that $d_1 \geq d_2$. We thus have

Theorem 4.8. [4, Theorem 4.21] *Let d_1 and d_2 be positive integers, $d_1 \geq d_2$. The following are equivalent.*
(i) *(d_1, d_2) is a downward optimal sequence.*
(ii) *Either d_1 is odd and $d_1 - d_2 \leq 2$, or d_1 is even.*
(iii) *For every two consecutive arcs of a maximal cycle in $\Delta(d_1, d_2)$, a path in $\tilde{\Delta}(d_1, d_2)$ corresponding to those arcs does not lie in the interior of the union of any two arcs of $\tilde{\Delta}(d_1, d_2)$.*

The following example illustrates the equivalence (i) $\Longleftrightarrow$ (ii) in Theorem 4.8.

Example 4.9. Let $n = 5$ and choose $c = 0.5$. For all possible sequences of two elements we have

$$\begin{cases} \rho((1,1),0.5) = 0.25 \\ \rho((2,1),0.5) = 0.5 \\ \rho((2,2),0.5) = 0.5 \\ \rho((3,1),0.5) = 0.5 \\ \rho((3,2),0.5) = 0.5 \\ \rho((3,3),0.5) = 0.63 \\ \rho((4,1),0.5) = 0.7071 \\ \rho((4,2),0.5) = 0.7071 \\ \rho((4,3),0.5) = 0.7071 \\ \rho((4,4),0.5) = 0.7071 \\ \rho((5,1),0.5) = 0.63 \\ \rho((5,2),0.5) = 0.5 \\ \rho((5,3),0.5) = 0.7071 \\ \rho((5,4),0.5) = 0.7071 \\ \rho((5,5),0.5) = 0.7579 \end{cases} . \tag{23}$$

By Theorem 4.8, the only sequences that are not downward optimal are $(5,1)$ and $(5,2)$. Indeed, from (23) we have $\rho((5,1),0.5) = 0.63 < 0.7071 = \rho((4,1),0.5)$ and

$$\rho((5,2),0.5) = 0.5 < \begin{cases} 0.7071 = \rho((4,1),0.5) \\ 0.7071 = \rho((4,2),0.5) \\ 0.63 = \rho((5,1),0.5) \end{cases} .$$

It is shown in [4] that the sufficient condition for downward optimality proven in Theorem 4.6 is not sufficient for upward optimality. In order to handle upward optimal sequences we define

Definition 4.10. A set S of arcs in $\tilde{\Delta}(\delta)$ is said to be *non-overlapping* if for every two arcs (t_1, t_2) and (t_3, t_4) in S we have either $t_2 \leq t_3$ or $t_4 \leq t_1$.

We then have

Theorem 4.11. [4, Theorem 5.6] *If there exists a maximal cycle γ in $\Delta(\delta)$ and a positive integer k such that no union of k non-overlapping arcs of $\tilde{\Delta}(\delta)$ lies in the interior of a path of k arcs in $\tilde{\Delta}(\delta)$ corresponding to k consecutive arcs of γ, then δ is an upward optimal sequence.*

Here too, he converse of Theorem 4.11 does not hold in general. However, it does hold in the case of sequences of two elements.

Theorem 4.12. [4, Theorem 5.11] *Let d_1 and d_2 be positive integers, $d_1 \geq d_2$. The following are equivalent.*

(i) *(d_1, d_2) is an upward optimal sequence.*
(ii) *Either d_1 is odd and d_2 is even, or d_2 is odd and $d_1 - d_2 \leq 2$, or d_1 is even and $d_2 = d_1$.*

(iii) *No union of two non-overlapping arcs of* $\tilde{\Delta}(d_1, d_2)$ *lies in the interior of a path of two arcs in* $\tilde{\Delta}(d_1, d_2)$ *corresponding to two consecutive arcs of a maximal cycle in* $\Delta(d_1, d_2)$.

The following example illustrates the equivalence (i) $\Longleftrightarrow$ (ii) in Theorem 4.12.

Example 4.13. As in Example 4.9, let $n = 5$ and choose $c = 0.5$. By Theorem 4.12, the only sequences that are not upward optimal are $(4, 1)$, (4.2) and $(5, 2)$. Indeed, from (23) we have

$$\rho((4.1), 0.5) = 0.7071 > \begin{cases} 0.63 = \rho((5,1), 0.5) \\ 0.5 = \rho((5,2), 0.5) \end{cases},$$

as well as $\rho((5, 1), 0.5) = 0.63 > 0.5 = \rho((5, 2), 0.5)$ and $\rho((4, 2), 0.5) = 0.63 > 0.5 = \rho((5, 2), 0.5)$.

A sufficient condition for a sequence to be both downward optimal and upward optimal is given in the following theorem.

Theorem 4.14. [4, Corollary 5.13] *If no arc of* $\tilde{\Delta}(\delta)$ *lies in the interior of another arc of* $\tilde{\Delta}(\delta)$, *then every cycle in* $\Delta(\delta)$ *is both a maximal cycle and a minimal cycle, and* δ *is both a downward optimal sequence and an upward optimal sequence.*

An immediate consequence of Theorem 4.14 is the following theorem, proven in [7]

Theorem 4.15. *If* $d_i \leq r + d_{i+r}$ *for all* $i, r \in \{1, \ldots, p\}$ *then* δ *is both a downward optimal sequence and an upward optimal sequence.*

Another issue discussed in [4] is order invariance of sequences.

Definition 4.16. The sequence δ is said to be order invariant (for the graph $\Delta(\delta)$) if $\mu(\Delta(\delta))$ is order invariant, that is, if $\mu(\Delta(d_{\hat{1}}, \ldots, d_{\hat{p}})) = \mu(\Delta(\delta))$ for every permutation $d_{\hat{1}}, \ldots, d_{\hat{p}}$ of $d_1, \ldots, d_p$.

We conclude this article with a bunch of conditions for order invariance of δ.

Theorem 4.17. [4, Theorems 6.8–6.12] *Let* $\delta = (d_1, \ldots, d_p)$ *be a sequence of positive integers. Then*

(i) *If all the* d_i*'s but one are the same, then* δ *is order invariant.*
(ii) *If there exists a positive integer* d, *relatively prime to* p, *such that* $d_i \equiv d(\text{mod} p)$, $i \in \{1, \ldots, p\}$, *then* δ *is order invariant.*
(iii) *If the largest* d_i *that is divisible by* p *is greater than or equal to the average of the two largest* d_i*'s that are not divisible by* p, *then* δ *is order invariant.*
(iv) *If the largest* d_i *is divisible by* p, *then* δ *is order invariant.*
(v) *If no partial sum of the set of* d_i*'s that are not divisible by* p *is divisible by* p, *then* δ *is order invariant.*

References

[1] A. Berman and R. Plemmons, *Nonnegative Matrices in Mathematical Sciences*, SIAM, Philadelphia, 1994.

[2] L. Elsner and D. Hershkowitz, Hadamard functions preserving nonnegative H-matrices, Linear Algebra Appl. 279 (1998), 13–19.

[3] L. Elsner, D. Hershkowitz and A. Pinkus, Functional inequalities for spectral radii of non-negative matrices, Linear Algebra Appl. 129 (1990), 103–130.

[4] L. Elsner, D. Hershkowitz and H. Schneider, Spectral radii of certain iteration matrices and cycle means of graphs, Linear Algebra Appl. 192 (1993), 61–81.

[5] L. Elsner and C.R. Johnson, Nonnegative matrices, zero patterns, and spectral inequalities, Linear Algebra Appl. 120 (1989), 225–236.

[6] L. Elsner, C.R. Johnson and J.A. Dias da Silva, The Perron root of a weighted geometric mean of nonnegative matrices. Linear and Multilinear Algebra 24 (1988), 1–13.

[7] L. Elsner and M. Neumann, Monotonic sequences and rates of convergence of asynchronized iterative methods, Linear Algebra Appl. 180 (1993), 17–33.

[8] G. Frobenius, Über Matrizen aus positiven Elementen, *S.–B. Preuss. Akad. Wiss.* (1909), 471–476.

[9] R.A. Horn and C.R. Johnson, *Matrix Analysis*, Cambridge University Press, Cambridge, 1985.

[10] S. Karlin and F. Ost, Some monotonicity properties of Schur powers of matrices and related inequalities, Linear Algebra Appl. 68 (1985), 47–65.

Daniel Hershkowitz
Department of Mathematics
Technion — Israel Institute of Technology
Haifa 32000
Israel

Operator Theory:
Advances and Applications, Vol. 134, 243–286

A Generic Schur Function is an Inner One

V. Katsnelson

Abstract. A *Schur function* s is a function which is holomorphic in an open unit disk $\mathbb{D}$ of the complex plane and is contractive there, i.e. $|s(z)| \leq 1$ for $z \in \mathbb{D}$. A Schur function is called exceptional if it is rational inner one. A *contractive sequence* ω is a sequence $\omega = \{\gamma_k\}_{0 \leq k < \infty}$ of complex numbers satisfying the condition $|\gamma_k| < 1$ for every k. The Schur algorithm establishes a one-to-one correspondence between the set Ω of all contractive sequences $\omega = \{\gamma_k\}_{0 \leq k < \infty}$ and the set of all non-exceptional Schur functions. A sequence $\omega \in \Omega$ is the sequence of the *Schur parameters* of the appropriate Schur function denoted as s_ω. Using this Schur correspondence, we introduce a probability measure on the set Ω, or, equivalently, on the set of all Schur functions. Namely, starting from an arbitrary probability measure μ on $\mathbb{D}$, we consider the set Ω as the set of sequences of independent identically distributed complex numbers from $\mathbb{D}$, with common distribution μ. (In other words, we introduce the product measure $p_\mu = \mu \otimes \mu \otimes \mu \otimes \cdots$ on $\Omega = \mathbb{D} \times \mathbb{D} \times \mathbb{D} \times \cdots$). We show that if the support of the measure μ consists of more than one point (otherwise there is no randomness), then p_μ almost every Schur function s_ω is inner. If, in addition, the logarithmic integral converges: $\int\limits_{\mathbb{D}} \ln(1-|\gamma|)\,\mu(d\gamma) > -\infty$, then for p_μ almost every Schur function the sequence of its *Schur approximants* converges pointwise almost everywhere (with respect to the Lebesgue measure) on the unit circle. The multiplicative ergodic theory is the main tool of investigation.

1. The Schur class, the Schur algorithm and the Schur parametrization

Definition 1. *A function* $s(\,.\,)$ *is said to be* a Schur function *if* $s(\,.\,)$ *is holomorphic in the open unit disk* $\mathbb{D}$ *and satisfies the inequality*

$$|s(z)| \leq 1 \qquad (z \in \mathbb{D}). \tag{1.1}$$

The set of all Schur functions is said to be the Schur class. *The Schur class is denoted by* $\mathfrak{S}$. □

Key words and phrases. Holomorphic contractive functions, Schur functions, Schur algorithm, Schur approximants, Schur continued fractions, random Schur functions, random orthogonal polynomials on the circle, random linear dynamic system, random matrices, random operators, singular spectrum, multiplicative ergodic theory.

If $s(\,.\,)$ is a Schur function then for almost every $t \in \mathbb{T}$ (with respect to the Lebesgue measure $m(dt)$), the radial limit

$$s(t) \stackrel{\text{def}}{=} \lim_{r \to 1-0} s(rt) \tag{1.2}$$

exists. The function $s(t)$, which is defined $m(dt)$-almost everywhere on the unit circle $\mathbb{T}$, is said to be *the boundary value of the function s.*

Definition 2. *A Schur function s is said to be* inner *if the absolute value of its boundary value is equal to one almost everywhere on $\mathbb{T}$ with respect to the Lebesgue measure:*

$$|s(t)| = 1 \qquad \textit{a.e. with respect to the Lebesgue measure } m(dt). \tag{1.3}$$

□

Definition 3. *A Schur function $s(\,.\,)$ is said to be* exceptional *if it is a rational inner function. The class of all exceptional Schur functions is denoted by $\mathfrak{S}_e$. The class $\mathfrak{S}_{ne}$ of* non-exceptional Schur functions *is defined as $\mathfrak{S}_{ne} \stackrel{def}{=} \mathfrak{S} \setminus \mathfrak{S}_e$.* □

The class of exceptional inner functions coincides with the class of functions which are finite Blaschke products. (This statement is a particular case of the Factorization Theorem for inner functions.)

Now we discuss the *Schur algorithm.* To introduce this algorithm we start from the linear fractional transformation

$$\zeta \to \frac{\zeta - \gamma}{1 - \zeta \overline{\gamma}} \tag{1.4}$$

where γ is a complex number, $|\gamma| < 1$. This transformation provides a one to one mapping of the open unit disk $\mathbb{D}$ onto itself. It also maps the unit circle $\mathbb{T}$, which is the boundary of $\mathbb{D}$, one to one onto itself. If $|\gamma| = 1$ the transformation (1.4) maps the set $\mathbb{C} \setminus \{\gamma\}$ into the point $\{-\gamma\}$ and is *not defined* at the point γ.

Let f be a Schur function *which is not a unitary constant.* Then $|f(0)| < 1$. Substituting $f(z)$ as ζ and $f(0)$ as γ into (1.4) we come to the function $\frac{f(z)-f(0)}{1-f(z)\overline{f(0)}}$ which is a Schur function as well. The latter function vanishes at the origin. According to the Schwarz Lemma the function $\frac{f(z)-f(0)}{1-f(z)\overline{f(0)}} \cdot \frac{1}{z}$ is a Schur function as well. If the function f is a unitary constant then the $\frac{f(z)-f(0)}{1-f(z)\overline{f(0)}} \cdot \frac{1}{z}$ is not defined. Thus the transformation

$$f(z) \to \frac{f(z) - f(0)}{1 - f(z)\,\overline{f(0)}} \cdot \frac{1}{z} \tag{1.5}$$

maps the class of all Schur functions which are not unitary constants into the class $\mathfrak{S}$ of all Schur functions. In particular if f is a non-exceptional Schur function then $|f(0)| < 1$ and the transformation (1.5) is well defined. It is easy to see that the function $\frac{f(z)-f(0)}{1-f(z)\overline{f(0)}} \cdot \frac{1}{z}$ also is a *non-exceptional* Schur function in this case. Thus we have

Proposition 1. *The transform (1.5) is well defined on the class $\mathfrak{S}_{ne}$ of all non-exceptional Schur functions and maps this class into itself.* □

If f is a rational inner function and $f(0) \neq 0$ then it is of the form $f(z) = \frac{z^n \overline{P(\overline{z}^{-1})}}{P(z)}$ where P is a polynomial of the degree n, $0 \leq n < \infty$. This n is determined uniquely from f and said to be *the degree of the rational inner function f.*

Proposition 2. *Let $f(z)$ be a rational inner function of degree n, $n \geq 1$. Then the transform (1.5) is defined for this f and the right-hand side of (1.5) is a rational inner function of degree $n-1$.* □

Description of the Schur algorithm. The Schur algorithm relates recursively a certain sequence of Schur functions $\{s_k(\,.\,)\}_{0 \leq k < \infty}$ to the given Schur function $s(.)$. We provide the given Schur function s with index zero

$$s_0(z) \stackrel{\text{def}}{=} s(z) \tag{1.6}$$

and then define

$$s_k(z) \stackrel{\text{def}}{=} \frac{s_{k-1}(z) - s_{k-1}(0)}{1 - s_{k-1}(z)\,\overline{s_{k-1}(0)}} \cdot \frac{1}{z} \quad (k = 1, 2, 3, \ldots). \tag{1.7}$$

If the starting function s is a non-exceptional Schur function, then, applying Proposition 1 to the functions s_0, s_1, s_2, ..., we deduce by induction that each of the functions s_k is well defined and is a non-exceptional Schur function as well. In this case the Schur algorithm can be continued and produces infinitely many Schur functions $s_k(z)$, $k = 0, 1, 2, 3, \ldots$ without any restrictions.

If the starting function s is an exceptional Schur function, i.e. it is rational inner, then Schur algorithm will end after finitely many steps. Namely, let s be a rational inner function of degree n. Then, following Proposition 2, the function s_k, $0 \leq k \leq (n-1)$ is a rational inner function of degree $n-k$. (It can be proved by induction.) The function s_n is a unitary constant. The function s_{n+1} is not defined because the transformation (1.5) is not applicable to the unitary constant. So in this case it is possible to perform only n steps of Schur algorithm. □

Schur parameters. *Let s be a Schur function and let $\{s_k\}$ be the sequence (infinite or finite) of functions generated by the function s according the Schur algorithm. The values*

$$\gamma_k \stackrel{\text{def}}{=} s_k(0) \tag{1.8}$$

play a crucial role in considerations related to the Schur class. The numbers $\gamma_k = \gamma_k(s)$ are said to be the Schur parameters of the function s. □

If s is a non-exceptional Schur function then it generates an infinite sequence of the Schur parameters $\{\gamma_k\}_{0 \leq k < \infty}$. In this case

$$|\gamma_k| < 1\,, \quad k = 0, 1, 2, \ldots\,. \tag{1.9}$$

If s is an exceptional Schur function, say a rational function of degree n, then it generates a finite sequence of Schur parameters $\{\gamma_k\}_{0\le k\le n}$. In this case

$$|\gamma_k| < 1\,, \quad k = 0, 1, 2, \dots, n-1; \quad |\gamma_n| = 1. \tag{1.10}$$

If

$$s(z) = \sum_{0\le k} c_k(s) z^k \tag{1.11}$$

is a Taylor expansion of a Schur function s (which surely converges in the open unit disk $\mathbb{D}$), then the Schur parameter $\gamma_k(s)$ depends only on the Taylor coefficients $c_0(s), c_1(s), \dots, c_k(s)$ of the function s:

$$\gamma_k(s) = \Phi_k(c_0(s), c_1(s), \dots, c_k(s))\,, \tag{1.12}$$

where $\Phi_k(c_0, c_1, \dots, c_k)$ is a rational function of the variables

$$c_0, \overline{c_0}, c_1, \overline{c_1}, \dots, c_{k-1}, \overline{c_{k-1}}, c_k\,.$$

Conversely, the Taylor coefficient $c_k(s)$ of the Schur function depends only on the Schur parameters $\gamma_0(s), \gamma_1(s), \dots, \gamma_k(s)$ of this function:

$$c_k(s) = \Psi_k(\gamma_0(s), \gamma_1(s), \dots, \gamma_k(s))\,, \tag{1.13}$$

where $\Psi_k(\gamma_0, \gamma_1, \dots, \gamma_k)$ is a polynomial in $\gamma_0, \overline{\gamma_0}, \gamma_1, \overline{\gamma_1}, \dots, \gamma_{k-1}, \overline{\gamma_{k-1}}, \gamma_k$.

In [S1] an explicit expression for Φ_k and Ψ_k has been given. There, the function Φ_k is represented as the quotients of determinants which are constructed from the coefficients $c_0, c_1, \dots, c_{n-1}, c_n$ and their conjugates $\overline{c_0}, \overline{c_1}, \dots, \overline{c_{n-1}}$. (See Theorem III in §3 of [S1].) In §2 of [S1] a recursion formula is provided which allows us to calculate all functions Ψ_k.

Definition 4. *A sequence $\{\gamma_k\}_{0\le k<\infty}$ of complex numbers is said to be* contractive *if it satisfies the condition $|\gamma_k| < 1$ for every k.*

We showed that the sequence of Schur parameters of a Schur function is contractive.

It turns out that every preassigned contractive sequence

$$\{\gamma_0, \gamma_1, \gamma_2, \dots, \gamma_k, \dots\}$$

is the sequence of the Schur parameters for some unique non-exceptional Schur function. Such a function can be constructed by means of a certain continued fractions algorithm.

Schur continued fractions. Given an arbitrary contractive sequence

$$\{\gamma_0, \gamma_1, \dots, \gamma_k, \dots\}$$

of complex numbers, we construct a sequence of rational Schur functions which converges to the Schur functions s which sequence of Schur parameters

$$\{\gamma_0(s), \gamma_1(s), \dots, \gamma_k(s), \dots\}$$

coincides with this preassigned sequence. The underlying reason for this construction is the following. The desired function s is sought in the form which may be considered as a special kind of continued fraction. The n-th rational function of the mentioned sequences can be considered as n-th convergent of this continued fraction. The transformation on which the elementary step (1.5) of the Schur algorithm is based is given by

$$f(z) \to \frac{f(z) - \gamma}{1 - f(z)\overline{\gamma}} \cdot \frac{1}{z} \cdot \tag{1.14}$$

The transformation inverse to (1.14) is of the form

$$f(z) \to \frac{\gamma + zf(z)}{1 + \overline{\gamma}zf(z)} \cdot \tag{1.15}$$

If f is a Schur function then the function $\frac{\gamma+zf(z)}{1+\overline{\gamma}zf(z)}$ is a Schur function as well. We use the '*inverse Schur algorithm*' in a recursive manner to construct *the n-th Schur approximant*, which we (following I. Schur) will *denote* by $[z;\, \gamma_0,\, \gamma_1,\, \dots,\, \gamma_n]$. Namely, we write

$$\begin{aligned} &[z;\, \gamma_n] = \gamma_n; \\ &[z; \gamma_k,\, \gamma_{k+1},\, \gamma_{k+2},\, \dots,\, \gamma_n] = \frac{\gamma_k + z \cdot [z;\, \gamma_{k+1},\, \gamma_{k+2},\, \dots,\, \gamma_n]}{1 + \overline{\gamma_k} \cdot z \cdot [z;\, \gamma_{k+1},\, \gamma_{k+2},\, \dots,\, \gamma_n]}\,, \\ &k = n-1,\, n-2,\, \dots,\, 1,\, 0\,. \end{aligned} \tag{1.16}$$

The function $[z;\, \gamma_0,\, \gamma_1,\, \dots,\, \gamma_n]$ is a rational Schur function whose Schur parameters $\gamma_k\big([z;\, \gamma_0,\, \gamma_1,\, \dots,\, \gamma_n]\big)$ are equal to

$$\begin{aligned} \gamma_k\big([z;\, \gamma_0,\, \gamma_1,\, \dots,\, \gamma_n]\big) &= \gamma_k \qquad \text{for} \qquad k = 0,\, 1,\, \dots,\, n; \\ \gamma_k\big([z;\, \gamma_0,\, \gamma_1,\, \dots,\, \gamma_n]\big) &= 0 \qquad \text{for} \qquad k > n\,. \end{aligned}$$

Let n_1 and n_2 be two non-negative integers. Since the Schur parameters with the indices $k : 0 \leq k \leq \min(n_1,\, n_2)$ for the functions $[z;\, \gamma_0,\, \gamma_1,\, \dots,\, \gamma_{n_1}]$ and $[z;\, \gamma_0,\, \gamma_1,\, \dots,\, \gamma_{n_2}]$ coincide, the Taylor coefficients c_k^1 and c_k^2 $(0 \leq k \leq \min(n_1, n_2))$ for these functions coincide as well. Hence

$$[z;\, \gamma_0,\, \gamma_1,\, \dots,\, \gamma_{n_1}] - [z;\, \gamma_0,\, \gamma_1,\, \dots,\, \gamma_{n_2}] = \sum_{\min(n_1,n_2)<k<\infty} (c_k^1 - c_k^2)\, z^k\,.$$

Using the estimates $[z;\, \gamma_0,\, \gamma_1,\, \dots,\, \gamma_{n_1}] \leq 1$, $[z;\, \gamma_0,\, \gamma_1,\, \dots,\, \gamma_{n_2}] \leq 1$ for $z \in \mathbb{D}$ and Schwarz Lemma we come to the inequality

$$\big|[z;\, \gamma_0,\, \gamma_1,\, \dots,\, \gamma_{n_1}] - [z;\, \gamma_0,\, \gamma_1,\, \dots,\, \gamma_{n_2}]\big| \leq 2\,|z|^{\,1+\min(\,n_1,\,n_2)} \quad \text{for } z \in \mathbb{D}\,. \tag{1.17}$$

From (1.17) it follows that the limit

$$[z;\, \gamma_0,\, \gamma_1,\, \dots,\, \gamma_k,\, \dots] \stackrel{\text{def}}{=} \lim_{n\to\infty} [z;\, \gamma_0,\, \gamma_1,\, \dots,\, \gamma_n] \tag{1.18}$$

exists in the unit disk. *The function* $[z;\, \gamma_0,\, \gamma_1,\, \dots,\, \gamma_k,\, \dots]$ *is said to be* the Schur continued fraction constructed from the sequence $\{\gamma_0,\, \gamma_1,\, \dots,\, \gamma_k,\, \dots\}$. □

The function $[z;\, \gamma_0,\, \gamma_1,\, \dots,\, \gamma_k,\, \dots]$ is a non-exceptional Schur function. Its Schur parameters $\gamma_k\big([z;\, \gamma_0,\, \gamma_1,\, \dots,\, \gamma_k,\, \dots]\big)$ coincide with the numbers γ_k:

$$\gamma_k\big([z;\, \gamma_0,\, \gamma_1,\, \dots,\, \gamma_k,\, \dots]\big) = \gamma_k, \quad 0 \leq k < \infty.$$

Given a non-exceptional Schur function s, we can form the sequence

$$\{\gamma_0(s),\, \gamma_1(s),\, \ldots,\, \gamma_k(s),\, \ldots\}$$

of its Schur parameters and then construct the Schur continued fraction

$$[z;\, \gamma_0(s),\, \gamma_1(s),\, \ldots\,,\, \gamma_k(s),\, \ldots].$$

The function represented by this fraction is a Schur function whose sequence of Schur parameters coincide with the sequence of Schur parameters of the original function s. Hence, Taylor coefficients of these two functions coincide as well.

Thus, *every non-exceptional Schur function s admits the Schur Continued Fraction Expansion*

$$s(z) = [z;\, \gamma_0(s),\, \gamma_1(s),\, \ldots\,,\, \gamma_k(s),\, \ldots\,]. \tag{1.19}$$

Definition 5. *Let $s(z)$ be a non-exceptional Schur function and n be a non-negative integer. Let* (1.19) *be the Schur continued fraction expansion of the function s. The function*

$$\mathrm{Ap}_n(s;\, z) \stackrel{\mathrm{def}}{=} [z; \gamma_0(s),\, \gamma_1(s),\, \ldots\,,\, \gamma_n(s)\,] \tag{1.20}$$

is said to be the n-th Schur approximant of the function s.

Remark 1. *The n-th Schur approximant is a rational function of z whose numerator and denominator are polynomials of degree not greater than n. In fact the n-th Schur approximant of a non-exceptional Schur function s is the n-th convergent of its Schur continued fraction expansion (1.19).*

The estimate

$$\left|s(z) - \mathrm{Ap}_n(s;\, z)\right| \le 2\,|z|^{n+1} \tag{1.21}$$

holds for every non-exceptional Schur function. (As in (1.17), this estimate can be obtained using Schwarz Lemma.) From the estimate (1.21) it follows that *the sequences of the Schur approximants of the non-exceptional Schur function s converges to this function locally uniformly in the open unit disk $\mathbb{D}$*. The problem of convergence of Schur approximants to s on the unit circle $\mathbb{T}$ is much more difficult. One of our main results is concerned with this problem.

Let us summarize results related to the Schur algorithm:

1. If for two non-exceptional Schur functions the sequences of their Schur parameters coincide then these functions coincide as well.
2. The Schur parameters of a non-exceptional Schur function form a contractive sequence.
3. Every non-exceptional Schur function s admits the Schur Continued Fraction Expansion (1.19).
4. The sequence of the Schur approximants of any non-exceptional Schur function s converges to this function locally uniformly in the unit disk $\mathbb{D}$, and the estimate (1.21) holds.
5. For an arbitrary preassigned contractive sequence $\{\gamma_0,\, \gamma_1,\, \ldots\,, \gamma_k,\, \ldots\}$ there exists a non-exceptional Schur function whose Schur parameters coincide with

these numbers γ_k. Such a function can be constructed as the Schur continued fraction (1.18).

In particular, from the above mentioned results about the Schur algorithm it follows that *the correspondence*

$$\{\gamma_0, \gamma_1, \dots, \gamma_k, \dots\} \Leftrightarrow [z; \gamma_0, \gamma_1, \dots, \gamma_k, \dots] \tag{1.22}$$

is a free parametrization of the class $\mathfrak{S}_{ne}$ of all non-exceptional Schur functions by means of the set of all contractive sequences $\{\gamma_0, \gamma_1, \gamma_2, \dots, \gamma_k, \dots\}$, where sequences serve as free parameters of this class.

The later fact is of principal importance for us. The point is that the parametrization (1.22) is an appropriate tool for probabilistic study of Schur class. The geometry of the set of all Taylor coefficients of functions of this class is rather complicated. Therefore Taylor coefficients are not suitable for our purpose. On the other hand, the geometry of the set of all Schur parameters is very simple: the latter set is the direct product of the unit disks. Such a geometry is well compatible with probabilistic structures and is very suitable for our purpose.

Remark 2. *It is not easy to express properties of a concrete Schur function s in terms of its Schur parameters $\gamma_k(s)$. In particular, it is not easy to recognize whether the function s is inner or not. Not much is known about this.*

If $\sum\limits_{0\leq k<\infty} |\gamma_k(s)| < \infty$, then the function s is continuous in the closed unit disk $\overline{\mathbb{D}}$, and $\max\limits_{z\in\mathbb{D}} |s(z)| < 1$. (Of course, s is not inner.) This result was obtained by I. Schur, [S2], *§15, Theorem XVIII.*

If $\sum\limits_{0\leq k<\infty} |\gamma_k(s)|^2 < \infty$, then the function s is also not inner. This follows from the identity

$$\prod_{0\leq k<\infty} (1-|\gamma_k(s)|^2) = \exp\Big\{\int\limits_{\mathbb{T}} \ln\left(1-|\,s(t)|^2\right) m(dt)\Big\}.$$

(*See* [Boy]. *See also the formula* (8.14) *in* [Gers3], *which expresses a similar result for polynomials orthogonal on* $\mathbb{T}$.)

If $\overline{\lim\limits_{k\to\infty}} |\gamma_k(s)| = 1$, then the function s is inner. This result was obtained by E.A. Rakhmanov, [Rakh] *and it is known as Rakhmanov's Lemma. We present a proof of Rakhmanov's Lemma in Section 4 of our paper* (*Lemma* 8).

If the sequence of the Schur parameters $\{\gamma_k(s)\}_{0\leq k<\infty}$ satisfies Maté-Nevai condition $\lim\limits_{k\to\infty} \gamma_k\gamma_{k+n} = 0$ for $n = 1, 2, 3, \dots$, but $\overline{\lim\limits_{k\to\infty}} |\gamma_k| > 0$, then s is an inner function. This is Theorem 5 *and* Corollary 9.1 *in* [Khru2].

It is also known that there exists infinite Blaschke product s such that

$$\sum_{0\leq k<\infty} |\gamma_k(s)|^p < \infty$$

for every $p > 2$. (*This is shown in* [Khru3].) □

2. The formulations of the main results

A random Schur function is a non-exceptional Schur function whose Schur parameters are independent identically distributed random variables in the open unit disk. We will show that such a function is inner almost surely. First we introduce necessary definitions.

Let μ be an arbitrary probability measure in the unit disk $\mathbb{D}$. The probability space, which is denoted by Ω, is the countable product

$$\Omega = \mathbb{D} \times \mathbb{D} \times \mathbb{D} \times \cdots \times \mathbb{D} \times \cdots. \tag{2.1}$$

Points ω of Ω are contractive sequences

$$\omega = \{\gamma_0, \gamma_1, \gamma_2, \ldots, \gamma_k, \ldots\} \tag{2.2}$$

of complex numbers γ_k. So the set Ω is the set of all contractive sequences ω. The space Ω, equipped with the product topology, becomes a topological space. In particular the notion of a Borel set in Ω is defined. The collection of all Borel sets in Ω forms a σ-algebra which is denoted by $\mathfrak{B}(\Omega)$ and is called the Borel σ-algebra of Ω.

The measure p_μ on $\mathfrak{B}(\Omega)$ is defined as the product measure

$$p_\mu = \mu \otimes \mu \otimes \mu \otimes \cdots \otimes \mu \otimes \cdots. \tag{2.3}$$

Of course, the measure p_μ is a probability measure on the Borel σ-algebra $\mathfrak{B}(\Omega)$. The completion of the measure p_μ is defined on the σ-algebra Σ generated by the σ-algebra $\mathfrak{B}(\Omega)$ and all subsets of all Borel sets of zero p_μ-measure.

Thus we have defined the *probability space* (Ω, Σ, p_μ). Ω is *a sample space*, Σ is *a σ-algebra of events*, p_μ is *a probability measure*:

$$p_\mu(\Omega) = 1. \tag{2.4}$$

As it was mentioned in Sction 3, to every sequence

$$\omega = \{\gamma_0, \gamma_1, \gamma_2, \ldots, \gamma_k, \ldots\} \in \Omega,$$

there corresponds a unique Schur function $s_\omega(z)$ such that the sequence of its Schur parameters coincides with this sequence ω. We denote this Schur function by $s_\omega(z)$. The function $s_\omega(z)$ can be constructed by means of the Schur continued fraction (1.18):

$$s_\omega(z) = [z; \gamma_0, \gamma_1, \ldots, \gamma_k, \ldots]. \tag{2.5}$$

It was also mentioned that every non-exceptional Schur function is of the form s_ω for some $\omega \in \Omega$ and if $\omega_1 \neq \omega_2$ then $s_{\omega_1} \neq s_{\omega_2}$. Thus the class $\mathfrak{S}_{ne}$ can be considered as the set $\{s_\omega\}_{\omega\in\Omega}$:

$$\mathfrak{S}_{ne} \simeq \{s_\omega\}_{\omega\in\Omega}. \tag{2.6}$$

The measure p_μ can be considered as a measure on the class $\mathfrak{S}_{ne}$.

Theorem I. *Let μ be a probability measure in the open unit disk $\mathbb{D}$ whose support consists of more than one point. Then almost every (with respect to the measure*

p_μ) non-exceptional Schur function is an inner one: there exists a set $R \subset \Omega$ such that

i. $p_\mu(R) = 0$;
ii. *For every $\omega \in \Omega \setminus R$, the function s_ω is an inner one.*

Theorem II. *Let μ be a probability measure in the open unit disk $\mathbb{D}$ whose support consists of more than one point and which satisfies the condition*

$$\int\limits_{\mathbb{D}} \ln \frac{1}{1-|\gamma|} \, \mu(d\gamma) < \infty . \tag{2.7}$$

Then for almost every (with respect to the measure p_μ) non-exceptional Schur function s, the sequence of its Schur approximants $\mathrm{Ap}_n(s;\,.)$ converges to $s(\,.\,)$ almost everywhere (with respect to Lebesgue measure $m(dt)$) on the unit circle $\mathbb{T}$: there exists a set $R \subset \Omega$ such that

i. $p_\mu(R) = 0$;
ii. *For every $\omega \in \Omega \setminus R$,*

$$\lim_{n\to\infty} \mathrm{Ap}_n(s_\omega;\, t) = s_\omega(t) \quad \text{for a.e. } t \in \mathbb{T} \text{ w.r.t. Lebesgue measure } m(dt) \tag{2.8}$$

($s_\omega(t)$ is the boundary value of the function s_ω).

Remark 3. *The rate of convergence in* (2.8) *is in some sense* exponential. *(See Remark* 8 *later.)*

Remark 4. *For every inner function s the sequence of its Schur approximants $\mathrm{Ap}_n(s)$ converges to s in $L^2(m(dt))$:*

$$\lim_{n\to\infty} \int\limits_{\mathbb{T}} \left|s(t) - \mathrm{Ap}_n(s;t)\right|^2 m(dt) = 0. \tag{2.9}$$

Let $c_k(s)$ and $c_k(\mathrm{Ap}_n(s))$ be Taylor coefficients of the functions s and $\mathrm{Ap}_n(s)$ respectively. Since $c_k(s) = c_k(\mathrm{Ap}_n(s))$ for $k = 0, 1, \dots, n$, the Parseval identity gives us:

$$\int\limits_{\mathbb{T}} \left|s(t) - \mathrm{Ap}_n(s;t)\right|^2 m(dt) = \sum_{n+1\le k<\infty} \left|c_k(s) - c_k(\mathrm{Ap}_n(s)\right|^2$$
$$\le 2 \sum_{n+1\le k<\infty} \left(|\, c_k(s)|^2 + |\, c_k(\mathrm{Ap}_n(s)|^2\right).$$

It is clear that $\sum_{n+1\le k<\infty} |c_k(s)|^2 \to 0$ as $n \to \infty$. (The later sum is the n-th remainder of a convergent series.) The sum $\sum_{n+1\le k<\infty} \left|c_k(\mathrm{Ap}_n(s))\right|^2$ is not the n-th remainder of a certain series: the series depends on n itself. Nevertheless, this sum also tends to zero as $n \to \infty$. Since $|\mathrm{Ap}_n(s;\, t)| \le 1$ on $\mathbb{T}$, the Parseval identity for the function $\mathrm{Ap}_n(s;\,.)$ gives us: $\sum_{0\le k<\infty} |\, c_k(\mathrm{Ap}_n(s))|^2 \le 1$. Since

$c_k(s) = c_k(\mathrm{Ap}_n(s))$ for $0 \le k \le n$ and $1 = \sum\limits_{0 \le k < \infty} |\, c_k(s)|^2$ (the later identity is the Parseval identity for the *inner* function s), we obtain: $\sum\limits_{n+1 \le k < \infty} |\, c_k(\mathrm{Ap}_n(s))|^2 \le \sum\limits_{n+1 \le k < \infty} |\, c_k(s)|^2$. Thus,

$$\int\limits_{\mathbb{T}} \left| s(t) - \mathrm{Ap}_n(s;t) \right|^2 m(dt) \le 4 \sum_{n+1 \le k < \infty} |\, c_k(s)|^2 . \qquad \square$$

According to Theorem I, p_μ-almost every Schur function is an inner one. Thus for p_μ-almost every Schur function s, the sequence of its Schur approximants converges to s in $L^2(m(dt))$. However, Theorem II claims that under the condition (2.7), the convergence is also pointwise for p_μ-almost every Schur function.

3. The matrices related to the Schur algorithm

It is very useful to present the Schur algorithm in matrix form. Given a complex number c, we associate the vector $\begin{bmatrix} c \\ 1 \end{bmatrix}$ with this number. In this agreement, the k-th elementary step (1.7) of the Schur algorithm can be presented in the form

$$\begin{bmatrix} s_k(z) \\ 1 \end{bmatrix} = \begin{bmatrix} z^{-1} & -\gamma_{k-1}\, z^{-1} \\ -\overline{\gamma_{k-1}} & 1 \end{bmatrix} \cdot \begin{bmatrix} s_{k-1}(z) \\ 1 \end{bmatrix} \cdot \frac{1}{-\overline{\gamma_{k-1}}\, s_{k-1}(z) + 1},$$

where $\gamma_{k-1} = s_{k-1}(0)$. So, it is natural to associate the matrix

$$\begin{bmatrix} z^{-1} & -\gamma_{k-1}\, z^{-1} \\ -\overline{\gamma_{k-1}} & 1 \end{bmatrix}$$

with the k-th elementary step of Schur algorithm. However, it turns out that it is much more fruitful to deal with a proportional matrix. Namely, instead of this matrix we consider the matrix $m_{\gamma_{k-1}}$, where the matrix m_γ is defined below.

For a complex number $\gamma : |\gamma| < 1$, let us introduce the matrix function

$$m_\gamma(z) = \begin{bmatrix} z^{-1} & -\gamma \cdot z^{-1} \\ -\overline{\gamma} & 1 \end{bmatrix} \cdot \frac{1}{\sqrt{1 - |\gamma|^2}}. \tag{3.1}$$

By a direct calculation we obtain that

$$m_\gamma(z)^{-1} = \begin{bmatrix} z & \gamma \\ z \cdot \overline{\gamma} & 1 \end{bmatrix} \cdot \frac{1}{\sqrt{1 - |\gamma|^2}}, \tag{3.2}$$

$$m_\gamma(z)^* = \begin{bmatrix} (\overline{z})^{-1} & -\gamma \\ -\overline{\gamma} \cdot (\overline{z})^{-1} & 1 \end{bmatrix} \cdot \frac{1}{\sqrt{1 - |\gamma|^2}} \tag{3.3}$$

and

$$(m_\gamma(z)^*)^{-1} = \begin{bmatrix} \overline{z} & \overline{z}\cdot\gamma \\ \overline{\gamma} & 1 \end{bmatrix} \cdot \frac{1}{\sqrt{1-|\gamma|^2}} \cdot \tag{3.4}$$

The matrix m_γ is a matrix of the linear fractional transformation (1.14) which is the elementary step (1.5) of the Schur algorithm.

The matrix of a linear fractional transformation is determined up to a nonzero scalar factor. We choose the matrix of the linear fractional transformation (1.14) in the form (3.1) for the following reason. The matrix m_γ of the form (3.1) satisfies j - properties. Let j be the matrix

$$j = \begin{bmatrix} -1 & 0 \\ 0 & 1 \end{bmatrix} \tag{3.5}$$

Then

$$(m_\gamma(z)^*)^{-1} j (m_\gamma(z))^{-1} - j = (1-|z|^2) \cdot \begin{bmatrix} 1 \\ 0 \end{bmatrix} \begin{bmatrix} 1 & 0 \end{bmatrix}. \tag{3.6}$$

Definition 6. *Let $\omega = \{\gamma_k\}_{0\le k<\infty}$ be a contractive sequence of complex numbers. Let us define a sequence of matrices (or matrix-functions) $\{M_{\omega,n}(z)\}_{0\le n<\infty}$:*

$$M_{\omega,n}(z) \overset{def}{=} m_{\gamma_n}(z)\cdot m_{\gamma_{n-1}}(z)\cdot \ \ldots \ \cdot m_{\gamma_1}(z)\cdot m_{\gamma_0}(z), \quad n = 0, 1, 2, \ldots$$

$$M_{\omega,-1}(z) \overset{def}{=} I \ \text{ – the identity matrix}. \tag{3.7}$$

The entries of the matrix $M_{\omega,n}$ are denoted as

$$M_{\omega,n}(z) = \begin{bmatrix} a_{\omega,n}(z) & b_{\omega,n}(z) \\ c_{\omega,n}(z) & d_{\omega,n}(z) \end{bmatrix}. \tag{3.8}$$

The entries of the inverse matrix are denoted as

$$M^{-1}_{\omega,n}(z) = \begin{bmatrix} A_{\omega,n}(z) & B_{\omega,n}(z) \\ C_{\omega,n}(z) & D_{\omega,n}(z) \end{bmatrix}. \tag{3.9}$$

Remark 5. *For a certain n, the matrix $M_{\omega,n}$ depends not on the whole sequence $\omega = \{\gamma_k\}_{0\le k<\infty}$, but only on its "initial interval" $\{\gamma_k\}_{0\le k\le n}$: if $\omega' = \{\gamma'_k\}_{0\le k<\infty}$ and $\omega'' = \{\gamma''_k\}_{0\le k<\infty}$ are two sequences for which $\gamma'_k = \gamma''_k$ by $0 \le k \le n$, then $M_{\omega',n} = M_{\omega'',n}$.* □

We use the sequence of matrices $M_{\omega,n}$ in two ways:

1. *First, we can study* a given *Schur function s starting from the sequence $\omega = \{\gamma_k\}$ of* its *Schur parameters, with an aim to express the properties of s in terms of the behavior of the sequence $\{M_{\omega,n}\}$.*

2. *Second, we can try to* specify *a Schur function* s with prescribed properties *choosing a sequence* $\omega = \{\gamma_k\}$ *for which the sequence* $\{M_{\omega,n}\}$ *has an appropriate behavior.*

If $s(z)$ is a non-exceptional Schur function, $\omega = \{\gamma_k\}_{0\leq k<\infty}$ is the sequence of its Schur parameters and $\{s_k(z)\}_{0\leq k<\infty}$ is the sequence of Schur functions generated by the Schur algorithm (so $\gamma_k = s_k(0)$) , then

$$M_{\omega,n}(z)\begin{bmatrix} s(z) \\ 1 \end{bmatrix} = \begin{bmatrix} s_{n+1}(z) \\ 1 \end{bmatrix} \cdot \left(c_{\omega,n}(z)\, s(z) + d_{\omega,n}(z)\right), \tag{3.10}$$

or

$$\frac{a_{\omega,n}(z)\, s(z) + b_{\omega,n}(z)}{c_{\omega,n}(z)\, s(z) + d_{\omega n}(z)} = s_{n+1}(z)\,. \tag{3.11}$$

Hence

$$s(z) = \frac{A_{\omega,n}(z)\, s_{n+1}(z) + B_{\omega,n}(z)}{C_{\omega,n}(z)\, s_{n+1}(z) + D_{\omega,n}(z)}\,. \tag{3.12}$$

If we replace $s_{n+1}(z)$ by 0 in (3.12) we obtain the n-th Schur approximant of the function s:

$$\mathrm{Ap}_n(s;\, z) = \frac{B_{\omega,n}(z)}{D_{\omega,n}(z)}\,. \tag{3.13}$$

This follows from the definition of Schur approximant (see Definition 5). In fact, the Schur approximant of the function s was defined from its Schur parameters using the inverse Schur algorithm (see (1.16)). Since an "elementary step" (1.15) of the inverse Schur algorithm is described by the matrix (3.2), the n-th Schur approximant can be expressed in terms of the entries of the matrix $M^{-1}_{\omega,n}(z)$ (see (3.13)). However it is more convenient to express it in terms of the entries of the matrix $M_{\omega,n}(z)$. The inverse matrix can be easily calculated:

$$\begin{bmatrix} A_{\omega,n}(z) & B_{\omega,n}(z) \\ C_{\omega,n}(z) & D_{\omega,n}(z) \end{bmatrix} = \begin{bmatrix} a_{\omega,n}(z) & b_{\omega,n}(z) \\ c_{\omega,n}(z) & d_{\omega,n}(z) \end{bmatrix}^{-1}$$

$$= \begin{bmatrix} d_{\omega,n}(z) & -b_{\omega,n}(z) \\ -c_{\omega,n}(z) & a_{\omega,n}(z) \end{bmatrix} \cdot \frac{1}{a_{\omega,n}(z) d_{\omega,n}(z) - b_{\omega,n}(z) c_{\omega n}(z)}\,.$$

Thus, using (3.13), we obtain

Lemma 1. *For a non-exceptional Schur function s with the sequence of Schur parameters $\omega = \{\gamma_k\}_{0\leq k<\infty}$, the n-th Schur approximant $\mathrm{Ap}_n(s, z)$ can be expressed in terms of the entries of the matrix-function $M_{\omega,n}$:*

$$\mathrm{Ap}_n(s;\, z) = -\frac{b_{\omega,n}(z)}{a_{\omega,n}(z)}\,. \tag{3.14}$$

Of course, we can also derive the expression (3.14) for $\mathrm{Ap}_n(s;\, z)$ in the following way. Replacing $s_{n+1}(z)$ with zero in (3.11), we obtain the following equation for $\mathrm{Ap}_n(s;\, z)$:

$$\frac{a_{\omega,n}(z)\,\mathrm{Ap}_n(s;\, z)+b_{\omega,n}(z)}{c_{\omega,n}(z)\,\mathrm{Ap}_n(s;\, z)+d_{\omega n}(z)}=0\,.$$

Solving this equation for $\mathrm{Ap}_n(s;\, z)$, we obtain (3.14).

Let us derive the so-called *Christoffel-Darboux formula.* From (3.7) it follows that

$$\begin{aligned}&M^*_{\omega,k-1}(z)\,j\,M_{\omega,k-1}(z)-M^*_{\omega,k}(z)\,j\,M_{\omega,k}(z)\\&=M^*_{\omega,k}(z)\cdot\{m^*_{\gamma_k}(z)^{-1}j\;m_{\gamma_k}(z)^{-1}-j\}\cdot M_{\omega,k}(z).\end{aligned}\tag{3.15}$$

Using (3.6) we obtain,

$$\begin{aligned}&M^*_{\omega,k-1}(z)\,j\,M_{\omega,k-1}(z)-M^*_{\omega,k}(z)\,j\,M_{\omega,k}(z)\\&=(1-|z|^2)\,M^*_{\omega,k}(z)\begin{bmatrix}1\\0\end{bmatrix}\cdot\begin{bmatrix}1&0\end{bmatrix}M_{\omega,k}(z)\,.\end{aligned}\tag{3.16}$$

In view of (3.8), the last expression can be written in terms of matrix entries

$$\begin{aligned}&M^*_{\omega,k-1}(z)\,j\,M_{\omega,k-1}(z)-M^*_{\omega,k}(z)\,j\,M_{\omega,k}(z)\\&=(1-|z|^2)\begin{bmatrix}\overline{a_{\omega,k}(z)}\\\overline{b_{\omega,k}(z)}\end{bmatrix}\cdot\begin{bmatrix}a_{\omega,k}(z)&b_{\omega,k}(z)\end{bmatrix}\end{aligned}\tag{3.17}$$

Summing over k from $k=0$ to $k=n$, we obtain

$$\begin{aligned}&j-M^*_{\omega,n}(z)\,j\,M_{\omega,n}(z)=\\&=(1-|z|^2)\sum_{0\le k\le n}\begin{bmatrix}\overline{a_{\omega,k}(z)}\\\overline{b_{\omega,k}(z)}\end{bmatrix}\cdot\begin{bmatrix}a_{\omega,k}(z)&b_{\omega,k}(z)\end{bmatrix}\end{aligned}\tag{3.18}$$

Multiplying by $\begin{bmatrix}\overline{s(z)}&1\end{bmatrix}$ on the left and by $\begin{bmatrix}s(z)\\1\end{bmatrix}$ on the right, we obtain:

$$\begin{aligned}&(1-|\,s(z)|^2)-(1-|\,s_{n+1}(z)|^2)\cdot\,|c_{\omega,n}(z)\,s(z)+d_{\omega,n}(z)|^2\\&=(1-|z|^2)\sum_{0\le k\le n}|a_{\omega,k}(z)\,s(z)+b_{\omega,k}(z)\,|^2\end{aligned}\tag{3.19}$$

The formula (3.19) is called the Christoffel-Darboux formula.

Since $1-|s_{n+1}(z)|^2\ge 0$ the inequality

$$\sum_{0\le k\le n}|a_{\omega,k}(z)\,s(z)+b_{\omega,k}(z)\,|^2\le\frac{1-|s(z)|^2}{1-|z|^2}\tag{3.20}$$

holds. Letting n tend to infinity, we get

$$\sum_{0\le k<\infty}|a_{\omega,k}(z)\,s(z)+b_{\omega,k}(z)\,|^2\le\frac{1-|s(z)|^2}{1-|z|^2}.\tag{3.21}$$

Remark 6. *In particular,*

$$\sum_{0\leq k<\infty} |\, a_{\omega,\,k}(z)\, s(z) + b_{\omega,\,k}(z)\,|^2 < \infty \quad \text{for} \quad |z| < 1\,. \tag{3.22}$$

For $|\, z| < 1$, *the series* $\sum\limits_{0\leq k<\infty} |\, a_{\omega,\,k}(z)|^2$ *and* $\sum\limits_{0\leq k<\infty} |\, b_{\omega,\,k}(z)|^2$ *can diverge even exponentially. Despite this, the series (3.22) converges for* $|\, z| < 1$. *The linear combination* $\{a_{\omega,\,k}(z)s(z) + b_{\omega,\,k}(z)\}$ *is analogous to the so-called Weyl solution in the theory of the singular Sturm-Liouville differential equation.*

Now we would like to discuss the j-*properties* of the matrix-functions $M_{\omega,n}(\,.\,)$ and derive consequences of these j-properties. By j-properties, we mean such properties as the j-contractivity and the j-unitarity.

Definition 7. *Let* j *be a* $d \times d$ *matrix such that*

$$j = j^{\,*}, \quad j^{\,2} = I\,. \tag{3.23}$$

i. *A* $d \times d$ *matrix* M *is said to be* j-*contractive if the inequality*

$$j - M^{\,*} j M \geq 0 \tag{3.24}$$

or, equivalently, the inequality

$$j - M j M^{\,*} \geq 0 \tag{3.25}$$

holds. (It is possible to prove that (3.24) and (3.25) are equivalent.)

ii *A* $d \times d$ *matrix* M *is said to be* j-*unitary if the equality*

$$j - M^{\,*} j M = 0 \tag{3.26}$$

or, equivalently, the equality

$$j - M j M^{\,*} = 0 \tag{3.27}$$

holds. □

In what follows we consider only the 2×2 matrix j of the form (3.5) and discuss j-properties of 2×2 matrices with respect to this matrix j.

Lemma 2. *Given a contractive sequence* $\omega = \{\gamma_k\}$, *the value of the appropriate matrix-function* $M_{\omega,\,n}(\,.\,)$ *is* j-*contractive in the unit disk and is* j-*unitary on its boundary:*

$$j - M^{\,*}_{\omega,\,n}(z) j M_{\omega,\,n}(z) \geq 0, \ \textit{or equivalently,}\ j - M_{\omega,\,n}(z) j M^{\,*}_{\omega,\,n}(z) \geq 0 \tag{3.28}$$

for $|z| \leq 1$,

$$j - M^{\,*}_{\omega,\,n}(t) j M_{\omega,\,n}(t) = 0\,, \quad \textit{or equivalently,}\ j - M_{\omega,\,n}(t) j M^{\,*}_{\omega,\,n}(t) = 0 \tag{3.29}$$

for $|t| = 1$.

Proof. The "left" inequality (3.28) and the "left" equality (3.29) are consequences of the identity (3.18). The equivalency of the "left" and the "right" relations is a general fact of the linear j-algebra. □

Now we derive some inequalities for entries of j-contractive and j-unitary matrices, with an aim to apply these inequalities to the values of the matrix-function $M_{\omega,n}(\,.\,)$.

Lemma 3. *Let* $M = \begin{bmatrix} a & b \\ c & d \end{bmatrix}$ *be a* 2×2 *matrix, which is* j*-contractive with respect to* j *of the form* (3.5)*. Then the following inequalities hold:*

$$\text{i).}\ \ |a| \geq 1\,; \qquad \text{ii).}\ \ |b\cdot a^{-1}| < 1\,; \qquad \text{iii).}\ \ |c\cdot a^{-1}| < 1\,. \tag{3.30}$$

If, in addition, the matrix M *is* j*-unitary, then*

$$\text{i).}\ \ |a| = |d|\,; \qquad \text{ii).}\ \ |b| = |c|\,; \qquad \text{iii).}\ \ |ad-bc| = 1\,, \tag{3.31}$$

and, in addition to (3.30)*, the following inequalities hold:*

$$\text{i).}\ \ 1-|b\cdot a^{-1}| \leq |a|^{-1}\,; \qquad \text{ii).}\ \ 1-|c\cdot a^{-1}| \leq |a|^{-1}\,. \tag{3.32}$$

Proof. The inequalities $j - M^{*}jM \geq 0$ and $j - MjM^{*} \geq 0$, written in terms of matrix entries, mean

$$\begin{bmatrix} |a|^2-|c|^2-1 & \bar{a}\,b-\bar{c}\,d \\ \bar{b}\,a-\bar{d}\,c & |b|^2-|d|^2+1 \end{bmatrix} \geq 0 \tag{3.33}$$

and

$$\begin{bmatrix} |a|^2-|b|^2-1 & a\,\bar{c}-b\,\bar{d} \\ c\,\bar{a}-d\,\bar{b} & |c|^2-|d|^2+1 \end{bmatrix} \geq 0\,. \tag{3.34}$$

Since the diagonal entries of a non-negative matrix are non-negative, the inequalities

$$|a|^2-|b|^2-1 \geq 0 \qquad \text{and} \qquad |a|^2-|c|^2-1 \geq 0 \tag{3.35}$$

hold. From here (3.30) follows. If the matrix M is j-unitary, then equalities hold in (3.33) and (3.34). In particular,

$$\text{i).}\ \ |a|^2-|c|^2-1 = 0\,; \qquad \text{ii).}\ \ |d|^2-|b|^2-1 = 0\,; \tag{3.36}$$

and

$$\text{i).}\ \ |a|^2-|b|^2-1 = 0\,; \qquad \text{ii).}\ \ |d|^2-|c|^2-1 = 0\,. \tag{3.37}$$

From here the equalities (3.31.i) and (3.31.ii) follow. To derive the equality (3.31.iii) we observe that the equality $j - M^{*}jM = 0$ implies the equality $|\det M| = 1$, or $|a\,d-b\,c| = 1$. To derive the equalities (3.32.i) and (3.32.ii) we use the equalities (3.37.i) and (3.36.i) which can be written in the form

$$|a|-|b| = \frac{1}{|a|+|b|} \qquad \text{and} \qquad |a|-|c| = \frac{1}{|a|+|c|}\,.$$

Since $|a| \geq 1$, $|a|-|b| \leq 1$ and $|a|-|c| \leq 1$. The later inequalities are the inequalities (3.32.i) and (3.32.ii). □

Lemma 4. *Let* $M = \begin{bmatrix} a & b \\ c & d \end{bmatrix}$ *be a* 2×2 *matrix, which is* j*-unitary with respect to* j *of the form* (3.5). *Then the following norm estimate holds:*

$$\text{i).}\quad \left\| \begin{bmatrix} a & b \\ c & d \end{bmatrix} \right\| = |\,a| + \sqrt{|\,a|^2 - 1}\,, \quad \textit{and hence,} \quad \text{ii).} \quad \left\| \begin{bmatrix} a & b \\ c & d \end{bmatrix} \right\| < 2|\,a|\,. \tag{3.38}$$

Proof. It is clear that $\left\| \begin{bmatrix} a & b \\ c & d \end{bmatrix} \right\| \leq \left\| \begin{bmatrix} |a| & |b| \\ |c| & |d| \end{bmatrix} \right\|$. In view of (3.31.i), (3.31.ii) and (3.36.i), $\begin{bmatrix} |a| & |b| \\ |c| & |d| \end{bmatrix} = \begin{bmatrix} |\,a| & \sqrt{|\,a|^2 - 1} \\ \sqrt{|\,a|^2 - 1} & |\,a| \end{bmatrix}$. The later matrix is symmetric, and its norm is equal to its largest eigenvalue. This eigenvalue is equal to $|\,a| + \sqrt{|\,a|^2 - 1} < 2|\,a|$. □

Lemma 5. *Let* $\omega = (\gamma_0, \gamma_1, \dots, \gamma_k, \dots)$ *be a contractive sequence of complex numbers and let* $M_{\omega,n}(\,.\,)$ *be the matrix function, with the entries* (3.8), *which is constructed from* ω *as the product* (3.7).

Then

1. *The inequalities*

$$\text{i).}\ |\,a_{\omega,n}(z)| \geq 1\,; \qquad \text{ii).}\ \left| \frac{b_{\omega,n}(z)}{a_{\omega,n}(z)} \right| < 1\,; \qquad \text{iii).}\ \left| \frac{c_{\omega,n}(z)}{a_{\omega,n}(z)} \right| < 1 \tag{3.39}$$

hold in the closed unit disk.

2. *The inequalities*

$$\text{i).}\ 1 - \left| \frac{b_{\omega,n}(t)}{a_{\omega,n}(t)} \right| \leq \frac{1}{|\,a_{\omega,n}(t)|}\,; \qquad \text{ii).}\ 1 - \left| \frac{c_{\omega,n}(t)}{a_{\omega,n}(t)} \right| \leq \frac{1}{|\,a_{\omega,n}(t)|} \tag{3.40}$$

and the equality

$$\left| a_{\omega,n}(t)\, d_{\omega,n}(t) - b_{\omega,n}(t) c_{\omega,n}(t) \right| = 1 \tag{3.41}$$

hold on the boundary of the unit disk.

3. *The estimate for* $\|M_{\omega,t}(t)\|$:

$$\left\| \begin{bmatrix} a_{\omega,n}(t) & b_{\omega,n}(t) \\ c_{\omega,n}(t) & d_{\omega,n}(t) \end{bmatrix} \right\| < 2|\,a_{\omega,n}(t)| \tag{3.42}$$

holds on the boundary of the unit disk.

Proof. The lemma is an immediate consequence of Lemma 2, Lemma 3 and Lemma 4. □

4. Some deterministic considerations

Lemma 6. *Let* s *be a Schur function,* $\omega = \{\gamma_k\}_{0 \leq k < \infty}$ *be a sequence of its Schur parameters (so* $\gamma_k = \gamma_k(s)$, $k = 0, 1, 2, \dots$) *and let* $\{M_{\omega,n}(\,.\,)\}_{0 \leq n < \infty}$ *be the sequence of the matrix functions, with the entries (3.8), which is constructed from*

ω as the product (3.7) of the matrices $m_{\gamma_k}(t)$ where the matrix $m_\gamma(t)$ is defined by (3.1).

For this ω, let T_ω be the set of all those $t \in \mathbb{T}$ for which the following two conditions are satisfied:

i.
$$|a_{\omega,n}(t)| \to \infty \quad \text{as} \quad n \to \infty; \tag{4.1}$$

ii. *there exists a linear combination of $a_{\omega,n}(t)$ and $b_{\omega,n}(t)$, with an appropriate coefficient $\zeta_\omega(t)$, which vanishes asymptotically:*
$$a_{\omega,n}(t)\,\zeta_\omega(t) + b_{\omega,n}(t) \to 0 \quad \textit{as} \quad n \to \infty. \tag{4.2}$$

Assume that the set T_ω is a set of full Lebesgue measure:
$$m(T_\omega) = 1. \tag{4.3}$$

Then the function $\zeta_\omega(t)$ (which is defined on the set T_ω) coincides with the boundary value $s(t)$ of the function $s(t)$ for $m(dt)$-almost every $t \in T_\omega$. Moreover,

i. *The function s is inner:*
$$|s(t)| = 1 \ \text{ for } \ m(dt) \textit{ almost every } \ t \in \mathbb{T}. \tag{4.4}$$

ii. *The sequence $\mathrm{Ap}_n(s;\,.)$ of the Schur approximants of the function s converges to its boundary value s pointwise almost everywhere on $\mathbb{T}$:*
$$\mathrm{Ap}_n(s;\,t) \to s(t) \ \text{ as } \ n \to \infty \ \text{ for } \ m(dt) \textit{ almost every } \ t \in \mathbb{T}. \tag{4.5}$$

Lemma 7. *The set T_ω is a Borel subset of $\mathbb{T}$.*

Proof. Note that the functions $a_{\omega,n}(t)$ and $b_{\omega,n}(t)$ are continuous on $\mathbb{T}$. Therefore, the set $E_1 = \{t \in \mathbb{T} : |a_{\omega,n}(t)| \to \infty\}$ is a Borel set. The set $E_3 = \{t \in \mathbb{T} : \exists \lim_{n\to\infty} b_{\omega,n}(t)/a_{\omega,n}(t) \stackrel{\text{def}}{=} -\tilde{\zeta}_\omega(t)\}$ is a Borel set and the function $\tilde{\zeta}_\omega(t)$ is defined and Borel measurable on E_3. Finally, the set $E_2 = \{t \in E_3 : \lim_{n\to\infty} (a_{\omega,n}(t)\,\tilde{\zeta}_\omega(t) + b_{\omega,n}(t)) = 0\}$ is a Borel set as well. It is clear that $\tilde{\zeta}_\omega(t) = \zeta_\omega(t)$ for $t \in E_2$ and that $T_\omega = E_1 \cap E_2$. □

Proof of Lemma 6. Since according to $|a_{n,\omega}(t)| \geq 1$ on $\mathbb{T}$, it follows from (4.2) that
$$\zeta_\omega(t) = \lim_{n\to\infty} -\frac{b_{\omega,n}(t)}{a_{\omega,n}(t)} \quad \text{for} \quad t \in T_\omega\,.$$
From (4.1) and from (3.39.ii), (3.40.i) it follows that
$$\left|\frac{b_{\omega,n}(t)}{a_{\omega,n}(t)}\right| \to 1 \ \text{ as } \ n \to \infty \ \text{ for } \ t \in T_\omega\,.$$
Thus,
$$|\zeta_\omega(t)| = 1 \quad \text{for} \quad t \in T_\omega\,. \tag{4.6}$$
We will now prove that $\zeta_\omega(t)$ coincides with $s(t)$ for $m(dt)$ - a. e. $t \in \mathbb{T}$. First of all we remark that the function $\zeta_\omega(t)$ can be "continued analytically" into the unit

disk. By (3.39.ii) $\mathrm{Ap}_n(s;\, z) = -\dfrac{b_{\,\omega,n}(z)}{a_{\,\omega,n}(z)}$ is a contractive rational function in the unit disk. Hence, for each n

$$\int\limits_{T} \mathrm{Ap}_n(s;\, t)\cdot t^k \cdot m(dt) = 0, \qquad k = 1,\, 2,\, 3,\, \dots$$

By (3.14),

$$\mathrm{Ap}_n(s;\, t) \to \zeta_\omega(t) \quad \text{for} \quad t \in T_\omega\,. \tag{4.7}$$

Letting n tend to infinity and applying the Lebesgue dominated convergence theorem we conclude that

$$\int\limits_{\mathbb{T}} \zeta_\omega(t)\cdot t^k \cdot m(dt) = 0, \qquad k = 1,\, 2,\, 3,\, \dots \quad .$$

From the last condition it follows that the function

$$\zeta_\omega(z) \stackrel{\text{def}}{=} \int\limits_{\mathbb{T}} \zeta_\omega(t)\, \frac{1}{1-\bar{t}z}\, m(dt) \qquad (|z|<1),$$

defined by means of the Cauchy integral, is also representable by the Poisson integral:

$$\zeta_\omega(z) = \int\limits_{\mathbb{T}} \zeta_\omega(t)\, \frac{1-|z|^2}{|t-z|^2}\, m(dt) \qquad (|z|<1)\,.$$

Indeed, for $t \in \mathbb{T}$ and $z \in \mathbb{D}$,

$$\frac{1-|z|^2}{|t-z|^2} = \frac{1}{1-\bar{t}z} + \frac{t\bar{z}}{1-t\bar{z}} = \frac{1}{1-\bar{t}z} + \sum_{1\le k<\infty} t^k \bar{z}^k.$$

From the Cauchy integral representation it follows that the function $\zeta_\omega(z)$ is holomorphic in the open unit disk $\mathbb{D}$ and from the Poisson integral representation it follows that the boundary values $\lim\limits_{\rho\to 1-0} \zeta_\omega(\rho t)$ of the function $\zeta_\omega(z)$ coincide with the original function $\zeta_\omega(t)$ for m-a.e. $t \in \mathbb{T}$ and that $|\zeta_\omega(z)| \le 1 \quad (z \in \mathbb{D})$. It is clear that

$$\zeta_\omega(z) - \mathrm{Ap}_n(s;\, z) = \int\limits_{\mathbb{T}} (\zeta_\omega(t) - \mathrm{Ap}_n(s;\, t))\, \frac{1-|z|^2}{|t-z|^2}\, m(dt) \qquad (|z|<1)\,.$$

Since $|\,\zeta_\omega(t) - \mathrm{Ap}_n(s;\, t)| \le 2$ on $\mathbb{T}$ and $\zeta_\omega(t) - \mathrm{Ap}_n(s;\, t) \to 0$ for m-a.e. $t \in \mathbb{T}$, by the Lebesgue dominated convergence theorem

$$\mathrm{Ap}_n(s;\, z) \to \zeta_\omega(z) \quad \text{locally uniformly in } \mathbb{D}\,. \tag{4.8}$$

According to (1.21),

$$\mathrm{Ap}_n(s;\, z) \to s(z) \quad \text{locally uniformly in } \mathbb{D}\,. \tag{4.9}$$

From (4.9) and (4.8) it follows that $\zeta_\omega(z) = s(z)$ for all $z \in \mathbb{D}$. Hence, $\zeta_\omega(t) = s(t)$ for m-a.e. $t \in \mathbb{T}$. This completes the proof of Lemma 6. □

Lemma 8. *Let the sequence* $\{\gamma_n\}_{0\le n<\infty}$ *of the Schur parameters of a non-exceptional Schur function* s *satisfy the condition*

$$\overline{\lim_{n\to\infty}} |\gamma_n| = 1\,. \tag{4.10}$$

Then s *is an inner function.*

Proof. Let $\{s_n(\,.\,)\}_{1\le n<\infty}$ be the sequence of Schur functions generated by the Schur algorithm applied to the function $s(\,.\,)$: (1.6) and (1.7). From the Parseval identity for the function s_{n+1} we derive that $|\,s_{n+1}(0)|^2 \le \int_{\mathbb{T}} |\,s_{n+1}(t)|^2\, m(dt)$ which implies

$$\int_{\mathbb{T}} (1-|s_{n+1}(t)|^2)\, m(dt) \le 1-|\gamma_{n+1}|^2. \tag{4.11}$$

Let $\{M_{\omega,n}(\,.\,)\}_{0\le n<\infty}$ be the sequence of the matrix functions, with the entries (3.8), which is constructed from the sequence $\omega = (\gamma_0,\, \gamma_1,\, \dots,\, \gamma_k,\, \dots)$ of Schur parameters of the function s as the product (3.7) of matrices $m_{\gamma_k}(\,.\,)$ where the matrix $m_\gamma(\,.\,)$ is defined by (3.1). Using the linear fractional transformation (3.11), the function s_{n+1} can be expressed in the matrix form (3.10). Since the matrix function $M_{\omega,n+1}$ takes j-unitary values on $\mathbb{T}$ (Lemma 2: identities (3.29)), it follows from (3.10) that

$$1-|\,s(t)|^2 = (1-|\,s_{n+1}(t)|^2)\,\left|c_{\omega,n}(t)\,s(t)+d_{\omega,n}(t)\right|^2 \quad \text{for} \quad t\in\mathbb{T}. \tag{4.12}$$

From (3.11) we obtain:

$$a_{\omega,n}(t) - c_{\omega,n}(t)\,s_{n+1}(t) = \frac{a_{\omega,n}(t)\,d_{\omega,n}(t) - b_{\omega,n}(t)\,c_{\omega,n}(t)}{c_{\omega,n}(t)\,s(t)+d_{\omega,n}(t)}\,.$$

The determinant relation (3.41) implies the identity

$$|\,a_{\omega,n}(t) - c_{\omega,n}(t)\,s_{n+1}(t)|^2 = \frac{1}{|c_{\omega,n}(t)\,s(t)+d_{\omega,n}(t)|^2}\,. \tag{4.13}$$

From (4.12) and (4.13) it follows that

$$(1-|s(t)|^2)\cdot|\,a_{\omega,n}(t) - c_{\omega,n}(t)\,s_{n+1}(t)|^2 = 1-|\,s_{n+1}(t)|^2 \quad \text{for} \quad t\in\mathbb{T}\,. \tag{4.14}$$

Since $|a_{\omega,n}(t)|\ge 1$ (Lemma 5: (3.39.i)), we obtain

$$(1-|s(t)|^2)\cdot\left|1-\frac{c_{\omega,n}(t)}{a_{\omega,n}(t)}\,s_{n+1}(t)\right|^2 \le 1-|\,s_{n+1}(t)|^2 \quad \text{for} \quad t\in\mathbb{T}\,. \tag{4.15}$$

Integrating over $\mathbb{T}$ with respect to the Lebesgue measure and using (4.11), we get

$$\int_{\mathbb{T}} (1-|s(t)|^2)\left|1-\frac{c_{\omega,n}(t)}{a_{\omega,n}(t)}\,s_{n+1}(t)\right|^2 m(dt) \le 1-|\,\gamma_{n+1}|^2. \tag{4.16}$$

If the function s is not inner, then there exists a set E of positive Lebesgue measure and a positive constant ϵ such that

$$1-|\,s(t)|^2 \ge \epsilon \quad \text{for} \quad t\in E\,. \tag{4.17}$$

From (4.16) and (4.17) we derive the crucial estimate

$$\int_E \left|1-\frac{c_{\omega,n}(t)}{a_{\omega,n}(t)}s_{n+1}(t)\right|^2 m(dt) \leq \epsilon^{-1}(1-|\gamma_{n+1}|^2) \qquad (n=0,1,2,\dots). \tag{4.18}$$

According to Lemma 5: (3.39.iii), the quotient $\frac{c_{\omega,n}}{a_{\omega,n}}$, which in particular appears in (4.18), represents a contractive holomorphic function in $\mathbb{D}$. It is important to know the value of this function at the origin. However, it is not possible to calculate this value by substituting $z=0$ into $c_{\omega,n}(z)$ and $a_{\omega,n}(z)$: the function $c_{\omega,n}$ and $a_{\omega,n}$ themselves are not holomorphic at the origin. They have a pole there, only their quotient is holomorphic.

To "resolve" the singularity at $z=0$, we consider $z^n a_{\omega,n}(z)$ and $z^n c_{\omega,n}(z)$ which have the same quotient but are polynomials. By (3.7) and (3.1) the matrix $z^n M_{\omega,n}(z)$ has the representation

$$\begin{bmatrix} z^n a_{\omega,n}(z) & z^n b_{\omega,n}(z) \\ z^n c_{\omega,n}(z) & z^n d_{\omega,n}(z) \end{bmatrix} = \overset{\curvearrowleft}{\prod_{0\leq k\leq n}} \begin{bmatrix} 1 & -\gamma_k \\ -\overline{\gamma_k}\, z & z \end{bmatrix} \cdot \frac{1}{\sqrt{1-|\gamma_k|^2}}. \tag{4.19}$$

For $z=0$,

$$\begin{bmatrix} z^n a_{\omega,n}(z)_{|z=0} & z^n b_{\omega,n}(z)_{|z=0} \\ z^n c_{\omega,n}(z)_{|z=0} & z^n d_{\omega,n}(z)_{|z=0} \end{bmatrix}$$

$$= \overset{\curvearrowleft}{\prod_{0\leq k\leq n}} \begin{bmatrix} 1 & -\gamma_k \\ 0 & 0 \end{bmatrix} \cdot \frac{1}{\sqrt{1-|\gamma_k|^2}} = \begin{bmatrix} 1 & -\gamma_0 \\ 0 & 0 \end{bmatrix} \cdot \overset{\curvearrowleft}{\prod_{0\leq k\leq n}} \frac{1}{\sqrt{1-|\gamma_k|^2}} \tag{4.20}$$

In particular, $\left|z^n a_{\omega,n}(z)_{|z=0}\right| > 1$, $\quad z^n c_{\omega,n}(z)_{|z=0} = 0$, and

$$\left.\frac{c_{\omega,n}(z)}{a_{\omega,n}(z)}\right|_{z=0} = 0. \tag{4.21}$$

Let us introduce the notation

$$f_n(z) \overset{\text{def}}{=} 1-\frac{c_{\omega,n}(z)}{a_{\omega,n}(z)}s_{n+1}(z). \tag{4.22}$$

Since both s_{n+1} and $c_{\omega,n+1}/a_{\omega,n+1}$ are holomorphic and contractive in $\mathbb{D}$, the sequence $\{f_n\}$ is uniformly bounded in $\mathbb{D}$:

$$|f_n(z)| \leq 2 \quad \text{for} \quad z\in\mathbb{D}, \qquad (n=0,1,2,\dots). \tag{4.23}$$

By (4.21),

$$f_n(0)=1, \qquad (n=0,1,2,\dots). \tag{4.24}$$

The inequality (4.18) can now be written as

$$\int_E |f_n(t)|^2 m(dt) \leq \epsilon^{-1}(1-|\gamma_n|^2), \qquad (n=0,1,2,\dots). \tag{4.25}$$

Here E is a set of positive Lebesgue measure and ϵ is a positive number which do not depend on n.

However, the totality of the conditions (4.23), (4.24), (4.25) for a holomorphic function f_n is incompatible if $1-|\gamma_n|^2$ is small enough and E and ϵ are fixed. Indeed, this contradict to the *Jensen inequality.* The Jensen inequality for a bounded holomorphic function Φ in the unit disk is given by

$$|\Phi(0)| \leq \exp\Big\{ \int_{\mathbb{T}} \ln|\Phi(t)|\, m(dt) \Big\},$$

where $\Phi(t)$ is the boundary values of the function Φ. Applying (4.24) and the Jensen inequality to the function f_n, we obtain

$$\begin{aligned} 1 &\leq \exp\Big\{ \int_{\mathbb{T}} \ln|f_n(t)|\, m(dt) \Big\} \\ &= \exp\Big\{ \int_{\mathbb{T}\setminus E} \ln|f_n(t)|\, m(dt) \Big\} \cdot \exp\Big\{ \int_{E} \ln|f_n(t)|\, m(dt) \Big\}, \end{aligned} \tag{4.26}$$

where E is the set on which the inequality (4.17) holds. By (4.23), $|f_n(t)| \leq 2$ on $\mathbb{T}$, thus

$$\exp\Big\{ \int_{\mathbb{T}\setminus E} \ln|f_n(t)|\, m(dt) \Big\} \leq 2. \tag{4.27}$$

Using inequality between arithmetic and geometric mean (for the probability measure $m(dt)/m(E)$ on E) we obtain

$$\begin{aligned} \exp\Big\{ \int_E \ln|f_n(t)|\, m(dt) \Big\} &= \Big(\exp\Big\{ \int_E \ln|f_n(t)|^2 \frac{m(dt)}{m(E)} \Big\} \Big)^{\frac{m(E)}{2}} \\ &\leq \Big(\int_E |f_n(t)|^2 \frac{m(dt)}{m(E)} \Big)^{\frac{m(E)}{2}} = \Big(\frac{1}{m(E)} \Big)^{\frac{m(E)}{2}} \cdot \Big(\int_E |f_n(t)|^2\, m(dt) \Big)^{\frac{m(E)}{2}} \end{aligned} \tag{4.28}$$

From (4.26), (4.27) and (4.28) we obtain the inequality:

$$m(E) \cdot 2^{-\frac{2}{m(E)}} \leq \int_E |f_n(t)|^2\, m(dt). \tag{4.29}$$

Finally, the inequality (4.25) implies

$$\epsilon \cdot m(E) \cdot 2^{-\frac{2}{m(E)}} \leq 1 - |\gamma_n|^2, \qquad (n = 0, 1, 2, \dots), \tag{4.30}$$

where E and ϵ are same as in (4.17) and γ_n is the n-th Schur parameter of the function s. Thus, if the function s is not inner, its Schur parameters $\gamma_n(s)$ are separated from 1: $(\sup_k |\gamma_k(s)| < 1)$, which contradicts (4.10). □

In the framework of polynomials orthogonal on the unit circle a result similar to our Lemma 8 was obtained by E.A. Rakhmanov (see [Rakh], Lemma 4 there).

This result by E.A. Rakhmanov was adopted for Schur functions by L. Golinskii (see [Gol2], Theorem 2). Our proof of Lemma 8 is different from the proof by E.A. Rakhmanov. Another proof of the lemma was given by S.V. Khrushchev in [Khru1] (see Theorem 9.3 and Corollary 9.5).

To apply Lemma 8, we need the following

Lemma 9. *Let μ be a probability measure on the open unit disk $\mathbb{D}$, and let Ω be the product space (2.1), i.e. the set of all contractive sequences ω of complex numbers. Let p_μ be the product measure on Ω generated by μ. If the support of the measure μ is* not *separated from $\mathbb{T}$, i.e. if the condition*

$$\sup_{\gamma \in \operatorname{supp} \mu} |\gamma| = 1 \tag{4.31}$$

is satisfied, then for p_μ-almost every sequence $\omega = (\gamma_0, \gamma_1, \gamma_2, \ldots, \gamma_k, \ldots)$ the condition

$$\overline{\lim_{k \to \infty}} |\gamma_k| = 1 \tag{4.32}$$

holds.

Proof. Proving the Lemma is equivalent to showing that (4.31) implies

$$p_\mu(L_1) = 0\,, \quad \text{where} \;\; L_1 = \{\omega : \overline{\lim_{k\to\infty}} |\gamma_k| < 1\}\,.$$

Since

$$L_1 = \bigcup_{1 \leq n < \infty} L_{1-1/n}\,, \text{where} \;\; L_{1-1/n} = \{\omega : \overline{\lim_{k\to\infty}} |\gamma_k| < 1 - 1/n\}\,,$$

it is enough to prove that $p_\mu(L_{1-1/n}) = 0$. However $L_{1-1/n} \subseteq \bigcup_{1 \leq m < \infty} C_{1-1/n,\, m}$,

where

$$C_{1-1/n,\, m} = \{\omega : |\gamma_k < 1 \text{ for } k = 0, 1, \cdots, m-1 \\ \text{and } |\gamma_k| < 1 - 1/n \;\; \text{for} \;\; k \geq m\}\,.$$

Since the set $C_{1-1/n,\, m}$ is a cylindric set (direct product),

$$p_\mu(C_{1-1/n\,,m}) = \left(\prod_{0 \leq k \leq m-1} \mu(\mathbb{D}) \right) \cdot \left(\prod_{m \leq k < \infty} \mu((1 - 1/n)\, \mathbb{D}) \right),$$

where

$$(1 - 1/n) \cdot \mathbb{D} = \{\gamma \in \mathbb{C} : |\gamma| < 1 - 1/n\}\,.$$

In view of (4.31),

$$\mu\,((1 - 1/n)\, \mathbb{D}) < 1 \;\; \text{for every} \;\; n\,.$$

Thus,

$$p_\mu(C_{1-1/n,\, m}) = 0 \;\; \text{for every} \;\; m, n\,, \;\; p_\mu\,(L_{1-1/n}) = 0 \;\; \text{for every} \;\; n$$

and hence $p_\mu\,(L_1) = 0$. □

5. Furstenberg–Kesten, Furstenberg and Oseledets ergodic theorems

To prove that some Schur function s is inner we use Lemma 6. For an individual function s, verification of conditions (4.1) and (4.2) is extremely difficult. However the situation is changed by a probabilistic consideration. It turns out that for vast majority of sequences $\omega = \{\gamma_0, \gamma_1, \gamma_2, \dots, \gamma_k, \dots\}$ the family (3.7) of matrices $m_{\gamma_n}(t) \cdot m_{\gamma_{n-1}}(t) \cdot \dots \cdot m_{\gamma_1}(t) \cdot m_{\gamma_0}(t)$ behaves appropriately. The probabilistic basis for our consideration are results on noncommutative random products which are known as Furstenberg-Kesten theorem, Furstenberg theorem and Oseledets theorem. We will formulate these theorems in generality which is sufficient for our consideration. These theorems are related to arbitrary $d \times d$ matrices. However for simplicity we formulate these theorems for 2×2 matrices only.

Definition 8. $\mathsf{PSL}(2, \mathbb{C})$ *is the set of all* 2×2 *matrices* g *with complex entries which satisfy the condition* $|\det g| = 1$.

$\mathsf{PSL}(2, \mathbb{C})$ is a noncompact Lie group.

The above mentioned theorems deal with products of independent identically distributed (i.i.d.) matrices. The common distribution of these matrices is described by a probability measure μ on the σ-algebra of Borel sets of $\mathsf{PSL}(2, \mathbb{C})$. The condition

$$\int\limits_{\mathsf{PSL}(2,\mathbb{C})} (\ln \|g\|)\, \mu(dg) < \infty \tag{5.1}$$

is usually imposed on the measure μ. (Since $|\det g| = 1$, $\|g\| \geq 1$ for $g \in \mathsf{PSL}(2, \mathbb{C})$. Thus, $\ln \|g\| \geq 0$ in (5.1).) For every sequence

$$\omega_g = \{g_0, g_1, g_2, \dots, g_k, \dots\} \tag{5.2}$$

we consider the sequence of products

$$M_{\omega_g, n} = g_n \cdot g_{n-1} \cdot g_{n-2} \cdot \dots \cdot g_1 \cdot g_0\,, \quad n = 0, 1, 2, \dots . \tag{5.3}$$

We consider the sequence $\{M_{\omega_g, n}\}_{0 \leq n < \infty}$ as a sequence of matrix valued random functions. The appropriate sample space, which we denote by Ω_G, is the space of all sequences ω of the form (5.2). On other words, Ω_G is the countable product

$$\Omega_G = \mathsf{PSL}(2, \mathbb{C}) \times \mathsf{PSL}(2, \mathbb{C}) \times \cdots \times \mathsf{PSL}(2, \mathbb{C}) \times \cdots . \tag{5.4}$$

Being the product of the topological spaces $\mathsf{PSL}(2, \mathbb{C})$, Ω_G itself is a topological space. The σ-algebra of events Σ is the algebra of all Borel sets in Ω_G. The probability measure p_μ is the product measure

$$p_\mu = \mu \otimes \mu \otimes \cdots \otimes \mu \otimes \cdots . \tag{5.5}$$

Of course, the measure p_μ is generated from the original probability measure μ on $\mathsf{PSL}(2, \mathbb{C})$.

Thus we have constructed the probability space $(\Omega_G, \Sigma, p_\mu)$. This space is used for the probabilistic study of the products $M_{\omega, n}$ of independent identically distributed random matrices (with the common distribution described by the measure μ.)

The following theorem can be considered as a matrix generalization of the strong law of large numbers:

Theorem (Furstenberg-Kesten). *Let μ be a probability measure on* $\mathsf{PSL}(2, \mathbf{C})$ *which satisfies the condition* (5.1). *Let* $(\Omega_G, \Sigma, p_\mu)$ *be the above probability space. Let* $M_{\omega_g, n}$ *be the sequence of successive products for a sequence* ω_g *of matrices, as it is described in* (5.2)–(5.3).

Then there exist an "exceptional" set R, $R \subseteq \Omega_G$, of zero p_μ-measure: $p_\mu(R) = 0$, and a number $\lambda \geq 0$ such that for every sequence $\omega_g \in \Omega_G \setminus R$, the limit

$$\lim_{n\to\infty} \frac{\ln \|M_{\omega_g, n}\|}{n+1} = \lambda . \tag{5.6}$$

exists. (This limit does not depend on $\omega_g \in \Omega_G \setminus R$.) □

This value λ of the limit (5.6) (which is common for almost every $\omega_g \in \Omega$) is said to be *the upper Lyapunov exponent* of the probability measure μ. There is an expression for the upper Lyapunov exponent:

$$\begin{aligned}\lambda = \lim_{n\to\infty} \frac{1}{n} \cdot \int\limits_{\Omega} \ln \|M_{\omega, n-1}\| \, dp_\mu(\omega) \\ = \lim_{n\to\infty} \frac{1}{n} \cdot \iint \cdots \int \left(\ln \|g_{n-1} \cdot g_{n-2} \cdot \cdots \cdot g_1 \cdot g_0\|\right) \\ d\mu(g_0)\, d\mu(g_1) \cdots d\mu(g_{n-1}) .\end{aligned} \tag{5.7}$$

The limit in (5.7) exists because the sequence $\int\limits_{\Omega} \ln \|M_{\omega, n-1}\, dp_\mu(\omega)\|$ is subadditive with respect to n. However, the expression (5.7) is practically useless for our purpose. We need conditions which ensure that the upper Lyapunov exponent is strictly positive: $\lambda > 0$. (Since $\ln \|M_{\omega_g, n}\| \geq 0$, $\lambda \geq 0$ always.) The following remarkable theorem of Furstenberg gives conditions which guarantee that the Lyapunov exponent λ is strictly positive.

To formulate Furstenberg's theorem we need to introduce some notion. Every non-degenerate 2×2 matrix $g = \begin{bmatrix} a & b \\ c & d \end{bmatrix}$ acts on the projective space $\mathbb{CP}^1$ (which can be identified with the Riemann sphere $\overline{\mathbb{C}} = \mathbb{C} \cup \{\infty\}$). If $\xi : \eta$ is a point of the space $\mathbb{CP}^1$, then the point $g(\xi : \eta)$ is: $g(\xi : \eta) = (a\xi + b\eta) : (c\xi + d\eta)$. If we identify the point $\xi : \eta \in \mathbb{CP}^1$ with the point $\zeta = \xi/\eta \in \overline{\mathbb{C}}$, then the point $g(\zeta)$ is: $g(\zeta) = \dfrac{a\zeta + b}{c\zeta + d}$. As g acts on the space $\mathbb{CP}^1 = \overline{\mathbb{C}}$, g acts also on measures on this space: if ν is a measure on $\overline{\mathbb{C}}$, then $g(\nu)(E) \stackrel{\text{def}}{=} \nu(g^{-1}(E))$ for the set E. (Of course, we assume that the domain of definition of the measure ν is invariant under the action of g.)

Definition 9. *The measure ν on $\overline{\mathbb{C}}$ is said to be* invariant for a (non-degenerated) matrix g *if* $g(\nu)=\nu$. *The measure ν on $\overline{\mathbb{C}}$ is said to be* invariant for a family G of non-degenerated matrices *if ν is invariant for every matrix g from this family.* □

Definition 10. *For a Borelian measure μ, $G(\operatorname{supp}\mu)$ is the subgroup of* $\mathsf{PSL}(2,\mathbb{C})$ *generated by matrices g from the set $\operatorname{supp}\mu$. In other words, $G(\operatorname{supp}\mu)$ is the set of all the matrices of the form $g_1^{\varepsilon_1}\cdot g_2^{\varepsilon_2}\cdot\cdots g_l^{\varepsilon_m}$, where $g_1, g_2, \dots, g_l \in \operatorname{supp}\mu$, l is an arbitrary natural number and $\varepsilon_k=\pm 1$.* □

Theorem (Furstenberg). *Let μ be a Borel probability measure on* $\mathsf{PSL}(2,\mathbb{C})$, *which satisfies* (5.1) *and the following additional condition: there is no Borel probability measure ν on $\overline{\mathbb{C}}$ which is invariant with respect to all the matrices from the subgroup $G(\operatorname{supp}\mu)$.*

Then the upper Lyapunov exponent λ of the measure μ is strictly positive: $(\lambda>0)$. □

The following result can be considered as a generalization of Furstenberg-Kesten theorem.

Multiplicative ergodic theorem (Oseledets). *Let μ be a Borel probability measure on* $\mathsf{PSL}(2,\mathbb{C})$ *which satisfies the condition* (5.1). *Let $(\Omega_G, \Sigma, p_\mu)$ be the above probability space. Let $M_{\omega_g,n}$ be the sequence of successive products (for a sequence ω_g of matrices) as it is described in* (5.2)–(5.3).

Then there exists an "exceptional" set R, $R\subseteq\Omega_G$, of zero p_μ-measure such that for every $\omega_g\in\Omega_G\setminus R$ the following holds:

i. *The limit*
$$\lim_{n\to\infty}\left\{M^*_{\omega_g,n}M_{\omega_g,n}\right\}^{1/2n}\overset{\text{def}}{=}\Psi(\omega_g), \tag{5.8}$$
exists. Here $\Psi(\omega_g)$ is a 2×2 Hermitian matrix with eigenvalues e^{λ} and $e^{-\lambda}$, where λ is the Lyapunov exponent of the measure μ. (So, the eigenvalues of the matrix Ψ do not depend on $\omega_g\in\Omega_G\backslash R$, but the eigenspaces may depend).

ii. *If $\lambda>0$, which is the case of different eigenvalues of the matrix $\Psi(\omega_g)$, then*
$$\lim_{n\to\infty}\frac{\ln\|M_{\omega_g,n}\,y\|}{n}=-\lambda \tag{5.9}$$
for every non-zero y in the eigenspace $V_{-\lambda,\omega_g}$ of the matrix $\Psi(\omega_g)$ corresponding to the eigenvalue $e^{-\lambda}$ and
$$\lim_{n\to\infty}\frac{\ln\|M_{\omega_g,n}\,x\|}{n}=\lambda \tag{5.10}$$
for every non-zero x which does not belong to this eigenspace (i.e. for $x\in\mathbb{C}^2\setminus V_{-\lambda,\omega_g}$). □

The multiplicative ergodic theorem of Oseledets is informative in the case of different eigenvalues of $\Psi(\omega)$, that is in the case of strictly positive upper Lyapunov exponent of the measure μ. Because of this, Oseledets' theorem is usually used together with Furstenberg's theorem.

The Furstenberg-Kesten theorem first appeared in [FuKe]. The original proof is based on Birkgoff ergodic theorem. A good presentation (also based on Birkgoff ergodic theorem) is contained in [BoLa]. Another proof (which appeared as a simple corollary the so-called subadditive ergodic theory) is contained in [King] (Theorem 6). The Furstenberg theorem first appeared in [Fur]. This is Theorem 8.6 in [Fur] [1]. A presentation of the Furstenberg theorem is also contained in [BoLa]. [BoLa] contains not only a presentation for the general case (of $d \times d$ matrices) but also a simplified presentation for 2×2 matrices. (For 2×2 matrices, the Furstenberg theorem appears as Theorem 4.4 of the part A of [BoLa].) The multiplicative ergodic theorem first appeared in [Os]. This theorem was motivated by smooth ergodic theory. It was first applied to smooth dynamical system. Later on other important applications of this fundamental theorem had been found. One of these applications there is the application to random Schrödinger operator which is conceptually close to our consideration. The original proof in [Os] was given for matrices with real entries but in fact this proof is valid for matrices with complex entries as well. A new proof under very general assumptions (for matrices whose entries belong to a local field, archimedian or non-archimedian) was given by M.S. Raghunathan in [Ragh]. The book [Arn] contains proofs of the Furstenberg-Kesten and Oseledets theorem as well as a lot of other results about products of random matrices and a rich bibliography. The paper by I.Ya. Goldsheid and G.A. Margulis [GoMa] contains a good overview of the above mentioned classical results about products of random matrices. It also contains a new criterion for simplicity of the Lyapunov spectrum (the set of Lyapunov exponents) of the product of $d \times d$ random matrices in terms of algebraic geometry. (Namely, in terms of the Zariski closure of the support of the measure μ.) The books [CarLa] by R. Carmona and J. Lacroix and [PaFig] by L. Pastur and A. Figotin contain a lot of applications of results on products of random matrices to random differential and difference operators. Part B of the book [BoLa] is also dedicated to such applications.

It should be mentioned that we intentionally formulated the results for sequences of 2×2 matrices. We also assumed that the matrices in the sequences are independent. However, the classical papers [FuKe], [Fur] and [Os], as well as many other sources, deal with stationary (but not necessary independent) sequences of $d \times d$ matrices. In some sources, results of this kind are obtained even under weaker assumptions on dependence than the stationarity.

[1] In the formulation of Theorem 8.6, which is stated in [Fur], it is not assumed that the subgroup $G(\operatorname{supp}\mu)$ cannot leave a measure fixed, but some other assumptions are made. However, from these assumptions, it follows immediately that there is no Borel probability measure ν on $\overline{\mathbb{C}}$ which is invariant with respect to all the matrices from the subgroup $G(\operatorname{supp}\mu)$. The main part of the proof of Theorem 8.6 is dedicated to prove that if the Lyapunov exponent $\lambda(\mu)$ of the measure μ is equal to zero then there exists a probability measure ν on $\overline{\mathbb{C}}$ which is invariant for all the matrices from the subgroup $G(\operatorname{supp}\mu)$.

6. Checking the conditions for the positivity of the upper Lyapunov exponent

To prove our main results, *Theorems I* and *II*, we use Lemma 6 (Section 4). To verify the assumptions (4.1) and (4.2) of Lemma 6 (for almost all sequences $\{m_{\gamma_k}(t)\}$), we apply Oseledets' theorem (together with Furstenberg' theorem) to these random sequences. Let t be an arbitrary but fixed point of the unit circle. We introduce a mapping Γ_t which takes a point $\gamma \in \mathbb{D}$ to the matrix $m_\gamma(t)$:

$$\Gamma_t(\gamma) = m_\gamma(t)\,, \quad \text{where} \quad m_\gamma(t) = \begin{bmatrix} t^{-1} & -\gamma \cdot t^{-1} \\ -\overline{\gamma} & 1 \end{bmatrix} \cdot \frac{1}{\sqrt{1-|\gamma|^2}} \cdot \tag{6.1}$$

The mapping $\Gamma_t : \mathbb{D} \to \mathsf{PSL}\,(2, \mathbb{C})$ is injective and continuous. We define a new measure $\mu_t\,(= \mu \cdot \Gamma_t^{-1})$ by

$$\mu_t\,(\mathcal{G}) = \mu(\Gamma_t^{-1}(\mathcal{G}) \text{ for } \mathcal{G} \subseteq \mathsf{PSL}\,(2, \mathbb{C})\,. \tag{6.2}$$

μ_t is a Borel measure on $\mathsf{PSL}\,(2, \mathbb{C})$. The image $\Gamma_t(\mathbb{D})$ of the open unit disk $\mathbb{D}$ is the closed (but non-compact) subset of $\mathsf{PSL}\,(2, \mathbb{C})$. It is clear that

$$\operatorname{supp} \mu_t \subseteq \Gamma_t\,(\mathbb{D})\,. \tag{6.3}$$

In view of (6.1) and (6.2),

$$\int\limits_{\mathsf{PSL}\,(2,\,\mathbb{C})} (\ln \|\, g \|\, \mu_t)(dg) = \int\limits_{\mathbb{D}} (\ln \|m_\gamma(t)\|)\, \mu(d\gamma)\,. \tag{6.4}$$

Since $|t| = 1$ and $|\gamma| < 1$, the values $\|m_\gamma(t)\|$ and $\dfrac{1}{\sqrt{1-|\gamma|^2}}$ are comparable and we have the following lemma.

Lemma 10. *The conditions*

$$\int\limits_{\mathsf{PSL}\,(2,\,\mathbb{C})} (\ln \|\, g \|)\, \mu_t(dg) < \infty \quad \textit{and} \quad \int\limits_{\mathbb{D}} \ln \frac{1}{1-|\gamma|}\, \mu(d\gamma) < \infty \tag{6.5}$$

are equivalent.

Definition 11. *Let γ', $\gamma'' \in \mathbb{D}$, $t \in \mathbb{T}$. The subgroup of* **PSL** $(2, \mathbb{C})$ *generated by two matrices $m_{\gamma'}(t)$ and $m_{\gamma''}(t)$ will be denoted as $G(\gamma', \gamma'', t)$.* □

To apply Furstenberg's theorem, we need the following two results:

Lemma 11. *If $\gamma' \neq \gamma''$, then for every $t \in \mathbb{T}$*

$$\left\|\left(m_{\gamma'}(t) \cdot m_{\gamma''}(t)^{-1}\right)^n\right\| \to \infty\,, \quad \left\|\left(m_{\gamma'}(t)^{-1} \cdot m_{\gamma''}(t)\right)^n\right\| \to \infty \quad \text{as} \quad n \to \pm\infty \tag{6.6}$$

□

Lemma 12. *Let at least one of the following two conditions be satisfied:*

i. $\gamma' \neq \gamma''$ *and* $t \neq \pm 1$.
ii. $\gamma'\overline{\gamma''} \neq \overline{\gamma'}\gamma''$.

Then there exists no Borel probability measure ν on $\overline{\mathbb{C}}$ which is invariant with respect to all the matrices from the subgroup $G(\gamma', \gamma'', t)$. □

Proof of Lemma 11. Using (3.1) we obtain

$$m_{\gamma'}(t)\cdot(m_{\gamma''}(t))^{-1} = \begin{bmatrix} 1-\gamma'\overline{\gamma''} & (\gamma''-\gamma')\,t^{-1} \\ (\overline{\gamma''}-\overline{\gamma'})\,t & 1-\overline{\gamma'}\,,\gamma'' \end{bmatrix}\cdot\frac{1}{\sqrt{1-|\gamma'|^2}}\cdot\frac{1}{\sqrt{1-|\gamma''|^2}}\,. \tag{6.7}$$

and

$$(m_{\gamma'}(t))^{-1}\cdot m_{\gamma''}(t) = \begin{bmatrix} 1-\gamma'\overline{\gamma''} & -(\gamma''-\gamma') \\ -(\overline{\gamma''}-\overline{\gamma'}) & 1-\overline{\gamma'}\gamma'' \end{bmatrix}\cdot\frac{1}{\sqrt{1-|\gamma'|^2}}\cdot\frac{1}{\sqrt{1-|\gamma''|^2}}\,. \tag{6.8}$$

Since $\det m_\gamma(t) = t^{-1}$,

$$\det\,(m_{\gamma'}(t)\cdot m_{\gamma''}(t)^{-1}) = 1 \qquad \text{and} \qquad \det\,(m_{\gamma'}(t)^{-1}\cdot m_{\gamma''}(t)) = 1\,.$$

The characteristic equation for each of two matrices $(m_{\gamma'}(t))^{-1}\cdot m_{\gamma''}(t)$ and $m_{\gamma'}(t)\cdot(m_{\gamma''}(t))^{-1}$ is of the form

$$\lambda^2 - \frac{2-\gamma'\overline{\gamma''}-\overline{\gamma'}\gamma''}{\sqrt{1-|\gamma'|^2}\cdot\sqrt{1-|\gamma''|^2}}\cdot\lambda + 1 = 0\,. \tag{6.9}$$

Since $|\gamma'| < 1$ and $|\gamma''| < 1$, the coefficient of λ is negative. If the discriminant of the characteristic equation is positive, then both characteristic roots are positive, one of them, say λ', is larger than 1, the second, say λ'', is less than 1. In particular, the eigenvalues of the matrix $(m_{\gamma'}(t))^{-1}\cdot m_{\gamma''}(t)$ (as well as the eigenvalues of the matrix $m_{\gamma'}(t)\cdot(m_{\gamma''}(t))^{-1}$) are distinct. Moreover, $\lambda'\cdot\lambda'' = 1$. Let us prove that if $\gamma' \neq \gamma''$, $|\gamma'| < 1$, $|\gamma''| < 1$, then the discriminant of the characteristic equation (6.9) is strictly positive. The discriminant is positive if and only if the expression $(2-\gamma'\overline{\gamma''}-\overline{\gamma'}\gamma'')^2 - 4\,(1-|\gamma'|^2)\,(1-|\gamma''|^2)$ is strictly positive. However,

$$\begin{aligned}
&(2-\gamma'\overline{\gamma''}-\overline{\gamma'}\gamma'')^2 - 4\,(1-|\gamma'|^2)\,(1-|\gamma''|^2)\\
&\quad = (2-\gamma'\overline{\gamma''}-\overline{\gamma'}\gamma'')^2 - (2-2\,|\gamma'|\,|\gamma''|)^2\\
&\quad + 4\left((1-|\gamma'|\,|\gamma''|)^2 - (1-|\gamma'|^2)(1-|\gamma''|^2)\right)\\
&= \left(4-2|\gamma'|\,|\gamma''|-\gamma'\overline{\gamma''}-\overline{\gamma'}\gamma''\right)\left(2\,|\gamma'|\,|\gamma''|-\gamma'\overline{\gamma''}-\overline{\gamma'}\gamma''\right)\\
&\qquad + 4\left(|\gamma'|-|\gamma''|\right)^2.
\end{aligned} \tag{6.10}$$

It is clear that the last expression is non-negative and it vanishes if and only if $|\gamma'| = |\gamma''|$ and $|\gamma'|\,|\gamma''| = \gamma'\overline{\gamma''}$, that is if and only if $\gamma' = \gamma''$. Since $\gamma' \neq \gamma''$, the discriminant is strictly positive, and eigenvalues λ' and λ'' of each of two matrices

$m_{\gamma'}(t) \cdot m_{\gamma''}(t)^{-1}$ and $(m_{\gamma'}(t))^{-1} \cdot m_{\gamma''}(t)$ (these two matrices have the same eigenvalues) satisfy the condition

$$0 < \lambda'' < 1 < \lambda' \qquad \text{and} \qquad \lambda' \cdot \lambda'' = 1\,. \tag{6.11}$$

Since $(\lambda')^n$ is an eigenvalue for each of the matrices $\left(m_{\gamma'}(t) \cdot m_{\gamma''}(t)^{-1}\right)^n$ and $\left(m_{\gamma'}(t)^{-1} \cdot m_{\gamma''}(t)\right)^n$,

$$\left\|\left(m_{\gamma'}(t) \cdot m_{\gamma''}(t)^{-1}\right)^n\right\| \geq (\lambda')^n, \quad \left\|\left(m_{\gamma'}(t)^{-1} \cdot m_{\gamma''}(t)\right)^n\right\| \geq (\lambda')^n$$

$(n = 0, 1, 2, \dots)$. Since $(\lambda'')^{-n} = (\lambda')^n$ is an eigenvalue for each of the matrices $\left(m_{\gamma'}(t) \cdot m_{\gamma''}(t)^{-1}\right)^{-n}$ and $\left(m_{\gamma'}(t)^{-1} \cdot m_{\gamma''}(t)\right)^{-n}$,

$$\left\|\left(m_{\gamma'}(t) \cdot m_{\gamma''}(t)^{-1}\right)^{-n}\right\| \geq (\lambda')^n, \quad \left\|\left(m_{\gamma'}(t)^{-1} \cdot m_{\gamma''}(t)\right)^{-n}\right\| \geq (\lambda')^n$$

$(n = 0, 1, 2, \dots)$. Since $\lambda' > 1$,

$$\left\|\left(m_{\gamma'}(t) \cdot m_{\gamma''}(t)^{-1}\right)^n\right\| \to \infty, \quad \left\|\left(m_{\gamma'}(t)^{-1} \cdot m_{\gamma''}(t)\right)^n\right\| \to \infty$$

as $n \to \pm\infty$. □

Proof of Lemma 12. We will prove a stronger result in this case. Namely we will show that no probability measure ν exists which is invariant for all matrices from the group generated by A_1 and A_2, where

$$A_1 \stackrel{\text{def}}{=} m_{\gamma'}(t) \cdot (m_{\gamma''}(t))^{-1} \qquad \text{and} \qquad A_2 \stackrel{\text{def}}{=} m_{\gamma'}(t))^{-1} \cdot m_{\gamma''}(t) \tag{6.12}$$

(Since A_1 and A_2 belong to the group $G(\gamma', \gamma''; t)$, the group generated by A_1, A_2 is contained in $G(\gamma', \gamma''; t)$.)

If a matrix A is j - unitary: $A^* j A = j$, then the eigenvector x that corresponds to the eigenvalue λ with $|\lambda| \neq 1$ is j - neutral: $x^* j x = 0$. Indeed, the identity $Ax = \lambda x$ implies $x^* A^* j A x = |\lambda|^2 x^* j x$, and j - unitarity of the matrix A implies $x^* j x = |\lambda|^2 x^* j x$. Therefore, each eigenvector of such a matrix A is proportional to a vector of the form $\begin{bmatrix} \zeta \\ 1 \end{bmatrix}$ where ζ is an appropriate unimodular complex number (which depends on the choice of the matrix and its eigenvalue). If the matrix A is hermitian, i.e. $A^* = A$, and λ' and λ'' are two distinct eigenvalues of the matrix A, then the corresponding eigenvectors e' and e'' are orthogonal: $(e'')^* \cdot e' = 0$.

Thus if $A = \begin{bmatrix} a_{11} & a_{12} \\ a_{21} & a_{22} \end{bmatrix}$ is a 2×2 matrix which is both j - unitary as well as hermitian and if its eigenvalues λ' and λ'' satisfy the conditions $\lambda' \neq \lambda''$, $|\lambda'| \neq 1$, $|\lambda''| \neq 1$, then its eigenvectors can be chosen in the form $e' = \begin{bmatrix} \zeta \\ 1 \end{bmatrix}$ and $e'' = \begin{bmatrix} -\zeta \\ 1 \end{bmatrix}$ where ζ is an unimodular complex number. The eigenvectors e' and e'' of the matrix A correspond to the fixed points $\{\zeta\}$ and $\{-\zeta\}$ of the linear fractional transformation $z \to \dfrac{a_{11}z + a_{12}}{a_{21}z + a_{22}}$, and these are the only fixed points of this transformation on $\overline{\mathbb{C}}$.

Now we investigate whether a probability measure ν exists on $\overline{\mathbb{C}}$ which is invariant under the actions of all the matrices from the group generated by the matrices A_1 and A_2. Both matrices are j-unitary and hermitian. First we consider the action of the matrix A_1. Let λ', λ'' be the eigenvalues of the matrix A_1, $\lambda' \in (1, \infty)$, $\lambda'' \in (0, 1)$. Let

$$e' = \begin{bmatrix} \zeta \\ 1 \end{bmatrix} \quad \text{and} \quad e'' = \begin{bmatrix} -\zeta \\ 1 \end{bmatrix},$$

where ζ is a unimodular complex number, be the corresponding eigenvectors. The points $\zeta \in \overline{\mathbb{C}}$ and $-\zeta \in \overline{\mathbb{C}}$ are the fixed points for the linear fractional transformation

$$z \rightarrow \frac{(1-\gamma'\overline{\gamma''})\, z + (\gamma'' - \gamma')\, t^{-1}}{(\overline{\gamma''} - \overline{\gamma'})\, t\, z + (1 - \overline{\gamma'}\gamma'')}, \tag{6.13}$$

which corresponds to the matrix A_1 (see (6.12) and (6.7)), as well as for all integer powers of this transformation. The point ζ is an attracting fixed point for the transformation (6.13), the point $-\zeta$ is a repelling one.

The measure ν is also invariant under the action of all matrices A_1^n, $n \in \mathbb{Z}$. Under the action of the matrix A_1^n, the punctured Riemann sphere $\overline{\mathbb{C}} \setminus (-\zeta)$ shrinks to the point ζ as $n \rightarrow +\infty$. Hence,

$$\operatorname{supp} \nu \subseteq \{\zeta\} \cup \{-\zeta\}. \tag{6.14}$$

Now we consider the action of the matrix A_2. We will show that each of the fixed points ζ and $-\zeta$ for the linear fractional transformation (6.13) is also a fixed point for the linear fractional transformation

$$z \rightarrow \frac{(1-\gamma'\overline{\gamma''})\, z - (\gamma'' - \gamma')}{-(\overline{\gamma''} - \overline{\gamma'})\, z + (1 - \overline{\gamma'}\gamma'')}, \tag{6.15}$$

which corresponds to the matrix A_2, or, equivalently, the vectors $\begin{bmatrix} \zeta \\ 1 \end{bmatrix}$ and $\begin{bmatrix} -\zeta \\ 1 \end{bmatrix}$, which originally appear as eigenvectors of the matrix A_1, are eigenvectors of the matrix A_2 as well.

Since the measure ν is invariant for the matrices A_2, A_2^{-1}, its support supp ν is A_2-invariant.

If supp ν consists of one point, say ζ, then the one-point set $\{\zeta\}$ must be invariant under the transformation (6.15), i.e. the vector $\begin{bmatrix} \zeta \\ 1 \end{bmatrix}$ is an eigenvector of the matrix A_2. Since A_2 is selfadjoint, the orthogonal vector $\begin{bmatrix} -\zeta \\ 1 \end{bmatrix}$ must be an eigenvector of the matrix A_2, i.e. the point $\{-\zeta\}$ must be a fixed point of the linear fractional transformation (6.15).

If supp ν consists of both points ζ and $-\zeta$ but $\nu(\{\zeta\}) \neq \nu(\{-\zeta\})$, then both sets $\{\zeta\}$ and $\{-\zeta\}$ are invariant under the actions of transformatios (6.13) and (6.15).

Finally consider the case where supp ν consists of both points $\{\zeta\}$ and $\{-\zeta\}$ and $\nu(\zeta) = \nu(-\zeta)$. In this case the linear fractional transformations (6.13) preserves ζ and $-\zeta$ and the linear fractional transformation (6.15) either preserves these points or permutes them. In any case the points ζ and $-\zeta$ are fixed points for the square of the transformation(6.15). Thus $\begin{bmatrix} \zeta \\ 1 \end{bmatrix}$ and $\begin{bmatrix} -\zeta \\ 1 \end{bmatrix}$ are eigenvectors of the matrix A_2^2. However, the mapping $\lambda \to \lambda^2$ is one-to-one on the spectrum of matrix A_2. Therefore these vectors are eigenvectors for the matrix A_2 as well. Hence, both ζ and $-\zeta$ are fixed points of the transformation (6.15).

Thus, if a probability measure ν exists which is invariant for the group generated by the matrices A_1 and A_2 then a unimodular complex number ζ exists such that the points ζ and $-\zeta$ are fixed points for both the linear fractional transformations (6.13) and (6.15). The equations

$$\begin{aligned} \frac{(1-\gamma'\overline{\gamma''})\,\zeta + (\gamma''-\gamma')\,t^{-1}}{(\overline{\gamma''}-\overline{\gamma'})\,t\,\zeta + (1-\overline{\gamma'}\gamma'')} &= \zeta, \\ \frac{(1-\gamma'\overline{\gamma''})\,(-\zeta) + (\gamma''-\gamma')\,t^{-1}}{(\overline{\gamma''}-\overline{\gamma'})\,t\,(-\zeta) + (1-\overline{\gamma'}\gamma'')} &= -\zeta, \end{aligned}$$

which express that ζ and $-\zeta$ are fixed points for the transformation (6.13), are equivalent to the equations

$$\begin{aligned} (1-\gamma'\overline{\gamma''}) + (\gamma''-\gamma')\,t^{-1}\overline{\zeta} &= (\overline{\gamma''}-\overline{\gamma'})\,t\,\zeta + (1-\overline{\gamma'}\gamma''), \\ (1-\gamma'\overline{\gamma''}) - (\gamma''-\gamma')\,t^{-1}\overline{\zeta} &= -(\overline{\gamma''}-\overline{\gamma'})\,t\,\zeta + (1-\overline{\gamma'}\gamma'') \end{aligned} \tag{6.16}$$

respectively. Similarly, the equations

$$\begin{aligned} \frac{(1-\gamma'\overline{\gamma''})\,\zeta - (\gamma''-\gamma')}{-(\overline{\gamma''}-\overline{\gamma'})\zeta + (1-\overline{\gamma'}\gamma'')} &= \zeta, \\ \frac{(1-\gamma'\overline{\gamma''})\,(-\zeta) - (\gamma''-\gamma')}{-(\overline{\gamma''}-\overline{\gamma'})\,(-\zeta) + (1-\overline{\gamma'}\gamma'')} &= -\zeta, \end{aligned}$$

which express that ζ and $-\zeta$ are fixed points for the transformation (6.15), are equivalent to the equations

$$\begin{aligned} (1-\gamma'\overline{\gamma''}) - (\gamma''-\gamma')\overline{\zeta} &= -(\overline{\gamma''}-\overline{\gamma'})\zeta + (1-\overline{\gamma'}\gamma''), \\ (1-\gamma'\overline{\gamma''}) + (\gamma''-\gamma')\overline{\zeta} &= (\overline{\gamma''}-\overline{\gamma'})\zeta + (1-\overline{\gamma'}\gamma'') \end{aligned} \tag{6.17}$$

respectively.

From (6.16) and (6.17) it follows that

$$\gamma'\overline{\gamma''} = \overline{\gamma'}\gamma'', \tag{6.18}$$

$$(\gamma''-\gamma')\overline{\zeta} = (\overline{\gamma''}-\overline{\gamma'})\zeta \tag{6.19}$$

and

$$t^2 = 1. \tag{6.20}$$

The equation (6.21) gives us the possible values for common fixed points $\pm\zeta$ of the linear fractional transformations (6.13) and (6.15):

$$\zeta = \pm\frac{\gamma'' - \gamma'}{|\gamma'' - \gamma'|}. \tag{6.21}$$

The system of equations (6.16)–(6.17) is compatible only under the conditions (6.18) and $t = \pm 1$.

Therefore if either one of the conditions (6.18) or (6.20) is violated, the linear fractional transformations (6.13) and (6.15) will have no common fixed point. In this case no probability measure ν exists which is invariant for the group generated by A_1 and A_2. □

7. Proofs of the main results

In this section we prove our main results – Theorems I and II. For a contractive sequence $\omega = (\gamma_0, \gamma_1, \gamma_2, \ldots, \gamma_k, \ldots)$ of complex numbers γ_k, we consider the sequence of matrices $\{M_{\omega,n}(t)\}_{n=0,1,2,\ldots}$, where

$$M_{\omega,n}(t) = \begin{bmatrix} a_{\omega,n}(t) & b_{\omega,n}(t) \\ c_{\omega,n}(t) & d_{\omega,n}(t) \end{bmatrix} \tag{7.1}$$

is defined as the product

$$M_{n,\omega}(t) \stackrel{\text{def}}{=} m_{\gamma_n}(t)\cdot m_{\gamma_{n-1}}(t)\cdot \;\ldots\; \cdot m_{\gamma_1}(t)\cdot m_{\gamma_0}(t)\,, \quad n = 0, 1, 2, \ldots. \tag{7.2}$$

The matrix $m_\gamma(t)$ is defined in (3.1):

$$m_\gamma(t) = \begin{bmatrix} t^{-1} & -\gamma\cdot t^{-1} \\ -\overline{\gamma} & 1 \end{bmatrix} \cdot \frac{1}{\sqrt{1-|\gamma|^2}}.$$

As before, Ω is the product space (2.1), i.e. the set of all sequences

$$\omega = (\gamma_0, \gamma_1, \gamma_2, \ldots)$$

of complex numbers γ_k satisfying the condition $|\gamma_k| < 1$, $k = 0, 1, 2, \ldots$

Lemma 13. *For a given t on the unit circle, let Ω_t be the set of all those $\omega \in \Omega$ for which the following two conditions are satisfied:*

i.

$$|a_{n,\omega}(t)| \to \infty \quad as \quad n \to \infty\,, \tag{7.3}$$

ii. *There exists a linear combination of $a_{n,\omega}(t)$ and $b_{n,\omega}(t)$, with an appropriate coefficient $\zeta_\omega(t)$, which vanishes asymptotically:*

$$a_{n,\omega}(t)\,\zeta_\omega(t) + b_{n,\omega}(t) \to 0 \quad as \quad n\to\infty\,, \tag{7.4}$$

where $a_{\omega,n}(t)$ and $b_{\omega,n}(t)$ are the entries of the matrix $M_{\omega,n}(t)$.

Let μ be a probability measure on the open unit disk $\mathbb{D}$ and p_μ be the probability product measure on Ω generated by μ : $p_\mu = \mu\otimes\mu\otimes\mu\otimes \;\cdots\; \otimes\mu\otimes\cdots$.

Assume *that the measure* μ *satisfies the following two conditions:*

i. *The logarithmic integral converges:*

$$\int_{\mathbb{D}} \ln \frac{1}{1-|\gamma|}\ \mu(d\gamma) < \infty\,; \tag{7.5}$$

ii. *The support of the measure* μ *consists of more than one point:*

$$\exists\, \gamma',\, \gamma'' \in \mathbb{D}:\ \gamma' \in \operatorname{supp}\mu,\ \gamma'' \in \operatorname{supp}\mu,\ \gamma' \neq \gamma''\,. \tag{7.6}$$

Then Ω_t *is a Borel set of full* p_μ*-measure in* Ω*:*

$$p_\mu(\Omega_t) = 1 \tag{7.7}$$

for every $t \in \mathbb{T}$ *except possibly* $t = \pm 1$.

ii$'$. *If instead* (7.6), *the following stronger condition is satisfied:*

$$\exists\, \gamma',\, \gamma'' \in \mathbb{D}:\ \gamma' \in \operatorname{supp}\mu,\ \gamma'' \in \operatorname{supp}\mu,\ \ \gamma'\overline{\gamma''} \neq \overline{\gamma'}\gamma''\,, \tag{7.8}$$

then (7.7) *holds for every* $t \in \mathbb{T}$.

Proof. The fact that Ω_t is a Borel set can be established by the same reasoning that we have already used in the proof of Lemma 7. Now we are going to apply the Furstenberg theorem and the Oseledets theorem to the family of the matrices $\{m_\gamma(t)\}_{\gamma\in\mathbb{D}}$. ($\gamma$ is the parameter which enumerates the family, $t \in \mathbb{T}\setminus\{\pm 1\}$ is fixed.) We consider matrices $m_\gamma(t)$ as members of the group **PSL** $(2, \mathbb{C})$. The measure μ_t on **PSL**$(2, \mathbb{C})$ is the image of the original measure μ on $\mathbb{D}$ under the embedding $\Gamma_t : \mathbb{D} \to$ **PSL**$(2, \mathbb{C})$, as it was explained at the beginning of Section 6 (see (6.1) and (6.2)). The measure μ_t is supported on the set $\Gamma_t(\mathbb{D})$, and only those $g \in \mathsf{PSL}(2, \mathbb{C})$ can belong to the support of the measure μ_t which are of the form $g = m_\gamma$ for some $\gamma \in \mathbb{D}$. Therefore, applying theorems on products of random matrices which were formulated in Section 5, we can restrict our consideration only to those sequences $\omega_G = (g_0, g_1, g_2, \dots)$ which are of the form $\omega_G = (m_{\gamma_0}, m_{\gamma_1}, m_{\gamma_2}, \dots)$: the sequences which are not of this form constitute a set of zero probability (i.e. of zero p_{μ_t}-measure). By the assumptions on the support of μ, there exist $\gamma', \gamma'' \in \mathbb{D}$, $\gamma' \neq \gamma''$ such that $m_{\gamma'} \in \operatorname{supp}\mu_t$, $m_{\gamma''} \in \operatorname{supp}\mu_t$. By Lemma 11, the group generated by the matrices $m_{\gamma'}$ and $m_{\gamma''}$ is non-compact. By Lemma 12, this group is irreducible as well. By Furstenberg's theorem, the upper Lyapunov exponent λ_t of the measure μ_t is strictly positive : $\lambda_t > 0$. This implies that

$$\lim_{n\to\infty} \frac{\ln \|M_{\omega,n}(t)\|}{n+1} = \lambda_t > 0 \tag{7.9}$$

for p_{μ_t}-almost every sequences $\omega_g = (m_{\gamma_0}, m_{\gamma_1}, \cdots, m_{\gamma_k}, \cdots) \in \Omega_G$, or, *equivalently, for* p_μ*-almost every sequences* $\omega = (\gamma_0, \gamma_1, \cdots, \gamma_k, \cdots) \in \Omega$. (The sample spaces Ω and Ω_G were defined in section 2 and 5 respectively.) Using (3.42) we obtain

$$\lim_{n\to\infty} \frac{\ln |a_{\omega,n}(t)|}{n+1} = \lambda_t > 0 \tag{7.10}$$

All the more, the claim (7.3) of Lemma 13 holds.

Since $M_{\omega,n}(t)\,x = \begin{bmatrix} a_{\omega,n}(t) \\ b_{\omega,n}(t) \end{bmatrix}$ for $x = \begin{bmatrix} 1 \\ 0 \end{bmatrix}$, it follows from (7.10) that

$$\lim_{n\to\infty} \frac{\ln \| M_{\omega,n}(t)\, x\|}{n+1} = \lambda_t$$

In particular, this vector x does not belong to the eigenspace $V_{-\lambda_t,\omega}(t)$ of the matrix $\Psi(\omega, t)$:

$$\Psi(\omega, t) \stackrel{\text{def}}{=} \lim_{n\to\infty} \{M^*_{\omega,n}(t)\, M_{\omega,n}(t)\}^{1/2n}, \tag{7.11}$$

which corresponds to its eigenvalue $e^{-\lambda_t}$. (See the formulation of the multiplicative ergodic theorem by Oseledets in Section 5.) For any vector $y = \begin{bmatrix} \xi \\ \eta \end{bmatrix} \neq 0$ in the eigenspace $V_{-\lambda_t,\omega}(t)$, the limiting relation

$$\lim_{n\to\infty} \frac{\ln \| M_{\omega,n}(t)\, y\|}{n+1} = -\lambda_t \tag{7.12}$$

holds. The second entry η of such a vector y is not equal to zero: For, if η were equal to zero then y would be proportional to the vector $x = \begin{bmatrix} 1 \\ 0 \end{bmatrix}$. Therefore the eigenspace $V_{-\lambda_t,\omega}(t)$ contains the "normalized" eigenvector of the form

$$y_\omega(t) = \begin{bmatrix} \zeta_\omega(t) \\ 1 \end{bmatrix}, \tag{7.13}$$

where ζ_ω is a complex number: we can start from an arbitrary eigenvector y with entries ξ and η and then form the vector $y_\omega(t)$ with $\zeta_\omega(t) = \xi\eta^{-1}$. From the relation (7.12) for the vector $y = y_\omega(t)$ it follows that

$$\lim_{n\to\infty} \frac{\ln | a_{\omega,n}(t)\, \zeta_\omega(t) + b_{\omega,n}(t)|}{n+1} \leq -\lambda_t. \tag{7.14}$$

(The expression

$$a_{\omega,n}(t)\, \zeta_\omega(t) + b_{\omega,n}(t)$$

is the first entry of the vector $M_{\omega,n}(t)\, y_\omega(t)$.) In particular, the claim (7.4) of Lemma 13 holds. □

Remark 7. *It is possible to prove that for arbitrary probability measure μ on $\mathbb{D}$ the Lyapunov exponent λ_t of the matrix family $\{m_\gamma(t)\}_{\gamma\in\mathbb{D}}$ (with respect to this measure) depends continuously on t (on the unit circle). By Lemmas 11 and 12 and Furstenberg theorem, under the condition* (7.6) $\lambda_t > 0$ *for $t \in \mathbb{T} \setminus \{\pm 1\}$, and under the condition* (7.8) $\lambda_t > 0$ *for all $t \in \mathbb{T}$. Thus, under the condition (7.8),* $\min\limits_{t\in\mathbb{T}} \lambda_t > 0$. □

Proofs of Theorems I and II. We consider the following two cases. Either the logarithmic integral (7.5) converges, or it diverges. If this integral diverges, then the support of the measure μ is not separated from $\mathbb{T}$, i.e. the condition (4.31 holds. In this case, using Lemma 9, we conclude that p_μ almost every sequence $\omega = (\gamma_0, \gamma_1, \gamma_2, \dots)$ satisfies the condition (4.32): $\overline{\lim\limits_{k\to\infty}} |\gamma_k| = 1$. By Lemma 8, for *every* sequence $\omega = (\gamma_0, \gamma_1, \gamma_2, \dots)$ which satisfies the condition (4.32), the corresponding function s_ω is inner. So, if the logarithmic integral, corresponding to the measure μ, diverges, then almost every (with respect to the measure p_μ) Schur function s_ω is inner, i.e. Theorem I holds.

In the case of divergence of the logarithmic integral, the proof of Theorem *I* is essentially deterministic. It is mainly based on Lemma 8 which is purely deterministic. The only probabilistic fact that we use is Lemma 9 which is quite elementary. The case of the convergence of the logarithmic integral is much more complicated. In this case, our reasoning are essentially probabilistic. It is based on Furstenberg theorem and in the end on Oseledets theorem. So, let the logarithmic integral converge. We will derive Theorems *I* and *II* from Lemma 6. To apply this Lemma, we have to find a full p_μ measure set Ω_{gen}, $\Omega_{\text{gen}} \subseteq \Omega$ such that for every $\omega \in \Omega_{\text{gen}}$ the conditions (4.1), (4.2) are satisfied for *almost every* (with respect to $m(dt)$) $t \in \mathbb{T}$. For each *fixed* $t \in \mathbb{T}$, the existence of a full p_μ measure set Ω_t such that for every $\omega \in \Omega_t$ the conditions (4.1) and (4.2) are satisfied is provided by Lemma 13. If $\bigcap\limits_{t\in\mathbb{T}} \Omega_t$ is a set of full measure, it would serve as the set Ω_{gen}. Unfortunately, we cannot assert that this is a full measure set because we have an uncountable intersection of full measure sets. However this obstacle can be overcome using Fubini's theorem.

Let us consider the product space $\Omega \times \mathbb{T}$ and the product measure

$$P_\mu \stackrel{\text{def}}{=} p_\mu \otimes m \quad \text{on} \quad \Omega \times \mathbb{T}. \tag{7.15}$$

We equip the set $\Omega \times \mathbb{T}$ with the product topology. The measure P_μ is defined on the Borel σ-algebra of the topological space $\Omega \times \mathbb{T}$ and is a probability measure:

$$P_\mu(\Omega \times \mathbb{T}) = 1. \tag{7.16}$$

Again, we consider the sequence of matrix functions $\{M_{\omega,n}(t)\}_{0\le n<\infty}$ which is constructed from ω and t as the product (3.7) of the matrices $m_{\gamma_k}(t)$ where the matrix $m_\gamma(t)$ is defined by (3.1).

Let us define the set $\mathcal{M}$, $\mathcal{M} \subseteq \Omega\times\mathbb{T}$ in a similar manner as we have introduced the sets T_ω and Ω_t (see formulations of Lemmas 6 and 13):

Thus, let $\mathcal{M}$ be the set of all those $(\omega, t) \in \Omega \times \mathbb{T}$ for which the following two conditions are satisfied:

i.

$$|a_{n,\omega}(t)| \to \infty \quad as \quad n \to \infty, \tag{7.17}$$

ii. *There exists a linear combination of $a_{n,\omega}(t)$ and $b_{n,\omega}(t)$, with an appropriate coefficient $\zeta_\omega(t)$, which vanishes asymptotically:*

$$a_{n,\omega}(t)\,\zeta_\omega(t) + b_{n,\omega}(t) \to 0 \quad as \quad n \to \infty, \tag{7.18}$$

where $a_{\omega,n}(t)$ and $b_{\omega,n}(t)$ are matrix entries of the matrix $M_{\omega,n}(t)$.

The set $\mathcal{M}$ is a Borel set of $\Omega \times \mathbb{T}$. The proof of this statement is similar to the proof of Lemma 7. Let $\mathcal{M}_t$ and $\mathcal{M}_\omega$ be the sections of the set $\mathcal{M}$:

$$\text{For fixed } t \in \mathbb{T}, \quad \mathcal{M}_t = \{\omega \in \Omega : (\omega, t) \in \mathcal{M}\}, \tag{7.19}$$

$$\text{For fixed } \omega \in \Omega, \quad \mathcal{M}_\omega = \{t \in \mathbb{T} : (\omega, t) \in \mathcal{M}\}. \tag{7.20}$$

By Fubini's theorem and the definition (7.15) of the product measure P_μ,

$$P_\mu(\mathcal{M}) = \int_{\mathbb{T}} p_\mu(\mathcal{M}_t)\, m(dt) \tag{7.21}$$

and

$$P_\mu(\mathcal{M}) = \int_{\Omega} m(\mathcal{M}_\omega)\, p_\mu(d\omega). \tag{7.22}$$

It is clear that the sections $\mathcal{M}_t$ and $\mathcal{M}_\omega$ of the set $\mathcal{M}$ coincide with the sets Ω_t and T_ω which were considered in Lemmas 6 and 13 respectively. By Lemma 13, $p_\mu(\mathcal{M}_t) = 1$ for every $t \in \mathbb{T} \setminus \{\pm 1\}$. From here and from (7.21) it follows that

$$P_\mu(\mathcal{M}) = 1. \tag{7.23}$$

It is clear that

$$0 \le m(\mathcal{M}_\omega) \le 1 \quad \text{for every} \quad \omega \in \Omega. \tag{7.24}$$

Since $p_\mu(\Omega) = 1$, it follows from (7.22) and (7.24) that

$$m(\mathcal{M}_\omega) = 1 \quad \text{for} \quad p_\mu \quad \text{almost every} \quad \omega \in \Omega. \tag{7.25}$$

Let us define

$$\Omega_{\text{gen}} = \{\omega \in \Omega : m(\mathcal{M}_\omega) = 1\}. \tag{7.26}$$

(7.25) means that Ω_{gen} is a set of full p_μ measure:

$$p_\mu(\Omega_{\text{gen}}) = 1 \tag{7.27}$$

and that

$$m(T_\omega) = 1 \quad \text{for every} \quad \omega \in \Omega_{\text{gen}}. \tag{7.28}$$

(Here we recall that the section $\mathcal{M}_\omega$ of the set $\mathcal{M}$ is the same as the set T_ω, considered in Lemma 6.)

Let us consider an arbitrary $\omega \in \Omega_{\text{gen}}$. Since (7.28) holds, Lemma 6 is applicable to the sequence of the matrix functions $\{M_{\omega,n}(t)\}_{0 \le n < \infty}$. (Now ω is fixed, t runs over $\mathbb{T}$.) By this lemma, the function s_ω, whose sequence of Schur parameters is ω, is inner, and the sequence of the Schur approximants converges pointwise almost everywhere on $\mathbb{T}$. The pointwise limit of this sequence coincide with the boundary value of the function s_ω almost everywhere on $\mathbb{T}$. Thus, Theorems *I* and *II* hold with $R = \Omega \setminus \Omega_{\text{gen}}$. □

Remark 8. *In fact we have proved more than what was formulated in Theorem II. Namely we proved that if the assumption of Theorem II is satisfied then for every*

$\omega \in \Omega \setminus R$, where $p_\mu(R) = 0$, the sequence $\mathrm{Ap}_n(s_\omega, .)$ of the Schur approximants of the function $s_\omega(.)$ satisfies the condition

$$\overline{\lim_{n\to\infty}} \frac{|s_\omega(t) - \mathrm{Ap}_n(s_\omega, t)|}{n+1} \leq -2\lambda_t \quad \forall t \in \mathbb{T} \setminus R_\omega , \tag{7.29}$$

where R_ω is a subset of $\mathbb{T}$ of zero Lebesgue measure (which depends on ω, i.e. on function s_ω). Since $\lambda_t > 0$ for $t \in \mathbb{T} \setminus \{\pm 1\}$, the sequence $\mathrm{Ap}_n(s_\omega, t))$ converges to $s_\omega(t)$ exponentially for almost all t. However, since the convergence is not uniform with respect to t, we cannot say anything about the rate of convergence in L^2 even if $\min_{t\in\mathbb{T}} \lambda_t > 0$. This holds under the condition (7.8). *(See Remark 7.)* □

8. Concluding remarks

In this section we discuss the relation between our results and numerous results on random differential and difference equations. Note that

$$\zeta \to \frac{1+\zeta}{1-\zeta}$$

is a one-to-one mapping of the unit disk $\{\zeta : |\zeta| < 1\}$ onto the right half-plane $\{\zeta : \mathrm{Re}\,\zeta > 0\}$. Therefore, if $s(z)$ is a Schur function, then the function

$$w(z) = \frac{1 + zs(z)}{1 - zs(z)} \tag{8.1}$$

is a Carathéodory function, i.e. a function which is holomorphic and has a non-negative real part in the unit disk:

$$\mathrm{Re}\, w(z) \geq 0 \quad \text{for} \quad z \in \mathbb{D} . \tag{8.2}$$

The factor z in (8.1) leads to the normalizing condition

$$w(0) = 1 . \tag{8.3}$$

Conversely, if $w(z)$ is a Carathéodory function satisfying the normalizing condition (8.3) then it can be uniquely represented in the form (8.1) where $s(z)$ is a Schur function. Every Carathéodory function $w(z)$ which satisfies the normalizing condition (8.3) admits the Herglotz representation

$$w(z) = \int_{\mathbb{T}} \frac{t+z}{t-z}\, \sigma(dt) , \tag{8.4}$$

where σ is a probability measure on $\mathbb{T}$. Conversely, if σ is a probability measure on $\mathbb{T}$, the formula (8.4) defines a normalized Carathéodory function $w(z)$. Thus, the transformation (8.1) together with the representation (8.4), establishes a one-to-one correspondence between Schur functions and probability measures on $\mathbb{T}$. It is easy to see that the Schur function is exceptional (i.e. a rational inner one) if and only if the corresponding probability measure σ on $\mathbb{T}$ is exceptional. (We call a probability measure on $\mathbb{T}$ *exceptional* if its support is a finite subset of $\mathbb{T}$.) A Carathéodory function $w(z)$ is called *exceptional* if it is rational and its poles are situated on $\mathbb{T}$. It is clear that a probability measure σ on $\mathbb{T}$ is exceptional if and only if the Carathéodory function $w(z)$ represented by (8.4) with *this* σ is exceptional. Thus, *the formula* (8.1), *together with the representation* (8.4), *establishes a one-to-one correspondence between the set of all non-exceptional Schur functions* $s(z)$ *and the set of all non-exceptional normalized Carathéodory functions* $w(z)$, *or equivalently, the set of all non-exceptional probability measures* σ *on* $\mathbb{T}$.

Given a non-exceptional probability measure σ on $\mathbb{T}$, we can relate this measure to the sequence $\{\varphi_k\}_{0\leq k<\infty}$ of polynomials orthonormal, with respect to σ. Such a sequence of polynomials can be obtained by applying of Gram-Schmidt orthogonalization procedure to the sequence $\{z^k\}_{0\leq k<\infty}$. Since the support of the measure σ is not a finite set, the system $\{z^k\}$ is linearly independent in $L^2(\mathbb{T}, \sigma(dt))$ and the Gram-Schmidt procedure can be performed unrestrictively for all $k: 0 \leq k < \infty$. As usual, for each polynomial φ_k of degree k, the so-called reciprocal polynomial φ_k^* can be defined: $\varphi_k^*(z) \overset{\text{def}}{=} z^k\overline{\varphi_k(1/\overline{z})}$. It turns out that the system of polynomials $\{\varphi_k, \varphi_k^*\}_{0\leq k<\infty}$ satisfies a linear recurrence relation, which can be written in the matrix form:

$$\begin{bmatrix} \varphi_{k+1}(z) \\ \varphi_{k+1}^*(z) \end{bmatrix} = \frac{1}{\sqrt{1-|a_k|^2}} \begin{bmatrix} z & -\overline{a}_k \\ -z\,a_k & 1 \end{bmatrix} \cdot \begin{bmatrix} \varphi_k(z) \\ \varphi_k^*(z) \end{bmatrix}, \quad 0 \leq k < \infty, \tag{8.5}$$

with the initial condition

$$\begin{bmatrix} \varphi_0(z) \\ \varphi_0^*(z) \end{bmatrix} = \begin{bmatrix} 1 \\ 1 \end{bmatrix}. \tag{8.6}$$

Here $\{a_k\}_{0\leq k<\infty}$ is a contractive sequence of complex numbers. The sequence $\{a_k\}$ is determined uniquely from the probability measure σ which generates the sequence $\{\varphi_k\}_{0\leq k<\infty}$ of orthogonal polynomials: $a_k = a_k(\sigma)$, $0 \leq k < \infty$. Conversely, given a contractive sequence of complex numbers $\{a_k\}_{0\leq k<\infty}$, we can *define* the sequence of polynomials $\{\varphi_k\}_{0\leq k<\infty}$ using the linear recurrence relations (8.5). It turns out that these polynomials form an orthonormal sequence with respect to some non-exceptional probability measure σ on $\mathbb{T}$. Thus, there exists a one-to-one correspondence between non-exceptional probability measures on $\mathbb{T}$ and contractive sequences $\{a_k\}_{0\leq k<\infty}$ of complex numbers. These numbers a_k can be considered as *free parameters* in the set of all non-exceptional probability measures. Different names are used for the numbers $a_k(\sigma)$: one calls them *cyclic parameters* or *circular parameters* or *reflection coefficients* of the measure σ or of the sequence $\{\varphi_k\}$ of orthogonal polynomials. The theory of polynomials orthogonal on the unit circle was started by G. Szegő in the early twenties, in connection with the trigonometric moment problem and with the Theory of Toeplitz matrices. It was further elaborated in great details by Ya.L. Geronimus, [Gers1], [Gers2]. For us the following result is of great importance:

Theorem [Ya.L. Geronimus]. *Let a non-exceptional Schur function* $s(z)$ *and a non-exceptional normalized Carathéodory function* $w(z)$ *are related by* (8.1). *Let* $\{\gamma_k(s)\}_{0\leq k<\infty}$ *be the sequence of the Schur parameters of the function* s, *and*

$\{a_k(\sigma)\}_{0\leq k<\infty}$ *be the sequence of cyclic parameters (or, in other terminology, of reflection coefficients) of the probability measure* σ*, representing the function* $w(z)$*. Then*

$$\gamma_k(s) = a_k(\sigma)\,, \quad 0 \leq k < \infty\,. \tag{8.7}$$

This theorem first appears in [Gers1], (Theorem IX.) It also appears in [Gers2], (Theorem 18.2 there). The paper [Gers1] is not easily available and it is not translated into English. The original paper [Gers2] is even more difficult to avail, but its English translation is available. The simplified presentation of this Geronimus result can be found in [Khru1] and in [PiNe].

Many (but not all) properties of a Schur function $s(z)$ can be naturally reformulated in terms of the related function $w(z)$, i.e. in terms of the related sequence of orthogonal polynomials. In particular: *a Schur function* $s(z)$ *is inner if and only if the related measure* $\sigma(dt)$ *is singular*. Indeed, $|\,s(t)| = 1$ if and only if $\operatorname{Re} w(t) = 0$. On the other hand, $\operatorname{Re} w(t) = \sigma'(t)$ for m almost every $t \in \mathbb{T}$. (Here $s(t)$ and $w(t)$ are the boundary values of the appropriate functions. $\sigma'(t)$ is the derivative of the measure σ with respect to the normalized Lebesgue measure m.) Some other connections between Schur functions and orthogonal polynomials can be found in [Gol1], [Gol2] and [Khru2].

The quoted theorem of Ya.L. Geronimus built a bridge between the theory of Schur function and the theory of polynomials orthogonal on the circle. In particular this theorem relates a study of random Schur functions and a study of random orthogonal polynomials and allied linear difference systems of the form (8.5). Especially spectral properties of such linear systems are important.

A study of spectral properties of random linear systems was initiated by physicists in connection with disordered physical structures. Starting from the fifties both numerical experiments and theoretical researches were carried out. Some of these investigations are summarized in the overviews [Lif] , [MaIsh] , [Ish] . (See also monographs [Hor] and [LGP], especially Chapter III of the later.) Around the end of the sixties, the role of the positivity of Lyapunov exponents (i.e. the role of exponentially growing and exponentially decaying solutions) in the study of spectral properties of randomly disordered systems was already clear to physicists. This role was mentioned in [MaIsh] where the Furstenberg theorem was invoked to prove the positivity of Lyapunov exponent. In [CaLeb] and in [Ish] some mathematical models of concrete physical disordered systems were studied and it was proved (on the "physical level of rigorousness") that the corresponding linear random systems don't have absolutely continuous spectrum, i.e. their spectrum is purely singular (see § 9 of the paper [Ish]). At the beginning of the seventies mathematicians joined these investigations. A lot of research papers, many overviews and several monographs appeared which are dedicated to spectral properties of random linear systems. See the monographs [BoLa], [CarLa], [PaFig] and the overview [Pas2], [Pas3]. In particular, the paper [Pas1] of L. Pastur contains the rigorous proof of the fact that the spectrum of the one-dimensional Schrödinger operator, continuous or discrete, with random potential, is purely singular. (See

Theorems 9 and 10 of [Pas1] and Appendix.) This result is commonly known as theorem of Ishii–Pastur. As it was mentioned in [Pas1], the main ideas and results of this paper were first published in his preprint in 1974. (See the reference in [Pas1].) The proof of the positivity of the Lyapunov exponent, which was given by L. Pastur in Appendix to [Pas1], did not use the Furstenberg theorem but did use peculiar features of a Sturm–Liouville equation. A deep study of the Lyapunov exponents and their connection to the structure of the spectrum of the Sturm–Liouville (Schródinger) operator was done by S. Kotani. (See his papers [Kot1], [Kot2] and his other papers quoted there.) The results by S. Kotany are based on the ingenious homology-like identities which was first discovered and used by R. Johnson and J. Moser in [JohMo] in the almost periodic situation. In particular, S. Kotani proved that for non-deterministic potential, the Lyapunov exponent of the Sturm–Liouville operator is strictly positive almost everywhere on $\mathbb{R}$ (with respect to the Lebesgue measure) (Theorem 4.5 and Corollary 1 in [Kot2]). S. Kotani did not use the Furstenberg theorem. He proved that if the Lyapunov exponent were to vanish on some subset of $\mathbb{R}$ of positive Lebesgue measure, the potential would be predictable. The reasoning and the results of S. Kotani were reproduced to some extent in [CarLa] and [PaFig]. A study of random polynomials orthogonal on the circle was initiated by E.M. Nikishin, [Nik]. Methods and results related to random Schrödinger operators, discrete and continuous, was later adopted to polynomials orthogonal on the unit circle. This was done by J.S. Geronimo [Germo], and by J.S. Geronimo and A. Teplyaev [GerTe]. (Some results in this direction were obtained in [Tep].) In particular, in [Germo], results of Kotani [Kot2] on relations between Lyapunov exponent and the structure of the spectrum of Schrödinger operator were extended to polynomials orthogonal on the circle. In the paper [GerTe] this study was continued. In particular, it was proved in [GerTe] (Theorem 6.2 there) that if the reflection coefficients $\{a_k\}$ in the system (8.5) form a sequence of independent identically distributed random values, then under the condition (7.5), the Lyapunov exponent λ_t is strictly positive almost everywhere with respect to the Lebesgue measure and hence the measure σ is purely singular. Thus we can use these results of the paper [GerTe] instead of the Furstenberg theorem in proofs of our Theorems I and II. However, the use of the Furstenberg theorem allows us to prove a little bit more then the reasoning used by S. Kotani. Namely, using the Furstenberg theorem we proved that $\lambda_t > 0$ for *all* $t \in \mathbb{T} \setminus \{\pm 1\}$, and under the condition (7.8) for *all* $t \in \mathbb{T}$. Kotani's reasoning led to the result that $\lambda_t > 0$ for *almost all* t only. Moreover, the paper [GerTe] is a little bit bulky.

Finally we would like to emphasize the following remarkable circumstance. The ideas and the methods which originally appeared in theoretical physics to study the transport phenomena in disordered physical structure [MaIsh], [CaLeb], work successfully in a quite distant field: in complex function theory to study the convergence of the Schur algorithm on the boundary of the disk of the analyticity. We address the reader to the beautiful paper of the F. Dyson [Dys] where other examples of such kind were given.

References

[Arn] ARNOLD, L.: *Random Dynamical Systems.* Springer-Verlag, Berlin · Heidelberg New York 1998.

[BoLa] BOUGEROL, P. and J. LACROIX.: *Products of Random Matrices with Applications to Schrödinger Operators.* (Progress in Probability and Statistics, 8). Birkhäuser Boston, Boston MA 1985.

[Boy] BOYD, D.: *Schur's algorithm for bounded holomorphic functions.* Bull. London Math. Soc., **11** (1979), pp. 145–150.

[CarLa] CARMONA, R. and J. LACROIX. *Spectral Theory of Random Schrödinger Operators.* Birkhäuser, Boston · Basel · Berlin 1990.

[CaLeb] CASHER, A. and J. LEBOWITZ. *Heat flow in regular and disordered harmonic chains.* Journ. Math. Phys., **12**, (1970), pp. 1702–1711.

[FrKi] *Ausgewählte Arbeiten zu den Ursprüngen der Schur-Analysis.* (*Series: Teubner-Archiv zur Mathematik:* 016). (FRITZSCHE, B. and B. KIRSTEIN – editors.) B.G. Teubner Verlagsgesellschaft, Stuttgart · Leipzig 1991.

[Dys] DYSON, F.: *Missed opportunities.* Bull. Amer. Math. Soc. **78**, (1972), pp. 635–652. Russian translation: DAISON, F.. *Upushchennye vozmozhnosti*, Uspekhi Matem. Nauk, **35**:1, (1980), pp. 171–191.

[Fur] FURSTENBERG, H.: *Noncommuting random products.* Trans. Amer. Math. Soc. **108** (1963), pp. 377–428.

[FuKe] FURSTENBERG, H. and H. KESTEN.: *Products of random matrices.* Ann. of Math. Statist. **31** (1960), pp. 457–469.

[Germo] GERONIMO, J.S.: *Polynomials orthogonal on the unit circle with random reflection coefficients.* In: *"Methods of Appriximation Theory in Complex Analysis and Mathematical Physics. Selected papers from the international seminar held in Leningrad, May 13–26, 1991."* GONCHAR, A.A and E. SAFF – editors. Lecture Notes in Math., vol. **1550**. Reissued by Springer-Verlag, Berlin 1993, originally published by "Nauka", Moscow 1993, pp. 43–61.

[GerTe] GERONIMO, J.S. and A. TEPLYAEV: *A difference equation arising from the trigonometric moment problem having random reflection coefficient – An operator theoretic approach.* Journ. of Functional Analysis, **123** (1994), pp. 12–45.

[Gers1] GERONIMUS, YA.L.: *O polinomakh, ortogonal'nykh na kruge, o trigonometricheskoi probleme momentov i ob associirovannykh s neyu funktsiyakh tipa Carathéodory i Schur'a. (On polynomials orthogonal on the circle, on trigonometric moment-problem and on allied Carathéodory and Schur functions.* Russian, Résumé Engl.) Matem. Sbornik. Nov. ser. **15** (**42**), no. 1 (1944), pp. 99–130.

[Gers2] GERONIMUS, YA.L.: *Polinomy, ortogonal'nye na kruge, i ikh prilozheniya.* Zapiski nauchno-issledovatel'skogo instituta matematiki i mekhaniki i Khar'kovskogo matem. obshchestva, **19** (1948), pp. 35–120 (Russian). English transl.: *Polynomials orthogonal on a circle and their applications.* Amer. Math. Soc. Transl. (ser. **1**), vol. **3** (1962), pp. 1–78.

[Gers3] GERONIMUS, YA.L.: *Polinomy Ortogonal'nye na Okruzhnosti i na Otrezke.* Fizmatgiz, Moscow, 1958 (in Russian). English transl.: *Polynomials Orthogonal on a Circle and Interval.* Pergamon Press, New York, 1960.

[GoMa] GOLDSHEID, I.J. and G.A. MARGULIS: *Pokazateli Lyapunova proizvedeniya sluchainykh matrits.* Uspekhi Matematicheskikh Nauk. Vol. **44**, no. 5 (1989), pp. 13–60 (in Russian). English transl.: *Lyapunov indices of a product of random matrices.* Russian Mathematical Surveys, vol. **44** (1989), pp. 11–71.

[Gol1] GOLINSKII, L.B: *Schur functions, Schur parameters and orthogonal polynomials on the unit circle.* Zeitschrift für Analysis und ihre Anwendungen **12** (1993), pp. 457–469.

[Gol2] GOLINSKII, L.: *On Schur functions and Szegő orthogonal polonomials.* In: *Topics in Interpolation Theory* (Operator Theory: Advances and Applications, **OT 95**). (DYM, H., B. FRITZSCHE, V. KATSNELSON, B. KIRSTEIN – editors.) Birkhäuser Verlag, Basel · Boston · Berlin 1997.

[Hor] HORI, J.: *Spectral Properties of Disordered Chains and Lattices.* Pergamon Press, Oxford, 1968.

[Ish] ISHII, K.: *Localisation of eigenstates and transport phenomena in the one dimensional disordered system.* Supplement of the Progress of Theor. Physics, No.**53** (1973), pp. 77–138.

[JohMo] JOHNSON, R and J. MOSER: *The rotation number for almost periodic potentials.* Commun. Math. Phys. **84** (1982), pp. 403–438.

[Khru1] KHRUSHCHEV, S.: *Parameters of orthogonal polynomials.* In: *"Methods of Approximation Theory in Complex Analysis and Mathematical Physics. Selected papers from the international seminar held in Leningrad, May 13–26, 1991."* GONCHAR, A.A. and E. SAFF – editors. Lecture Notes in Math., vol. **1550**. Reissued by Springer-Verlag, Berlin 1993, originally published by "Nauka", Moscow 1993, pp. 185–191.

[Khru2] KHRUSHCHEV, S.V.: *Schur's algorithm, orthogonal polynomials and convergence of Wall's continued fractions in* $L^2(\mathbb{T})$. Journal of Approximation Theory, vol. **108**, no. 2, (2001), pp. 161–248.

[Khru3] KHRUSHCHEV, S.V.: *A singular Riesz product in the Nevai class and inner functions with the Schur parameters* $\bigcap_{p>2} l^p$. Journal of Approximation Theory, vol. **108**, no. 2, (2001), pp. 249–255.

[King] KINGMAN, J.F.C.: *Subadditive ergodic theory.* Annals of Probability, vol. **1**, No. 6, (1973), pp. 883–909.

[Kot1] KOTANI, S.: *Ljapunov indices determine absolutely continuous spectra of stationary random one-dimensional Schrödinger operators.* In: *Proc. Taneguchi Internat. Sympos. on Stochastic Analysis. Katata and Kyoto, 1982.* ITO, K. – editor. North Holland 1983, pp. 225–247.

[Kot2] KOTANI, S.: *One-dimensional random Schrödinger operators and Herglotz functions.* In: *Proc. Taneguchi Internat. Sympos. on Stochast. Processes and Math. Physics. Katata and Kyoto, 1985.* ITO, K. – editor. North Holland 1987, pp. 219–250.

[Kr] KRENGEL, U.: *Ergodic Theory.* Walter de Gruiter. Berlin · New York 1985.

[Lif] LIFSHITZ, I.M.: *O strukture energeticheskogo spektra i kvantovykh sostoyaniyakh neuporyadochennykh kondensirovannykh sistem.* Uspekhi Phizicheskikh Nauk, **VXXXIII**, 4 (1964), pp. 617–663 (Russian), English translation:

On the structure of the energy spectrum and quantum states of disordered condensed systems. Soviet. Phys. Uspekhi **7**, (1964/65), pp. 549–573.

[LGP] LIFSHITZ, I.M., S.A. GREDESKUL and L.A. PASTUR: *Vvedenie v Teoriyu Neuporyadochennykh sistem.* Nauka, Moscow, 1982. (Russian). English translation: *Introduction to the Theory of Disordered Systems*, Wiley, New York, 1988.

[MaIsh] MATSUDA, H. and K. ISHII. *Localization of normal modes and energy transport in the disordered harmonic chain.* Supplement of the Progress of Theor. Physics, No. **45** (1970), pp. 56–86.

[Nik] NIKISHIN, E.M. *Sluchainye ortogonal'nye polinomy na okruzhnosti.* Vestnik Moskovskogo Universiteta Ser. I Matematika i Mekhanika **42**, no. 1 (1987), pp. 52–55, 102. English translation: *Random orthogonal polynomials on a circle.* Moscow Univ. Math. Bull. **42**, no. 1 (1987), pp. 42–45.

[Nja] NJÅSTAD, O. *Convergence of the Schur algorithm.* Proc. Amer. Math. Soc. **110**, No. 4, 1990, pp. 1003–1007.

[Os] OSELEDETS, V.I.: *Mul'tiplikativnaya ergodicheskaya theorema. Kharakteristicheskie pokazateli Ljapunova dinamicheskikh sistem.* Trudy Moskovskogo Matematicheskogo Obshtshestva, **19** (1968), pp. 179–210. English translation: *A multiplicative ergodic theorem. Ljapunov characteristic numbers for dynamical systems.* Trans. Moscow Math. Soc. **19**, (1968), pp. 197–231.

[Pas1] PASTUR, L.A. *Spectral properties of disordered systems in the one-body approximations.* Communications in Math. Phys. **75**, pp. 179–196.

[Pas2] PASTUR, L.A. *Spectral'naya teoriya sluchajnykh samosopryazhennykh operatorov.* Itogi Nauki i Tekhniki. Ser. Teoriya Veroyatnosteĭ. Matematicheskaya statistika. Tekhnicheskaya kibernetika. Tom **25**, pp. 3–67. (Russian). English translation: *Spectral theory of random self-adjoint operators.* Journ. of Soviet Math. **46**:4, pp. 1979–2021.

[Pas3] PASTUR, L.A. *Spectral'nye svoistva sluchainykh samosopryazhennykh operatorov i matric.* Trudy St. Peterburg. Matem. Obshch. **4**, 1996, pp. 222–286. (Russian). English translation: *Spectral properties of random selfadjoint operators and matrices (A survey).* Amer. Math. Soc. Translations (Ser. **2**), vol. **188** (1999), pp. 153–195.

[PaFig] PASTUR, L. and A. FIGOTIN. *Spectra of Random and Almost-Periodic Operators.* Springer-Verlag, Berlin·Heidelberg·New York 1991.

[PiNe] PINTÉR, F. and P. NEVAI. *Schur functions and orthogonal polynomials on the unit circle.* In: *Approximation Theory and Function Series. Budapest, 1995*, pp. 293–306. VÉRTESI, P., L. LEINDLER, F. MÓRICZ, SZ. RÉVÉSZ, J. SZABADOS and V. TOTIK – editors. János Bolyai Mathematical Society, Budapest 1996.

[Ragh] RAGHUNATHJAN, M.S.: *A proof of Oseledec's multiplicative ergodic theorem.* Israel Journ. of Math., vol. **32** (1979), no. 4, pp. 356–362.

[Rakh] RAHMANOV, E.A.(=RAKHMANOV, E.A.) *Ob asimptotike otnosheniya ortogonal'nykh polinomov. II.* Matem. Sbornik. Nov.ser. **118** (**160**), no. 1 (1982), pp. 104–117 (in Russian). English transl.: *On the asymptotics of the ratio of orthogonal polynomials. II.* Math. USSR Sbornik **46** (1983), pp. 105–117.

[S1] SCHUR, I.: *Über Potenzreihen, die im Innern des Einheitskreises beschränkt sind, I.* J. reine und angewandte Math. **147** (1917), 205–232. Reprinted in [S3], pp. 137–164. Reprinted also in: [FrKi], pp. 22–49. English translation: *On power series which are bounded in the interior of the unit circle. I.* In: [Sch], pp. 31–59.

[S2] SCHUR, I. : *Über Potenzreihen, die im Innern des Einheitskreises beschränkt sind, II*. J. reine und angewandte Math. **148** (1918), 122–145. Reprinted in [S3], pp. 165–188. Reprinted also in: [FrKi], pp. 50–73. English translation: *On power series which are bounded in the interior of the unit circle. II.* In: [Sch], pp. 61–88.

[S3] SCHUR, I. : *Gesammelte Abhandlungen. Band* II. [*Collected Papers, Vol.* II], in German. Springer-Verlag, Berlin · Heidelberg · New York 1973.

[Sch] *I. Schur Methods in Operator Theory and Signal Processing* (Operator Theory: Advances and Applications, **OT 18**) (GOHBERG, I. – editor). Birkhäuser Verlag, Basel · Boston · Stuttgart 1986.

[Tep] TEPLYAEV, A.V. *The pure point spectrum of random orthogonal polynomials on the unit circle.* (Russian) Dokl. Akad. Nauk SSSR **320**, no. 1 (1991), pp. 49–53. English transl.: Soviet Math. Dokl. **44**, no. 2 (1992), 407–411.

V. Katsnelson
Department of Mathematics
The Weizmann Institute of Science
Rehovot 76100
Israel
e-mail: `katze@wisdom.weizmann.ac.il`

Operator Theory:
Advances and Applications, Vol. 134, 287–317

Abstract Interpolation Scheme for Harmonic Functions

A. Kheifets

Dedicated to Professor Harry Dym on the occasion of his 60th birthday with deep appreciation

Abstract. In Section 1 we recall the setting and solution of the Abstract Interpolation Problem (AIP) from [KKY]. In Section 2 we rephrase the AIP in terms of unitary scattering systems rather than in terms of unitary colligations. This allows us to give up the orthogonality assumption on the data scales and to formulate a more general setting of the AIP that corresponds to interpolation problems for harmonic functions also. (The original formulation of the AIP corresponded naturally to interpolating analytic functions only.) In Section 3 we give a complete solution to this more general AIP under an additional assumption regarding the data scale ρ_0. Solutions are the spectral functions of the feedback coupling with respect to the scale ρ_0. In Section 4 we give up the additional assumption of Section 3 regarding the data scale ρ_0, and define the scale ρ associated with any feedback coupling by means of the corresponding wave operator and develop the appropriate modification of the results of Section 3. In Section 5 a remark is given on the feedback coupling of the scattering systems. We plan to demonstrate applications of this approach to the General Commutant Lifting problem at another occasion.

1. Abstract Interpolation Problem

The following Abstract Interpolation Problem (AIP) was introduced and studied in [KKY] (see also [Kh1], [Kh2], [Kh3], [KhY] and [Kh4]). Define first the data of the problem. Let X be a linear space, $D(x, y)$ be a positive semidefinite sesquilinear form on X, T_1 and T_2 be linear operators on X, $M_1 : X \to E_1$, $M_2 : X \to E_2$ be two linear mappings from the space X into given Hilbert spaces E_1 and E_2 respectively. All the listed objects are related by the following identity

$$D(T_1x, T_1y) + \langle M_1x, M_1y\rangle_{E_1} = D(T_2x, T_2y) + \langle M_2x, M_2y\rangle. \tag{1.1}$$

Analytic on the unit disc $\mathbb{D} = \{\zeta \in \mathbb{C} : |\zeta| < 1\}$ contractive operator-valued function $w(\zeta) : E_1 \to E_2$ is said to be a solution of the Abstract Interpolation Problem if there exists a linear mapping

$$F : X \to H^w$$

such that

(i) $$\bar{t}((FT_1x)(t)+\begin{bmatrix} w(t)\\ \mathbf{1}\end{bmatrix}M_1x)=(FT_2x)(t)+\bar{t}\begin{bmatrix}\mathbf{1}\\ w(t)^*\end{bmatrix}M_2X, \tag{1.2}$$

for *a.a.* t on the unit circle $\mathbb{T}=\{t\in\mathbb{C}:|t|=1\}$;

(ii) $$\|Fx\|^2_{H^w}\leq D(x,x). \tag{1.3}$$

We use the following notations throughout the paper:

$$L^w=\begin{bmatrix}\mathbf{1} & w\\ w^* & \mathbf{1}\end{bmatrix}^{1/2}\begin{bmatrix}L^2(E_2)\\ L^2(E_1)\end{bmatrix}, \tag{1.4}$$

endowed with the range norm;

$$H^w=L^w\cap\begin{bmatrix}H^2_+(E_2)\\ H^2_-(E_1)\end{bmatrix}, \tag{1.5}$$

with the norm induced from L^w. L^2 here is the usual space of square integrable with respect to Lebesgue measure vector-functions on the unit circle $\mathbb{T}$; H^2_+, H^2_- are the correspondent vector Hardy spaces.

The description of the solution set depends on the following construction. The positive semidefinite form D endows the space X with a Hilbert space structure (after proper quotioning and completing). This Hilbert space will be denoted by H_0. Then identity (1.1) can be read as a definition of this isometry

$$V:E_1\oplus H_0\to H_0\oplus E_2,$$

with domain

$$d_V=\mathrm{Clos}\{M_1x\oplus[T_1x],x\in X\}\subseteq E_1\oplus H_0$$

and the range

$$\Delta_V=\mathrm{Clos}\{[T_2x]\oplus M_2x,x\in X\}\subseteq H_0\oplus E_2,$$

where $[T_1x]$ and $[T_2x]$ stand for correspondent equivalent classes.

Due to inequality (1.3) the range Fx depends on the equivalence class $[x]$ rather than on x itself and, hence, the mapping F can be extended by continuity to H_0, keeping the inequality

$$\|Fh_0\|^2_{H^w}\leq\|h_0\|^2_{H_0}. \tag{1.6}$$

The space L^w defined by (1.4) can be decomposed as follows

$$L^w=\begin{bmatrix}w\\ \mathbf{1}\end{bmatrix}H^2_+(E_1)\oplus H^w\oplus\begin{bmatrix}\mathbf{1}\\ w^*\end{bmatrix}H^2_-(E_2). \tag{1.7}$$

Define

$$F:E_1\to L^w,\ F:E_2\to L^w$$

this way

$$Fe_1=\begin{bmatrix}w\\ \mathbf{1}\end{bmatrix}e_1,\ Fe_2=\bar{t}\begin{bmatrix}\mathbf{1}\\ w^*\end{bmatrix}e_2 \tag{1.8}$$

observe that according to (1.7) $F(E_1)$, $F(H_0)$, $F(E_2)$ are mutually orthogonal in L^w, and

$$\begin{aligned} F &: E_1 \oplus H_0 \to \begin{bmatrix} w \\ \mathbf{1} \end{bmatrix} E_1 \oplus H^w, \\ F &: H_0 \oplus E_2 \to H^w \oplus \bar{t} \begin{bmatrix} \mathbf{1} \\ w^* \end{bmatrix} E_2. \end{aligned} \tag{1.9}$$

Under these notations (1.2) reads as

$$FV \Big|_{d_V} = \bar{t}F \Big|_{d_V}. \tag{1.10}$$

If

$$N_{d_V} = (E_1 \oplus H_0) \ominus d_V \quad \text{and} \quad N_{\Delta_V} = (H_0 \oplus E_2) \ominus \Delta_V. \tag{1.11}$$

then one can define the unitary colligation

$$A_0 : N_2 \oplus E_1 \oplus H_0 \to H_0 \oplus E_2 \oplus N_1 \tag{1.12}$$

where N_1 and N_2 are copies of N_{d_V} and N_{Δ_V} respectively, by letting

$$\begin{aligned} &A_0 \Big| d_V = V, \\ &A_0 : N_{d_V} \to N_1 \quad \text{identically}, \\ &A_0 : N_2 \to N_{\Delta_V} \quad \text{identically}. \end{aligned} \tag{1.13}$$

Denote by $S(\zeta)$ the characteristic function of A_0:

$$\begin{aligned} S(\zeta) &= P_{N_2 \oplus E_1} A_0 (\mathbf{1} - \zeta P_{H_0} A)^{-1} \Big|_{E_2 \oplus N_1}, \\ S(\zeta) &= \begin{bmatrix} s_2(\zeta) & s_0(\zeta) \\ s_1(\zeta) & s(\zeta) \end{bmatrix} : \begin{bmatrix} E_2 \\ N_1 \end{bmatrix} \to \begin{bmatrix} N_2 \\ E_1 \end{bmatrix}. \end{aligned} \tag{1.14}$$

Let $A_1 : N_1 \oplus H_1 \to H_1 \oplus N_2$ be an arbitrary unitary colligation with the same as the above spaces N_1 and N_2 as input and output respectively. Let $\omega(\zeta)$ be its characteristic function, $\omega(\zeta) : N_1 \to N_2$.

Feedback loadings of A_0 with A_1 produce all the unitary extensions A of V,

$$A : E_1 \oplus H \to H \oplus E_1,$$

where $H = H_0 \oplus H_1$. A is a minimal extension if and only if A_1 is a simple colligation. The principal fact is that the characteristic functions of the extensions A give exactly the solution set of the AIP. This leads to this formula

$$w = s_0 + s_2 \omega (\mathbf{1} - s\omega)^{-1} s_1 \tag{1.15}$$

that describes the solution set of the AIP. The correspondent mappings F are also described in this way. See references at the beginning of this section for more details.

2. From AIP to a more general setting

We are going to rephrase the setting of AIP first and to make it more general than in this section. The rephrasing goes in terms of scattering systems rather than in terms of unitary colligations.

Let $\tilde{X}$ be the space of vectors $\tilde{x}$:

$$\tilde{x} = (\cdots e_1^{(1)}, e_2^{(0)}, x, e_2^{(-1)}, e_2^{(-2)}, \cdots) \tag{2.1}$$

such that

$$x \in X,\ e_1^{(k)} \in E_1, \quad k \geq 0 \quad \text{and} \quad \sum_{k=0}^{\infty} \|e_1^{(k)}\|^2 < \infty,$$

$$e_2^{(k)} \in E_2, \quad k \leq -1 \quad \text{and} \quad \sum_{k=-\infty}^{-1} \|e_2^{(k)}\|^2 < \infty$$

define

$$\tilde{D}(\tilde{x}, \tilde{x}) = \sum_{k=0}^{\infty} \|e_1^{(k)}\|^2 + D(x,x) + \sum_{k=-\infty}^{-1} \|e_2^{(k)}\|^2, \tag{2.2}$$

$$\tilde{T}_1\tilde{x} = (\cdots, e_1^{(1)}, e_1^{(0)}, M_1x, T_1x, e_2^{(-1)}, e_2^{(-2)}, e_2^{(-3)}, \cdots), \tag{2.3}$$

and

$$\tilde{T}_2\tilde{x} = (\cdots, e_1^{(2)}, e_1^{(1)}\ ,\ e_1^{(0)}, T_2x, M_2x\ , e_2^{(-1)}, e_2^{(-2)}, \cdots). \tag{2.4}$$

Then (1.1) can be rephrased as

$$\tilde{D}(\tilde{T}_1\tilde{x}, \tilde{T}_1\tilde{y}) = \tilde{D}(\tilde{T}_2\tilde{x}, \tilde{T}_2\tilde{y}). \tag{2.5}$$

$\tilde{D}$ endows the space $\tilde{X}$ with a Hilbert space structure, which results with the space

$$\tilde{H}_0 = \cdots \oplus E_1^{(1)} \oplus E_1^{(0)} \oplus H_0 \oplus E_2^{(-1)} \oplus E_2^{(-2)} \oplus \cdots . \tag{2.6}$$

We denoted here by $E_1^{(k)}$, $k \geq 0$ ($E_2^{(k)}, k \leq -1$) the subspaces of vectors $\tilde{x}$, such that all the entries but $e_1^{(k)}$ ($e_2^{(k)}$, respectively) are zeros.
One can define the isometry $\tilde{V}$:

$$\tilde{V} : [\tilde{T}_1\tilde{x}] \to [\tilde{T}_2\tilde{x}] \tag{2.7}$$

with domain

$$d_{\tilde{V}} = \text{Clos}\{[\tilde{T}_1\tilde{x}], \tilde{x} \in \tilde{X}\} \subseteq \tilde{H}_0, \tag{2.8}$$

and range

$$\Delta_{\tilde{V}} = \text{Clos}\{[\tilde{T}_2\tilde{x}], \tilde{x} \in \tilde{X}\} \subseteq \tilde{H}_0. \tag{2.9}$$

It is easy to see that

$$\tilde{H}_0 \ominus d_{\tilde{V}} = N_{d_V} = H_0 \ominus d_V \tag{2.10}$$

and

$$\tilde{H}_0 \ominus \Delta_{\tilde{V}} = N_{\Delta_V} = H_0 \ominus \Delta_V \tag{2.11}$$

are the same spaces as in Section 1.

One can extend the mapping $F : X \to H^w$ of Section 1 to the mapping

$$\tilde{F} : \tilde{X} \to L^w \tag{2.12}$$

by letting

$$(\tilde{F}\tilde{X})(t) = \begin{bmatrix} w(t) \\ \mathbf{1} \end{bmatrix} \sum_{k=0}^{\infty} t^k e_1^{(k)} + Fx + \begin{bmatrix} \mathbf{1} \\ w(t)^* \end{bmatrix} \sum_{k=-\infty}^{-1} \bar{t}^{|k|} e_2^{(k)}, \tag{2.13}$$

$|t| = 1$, where convergence of the series is understood in L^2 sense. Thus,

$$\begin{aligned} &\tilde{F}\Big| X = F : X \to H^w, \\ &\tilde{F} : \times_{k=\infty}^{0} E_1^{(k)} \to \begin{bmatrix} w \\ \mathbf{1} \end{bmatrix} H_+^2(E_1) \\ &\tilde{F} : \times_{k=-1}^{-\infty} E_2^{(k)} \to \begin{bmatrix} \mathbf{1} \\ w^* \end{bmatrix} H_-^2(E_2), \end{aligned} \tag{2.14}$$

two last subspaces are mapped one-to-one and onto. Obviously, for vectors $\tilde{x}$ with zero x entry

$$\|\tilde{F}\tilde{x}\|_{L^w}^2 = \tilde{D}(\tilde{x}, \tilde{x}). \tag{2.15}$$

Property (1.3) of Section 1 implies now

$$\|\tilde{F}\tilde{x}\|_{L^w}^2 \le \tilde{D}(\tilde{x}, \tilde{x}). \tag{2.16}$$

Analogously to Section 1, the mapping $\tilde{F}$ can be viewed as a mapping from $\tilde{H}_0$,

$$\tilde{F} : \tilde{H}_0 \to L^w.$$

Then (2.16) can be read as

$$\|\tilde{F}\tilde{h}_0\|_{L^w}^2 \le \|\tilde{h}_0\|_{\tilde{H}_0}^2. \tag{2.17}$$

Property (1.2), combined with definitions of $\tilde{V}$ and $\tilde{F}$, reads

$$\tilde{F}\tilde{V}\Big| d_{\tilde{V}} = \bar{t}\tilde{F}\Big| d_{\tilde{V}}. \tag{2.18}$$

Emphasize also the property

$$\begin{aligned} &\tilde{F} \text{ maps } E_1^{(0)} \text{ onto } \begin{bmatrix} w \\ \mathbf{1} \end{bmatrix} E_1 \text{ isometrically}, \\ &\tilde{F} \text{ maps } E_2^{(-1)} \text{ onto } \begin{bmatrix} \mathbf{1} \\ w \end{bmatrix} \bar{t}E_2 \text{ isometrically}, \end{aligned} \tag{2.19}$$

that is crucial in what follows. It turns out that any mapping $\tilde{F} : \tilde{X} \to L^w$ possessing properties (2.17)–(2.19) is of the structure described above. Namely, the following proposition holds true.

Proposition 2.1. Let $\tilde{X}$ and $\tilde{D}$ be the same as above. Let the mapping $\tilde{F} : \tilde{X} \to L^w$ possess the properties (2.17)–(2.19). Let $F = \tilde{F}\Big| X$. Then

(i) $F : X \to H^w$, and F possesses (1.2);
(ii) $\tilde{F} : E_1^{(k)} \to \left[\begin{smallmatrix} w \\ \mathbf{1} \end{smallmatrix}\right] t^k E_1,\ k \ge 0, \qquad \tilde{F} : E_2^{(k)} \to \left[\begin{smallmatrix} \mathbf{1} \\ w^* \end{smallmatrix}\right] \bar{t}^{|k|} E_2,\ k \le -1,$

and

$$\|\tilde{F}\tilde{x}\|^2_{L^w} = \tilde{D}(\tilde{x}, \tilde{x}).$$

for

$$\tilde{x} \in \left(\times_{k=\infty}^{0} E_1^{(k)}\right) \times \{0\} \times \left(\times_{k=-1}^{-\infty} E_2^{(k)}\right)$$

Proof. Properties (ii) follow immediately from (2.19), (2.18) and the definition of $\tilde{V}$. The main point one has to check is

$$\tilde{F} : X \to H^w.$$

According to property (2.17)

$$\begin{bmatrix} \|\tilde{F}\tilde{x}\|^2_{L^w} & \langle \tilde{F}\tilde{y}, \tilde{F}\tilde{x}\rangle_{L^w} \\ \langle \tilde{F}\tilde{x}, \tilde{F}\tilde{y}\rangle_{L^w} & \|\tilde{F}\tilde{y}\|^2_{L^w} \end{bmatrix} \leq \begin{bmatrix} \tilde{D}(\tilde{x}, \tilde{x}) & \tilde{D}(\tilde{y}, \tilde{x}) \\ \tilde{D}(\tilde{x}, \tilde{y}) & \tilde{D}(\tilde{y}, \tilde{y}) \end{bmatrix} \tag{2.20}$$

for any $\tilde{x}, \tilde{y} \in \tilde{X}$. Pick up $\tilde{x} = x \in X$, $\tilde{y} \in \left(\times_{k=\infty}^{0} E_1^{(k)}\right) \times \{0\} \times \left(\times_{k=-1}^{-\infty} E_2^{(k)}\right)$. Then according to the definition of $\tilde{D}$:

$$\tilde{D}(\tilde{x}, \tilde{y}) = 0, \quad \tilde{D}(\tilde{x}, \tilde{x}) = D(x, x). \tag{2.21}$$

As we have already proved

$$\|\tilde{F}\tilde{y}\|^2_{L^w} = \tilde{D}(\tilde{y}, \tilde{y}). \tag{2.22}$$

Substituting (2.21) and (2.22) into (2.20) we obtain

$$\begin{bmatrix} \|\tilde{F}x\|^2_{L^w} & \langle \tilde{F}\tilde{y}, \tilde{F}x\rangle_{L^w} \\ \langle \tilde{F}x, \tilde{F}\tilde{y}\rangle_{L^w} & \tilde{D}(\tilde{y}, \tilde{y}) \end{bmatrix} \leq \begin{bmatrix} D(x, x) & 0 \\ 0 & \tilde{D}(\tilde{y}, \tilde{y}) \end{bmatrix}.$$

In other words

$$\begin{bmatrix} D(x, x) - \|\tilde{F}x\|^2_{L^w} & \langle \tilde{F}\tilde{y}, \tilde{F}x\rangle_{L^w} \\ \langle \tilde{F}x, \tilde{F}\tilde{y}\rangle_{L^w} & 0 \end{bmatrix} \geq 0.$$

Which means that

$$\langle \tilde{F}x, \tilde{F}\tilde{y}\rangle_{L^w} = 0 \tag{2.23}$$

for any $x \in X$ and $\tilde{y}$ of chosen type. But as we know from property (ii) $\tilde{F}\tilde{y}$ runs through $\begin{bmatrix} w \\ \mathbf{1} \end{bmatrix} H^2_+(E_1) \oplus \begin{bmatrix} \mathbf{1} \\ w^* \end{bmatrix} H^2_-(E_2)$. Then (2.23) means that $\tilde{F}x \in H^w$, for any $x \in X$. The first part of assertion (i) follows. The rest of the assertion (i) follows immediately from (2.18) and definition of $\tilde{V}$. This completes the proof. □

We are going to impose now an additional assumption that we will get rid of in Section 4. Till the end of this section and in Section 3 we assume that

$$E_2^{(-1)} \subset \Delta_{\tilde{V}}. \tag{2.24}$$

Then

$$\tilde{V}^{-1} E_2^{(-1)} \subset d_{\tilde{V}} \subseteq \tilde{H}_0. \tag{2.25}$$

Define $\rho_0 : E_1 \oplus E_2 \to \tilde{H}_0$ by

$$\rho_0 : E_1 \to E_1^{(0)}\ ; \quad \rho_0 : E_2 \to \tilde{V}^{-1} E_2^{(-1)}\ . \tag{2.26}$$

By definition

$$\begin{aligned}\|\rho_0 e_1\|_{\tilde{H}_0} &= \|e_1\|_{E_1}\\ \|\rho_0 e_2\|_{\tilde{H}_0} &= \|e_2\|_{E_2}.\end{aligned} \tag{2.27}$$

The whole of ρ_0 need not be an isometry, but one can see that for $e = e_1 \oplus e_2$

$$\|\rho_0 e\|_{\tilde{H}_0} = \sqrt{2}\|e\|_{E_1 \oplus E_2}\ . \tag{2.28}$$

Denote by $\sigma(\zeta)$ the harmonic operator function

$$\sigma(\zeta) = \begin{bmatrix} \mathbb{1} & w(\zeta) \\ w(\zeta)^* & \mathbb{1} \end{bmatrix},\ |\zeta| < 1 \tag{2.29}$$

and by $\sigma(dt)$:

$$\sigma(dt) = \begin{bmatrix} \mathbb{1} & w(t) \\ w(t)^* & \mathbb{1} \end{bmatrix} m(dt),\ |t| = 1 \tag{2.30}$$

the correspondent measure on $\mathbb{T}$, where $m(dt)$ is the normalized Lebesgue measure. The space L^w can be identified with the Hellinger space L^σ (see, e.g. [BDKh], and Appendix to [Kh5] for definitions and notations). Now we are ready to reformulate the AIP.

AIP˜. Let $\tilde{X}$ be a linear space, $\tilde{D}$ be a positive semidefinite form on $\tilde{X}, \tilde{T}_1$ and $\tilde{T}_2$ be two linear operators on $\tilde{X}$, related by the identity

$$\tilde{D}(\tilde{T}_1\tilde{x}, \tilde{T}_1\tilde{y}) = \tilde{D}(\tilde{T}_2\tilde{x}, \tilde{T}_2\tilde{y}) \tag{2.31}$$

for all $\tilde{x}, \tilde{y} \in \tilde{X}$. Let E be a Hilbert space and

$$\rho_0 : E \to \tilde{H}_0 \tag{2.32}$$

be a bounded linear operator, where $\tilde{H}_0$ is the Hilbert space associated to $\tilde{X}$ and to the form $\tilde{D}$ (as it was described above). A positive semidefinite harmonic in the unit disc $\mathbb{D}$ operator-valued (on E) function $\sigma(\zeta)$ (or correspondent operator-valued measure $\sigma(dt)$ on the unit circle $\mathbb{T}$) is said to be a solution of the AIP˜ if there exists a linear mapping

$$\tilde{F} : \tilde{X} \to L^\sigma, \tag{2.33}$$

(where L^σ is the Hellinger space associated to σ, see [BDKh], Appendix to [Kh5]) such that

$$\begin{aligned} &\text{(i)} && \tilde{F}\tilde{T}_2\tilde{x} = \bar{t}\tilde{F}\tilde{T}_1 x\\ &\text{(ii)} && \|\tilde{F}\tilde{x}\|^2_{L^\sigma} \le \tilde{D}(\tilde{x}, \tilde{x})\\ &\text{(iii)} && \tilde{F}\rho_0 e = \sigma(dt)e,\ \forall e \in E. \end{aligned} \tag{2.34}$$

(In the latter property $\tilde{F}$ is understood as the continuation of $\tilde{F}$ to $\tilde{H}_0$).

Remark. If all the data of AIP˜ come from the AIP then these two problems are equivalent and any solution $\sigma(dt)$ of the AIP˜ is of the form (2.30). In fact, in this case

$$U\Big|E_2^{(-1)}\oplus E_2^{(-2)}\oplus\cdots=\tilde{V}\Big|E_2^{(-1)}\oplus E_2^{(-2)}\oplus\cdots$$

for any extension U of V, since $E_2^{(-1)}\oplus E_2^{(-2)}\oplus\cdots\subseteq d_{\tilde{V}}$. Hence, $E_2^{(-1)}$ is a wandering subspace for all U's also. Similarly,

$$U^*\Big|\cdots\oplus E_1^{(1)}\oplus E_1^{(0)}=V^{-1}\Big|\cdots\oplus E_1^{(1)}\oplus E_1^{(0)}$$

implying $E_1^{(0)}$ is also a wandering subspace for all U's which guarantees the desired structure of $\sigma(dt)$. See, e.g. [BDKh] for more details.

In Section 3 we will solve the AIP˜ without any special assumptions about the structure of $\tilde{X},\tilde{T}_1,\tilde{T}_2,E,\rho_0$.

3. Solving the AIP˜

We will omit ˜ throughout this section (although we are talking about AIP˜) to simplify notations. Hopefully this will not lead to any misunderstanding since we are not talking about AIP in this section but only about AIP˜.

3.1. We associate to the data of the AIP˜, the Hilbert space H_0 and the isometry V on it with domain d_V and range Δ_V. N_{d_V} and N_{Δ_V} are orthogonal complements of d_V and Δ_V, respectively. The subspaces N_{d_V} and N_{Δ_V} are called the defect subspaces. To formulate the first theorem of this section we need to remind of some definitions (see, e.g. [BDKh] for more details).

Let U be a unitary linear operator on a Hilbert space L. Let E be another Hilbert space and $\rho: E\to L$ be a bounded linear operator, called the scale. The triple (L,U,ρ) is called the scattering system.

The positive harmonic on $\mathbb{D}$ operator (on E)-valued function $\sigma(\zeta)$

$$\sigma(\zeta)=\rho^*\mathcal{P}_U(\zeta)\rho$$

is called the spectral function of U with respect to the scale ρ. Here $\mathcal{P}_U(\zeta)$ is the Poisson kernel:

$$\begin{aligned}\mathcal{P}_U(\zeta)&=(\mathbf{1}-\zeta U)^{-1}+(\mathbf{1}-\bar{\zeta}U^*)^{-1}-\mathbf{1}\\&=(1-|\zeta|^2)(\mathbf{1}-\zeta U)^{-1}(\mathbf{1}-\bar{\zeta}U^*)^{-1}.\end{aligned}$$

Theorem 3.1. The solution set of the AIP˜ coincides with the set of spectral functions $\sigma(\zeta)$ of unitary extensions U of V with respect to the scale $\rho=\rho_0$. (By unitary extension we understand a unitary operator $U: L\to L$, such that $L\supseteq H_0$ and $U\Big|d_V=V$.) The proof is similar to the correspondent theorem in the context of the AIP (see [KKY], [Kh1], [Kh2], [Kh3], [Kh4] and [BTr]).

3.2. We are interested in explicit parameterization of these functions $\sigma(\zeta)$. To this end we need more detailed information about the structure of unitary extensions. Define the unitary colligation

$$A_0 : N_2 \oplus H_0 \to H_0 \oplus N_1,$$

where N_1 and N_2 are copies of N_{d_V} and N_{Δ_V} respectively, and

$$A_0 \Big| d_V = V,$$

A_0 maps identically N_{d_V} onto N_1 and N_2 onto N_{Δ_V}. Then unitary extensions U of V are feedback couplings of A_0 with unitary colligations A_1 of the form

$$A_1 : N_1 \oplus H_1 \to H_1 \oplus N_2,$$

where N_1 and N_2 are the same spaces as above. The feedback coupling U of the colligation A_0 with the colligation A_1 is a unitary operator acting on the space $L = H_0 \oplus H_1$. We define U along with auxiliary operator $\Gamma : H_0 \oplus H_1 \to N_1 \oplus N_2$ as follows:

$$\begin{aligned} U\begin{pmatrix} h_0 \\ h_1 \end{pmatrix} &\overset{\text{def}}{=} \begin{pmatrix} h_0' \\ h_1' \end{pmatrix}, \\ \Gamma\begin{pmatrix} h_0 \\ h_1 \end{pmatrix} &\overset{\text{def}}{=} \begin{pmatrix} n_1 \\ n_2 \end{pmatrix}, \end{aligned} \tag{3.1}$$

where $h_0\prime, h_1\prime,\ n_1, n_2$ is the unique solution of the following system of linear equations

$$\begin{aligned} A_0(n_2 \oplus h_0) &= h_0' \oplus n_1, \\ A_1(n_1 \oplus h_1) &= h_1' \oplus n_2. \end{aligned} \tag{3.2}$$

Solvability of the system (3.2) and uniqueness of the solution are guaranteed by the property

$$A_0 : N_2 \to H_0. \tag{3.3}$$

Define $i : N_1 \oplus N_2 \to H_0 \oplus H_1$ as

$$i = \Gamma^*, \tag{3.4}$$

and will call it the defect scale correspondent to the coupling U.

Turning now to the data of AIP$^{\sim}$ one can see that it provides us with another scale (we will call it the data scale)

$$\rho = \rho_0 : E \to H_0 \subset L, \tag{3.5}$$

that does not depend on the operator U (according to assumption (2.32) that $\rho_0 : E \to H_0$).

Our goal (according to Theorem 3.1) is computing the spectral function of the scattering system (L, U, ρ). Consider also the scale

$$i + \rho : N_1 \oplus N_2 \oplus E \to L, \tag{3.6}$$

where the sum is defined as i on $N_1 \oplus N_2$ and ρ on E.

Denote the spectral function of the scattering system $(L, U, i+\rho)$ by $\Sigma(\zeta)$, then

$$\Sigma = \begin{bmatrix} \sigma_i & \sigma_{i\rho} \\ \sigma_{i\rho}^* & \sigma_\rho \end{bmatrix} \tag{3.7}$$

where σ_i is the spectral function of U with respect to the scale i, σ_ρ is the one with respect to the scale ρ. We are interested in σ_ρ, but we will compute all the other entries of (3.7) as well.

The Fourier transform is associated to any scattering system (see, e.g. [DBKh]). For a system (L, U, i) it is defined as

$$(\mathcal{F}_{U,i}\ell)(\zeta) = i^*\mathcal{P}_U(\zeta)\ell, \quad \ell \in L.$$

It turns out that (see, e.g. [DBKh])

$$\mathcal{F}_{U,i} : L \to L^{\sigma_i},$$

where L^{σ_i} is the Hellinger space associated to the spectral function σ_i (see Section 2).

3.3. We describe objects related to A_0 and to A_1 separately in this subsection, and will do coupling computations in the next one. Denote the characteristic function of the colligation A_1 by $\omega(\zeta)$. It can be an arbitrary Schur class operator-valued function on $\mathbb{D}$, $\omega(\zeta) : N_1 \to N_2$.

Define the scattering system (L_0, U_0, i_0) associated to the colligation A_0:

$$L_0 = \cdots \oplus N_2^{(1)} \oplus N_2^{(0)} \oplus H_0 \oplus N_1^{(-1)} \oplus N_1^{(-2)} \oplus \cdots, \tag{3.8}$$

where

$$N_2^{(0)} = N_2, \quad N_1^{(-1)} = N_1, \tag{3.9}$$

and other components of (3.8) are copies of N_1 and N_2 respectively, added orthogonally; U_0 acts on $N_2^{(0)} \oplus H_0$ in the same way as A_0 does,

$$U_0 : N_2^{(0)} \oplus H_0 \to H_0 \oplus N_1^{(0)} \tag{3.10}$$

and U_0 acts as shifts on the "tails":

$$\begin{aligned} U_0 &: N_1^{(k)} \to N_1^{(k-1)}, \quad k \leq -1, \\ U_0 &: N_2^{(k)} \to N_2^{(k-1)}, \quad k \geq 1; \end{aligned} \tag{3.11}$$

The scale i_0 is defined as follows:

$$\begin{aligned} &i_0 : N_1 \oplus N_2 \to L_0 \\ &i_0^{(2)} : N_2 \to N_2^{(0)} \quad \text{identically}, \\ &i_0^{(1)} : N_1 \to U_0^* N_1^{(-1)} \stackrel{def}{=} N_1^{(0)} \subseteq H_0. \end{aligned} \tag{3.12}$$

We remark here that the scattering system (L_0, U_0, i_0) coincides with the feedback loading of A_0 with a colligation A_1 whose characteristic function is zero, $\omega(\zeta) \equiv 0$.

Observe that the spectral function of the scattering system (L_0, U_0, i_0) is

$$\begin{bmatrix} \mathbf{1}_{N_1} & s(\zeta) \\ s(\zeta)^* & \mathbf{1}_{N_2} \end{bmatrix} : N_1 \oplus N_2 \to N_1 \oplus N_2, \tag{3.13}$$

where $s(\zeta)$ is the characteristic function of the colligation A_0. $s(\zeta) : N_2 \to N_1$ is a Schur class operator-valued function, and

$$s(0) = 0, \tag{3.14}$$

due to the property (3.3) of the colligation A_0.

We will also need the spectral function of U_0 with respect to the scale $i_0 + \rho_0$, that we denote by $\Sigma_0(\zeta)$:

$$\Sigma_0 = \begin{bmatrix} \mathbf{1}_{N_1} & s & r_1 \\ s^* & \mathbf{1}_{N_2} & r_2^* \\ r_1^* & r_2 & \sigma_0 \end{bmatrix} : N_1 \oplus N_2 \oplus E \to N_1 \oplus N_2 \oplus E. \tag{3.15}$$

Since $\rho_0 : E \to H$, $r_1(\zeta)$ is analytic on $\mathbb{D}$, $r_2(\zeta)^*$ is antianalytic on $\mathbb{D}$, $r_2(0)^* = 0$. Moreover, since $\Sigma_0(\zeta) \geq 0$,

$$\begin{aligned} \|r_1(\zeta)e\|^2_{N_1} &\leq < \sigma_0(\zeta)e, e >, \\ \|r_2(\zeta)^* e\|^2_{N_2} &\leq < \sigma_0(\zeta)e, e > . \end{aligned} \tag{3.16}$$

Since $\sigma_0(\zeta)$ is a harmonic function, (3.16) implies that r_1 is a strong H^2_+ operator function and r_2^* is a strong H^2_- operator function (see [NF]).

The Hellinger space $L^{\sigma_{i_0}}$ can be identified with L^s (see (1.4)). Moreover,

$$\mathcal{F}_{U_0, i_0} : H_0 \to H^s, \tag{3.18}$$

where $H^s = L^s \cap \begin{bmatrix} H^2_+(N_1) \\ H^2_-(N_2) \end{bmatrix}$. According to the definition of i_0 (3.12),

$$i_0^* = \begin{bmatrix} i_0^{(1)*} \\ i_0^{(2)*} \end{bmatrix}, \tag{3.19}$$

and, hence,

$$\mathcal{F}_{U_0, i_0} = \begin{bmatrix} \mathcal{F}_{U_0, i_0^{(1)}} \\ \mathcal{F}_{U_0, i_0^{(2)}} \end{bmatrix}. \tag{3.20}$$

Then (3.18) means that for $h_0 \in H_0$,

$$\begin{aligned} \mathcal{F}_{U_0, i_0^{(1)}} h_0 &\in H^2_+(N_1), \\ \mathcal{F}_{U_0, i_0^{(2)}} h_0 &\in H^2_-(N_2). \end{aligned} \tag{3.21}$$

Similarly, one can associate the scattering system (L_1, U_1, i_1), $i_1 : N_1 \oplus N_2 \to L_1$, to the unitary colligation

$$A_1 : N_1 \oplus H_1 \to H_1 \oplus N_2$$

this way

$$L_1 = \cdots \oplus N_1^{(1)} \oplus N_1^{(0)} \oplus H_1 \oplus N_2^{(-1)} \oplus N_2^{(-2)} \oplus \cdots ,$$
$$U_1 : N_1^{(0)} \oplus H_1 \to H_1 \oplus N_2^{(-1)} \quad \text{like} \quad A_1, \tag{3.22}$$

U_1 acts like shift on the "tails"; $i_1^{(1)} : N_1 \to N_1^{(0)}$, $i_1^{(2)} : N_2 \to U_1^* N_2^{(-1)}$.

The spectral function of the system (L_1, U_1, i_1) is equal to

$$\begin{bmatrix} \mathbb{1}_{N_1} & \omega^* \\ \omega & \mathbb{1}_{N_2} \end{bmatrix}, \tag{3.23}$$

where ω is the characteristic function of the colligation A_1. The correspondent Hellinger space can be identified with L^{ω^*},

$$\mathcal{F}_{U_1,i_1} : L_1 \to L^{\omega^*}.$$

According to definition (3.22) of i_1,

$$i_1^* = \begin{bmatrix} i_1^{(1)*} \\ i_1^{(2)*} \end{bmatrix}, \tag{3.24}$$

and, hence,

$$\mathcal{F}_{U_1,i_1} = \begin{pmatrix} \mathcal{F}_{U_1,i_1^{(1)}} \\ \mathcal{F}_{U_2,i_1^{(2)}} \end{pmatrix}. \tag{3.25}$$

For $h_1 \in H_1$,

$$\mathcal{F}_{U_1,i_1^{(1)}} h_1 \in H_-^2(N_1),$$
$$\mathcal{F}_{U_1,i_1^{(2)}} h_1 \in H_+^2(N_2).$$

Thus

$$\mathcal{F}_{U_1,i_1} : H_1 \to L^{\omega^*} \cap \begin{bmatrix} H_-^2(N_1) \\ H_+^2(N_2) \end{bmatrix} \overset{def}{=} H^{\omega^*}. \tag{3.26}$$

3.4. Now we are in the position to start computing:

Lemma 3.2.

$$\begin{aligned} \mathcal{P}_U(\zeta) &= \begin{bmatrix} \mathcal{P}_{U_0}(\zeta) & 0 \\ 0 & \mathcal{P}_{U_1}(\zeta) \end{bmatrix} + \\ &= \begin{bmatrix} \mathcal{P}_{U_0}^-(\zeta)^* & 0 \\ 0 & \mathcal{P}_{U_1}^-(\zeta)^* \end{bmatrix} \begin{bmatrix} -i_0^{(1)} & i_0^{(2)} \\ i_1^{(1)} & -i_1^{(2)} \end{bmatrix} i^* \mathcal{P}_U^+(\zeta) \\ &+ \begin{bmatrix} \mathcal{P}_{U_0}^+(\zeta)^* & 0 \\ 0 & \mathcal{P}_{U_1}^+(\zeta)^* \end{bmatrix} \begin{bmatrix} i_0^{(1)} & -i_0^{(2)} \\ -i_1^{(1)} & i_1^{(2)} \end{bmatrix} i^* \mathcal{P}_U^-(\zeta). \end{aligned} \tag{3.27}$$

Proof. According to definitions (3.1), (3.2), (3.4) of U and i, (3.12) of i_0, and (3.22) of i_1 one can write

$$\begin{cases} U_0(i_0^{(2)}i^{(2)^*} + P_{H_0})\begin{pmatrix} h_0 \\ h_1 \end{pmatrix} = P_{H_0}U\begin{pmatrix} h_0 \\ h_1 \end{pmatrix} + U_0 i_0^{(1)} i^{(1)^*}\begin{pmatrix} h_0 \\ h_1 \end{pmatrix} \\ U_1(i_1^{(1)}i^{(1)^*} + P_{H_1})\begin{pmatrix} h_0 \\ h_1 \end{pmatrix} = P_{H_1}U\begin{pmatrix} h_0 \\ h_1 \end{pmatrix} + U_1 i_1^{(2)} i^{(2)^*}\begin{pmatrix} h_0 \\ h_1 \end{pmatrix}. \end{cases} \tag{3.28}$$

Substitute in (3.28) $\mathcal{P}_U^+(\zeta)\binom{h_0}{h_1}$ instead of $\binom{h_0}{h_1}$. Since

$$U\mathcal{P}_U^+(\zeta) = \frac{\mathcal{P}_U^+(\zeta) - \mathbf{1}}{\zeta} \tag{3.29}$$

that leads to

$$\begin{cases} U_0(i_0^{(2)}i^{(2)^*}\mathcal{P}_U^+(\zeta) + P_{H_0}\mathcal{P}_U^+(\zeta)) = P_{H_0}\dfrac{\mathcal{P}_U^+(\zeta) - \mathbf{1}}{\zeta} + U_0 i_0^{(1)} i^{(1)^*}\mathcal{P}_U^+(\zeta) \\ U_1(i_1^{(1)}i^{(1)^*}\mathcal{P}_U^+(\zeta) + P_{H_1}\mathcal{P}_U^+(\zeta)) = P_{H_1}\dfrac{\mathcal{P}_U^+(\zeta) - \mathbf{1}}{\zeta} + U_1 i_1^{(2)} i^{(2)^*}\mathcal{P}_U^+(\zeta) \end{cases} \tag{3.30}$$

which can be rearranged as follows:

$$\begin{cases} (\mathbf{1} - \zeta U_0)P_{H_0}\mathcal{P}_U^+(\zeta) = P_{H_0} + \zeta U_0(i_0^{(2)}i^{(2)^*} - i_0^{(1)}i^{(1)^*})\mathcal{P}_U^+(\zeta) \\ (\mathbf{1} - \zeta U_1)P_{H_1}\mathcal{P}_U^+(\zeta) = P_{H_1} + \zeta U_1(i_1^{(1)}i^{(1)^*} - i_1^{(2)}i^{(2)^*})\mathcal{P}_U^+(\zeta). \end{cases} \tag{3.31}$$

Or,

$$\begin{cases} P_{H_0}\mathcal{P}_U^+(\zeta) = \mathcal{P}_{U_0}^+(\zeta)P_{H_0} + \mathcal{P}_{U_0}^-(\zeta)^*[-i_0^{(1)}, i_0^{(2)}]i^*\mathcal{P}_U^+(\zeta) \\ P_{H_1}\mathcal{P}_U^+(\zeta) = \mathcal{P}_{U_1}^+(\zeta)P_{H_1} + \mathcal{P}_{U_1}^-(\zeta)^*[i_1^{(1)}, -i_1^{(2)}]i^*\mathcal{P}_U^+(\zeta). \end{cases} \tag{3.32}$$

Finally,

$$\begin{aligned} \mathcal{P}_U^+(\zeta) &= \begin{bmatrix} \mathcal{P}_{U_0}^+(\zeta) & 0 \\ 0 & \mathcal{P}_{U_1}^+(\zeta) \end{bmatrix} \\ &+ \begin{bmatrix} \mathcal{P}_{U_0}^-(\zeta)^* & 0 \\ 0 & \mathcal{P}_{U_1}^-(\zeta)^* \end{bmatrix} \begin{bmatrix} -i_0^{(1)} & i_0^{(2)} \\ i_1^{(1)} & -i_1^{(2)} \end{bmatrix} i^*\mathcal{P}_U^+(\zeta). \end{aligned} \tag{3.33}$$

Similarly, substituting $\mathcal{P}_U^-(\zeta)\binom{h_0}{h_1}$ in (3.28) instead of $\binom{h_0}{h_1}$, and using

$$U\mathcal{P}_U^-(\zeta) = \bar{\zeta}(\mathcal{P}_U^-(\zeta) + \mathbf{1}), \tag{3.34}$$

one obtains

$$\begin{cases} U_0(i_0^{(2)}i^{(2)^*}\mathcal{P}_U^-(\zeta) + P_{H_0}\mathcal{P}_U^-(\zeta)) = \bar{\zeta}P_{H_0}(\mathcal{P}_U^-(\zeta) + \mathbf{1}) + U_0 i_0^{(1)} i^{(1)^*}\mathcal{P}_U^-(\zeta), \\ U_1(i_1^{(1)}i^{(1)^*}\mathcal{P}_U^- + P_{H_1}\mathcal{P}_U^-(\zeta)) = \bar{\zeta}P_{H_1}(\mathcal{P}_U^-(\zeta) + \mathbf{1}) + U_1 i_1^{(2)} i^{(2)^*}\mathcal{P}_U^-(\zeta). \end{cases} \tag{3.35}$$

Multiplying the first equation by U_0^* and the second by U_1^* we get to

$$\begin{cases} (\mathbf{1} - \bar{\zeta}U_0^*)P_{H_0}\mathcal{P}_U^-(\zeta) = \bar{\zeta}U_0^* P_{H_0} + \begin{bmatrix} i_0^{(1)} & -i_0^{(2)} \end{bmatrix} i^*\mathcal{P}_U^-(\zeta) \\ (\mathbf{1} - \bar{\zeta}U_1^*)P_{H_1}\mathcal{P}_U^-(\zeta) = \bar{\zeta}U_1^* P_{H_1} + \begin{bmatrix} -i_1^{(1)} & i_1^{(2)} \end{bmatrix} i^*\mathcal{P}_U^-(\zeta) \end{cases} \tag{3.36}$$

which leads to

$$\mathcal{P}_U^-(\zeta) = \begin{bmatrix} \mathcal{P}_{U_0}^-(\zeta) & 0 \\ 0 & \mathcal{P}_{U_1}^-(\zeta) \end{bmatrix} + \begin{bmatrix} \mathcal{P}_{U_0}^+(\zeta)^* & 0 \\ 0 & \mathcal{P}_{U_1}^+(\zeta)^* \end{bmatrix} \begin{bmatrix} i_0^{(1)} & -i_0^{(2)} \\ -i_1^{(1)} & i_1^{(2)} \end{bmatrix} i^*\mathcal{P}_U^-(\zeta). \tag{3.37}$$

Combining (3.33) and (3.37) we obtain (3.27). Thus, the Lemma follows. □

Lemma 3.3.

$$\begin{aligned} (\mathcal{F}_{U,i}\begin{bmatrix} h_0 \\ h_1 \end{bmatrix})(\zeta) &= \left(\mathbf{1} - \begin{bmatrix} 0 & s(\zeta) \\ \omega(\zeta) & 0 \end{bmatrix}\right)^{-1} \begin{bmatrix} (\mathcal{F}_{U_0,i_0^{(1)}}h_0)(\zeta) \\ (\mathcal{F}_{U_1,i_1^{(2)}}h_1)(\zeta) \end{bmatrix} \\ &+ \left(\mathbf{1} - \begin{bmatrix} 0 & \omega(\zeta)^* \\ s(\zeta)^* & 0 \end{bmatrix}\right)^{-1} \begin{bmatrix} (\mathcal{F}_{U_1,i_1^{(1)}}h_1)(\zeta) \\ (\mathcal{F}_{U_0,i_0^{(2)}}h_0)(\zeta) \end{bmatrix}. \end{aligned} \tag{3.38}$$

Proof. Apply $\mathcal{F}_{U_0,i_0^{(1)}}$ to the first entry of (3.33) and use the property that

$$\begin{aligned} (\mathcal{F}_{U_0,i_0}\mathcal{P}_{U_0}^+(\zeta)h_0)(t) &= \frac{(\mathcal{F}_{U_0,i_0}h_0)(t)}{1-\bar{t}\zeta} \\ (\mathcal{F}_{U_0,i_0}\mathcal{P}_{U_0}^-(\zeta)h_0)(t) &= \frac{(\mathcal{F}_{U_0,i_0}h_0)(t)}{1-t\bar{\zeta}}t\bar{\zeta}, \quad \text{a.e. } ,|t|=1. \end{aligned} \tag{3.39}$$

Then

$$\begin{aligned} &(\mathcal{F}_{U_0,i_0^{(1)}}P_{H_0}\mathcal{P}_{U_0}^+(\zeta)\begin{pmatrix} h_0 \\ h_1 \end{pmatrix})(t) \\ &= \frac{(\mathcal{F}_{U_0,i_0^{(1)}}h_0)(t)}{1-\bar{t}\zeta} + \frac{\bar{t}\zeta}{1-\bar{t}\zeta}\mathcal{F}_{U_0,i_0^{(1)}}(\begin{bmatrix} -i_0^{(1)} & i_0^{(2)} \end{bmatrix})(t)\cdot i^*\mathcal{P}_U^+(\zeta)\begin{pmatrix} h_0 \\ h_1 \end{pmatrix} \\ &= \frac{t(\mathcal{F}_{U_0,i_0^{(1)}}h_0)(t)}{t-\zeta} + \frac{\zeta}{t-\zeta}\begin{bmatrix} -\mathbf{1} & s(t) \end{bmatrix}\cdot i^*\mathcal{P}_U^+(\zeta)\begin{pmatrix} h_0 \\ h_1 \end{pmatrix}. \end{aligned} \tag{3.40}$$

The latter equality follows from (3.13). According to (3.21), the left-hand side of (3.40) is in $H_+^2(N_1)$, which implies

$$(\mathcal{F}_{U_0,i_0^{(1)}}h_0)(\zeta) + \begin{bmatrix} -\mathbf{1} & s(\zeta) \end{bmatrix}\cdot i^*\mathcal{P}_U^+(\zeta)\begin{pmatrix} h_0 \\ h_1 \end{pmatrix} = 0. \tag{3.41}$$

In the same way, applying $\mathcal{F}_{U_1,i_1^{(2)}}$ to the second entry of (3.33) we obtain

$$(\mathcal{F}_{U_1,i_1^{(2)}}h_1)(\zeta) + \begin{bmatrix} \omega(\zeta) & -\mathbf{1} \end{bmatrix}\cdot i^*\mathcal{P}_U^+(\zeta)\begin{pmatrix} h_0 \\ h_1 \end{pmatrix} = 0. \tag{3.42}$$

(3.41) and (3.42) together give

$$\begin{bmatrix} \mathbf{1} & -s(\zeta) \\ -\omega(\zeta) & \mathbf{1} \end{bmatrix} i^* \mathcal{P}_U^+(\zeta) \begin{bmatrix} h_0 \\ h_1 \end{bmatrix} = \begin{bmatrix} (\mathcal{F}_{U_0, i_0^{(1)}} h_0)(\zeta) \\ (\mathcal{F}_{U_1, i_1^{(2)}} h_1)(\zeta) \end{bmatrix}, \tag{3.43}$$

which leads to the first term in (3.38). The second term comes from (3.37) by similar reasons. Thus, the Lemma follows. □

Theorem 3.4 The following formulas hold true ($|\zeta| < 1$)

$$\sigma_i = \frac{1}{2}\left(\frac{\mathbf{1} + \begin{bmatrix} 0 & s \\ \omega & 0 \end{bmatrix}}{\mathbf{1} - \begin{bmatrix} 0 & s \\ \omega & 0 \end{bmatrix}} + \frac{\mathbf{1} + \begin{bmatrix} 0 & \omega^* \\ s^* & 0 \end{bmatrix}}{\mathbf{1} - \begin{bmatrix} 0 & \omega^* \\ s^* & 0 \end{bmatrix}} \right), \tag{3.44}$$

$$\sigma_{i\rho} = \left(\mathbf{1} - \begin{bmatrix} 0 & s \\ \omega & 0 \end{bmatrix}\right)^{-1} \begin{bmatrix} r_1 \\ 0 \end{bmatrix} + \left(\mathbf{1} - \begin{bmatrix} 0 & \omega^* \\ s^* & 0 \end{bmatrix}\right)^{-1} \begin{bmatrix} 0 \\ r_2^* \end{bmatrix} \tag{3.45}$$

$$\sigma_\rho = \sigma_0 + r_2 \omega (\mathbf{1} - s\omega)^{-1} r_1 + r_1^* (\mathbf{1} - \omega^* s^*)^{-1} \omega^* r_2^* \tag{3.46}$$

where $\sigma_1, \sigma_{i\rho}, \sigma_\rho$ are entries of (3.7), s, r_1, r_2, σ_0 are defined by (3.13), (3.15), ω by (3.23).

Proof. It suffices to substitute $\binom{h_0}{h_1} = i\binom{n_1}{n_2}$ and $\binom{h_0}{h_1} = \rho e$ in (3.38) to obtain (3.44) and (3.45), respectively. Substitute first

$$\begin{pmatrix} h_0 \\ h_1 \end{pmatrix} = \rho e. \tag{3.47}$$

According to our assumption $\rho = \rho_0 : E \to H_0$. Hence,

$$h_0 = \rho_0 e, \quad h_1 = 0$$

in (3.47). But now, according to Definition (3.15)

$$\begin{aligned} (\mathcal{F}_{U_0, i_0^{(1)}} \rho_0 e)(\zeta) &= i_0^{(1)^*} \mathcal{P}_{U_0}(\zeta) \rho_0 e = r_1(\zeta) e \\ (\mathcal{F}_{U_0, i_0^{(2)}} \rho_0 e)(\zeta) &= i_0^{(2)^*} \mathcal{P}_{U_0}(\zeta) \rho_0 e = r_2^*(\zeta) e. \end{aligned} \tag{3.48}$$

Substituting (3.48) in (3.38) we get (3.45). We need a new expression for i to compute (3.44). Letting $\zeta = 0$ in (3.38) one gets

$$i^* \begin{bmatrix} h_0 \\ h_1 \end{bmatrix} = \begin{bmatrix} \mathbf{1} & 0 \\ \omega(0) & \mathbf{1} \end{bmatrix} \begin{bmatrix} i_0^{(1)^*} h_0 \\ i_1^{(2)^*} h_1 \end{bmatrix}. \tag{3.49}$$

Therefore,

$$i \begin{bmatrix} n_1 \\ n_2 \end{bmatrix} = \begin{bmatrix} i_0^{(1)} & 0 \\ 0 & P_{H_1} i_1^{(2)} \end{bmatrix} \begin{bmatrix} \mathbf{1} & \omega(0)^* \\ 0 & \mathbf{1} \end{bmatrix} \begin{bmatrix} n_1 \\ n_2 \end{bmatrix}. \tag{3.50}$$

According to Definition (3.22)

$$i_1^{(2)} N_2 = U_1^* N_2^{(-1)},$$

hence,

$$i_1^{(2)} N_2 \in N_1^{(0)} \oplus H_1.$$

Then

$$P_{H_1} i_1^{(2)} = i_1^{(2)} - P_{N_1^{(0)}} i_1^{(2)}.$$

But again, according to Definition (3.22)

$$P_{N_1^{(0)}} = i_1^{(1)} i_1^{(1)*}.$$

Thus

$$P_{H_1} i_1^{(2)} = i_1^{(2)} - i_1^{(1)} i_1^{(1)*} i_1^{(2)}.$$

But according to (3.23),

$$i_1^{(1)*} i_1^{(2)} = \omega(0)^*.$$

Hence

$$P_{H_1} i_1^{(2)} = i_1^{(2)} - i_1^{(1)} \omega(0)^*.$$

Now (3.50) looks like

$$i^* \begin{bmatrix} n_1 \\ n_2 \end{bmatrix} = \begin{bmatrix} i_0^{(1)} \ (n_1 + \omega(0)^* n_2) \\ i_1^{(2)} \ n_2 - i_1^{(1)} \omega(0)^* n_2 \end{bmatrix}. \tag{3.51}$$

This we are going to substitute in (3.38). Remind that

$$\begin{aligned} &\mathcal{F}_{U_0, i_0^{(1)}} i_0^{(1)} = \mathbf{1}_{N_1} \\ &\mathcal{F}_{U_0, i_0^{(2)}} i_0^{(1)} = s^* \\ &\mathcal{F}_{U_1, i_1^{(1)}} (i_1^{(2)} - i_1^{(1)} \omega(0)^*) = \omega^* - \omega(0)^* \\ &\mathcal{F}_{U_1, i_1^{(2)}} (i_1^{(2)} - i_1^{(1)} \omega(0)^*) = \mathbf{1} - \omega\omega(0)^*. \end{aligned} \tag{3.52}$$

Now (3.38) turns into

$$\begin{aligned} \sigma_i(\zeta) \begin{bmatrix} n_1 \\ n_2 \end{bmatrix} = {} & \left(\mathbf{1} - \begin{bmatrix} 0 & s(\zeta) \\ \omega(\zeta) & 0 \end{bmatrix} \right)^{-1} \begin{bmatrix} n_1 + \omega(0)^* n_2 \\ (\mathbf{1} - \omega(\zeta)\omega(0)^*) n_2 \end{bmatrix} \\ & + \left(\mathbf{1} - \begin{bmatrix} 0 & \omega(\zeta)^* \\ s(\zeta)^* & 0 \end{bmatrix} \right)^{-1} \begin{bmatrix} \omega(\zeta)^* - \omega(0)^*) n_2 \\ s(\zeta)^* (n_1 + \omega(0)^* n_2) \end{bmatrix}. \end{aligned}$$

Meaning that

$$\begin{aligned}
\sigma_i(\zeta) = & \left(\mathbf{1} - \begin{bmatrix} 0 & s(\zeta) \\ \omega(\zeta) & 0 \end{bmatrix}\right)^{-1} \begin{bmatrix} \mathbf{1} & \omega(0)^* \\ 0 & \mathbf{1} - \omega(\zeta)\omega(0)^* \end{bmatrix} \\
& + \left(\mathbf{1} - \begin{bmatrix} 0 & \omega(\zeta)^* \\ s(\zeta)^* & 0 \end{bmatrix}\right)^{-1} \begin{bmatrix} 0 & \omega(\zeta)^* - \omega(0)^*) \\ s(\zeta)^* & s(\zeta)^*\omega(0)^* \end{bmatrix} \\
= & \left(\mathbf{1} - \begin{bmatrix} 0 & s(\zeta) \\ \omega(\zeta) & 0 \end{bmatrix}\right)^{-1} + \left(\mathbf{1} - \begin{bmatrix} 0 & \omega(\zeta)^* \\ s(\zeta)^* & 0 \end{bmatrix}\right)^{-1} \\
& \cdot \begin{bmatrix} 0 & \omega(\zeta)^* \\ s(\zeta)^* & 0 \end{bmatrix} \\
& + \left(\mathbf{1} - \begin{bmatrix} 0 & s(\zeta) \\ \omega(\zeta) & 0 \end{bmatrix}\right)^{-1} + \begin{bmatrix} 0 & \omega(0)^* \\ 0 & -\omega(\zeta)\omega(0)^* \end{bmatrix} \\
& - \left(\mathbf{1} - \begin{bmatrix} 0 & \omega(\zeta)^* \\ s(\zeta)^* & 0 \end{bmatrix}\right)^{-1} \begin{bmatrix} 0 & \omega(0)^* \\ 0 & -s(\zeta)^*\omega(0)^* \end{bmatrix} .
\end{aligned} \tag{3.53}$$

The third term in (3.53) equals to

$$\left(\mathbf{1} - \begin{bmatrix} 0 & s(\zeta) \\ \omega(\zeta) & 0 \end{bmatrix}\right)^{-1} \left(\mathbf{1} - \begin{bmatrix} 0 & s(\zeta) \\ \omega(\zeta) & 0 \end{bmatrix}\right) \begin{bmatrix} 0 & \omega(0)^* \\ 0 & 0 \end{bmatrix} = \begin{bmatrix} 0 & \omega(0)^* \\ 0 & 0 \end{bmatrix}.$$

Similarly the last term in (3.35) is equal to

$$\left(\mathbf{1} - \begin{bmatrix} 0 & \omega(\zeta)^* \\ s(\zeta)^* & 0 \end{bmatrix}\right)^{-1} \left(\mathbf{1} - \begin{bmatrix} 0 & \omega(\zeta)^* \\ s(\zeta)^* & 0 \end{bmatrix}\right) \begin{bmatrix} 0 & \omega(0)^* \\ 0 & 0 \end{bmatrix} = \begin{bmatrix} 0 & \omega(0)^* \\ 0 & 0 \end{bmatrix}.$$

Thus, the two last terms in (3.53) are cancelled. Then (3.53) can be transformed to the claimed from (3.44) by subtracting $\frac{1}{2}\mathbf{1}$ from the first term and adding it to the second one.

We turn now to proving formula (3.46). We have to compute (see (3.47))

$$\sigma_\rho(\zeta) = \rho^*\mathcal{P}_U(\zeta)\rho = [\rho_0^* \quad 0]\,\mathcal{P}_U(\zeta)\begin{bmatrix} \rho_0 \\ 0 \end{bmatrix}. \tag{3.54}$$

Then, according to (3.27) and (3.45)

$$\begin{aligned}
\rho^*\mathcal{P}_U(\zeta)\rho &= \rho_0^*\mathcal{P}_{U_0}(\zeta)\rho_0 \\
&+ \rho_0^*\mathcal{P}_{U_0}^-(\zeta)^* \begin{bmatrix}-i_0^{(1)} & i_0^{(2)}\end{bmatrix} i^*\mathcal{P}_U^+(\zeta)\begin{bmatrix}\rho_0\\0\end{bmatrix} \\
&+ \rho_0^*\mathcal{P}_{U_0}^+(\zeta)^* \begin{bmatrix}i_0^{(1)} & -i_0^{(2)}\end{bmatrix} i^*\mathcal{P}_U^-(\zeta)\begin{bmatrix}\rho_0\\0\end{bmatrix} \\
&= \sigma_0(\zeta) + \begin{bmatrix}0 & r_2(\zeta)\end{bmatrix}\begin{bmatrix}\mathbf{1} & -s(\zeta)\\ -\omega(\zeta) & \mathbf{1}\end{bmatrix}^{-1}\begin{bmatrix}r_1(\zeta)\\0\end{bmatrix} \\
&+ \begin{bmatrix}r_1(\zeta)^* & 0\end{bmatrix}\begin{bmatrix}\mathbf{1} & -\omega(\zeta)^*\\ -s(\zeta)^* & \mathbf{1}\end{bmatrix}^{-1}\begin{bmatrix}0\\ r_2(\zeta)^*\end{bmatrix} \\
&= \sigma_0(\zeta) + r_2(\zeta)\omega(\zeta)(\mathbf{1}-s(\zeta)\omega(\zeta))^{-1}r_1(\zeta) \\
&+ r_1(\zeta)^*(\mathbf{1}-\omega(\zeta)^*s(\zeta)^*)^{-1}\omega(\zeta)^*r_2(\zeta)^*.
\end{aligned}$$

At the last stage we used the formula

$$\begin{bmatrix}\mathbf{1} & -s\\ -\omega & \mathbf{1}\end{bmatrix}^{-1} = \begin{bmatrix}(\mathbf{1}-s\omega)^{-1} & s(\mathbf{1}-\omega s)^{-1}\\ \omega(\mathbf{1}-s\omega)^{-1} & (\mathbf{1}-\omega s)^{-1}\end{bmatrix}, \tag{3.55}$$

that is valid for all $|\zeta| < 1$, since $s(0) = 0$ (see (3.14)). □

Remark. In the case of AIP (see Section 2), $E = E_1 \oplus E_2$, $\rho_0(E_1)$ and $\rho_0(E_2)$ are wandering subspaces for U_0 with orthogonal semichannels. Then σ_0 is of the form

$$\sigma_0 = \begin{bmatrix}\mathbf{1} & s_0^*\\ s_0 & \mathbf{1}\end{bmatrix},$$

where s_0 is a Schur class function, r_1 and r_2 are of the form

$$r_2 = \begin{bmatrix}0\\ s_2\end{bmatrix}, \quad r_1 = \begin{bmatrix}s_1 & 0\end{bmatrix}$$

in this case. Then σ_ρ is also of the form

$$\begin{bmatrix}\mathbf{1} & w^*\\ w & \mathbf{1}\end{bmatrix},$$

where

$$w = s_0 + s_2\omega(\mathbf{1}-s\omega)^{-1}s_1,$$

and w is a Schur class function. The latter formula is known as the Arov-Grossman formula ([AG]).

4. Wave operators and induced scales

In this section we are going to get rid of the assumption

$$\rho_0 : E \to H_0 \ ,$$

imposed in Section 2 and used throughout Section 3. Now,

$$\rho_0 : E \to L_0 \ ,$$

where

$$L_0 = \cdots \oplus N_2^{(1)} \oplus N_2^{(0)} \oplus H_0 \oplus N_1^{(-1)} \oplus N_1^{(-2)} \oplus \cdots \tag{4.1}$$

(see Section 3.3). We cannot compute now the spectral function of U with respect to ρ_0 (as we did in Section 3.4) since the range of ρ_0 does not belong to the space L the extension U acts on. The problem is to realize what the natural scale $\rho : E \to L$ is, that we have to compute the spectral function of U with respect to. The scale ρ will, generically, depend on the extension U. We are going to demonstrate in another opportunity some examples where the very choice of the ρ corresponds to concrete problems of analysis.

Remark here, that for σ_i the same formula (3.44) remains valid, since it deals neither with ρ, nor with ρ_0. To define the scale ρ correspondent to the extension U we will need the wave operator.

4.1. Define the wave operator $W : L_0 \to L$ of an extension U with respect to the data scattering system U_0 this way:

$$W : H_0 \to H_0 \quad \text{identically,} \tag{4.2}$$

$$W|N_1^{(k)} = U^{-k} U_0^k | N_1^{(k)} \ , \ k \leq -1 \ , \tag{4.3}$$

$$W|N_2^{(k)} = U^{-k-1} U_0^{k+1} | N_2^{(k)} \ , \ k \geq 0 \ . \tag{4.4}$$

W extends then by linearity. Thus, W is at least densely defined on L_0, but generically it need not be a bounded operator.

Lemma 4.1.

$$i^{(1)} = i_0^{(1)} \tag{4.5}$$

$$i^{(2)} = U^* U_0 i_0^{(2)} \ . \tag{4.6}$$

Actually (4.5) is contained in (3.51), but we will give here an independent proof.

Proof. Let $\hat{n}_1 \in N_1$, then according to Definition (3.12),

$$i_0^{(1)} \hat{n}_1 = U_0^* (0 \oplus \hat{n}_1) \ . \tag{4.7}$$

For $h_0 \in H_0$ and $h_1 \in H_1$, denote

$$\begin{aligned} h_0' \oplus h_1' &= U(h_0 \oplus h_1) \\ n_1 \oplus n_2 &= i^*(h_0 \oplus h_1) \ . \end{aligned} \tag{4.8}$$

According to the definition of U and i^* ((3.1),(3.2)) system (4.8) is equivalent to

$$\begin{aligned} A_0(n_2 \oplus h_0) &= h_0' \oplus n_1 \\ A_1(n_1 \oplus h_1) &= h_1' \oplus n_2 \ . \end{aligned} \tag{4.9}$$

Consider now

$$\begin{aligned}\langle i\hat{n}_1, h_0 \oplus h_1\rangle_{H_0\oplus H_1} &= \langle \hat{n}_1 \oplus 0, i^*(h_0 \oplus h_1)\rangle_{N_1\oplus N_2}\\ &= \langle \hat{n}_1, n_1\rangle_{N_1} \quad (n_1 \text{ is from } (4.8))\\ &= \langle 0 \oplus \hat{n}_1, h_0' \oplus n_1\rangle_{H_0\oplus N_1}\\ &= \langle 0 \oplus \hat{n}_1, A_0(n_2 \oplus h_0)\rangle \quad (\text{because of } (4.9))\\ &= \langle A_0^*(0 \oplus \hat{n}_1), n_2 \oplus h_0\rangle_{N_2\oplus H_0}\\ &= \langle i_0\hat{n}_1, h_0\rangle_{H_0}\\ &= \langle i_0\hat{n}_1, h_0 \oplus h_1\rangle_{H_0\oplus H_1} .\end{aligned}$$

Hence

$$i\hat{n}_1 = i_0\hat{n}_1 ,$$

which is (4.5). Similarly, for $\hat{n}_2 \in N_2$,

$$i_0\hat{n}_2 = \hat{n}_2 \oplus 0 \in N_2 \oplus H_0 .$$

Then

$$\begin{aligned}&\langle i\hat{n}_2, h_0 \oplus h_1\rangle_{H_0\oplus H_1} = \langle 0 \oplus \hat{n}_2, i^*(h_0 \oplus h_1)\rangle_{N_1\oplus N_2}\\ &\quad= \langle \hat{n}_2, n_2\rangle_{N_2} = \langle \hat{n}_2 \oplus 0, n_2 \oplus h_0\rangle_{N_2\oplus H_0}\\ &\quad= \langle A_0(\hat{n}_2 \oplus 0), A_0(n_2 \oplus h_0)\rangle_{H_0\oplus N_1} = \langle A_0(\hat{n}_2 \oplus 0), h_0' \oplus n_1\rangle_{H_0\oplus N_1}\\ &\quad= \langle A_0(\hat{n}_2 \oplus 0), h_0'\rangle_{H_0} = \langle A_0 i_0\hat{n}_2, h_0' \oplus h_1'\rangle_{H_0\oplus H_1}\\ &\quad= \langle U_0 i_0\hat{n}_2, U(h_0 \oplus h_1)\rangle_{H_0\oplus H_1} = \langle U^*U_0 i_0\hat{n}_2, h_0 \oplus h_1\rangle_{H_0\oplus H_1} .\end{aligned}$$

Hence

$$i\hat{n}_2 = U^*U_0 i_0\hat{n}_2 ,$$

which is (4.6). The lemma follows. □

Remark. (4.5) and (4.6) mean that

$$W i_0 = i . \tag{4.10}$$

In fact, since $i_0^{(1)} : N_1 \to U_0^* N_1^{(-1)} \subseteq H_0$, then $W i_0^{(1)} = i_0^{(1)}$. Since $i_0^{(2)} : N_2 \to N_2^{(0)}$, then $W i_0^{(2)} = U^* U_0 i_0^{(2)}$.

A vector $\ell_0 \in L_0$ we will call a finite vector if its projection onto $N_1^{(-k)}$ and $N_2^{(k)}$ is zero except for a finite number of indexes. All finite vectors are obviously in the domain of W.

Lemma 4.2.

$$W = \sum_{k=\infty}^{0} U^{-k} i^{(2)} i_0^{(2)^*} U_0^k + P_{H_0} + \sum_{k=-1}^{-\infty} U^{-k} i^{(1)} i_0^{(1)^*} U_0^k . \tag{4.11}$$

(For finite vectors ℓ_0, only a finite number of terms appear in the formula.)

Proof. Consider orthogonal decomposition of identity of the space L_0:

$$\begin{aligned} I &= \sum_{k=\infty}^{0} P_{N_2^{(k)}} + P_{H_0} + \sum_{k=-1}^{\infty} P_{N_1^{(k)}} \\ &= \sum_{k=\infty}^{0} U_0^{-k} P_{N_2^{(0)}} U_0^k + P_{H_0} + \sum_{k=-1}^{-\infty} U_0^{-k} P_{N_1^{(0)}} U_0^k . \end{aligned} \tag{4.12}$$

Remind that $N_1^{(0)} = U_0^* N_1^{(-1)}$. According to Definitions (3.12),

$$\begin{aligned} P_{N_2^{(0)}} &= i_0^{(2)} i_0^{(2)*} \\ P_{N_1^{(0)}} &= i_0^{(1)} i_0^{(1)*} . \end{aligned} \tag{4.13}$$

Then

$$I = \sum_{k=\infty}^{0} U_0^{-k} i_0^{(2)} i_0^{(2)*} U_0^k + P_{H_0} + \sum_{k=-1}^{-\infty} U_0^{-k} i_0^{(1)} i_0^{(1)*} U_0^k . \tag{4.14}$$

Apply now W to (4.14):

$$\begin{aligned} W &= \sum_{k=\infty}^{0} U^{-k-1} U_0^{k+1} \cdot U_0^{-k} i_0^{(2)} i_0^{(2)*} U_0^k + P_{H_0} + \sum_{k=-1}^{-\infty} U^{-k} U_0^k \cdot U_0^{-k} i_0^{(1)} i_0^{(1)*} U_0^k \\ &= \sum_{k=\infty}^{0} U^{-k} i^{(2)} i_0^{(2)*} U_0^k + P_{H_0} + \sum_{k=-1}^{-\infty} U^{-k} i^{(1)} i_0^{(1)*} U_0^k , \end{aligned}$$

by Lemma 4.1. The lemma follows. □

4.2. In this section we obtain a formula that leads to a generalization of the formula (3.45). For $\ell_0 \in L_0$, denote

$$(\mathcal{F}_{U_0, i_0} \ell_0)(t) = \begin{bmatrix} n_1(t) \\ n_2(t) \end{bmatrix} \in L^s \tag{4.15}$$

(see Section 3.3). $\mathcal{F}_{U_0,i_0}$ sends H_0 onto H^s, that is for $\ell_0 \in H_0$, $n_1 \in H_+^2(N_1)$, $n_2 \in H_-^2(N_2)$. Remind that $\mathcal{F}_{U_0,i_0}$ is isometric on a part of H_0 and on $H_0^\perp$, and

$$\begin{aligned} \mathcal{F}_{U_0,i_0} &: \cdots \oplus N_2^{(1)} \oplus N_2^{(0)} \to \begin{bmatrix} s \\ \mathbf{1} \end{bmatrix} H_+^2(N_2) , \\ \mathcal{F}_{U_0,i_0} &: N_1^{(-1)} \oplus N_2^{(-2)} \oplus \cdots \to \begin{bmatrix} \mathbf{1} \\ s^* \end{bmatrix} H_-^2(N_1) . \end{aligned} \tag{4.16}$$

Any vector $\begin{bmatrix} n_1 \\ n_2 \end{bmatrix} \in L^s$ can be decomposed as

$$\begin{aligned} \begin{bmatrix} n_1 \\ n_2 \end{bmatrix} &= \begin{bmatrix} s \\ \mathbf{1} \end{bmatrix} n_2^+ \oplus \left(\begin{bmatrix} n_1 \\ n_2 \end{bmatrix} - \begin{bmatrix} \mathbf{1} & s \\ s^* & \mathbf{1} \end{bmatrix} \begin{bmatrix} n_1^- \\ n_2^+ \end{bmatrix} \right) \oplus \begin{bmatrix} \mathbf{1} \\ s^* \end{bmatrix} n_1^- \\ &= \begin{bmatrix} s \\ \mathbf{1} \end{bmatrix} n_2^+ \oplus \begin{bmatrix} n_1^+ - s n_2^+ \\ n_2^- - s^* n_1^- \end{bmatrix} \oplus \begin{bmatrix} \mathbf{1} \\ s^* \end{bmatrix} n_1^- \end{aligned} \tag{4.17}$$

in accordance with the decomposition

$$L^s = \begin{bmatrix} s \\ \mathbf{1} \end{bmatrix} H^2_+(N_2) \oplus H^s \oplus \begin{bmatrix} \mathbf{1} \\ s^* \end{bmatrix} H^2_-(N_1) \ . \tag{4.18}$$

Lemma 4.3.

$$\begin{aligned} \mathcal{F}_{U,i} W \ell_0 = &\left(\mathbf{1} - \begin{bmatrix} 0 & s \\ \omega & 0 \end{bmatrix} \right)^{-1} \begin{bmatrix} n_1^+ - s \cdot n_2^+ \\ 0 \end{bmatrix} \\ &+ \left(\mathbf{1} - \begin{bmatrix} 0 & \omega^* \\ s^* & 0 \end{bmatrix} \right)^{-1} \begin{bmatrix} 0 \\ n_2^- - s^* \cdot n_1^- \end{bmatrix} + \sigma_i \begin{bmatrix} n_1^- \\ n_2^+ \end{bmatrix} , \end{aligned} \tag{4.19}$$

where n_1, n_2 are defined by (4.15).

Remarks.

1. This is well defined at least for finite vectors ℓ_0.
2. The sum of the first two terms is understood as a vector-measure on $\mathbb{T}$ correspondent to the harmonic function on $\mathbb{D}$. The measure belongs to the Hellinger space L^{σ_i}.
3. The last term also belongs to L^{σ_i}.

Proof. Denote

$$\begin{aligned} n_{2,k} &= i_0^{(2)^*} U_0^k \ell_0 \in N_2 \ , \ k \geq 0 \\ n_{1,k} &= i_0^{(1)^*} U_0^k \ell_0 \in N_1 \ , \ k \leq -1 \ , \end{aligned} \tag{4.20}$$

then (see (4.14) and (4.17))

$$n_2^+ = \sum_{k=0}^{\infty} n_{2,k} t^k, \qquad n_1^- = \sum_{k=-\infty}^{-1} n_{1,k} t^k \ . \tag{4.21}$$

For a finite ℓ_0, there is only a finite number of nonvanishing terms. Using Lemma 4.2,

$$W \ell_0 = \sum_{k=\infty}^{0} U^{-k} i^{(2)} n_{2,k} + P_{H_0} \ell_0 + \sum_{k=-1}^{-\infty} U^{-k} i^{(1)} n_{1,k} \ . \tag{4.22}$$

Since

$$\mathcal{F}_{U,i} U = \bar{t} \mathcal{F}_{U,i} \qquad \text{and} \qquad \mathcal{F}_{U,i} i \begin{bmatrix} n_1 \\ n_2 \end{bmatrix} = \sigma_i \begin{bmatrix} n_1 \\ n_2 \end{bmatrix} ,$$

then

$$\mathcal{F}_{U,i} W \ell_0 = \sum_{k=\infty}^{0} t^k \sigma_i \begin{bmatrix} 0 \\ n_{2,k} \end{bmatrix} + \mathcal{F}_{U,i} P_{H_0} \ell_0 + \sum_{k=-1}^{-\infty} t^k \sigma_i \begin{bmatrix} n_{1,k} \\ 0 \end{bmatrix} = \mathcal{F}_{U,i} P_{H_0} \ell_0 + \sigma_i \begin{bmatrix} n_1^- \\ n_2^+ \end{bmatrix} . \tag{4.23}$$

According to (4.17),

$$\mathcal{F}_{U_0, i_0} P_{H_0} \ell_0 = \begin{bmatrix} n_1^+ - s \cdot n_2^+ \\ n_2^- - s^* \cdot n_1^- \end{bmatrix} .$$

(4.19) follows now from (4.23) by applying (3.38) to $h_0 = P_{H_0} \ell_0$ and $h_1 = 0$. The lemma follows. □

4.3. In this section we are going to obtain a formula that leads to a generalization of the formula (3.46). Namely, our goal is computing

$$W^*\mathcal{P}_U(\varsigma)W\ .$$

Denote by $L_0^{i_0}$ the minimal subspace of L_0 that contains $N_1^{(0)}, N_2^{(0)}$ and is invariant under U_0 and U_0^* (see, e.g. [BDKh]). Obviously, $(L_0^{i_0})^\perp \subseteq H_0$. Denote it by H_0'. It is known (see, e.g. [BDKh]) that H_0' is characterized by the property

$$\mathcal{F}_{U_0,i_0}\ell_0 = 0\ . \tag{4.24}$$

Similarly, define L^i to be the minimal subspace of L that contains the range of i and is invariant under U and U^*. Denote by

$$H' = (L^i)^\perp = L \ominus L^i\ . \tag{4.25}$$

Lemma 4.4. $H' = H_0'$.

Proof. The space H' is characterized by the property

$$\mathcal{F}_{U,i}\ell = 0\ . \tag{4.26}$$

According to formula (3.38), for $\ell = h_0 \oplus h_1$ the equality (4.26) holds if and only if

$$\mathcal{F}_{U_0,i_0}h_0 = 0 \tag{4.27}$$

and

$$\mathcal{F}_{U_1,i_1}h_1 = 0\ . \tag{4.28}$$

Since we are dealing with minimal extensions U, the colligation A_1 is simple. Then (4.28) implies $h_1 = 0$. But as it was mentioned above, (4.27) means that

$$h_0 \in H_0'\ .$$

The lemma follows. □

Remark. It follows from formula (4.11) that $W : H_0^\perp \to L^i$. It follows from the previous lemma that actually $W : (H_0')^\perp \to L^i$. If $\ell_0 = h_0 \oplus h_0^\perp$, $h_0 \in H_0$, $h_0^\perp \in H_0^\perp$, then, according to (4.15), (4.17),

$$\mathcal{F}_{U_0,i_0}h_0 = \begin{bmatrix} n_1 \\ n_2 \end{bmatrix} - \begin{bmatrix} \mathbf{1} & s \\ s^* & \mathbf{1} \end{bmatrix}\begin{bmatrix} n_1^- \\ n_2^+ \end{bmatrix}\ , \tag{4.29}$$

$$\mathcal{F}_{U_0,i_0}h_0^\perp = \begin{bmatrix} \mathbf{1} & s \\ s^* & \mathbf{1} \end{bmatrix}\begin{bmatrix} n_1^- \\ n_2^+ \end{bmatrix}\ , \tag{4.30}$$

according to (4.19) $\qquad \mathcal{F}_{U,i}Wh_0 = \mathcal{F}_{U,i}h_0$

$$= \left(\mathbf{1} - \begin{bmatrix} 0 & s \\ \omega & 0 \end{bmatrix}\right)^{-1}\begin{bmatrix} n_1^+ - s \cdot n_2^+ \\ 0 \end{bmatrix} + \left(\mathbf{1} - \begin{bmatrix} 0 & \omega^* \\ s^* & 0 \end{bmatrix}\right)^{-1}\begin{bmatrix} 0 \\ n_2^- - s^* \cdot n_1^- \end{bmatrix}\ , \tag{4.31}$$

and

$$\mathcal{F}_{U,i}Wh_0^\perp = \sigma_i \begin{bmatrix} n_1^- \\ n_2^+ \end{bmatrix}\ . \tag{4.32}$$

Theorem 4.4.

$$
\begin{aligned}
&\langle \mathcal{P}_U(\zeta)W\ell_0, W\ell_0\rangle = \langle \mathcal{P}_{U_0}(\zeta)\ell_0, \ell_0\rangle + \\
&+ \langle \omega(\zeta)(\mathbf{1} - s(\zeta)\omega(\zeta))^{-1}(n_1^+(\zeta) - s(\zeta)n_2^+(\zeta)), n_2^-(\zeta) - s(\zeta)^+ n_1^-(\zeta)\rangle + (\text{c.c.}) \\
&+ \int_{\mathbb{T}} \frac{1-|\zeta|^2}{|t-\zeta|^2} \left\langle \left\{ \begin{bmatrix} \mathbf{1} & -s \\ -\omega & \mathbf{1} \end{bmatrix}^{-1} \begin{bmatrix} 0 \\ \omega n_1^+ - n_2^+ \end{bmatrix} \right.\right. \\
&\qquad\qquad \left.\left. + \begin{bmatrix} \mathbf{1} & -\omega^* \\ -s^* & \mathbf{1} \end{bmatrix}^{-1} \begin{bmatrix} -n_1^- + \omega^* n_2^- \\ 0 \end{bmatrix} \right\} (dt), \begin{bmatrix} n_1^- \\ n_2^+ \end{bmatrix} \right\rangle + (\text{c.c.}) \\
&+ \int_{\mathbb{T}} \frac{1-|\zeta|^2}{|t-\zeta|^2} \left\langle \begin{bmatrix} \mathbf{1} & s(t) \\ s(t)^* & \mathbf{1} \end{bmatrix} \begin{bmatrix} n_1^-(t) \\ n_2^+(t) \end{bmatrix}, \begin{bmatrix} n_1^-(t) \\ n_2^-(t) \end{bmatrix} \right\rangle m(dt) \\
&+ \int_{\mathbb{T}} \frac{1-|\zeta|^2}{|t-\zeta|^2} \left\langle \sigma_i(dt) \begin{bmatrix} n_1^-(t) \\ n_2^+(t) \end{bmatrix}, \begin{bmatrix} n_1^-(t) \\ n_2^+(t) \end{bmatrix} \right\rangle .
\end{aligned}
\tag{4.33}
$$

Remind here that the expression in braces is understood as a vector-measure on $\mathbb{T}$ correspondent to the harmonic function on $\mathbb{D}$. Symbol (c.c) stands for the complex conjugate of the preceding term.

Proof.

$$
\begin{aligned}
\langle \mathcal{P}_U(\zeta)W\ell_0, W\ell_0\rangle &= \langle \mathcal{P}_U(\zeta)h_0, h_0\rangle + \langle \mathcal{P}_U(\zeta)h_0, Wh_0^\perp\rangle \\
&\quad + \langle \mathcal{P}_U(\zeta)Wh_0^\perp, h_0\rangle + \langle \mathcal{P}_U(\zeta)Wh_0^\perp, Wh_0^\perp\rangle .
\end{aligned}
\tag{4.34}
$$

Completely similar to the proof of the formula (3.46) in Theorem 3.4 one can show that the first term is computed as follows:

$$
\begin{aligned}
&\langle \mathcal{P}_U(\zeta)h_0, h_0\rangle = \langle \mathcal{P}_{U_0}(\zeta)h_0, h_0\rangle \\
&+ \left\langle \left(\mathbf{1} - \begin{bmatrix} 0 & s(\zeta) \\ \omega(\zeta) & 0 \end{bmatrix}\right)^{-1} \begin{bmatrix} n_1^+(\zeta) - s(\zeta)n_2^+(\zeta) \\ 0 \end{bmatrix}, \begin{bmatrix} 0 \\ n_2^-(\zeta) - s(\zeta)^* n_1^-(\zeta) \end{bmatrix} \right\rangle \\
&+ (\text{c.c.}) \\
&= \langle \mathcal{P}_{U_0}(\zeta)h_0, h_0\rangle \\
&+ \langle \omega(\zeta)(\mathbf{1} - s(\zeta)\omega(\zeta))^{-1}(n_1^+(\zeta) - s(\zeta)n_2^+(\zeta)), n_2^-(\zeta) - s(\zeta)^* n_1^-(\zeta)\rangle + (\text{c.c.}) .
\end{aligned}
\tag{4.35}
$$

In turn,

$$
\begin{aligned}
\langle \mathcal{P}_{U_0}(\zeta)h_0, h_0\rangle &= \langle \mathcal{P}_{U_0}(\zeta)\ell_0, \ell_0\rangle - \langle \mathcal{P}_{U_0}(\zeta)h_0, h_0^\perp\rangle \\
&\quad - \langle \mathcal{P}_{U_0}(\zeta)h_0^\perp, h_0\rangle - \langle \mathcal{P}_{U_0}(\zeta)h_0^\perp, h_0^\perp\rangle .
\end{aligned}
\tag{4.36}
$$

Since $h_0^\perp \in L_0^{io}$, then the last three terms in (4.36) can be replaced by their Fourier transforms, i.e.,

$$\langle \mathcal{P}_{U_0}(\zeta)h_0, h_0\rangle = \langle \mathcal{P}_{U_0}(\zeta)\ell_0, \ell_0\rangle - \left\langle \frac{1-|\zeta|^2}{|t-\zeta|^2}\mathcal{F}_{U_0,i_0}h_0, \mathcal{F}_{U_0,i_0}h_0^\perp \right\rangle_{L^s} - (\text{c.c.})$$
$$- \left\langle \frac{1-|\zeta|^2}{|t-\zeta|^2}\mathcal{F}_{U_0,i_0}h_0^\perp, \mathcal{F}_{U_0,i_0}h_0^\perp \right\rangle_{L^s} . \tag{4.37}$$

Substituting (4.29) and (4.30) we get

$$\begin{aligned}\langle \mathcal{P}_{U_0}(\zeta)h_0, h_0\rangle &= \langle \mathcal{P}_{U_0}(\zeta)\ell_0, \ell_0\rangle \\ &- \int_{\mathbb{T}} \frac{1-|\zeta|^2}{|t-\zeta|^2}\left\langle \begin{bmatrix} n_1^+(t) - s(t)\cdot n_2^+(t) \\ n_2^-(t) - s(t)^*\cdot n_1^-(t)\end{bmatrix}, \begin{bmatrix} n_1^-(t) \\ n_2^+(t)\end{bmatrix}\right\rangle m(dt) - (\text{c.c.}) \\ &- \int_{\mathbb{T}} \frac{1-|\zeta|^2}{|t-\zeta|^2}\left\langle \begin{bmatrix} \mathbf{1} & s(t) \\ s(t)^* & \mathbf{1}\end{bmatrix}\begin{bmatrix} n_1^-(t) \\ n_2^+(t)\end{bmatrix}, \begin{bmatrix} n_1^-(t) \\ n_2^+(t)\end{bmatrix}\right\rangle m(dt) .\end{aligned} \tag{4.38}$$

Similarly, since $Wh_0^\perp \in L^i$, the last three terms in (4.34) can be replaced by their Fourier transforms:

$$\langle \mathcal{P}_U(\zeta)h_0, Wh_0^\perp\rangle = \left\langle \frac{1-|\zeta|^2}{|t-\zeta|^2}\mathcal{F}_{U,i}h_0, \mathcal{F}_{U,i}h_0^\perp \right\rangle_{L^{\sigma_i}}$$

(according to (4.31), (4.32))

$$\begin{aligned}&= \int_{\mathbb{T}} \frac{1-|\zeta|^2}{|t-\zeta|^2}\left\langle \left\{ \begin{bmatrix} \mathbf{1} & -S \\ -\omega & \mathbf{1}\end{bmatrix}^{-1}\begin{bmatrix} n_1^+ - S\cdot n_2^+ \\ 0\end{bmatrix}\right.\right. \\ &\qquad\left.\left. + \begin{bmatrix} \mathbf{1} & -\omega^* \\ -S^* & \mathbf{1}\end{bmatrix}^{-1}\begin{bmatrix} 0 \\ n_2^- - S^*\cdot n_1^-\end{bmatrix}\right\}(dt), \begin{bmatrix} n_1^-(t) \\ n_2^+(t)\end{bmatrix}\right\rangle ,\end{aligned} \tag{4.39}$$

$$\langle \mathcal{P}_U(\zeta)Wh_0^\perp, Wh_0^\perp\rangle = \int_{\mathbb{T}} \frac{1-|\zeta|^2}{|t-\zeta|^2}\left\langle \sigma_i(dt)\begin{bmatrix} n_1^-(t) \\ n_2^+(t)\end{bmatrix}, \begin{bmatrix} n_1^-(t) \\ n_2^+(t)\end{bmatrix}\right\rangle . \tag{4.40}$$

Putting together (4.34), (4.35), (4.38), (4.39) and (4.40), we obtain

$$\begin{aligned}&\langle \mathcal{P}_U(\zeta)W\ell_0, W\ell_0\rangle \\ &= \langle \mathcal{P}_{U_0}(\zeta)\ell_0\ell_0\rangle \\ &\quad + \langle \omega(\zeta)(\mathbf{1} - s(\zeta)\omega(\zeta))^{-1}(n_1^+(\zeta) - s(\zeta)n_2^+(\zeta)), n_2^-(\zeta) - s(\zeta)^* n_1^-(\zeta)\rangle + \quad (\text{c.c.}) \\ &\quad - \int_{\mathbb{T}} \frac{1-|\zeta|^2}{|t-\zeta|^2}\left\langle \begin{bmatrix} n_1^+(t) - s(t)n_2^+(t) \\ n_2^-(t) - s(t)n_1^-(t)\end{bmatrix}, \begin{bmatrix} n_1^-(t) \\ n_2^+(t)\end{bmatrix}\right\rangle m(dt) - \quad (\text{c.c.}) \\ &\quad - \int_{\mathbb{T}} \frac{1-|\zeta|^2}{|t-\zeta|^2}\left\langle \begin{bmatrix} \mathbf{1} & s(t) \\ s(t)^* & \mathbf{1}\end{bmatrix}\begin{bmatrix} n_1^-(t) \\ n_2^+(t)\end{bmatrix}, \begin{bmatrix} n_1^-(t) \\ n_2^+(t)\end{bmatrix}\right\rangle m(dt)\end{aligned}$$

$$
\begin{aligned}
&+\int_{\mathbb{T}} \frac{1-|\zeta|^2}{|t-\zeta|^2}\left\langle \left\{ \begin{bmatrix} \mathbf{1} & -s \\ -\omega & \mathbf{1} \end{bmatrix} \begin{bmatrix} n_1^+ - s\cdot n_2^+ \\ 0 \end{bmatrix} \right.\right. \\
&\qquad\qquad \left.\left. + \begin{bmatrix} \mathbf{1} & -\omega^* \\ -s^* & \mathbf{1} \end{bmatrix}^{-1} \begin{bmatrix} 0 \\ n_2^- - s^*\cdot n_1^- \end{bmatrix} \right\} (dt), \begin{bmatrix} n_1^-(t) \\ n_2^+(t) \end{bmatrix} \right\rangle + \quad (\text{c.c.}) \\
&+\int_{\mathbb{T}} \frac{1-|\zeta|^2}{|t-\zeta|^2}\left\langle \sigma_i(dt) \begin{bmatrix} n_1^-(t) \\ n_2^+(t) \end{bmatrix}, \begin{bmatrix} n_1^-(t) \\ n_2^+(t) \end{bmatrix} \right\rangle .
\end{aligned} \tag{4.41}
$$

Since

$$
\begin{bmatrix} n_1^+(t) - s(t)n_2^+(t) \\ n_2^-(t) - s(t)n_1^-(t) \end{bmatrix} = \begin{bmatrix} n_1(t) \\ n_2(t) \end{bmatrix} - \begin{bmatrix} \mathbf{1} & s(t) \\ s(t)^* & \mathbf{1} \end{bmatrix} \begin{bmatrix} n_1^-(t) \\ n_2^+(t) \end{bmatrix},
$$

(4.41) reads as

$$
\begin{aligned}
&\langle \mathcal{P}_U(\zeta)W\ell_0, W\ell_0\rangle = \langle \mathcal{P}_{U_0}(\zeta)\ell_0\ell_0\rangle \\
&\quad + \langle \omega(\zeta)(\mathbf{1}-s(\zeta)\omega(\zeta))^{-1}(n_1^*(\zeta)-s(\zeta)n_2^+(\zeta)), n_2^-(\zeta)-s(\zeta)^*n_1^-(\zeta)\rangle + (\text{c.c.}) \\
&\quad - \int_{\mathbb{T}} \frac{1-|\zeta|^2}{|t-\zeta|^2}\left\langle \begin{bmatrix} n_1(t) \\ n_2(t) \end{bmatrix}, \begin{bmatrix} n_1^-(t) \\ n_2^+(t) \end{bmatrix} \right\rangle m(dt) - (\text{c.c.}) \\
&\quad + \int_{\mathbb{T}} \frac{1-|\zeta|^2}{|t-\zeta|^2}\left\langle \begin{bmatrix} \mathbf{1} & s(t) \\ s(t)^* & \mathbf{1} \end{bmatrix} \begin{bmatrix} n_1^-(t) \\ n_2^+(t) \end{bmatrix}, \begin{bmatrix} n_1^-(t) \\ n_2^+(t) \end{bmatrix} \right\rangle m(dt) \\
&\quad + \int_{\mathbb{T}} \frac{1-|\zeta|^2}{|t-\zeta|^2}\left\langle \left\{ \begin{bmatrix} \mathbf{1} & -s \\ -\omega & \mathbf{1} \end{bmatrix}^{-1} \begin{bmatrix} n_1^+ - s\cdot n_2^+ \\ 0 \end{bmatrix} \right.\right. \\
&\qquad\qquad \left.\left. + \begin{bmatrix} \mathbf{1} & -s^* \\ -\omega^* & \mathbf{1} \end{bmatrix}^{-1} \begin{bmatrix} 0 \\ n_2^- - s^*\cdot n_1^- \end{bmatrix} \right\} (dt), \begin{bmatrix} n_1^- \\ n_2^+ \end{bmatrix} \right\rangle + (\text{c.c.}) \\
&\quad + \int_{\mathbb{T}} \frac{1-|\zeta|^2}{|t-\zeta|^2}\left\langle \sigma_i(dt) \begin{bmatrix} n_1^-(t) \\ n_2^+(t) \end{bmatrix}, \begin{bmatrix} n_1^-(t) \\ n_2^+(t) \end{bmatrix} \right\rangle .
\end{aligned} \tag{4.42}
$$

Putting together the third and the fifth lines, we get (4.33). □

4.4. Define now the scale ρ correspondent to the extension U. We assume here that there is a dense set $E_0 \subseteq E$ such that for any $e \in E_0$, $\rho_0 e$ is a finite vector. Then define $\rho : E \to L$ by the formula

$$\rho = W\rho_0 . \tag{4.43}$$

It is defined at least on E_0. Let $e \in E_0$, then

$$\ell_0 = \rho_0 e \tag{4.44}$$

is a finite vector. According to the notations (3.15) and (4.15) we will get for $\ell_0 = \rho e$,

$$(\mathcal{F}_{U_0,i_0}\rho e)(t) = \begin{bmatrix} n_1(t) \\ n_2(t) \end{bmatrix} = \begin{bmatrix} r_1(t)e \\ r_2(t)^*e \end{bmatrix} . \tag{4.45}$$

Applying now (4.33) to $\ell_0 = \rho e$, we get

Theorem 4.5.

$$\begin{aligned}
&\langle \mathcal{P}_U(\zeta)\rho e, \rho e\rangle = \langle \mathcal{P}_U(\zeta)W\rho_0 e, W\rho_0 e\rangle = \langle \sigma_0(\zeta)e, e\rangle \\
&+ \langle \omega(\zeta)(\mathbf{1} - s(\zeta)\omega(\zeta))^{-1}(r_1^+(\zeta) - s(\zeta)(r_2^*)^+(\zeta))e, ((r_2^*)^-(\zeta) - s(\zeta)^* r_1^-(\zeta))e\rangle \\
&+ (c.c.) \\
&+ \int_{\mathbb{T}} \frac{1-|\zeta|^2}{|t-\zeta|^2} \left\langle \left\{ \begin{bmatrix} \mathbf{1} & -s \\ -\omega & \mathbf{1} \end{bmatrix}^{-1} \begin{bmatrix} 0 \\ \omega r_1^+ - (r_2^*)^+ \end{bmatrix} \right.\right. \\
&\qquad \left.\left. + \begin{bmatrix} \mathbf{1} & -\omega^* \\ -s^* & \mathbf{1} \end{bmatrix}^{-1} \begin{bmatrix} -r_1^- + \omega^*(r_2^*)^- \\ 0 \end{bmatrix} \right\} (dt)e, \begin{bmatrix} r_1^-(t) \\ (r_2^*)^+(t) \end{bmatrix} e \right\rangle + (c.c.) \\
&+ \int_{\mathbb{T}} \frac{1-|\zeta|^2}{|t-\zeta|^2} \left\langle \begin{bmatrix} \mathbf{1} & s(t) \\ s(t)^* & \mathbf{1} \end{bmatrix} \begin{bmatrix} r_1^-(t) \\ (r_2^*)^+(t) \end{bmatrix} e, \begin{bmatrix} r_1^-(t) \\ (r_2^*)^+(t) \end{bmatrix} e \right\rangle \\
&+ \int_{\mathbb{T}} \frac{1-|\zeta|^2}{|t-\zeta|^2} \left\langle \sigma_i(dt) \begin{bmatrix} r_1^-(t) \\ (r_2^*)^+(t) \end{bmatrix} e, \begin{bmatrix} r_1^-(t) \\ (r_2^*)^+(t) \end{bmatrix} e \right\rangle .
\end{aligned} \tag{4.46}$$

The latter formula is a generalization of the formula (3.45) for the case when $r_1(\zeta)e$ and $r_2^*(\zeta)e$ need not be $H_+^2(N_1)$ and $H_-^2(N_2)$ functions respectively. In this special case (4.46) turns into (3.45).

5. One remark on the feedback coupling

We are going to give here one more perspective on the feedback coupling discussed above, but now in terms of the scattering systems rather than in terms of unitary colligations. The main point of this section is Lemma 5.2.

5.1. Consider the model unitary colligations A_0 and A_1 (after performing Fourier transforms $\mathcal{F}_{U_0,i_0}$ and $\mathcal{F}_{U_1,i_1}$ respectively, see Section 3.3):

$$\begin{aligned} A_0 &: N_2 \oplus H^s \to H^s \oplus N_1 \ , \\ A_1 &: N_1 \oplus H^\omega \to H^\omega \oplus N_2 \ , \end{aligned} \tag{5.1}$$

where

$$H^s = \mathcal{F}_{U_0,i_0} H_0 = L^s \cap \begin{bmatrix} H_+^2(N_1) \\ H_-^2(N_2) \end{bmatrix} ,$$

$$H^{\omega^*} = \mathcal{F}_{U_1,i_1} H_1 = L^{\omega^*} \cap \begin{bmatrix} H_-^2(N_1) \\ H_+^2(N_2) \end{bmatrix} ,$$

(see (3.18), (3.26)). Remind that

$$L^s = \begin{bmatrix} s \\ \mathbf{1} \end{bmatrix} H^2_+(N_2) \oplus H^s \oplus \begin{bmatrix} \mathbf{1} \\ s^* \end{bmatrix} H^2_-(N_1)$$
$$L^{\omega^*} = \begin{bmatrix} \mathbf{1} \\ \omega \end{bmatrix} H^2_+(N_1) \oplus H^{\omega^*} \oplus \begin{bmatrix} \omega^* \\ \mathbf{1} \end{bmatrix} H^2_-(N_2)\,. \tag{5.2}$$

For $h^s \in H^s$, denote

$$h^s = \begin{bmatrix} h^s_+ \\ h^s_- \end{bmatrix} \tag{5.3}$$

and for $h^\omega \in H^{\omega^*}$ denote

$$h^\omega = \begin{bmatrix} h^\omega_- \\ h^\omega_+ \end{bmatrix}\,. \tag{5.4}$$

A_0 and A_1 act as multiplications by $\bar{t}$ in these models:

$$\bar{t} : \begin{bmatrix} s \\ \mathbf{1} \end{bmatrix} N_2 \oplus H^s \to H^s \oplus \bar{t} \begin{bmatrix} \mathbf{1} \\ s^* \end{bmatrix} N_1$$
$$\bar{t} : \begin{bmatrix} \mathbf{1} \\ \omega \end{bmatrix} N_1 \oplus H^\omega \to H^\omega \oplus \bar{t} \begin{bmatrix} \omega^* \\ \mathbf{1} \end{bmatrix} N_2\,. \tag{5.5}$$

Denote by U the feedback coupling of A_0 and A_1 (see Definitions (3.1), (3.2)):

$$U : H^s \oplus H^\omega \to H^s \oplus H^\omega\,. \tag{5.6}$$

Lemma 5.1. Let

$$\begin{bmatrix} g^s \\ g^\omega \end{bmatrix} = U \begin{bmatrix} h^s \\ h^\omega \end{bmatrix}, \tag{5.7}$$

then

$$g^s = \bar{t}\left(h^s + \begin{bmatrix} \mathbf{1} & s \\ s^* & \mathbf{1} \end{bmatrix} \begin{bmatrix} -h^s_+(0) \\ \omega(0)h^s_+(0) + h^\omega_+(0) \end{bmatrix} \right)$$
$$g^\omega = \bar{t}\left(h^\omega - \begin{bmatrix} \mathbf{1} & \omega^* \\ \omega & \mathbf{1} \end{bmatrix} \begin{bmatrix} -h^s_+(0) \\ \omega(0)h^s_+(0) + h^\omega_+(0) \end{bmatrix} \right)\,. \tag{5.8}$$

Proof. By definition of the coupling ((3.1),(3.2)) the operator U in (5.7) is computed from the system of equations (see (5.5))

$$\bar{t}\left(\begin{bmatrix} s \\ \mathbf{1} \end{bmatrix} n_2 \oplus h^s \right) = g^s \oplus \bar{t} \begin{bmatrix} \mathbf{1} \\ s^* \end{bmatrix} n_1$$
$$\bar{t}\left(\begin{bmatrix} \mathbf{1} \\ \omega \end{bmatrix} n_1 \oplus h^\omega \right) = g^\omega \oplus \bar{t} \begin{bmatrix} \omega^* \\ \mathbf{1} \end{bmatrix} n_2\,. \tag{5.9}$$

Remind that the coupling condition means that n_1 from the first equation and n_1 from the second one is the same vector, the same holds for n_2. Since

$$\bar{t}\left(\begin{bmatrix} s \\ \mathbf{1} \end{bmatrix} n_2 \oplus h^s \right) = \left(\begin{bmatrix} \bar{t}s \\ \bar{t} \end{bmatrix} n_2 + \bar{t}\left(h^s - \begin{bmatrix} \mathbf{1} \\ s^* \end{bmatrix} h^s_+(0) \right) \right) \oplus \bar{t} \begin{bmatrix} \mathbf{1} \\ s^* \end{bmatrix} h^s_+(0)\,, \tag{5.10}$$

then

$$\begin{aligned} g^s &= \begin{bmatrix} \bar{t}s \\ \bar{t} \end{bmatrix} n_2 + \bar{t}\left(h^s - \begin{bmatrix} \mathbf{1} \\ s^* \end{bmatrix} h_+^s(0) \right) \\ n_1 &= h_+^s(0) \,. \end{aligned} \tag{5.11}$$

Similarly,

$$\begin{aligned} \bar{t}\left(\begin{bmatrix} \mathbf{1} \\ \omega \end{bmatrix} n_1 \oplus h^\omega \right) = \bar{t} \begin{bmatrix} \mathbf{1} & \omega^* \\ \omega & \mathbf{1} \end{bmatrix} \begin{bmatrix} n_1 \\ -\omega(0)n_1 \end{bmatrix} + \bar{t}\left(h^\omega - \begin{bmatrix} \omega^* \\ \mathbf{1} \end{bmatrix} h_+^\omega(0) \right) \\ \oplus \bar{t} \begin{bmatrix} \omega^* \\ \mathbf{1} \end{bmatrix} (\omega(0)n_1 + h_+^\omega(0)) \,, \end{aligned} \tag{5.12}$$

which means

$$\begin{aligned} g^\omega &= \bar{t} \begin{bmatrix} \mathbf{1} & \omega^* \\ \omega & \mathbf{1} \end{bmatrix} \begin{bmatrix} n_1 \\ -\omega(0)n_1 \end{bmatrix} + \bar{t}\left(h^\omega - \begin{bmatrix} \omega \\ \mathbf{1} \end{bmatrix} h_+^\omega(0) \right) \\ n_2 &= \omega(0)n_1 + h_+^\omega(0) \,. \end{aligned} \tag{5.13}$$

Or, taking into account that $n_1 = h_+^s(0)$ (see (5.11)), we obtain

$$\begin{aligned} g^\omega &= \bar{t}\left(h^\omega + \begin{bmatrix} \mathbf{1} & \omega^* \\ \omega & \mathbf{1} \end{bmatrix} \begin{bmatrix} h_+^s(0) \\ -\omega(0)h_+^s(0) - h_+^\omega(0) \end{bmatrix} \right) \\ n_2 &= \omega(0)h_+^s(0) + h_+^\omega(0) \,. \end{aligned} \tag{5.14}$$

Substituting n_2 from (5.14) into (5.11), we get

$$\begin{aligned} g^s &= \bar{t}\left(h^s + \begin{bmatrix} \mathbf{1} & s \\ s^* & \mathbf{1} \end{bmatrix} \begin{bmatrix} -h_+^s(0) \\ \omega(0)h_+^s(0) + h_+^\omega(0) \end{bmatrix} \right) \\ n_1 &= h_+^s(0) \,. \end{aligned} \tag{5.15}$$

(5.15) and (5.14) give us (5.8). The lemma follows. □

Remark. By the way we obtained that in this model,

$$i^* \begin{bmatrix} h^s \\ h^\omega \end{bmatrix} = \begin{bmatrix} h_+^s(0) \\ \omega(0)h_+^s(0) + h_+^\omega(0) \end{bmatrix} \,. \tag{5.16}$$

5.2. Discuss now the Fourier representation

$$\mathcal{F}_{U,i} : H^s \oplus H^\omega \to L^{\sigma_i} \,,$$

where (see (3.44))

$$\sigma_i(\zeta) = \frac{1}{2} \left(\frac{\mathbf{1} + \begin{bmatrix} 0 & s(\zeta) \\ \omega(\zeta) & 0 \end{bmatrix}}{\mathbf{1} - \begin{bmatrix} 0 & s(\zeta) \\ \omega(\zeta) & \mathbf{1} \end{bmatrix}} + \frac{\mathbf{1} + \begin{bmatrix} 0 & \omega(\zeta)^* \\ s(\zeta)^* & 0 \end{bmatrix}}{\mathbf{1} - \begin{bmatrix} 0 & \omega(\zeta)^* \\ s(\zeta)^* & 0 \end{bmatrix}} \right) \,, \quad |\zeta| < 1.$$

According to (3.38) for $|\zeta| < 1$,

$$\mathcal{F}_{U,i}\begin{bmatrix} h^s \\ h^\omega \end{bmatrix} = \left(\mathbf{1} - \begin{bmatrix} 0 & s \\ \omega & 0 \end{bmatrix}\right)^{-1} \begin{bmatrix} h^s_+ \\ h^\omega_+ \end{bmatrix} + \left(\mathbf{1} - \begin{bmatrix} 0 & \omega^* \\ s^* & 0 \end{bmatrix}\right)^{-1} \begin{bmatrix} h^\omega_- \\ h^s_- \end{bmatrix}. \tag{5.17}$$

Let us denote it by

$$\begin{bmatrix} \nu_1 \\ \nu_2 \end{bmatrix} \equiv \left(\mathbf{1} - \begin{bmatrix} 0 & s \\ \omega & 0 \end{bmatrix}\right)^{-1} \begin{bmatrix} h^s_+ \\ h^\omega_+ \end{bmatrix} + \left(\mathbf{1} - \begin{bmatrix} 0 & \omega^* \\ s^* & 0 \end{bmatrix}\right)^{-1} \begin{bmatrix} h^\omega_- \\ h^s_- \end{bmatrix}. \tag{5.18}$$

By $\begin{bmatrix} \nu_1^+ \\ \nu_2^+ \end{bmatrix}$ we denote the analytic term

$$\begin{bmatrix} \nu_1^+ \\ \nu_2^+ \end{bmatrix} \equiv \left(\mathbf{1} - \begin{bmatrix} 0 & s \\ \omega & 0 \end{bmatrix}\right)^{-1} \begin{bmatrix} h^s_+ \\ h^\omega_+ \end{bmatrix}, \tag{5.19}$$

and by $\begin{bmatrix} \nu_1^- \\ \nu_2^- \end{bmatrix}$ the antianalytic term

$$\begin{bmatrix} \nu_1^- \\ \nu_2^- \end{bmatrix} \equiv \left(\mathbf{1} - \begin{bmatrix} 0 & \omega^* \\ s^* & 0 \end{bmatrix}\right)^{-1} \begin{bmatrix} h^\omega_- \\ h^s_- \end{bmatrix}. \tag{5.20}$$

Lemma 5.2. Formulas (5.18)–(5.20) are equivalent to

$$\begin{bmatrix} \nu_1 \\ \nu_2 \end{bmatrix} = \begin{bmatrix} s \\ \mathbf{1} \end{bmatrix} \nu_2^+ + h^s + \begin{bmatrix} \mathbf{1} \\ s^* \end{bmatrix} \nu_1^- = \begin{bmatrix} \mathbf{1} \\ \omega \end{bmatrix} \nu_1^+ + h^\omega + \begin{bmatrix} \omega^* \\ \mathbf{1} \end{bmatrix} \nu_2^-, \tag{5.21}$$

for $|\zeta| < 1$, where

$$h^s = \begin{bmatrix} h^s_+ \\ h^s_- \end{bmatrix}, \quad h^\omega = \begin{bmatrix} h^\omega_- \\ h^\omega_+ \end{bmatrix} \tag{5.22}$$

(see (5.3), (5.4)).

Proof. By straightforward computation, separating entries, analytic and antianalytic parts. □

Comment. The sides of (5.21) look like vectors from L^s and L^{ω^*} respectively (see (5.2)), but the functions $\nu_1^+, \nu_1^-, \nu_2^+, \nu_2^-$ need not be H^2_+ or H^2_- functions. Thus, in terms of scattering systems, the feedback coupling means (in a sense) that we add to the state vectors h^s and h^ω the "channel" terms in order to make the sums in (5.21) equal.

Acknowledgements. The author is grateful to V. Dubovoy for our many discussions which stimulated this work to a large extent. The author is thankful to J. Ball, C. Sadoski and V. Vinnikov for fruitful discussions and encouraging him to writing this paper. The author is thankful to Miriam Abraham and Diana Mandelik for careful typing of the manuscript. The work was done under support of the Ministry of Absorption of Israel.

References

[KKY] V.E. Katsnelson, A.Ya. Kheifets, P.M. Yuditskii, *Abstract Interpolation Problem and Isometric Operators Extension Theory*, Operators in Functional Spaces and Questions of Function Theory, Kiev (1987) 83–96, Russian. English transl. in Topics in Interpolation Theory (Leipzig,1994), Operator Theory: Advances and Applications, 95(1997) 283–298 , Birkhäuser Verlag, Basel.

[Kh1] A.Ya. Kheifets, *Parseval equality in abstract interpolation problem and coupling of open systems (I)*, Teor. Funk., Funk. Anal. i ikh Prilozhen, 49 (1988) 112–120, Russian. English transl.: J. Sov. Math. 49, 4 (1990) 1114–1120.

[Kh2] A.Ya. Kheifets, *Parseval equality in abstract interpolation problem and coupling of open systems (II)*, Teor. Funk., Funk. Anal. I ikh Prilozhen, 50 (1988) 98–103, Russian. English transl.: J. Sov. Math. 49, 6 (1990) 1307–1310.

[Kh3] A. Kheifets, *Scattering matrices and Parseval equality in Abstract Interpolation problem*, Ph.D. Thesis, Kharkov State University, 1990.

[KhY] A.Ya. Kheifets, P.M. Yuditskii, *An analysis and extension of V.P. Potapov's approach to interpolation problems with applications to the generalized bi-tangential Schur-Nevanlinna-Pick problem and J-inner-outer factorization*, In Operator Theory: Advances and Applications, 72 (1994) 133–161, Birkhäuser Verlag, Basel.

[Kh4] A. Kheifets, *Abstract interpolation problem and some applications*, in: Holomorphic Spaces (S. Axler, J. McCarthy, D. Sarason editors), MSRI Publications, 33 (1998) 351–381, Cambridge University Press.

[BDKh] S.S. Boiko, V.K. Dubovoy, A.Ya. Kheifets, *Measure Schur complements and spectral functions of unitary operators with respect to different scales*, in D. Alpay and V. Vinikov (editors), Operator Theory, System Theory and Related Topics (The Moshe Livšic Anniversary Volume), Operator Theory: Advances and Applications, 123(2001) 89–138, Birkhäuser, Basel.

[Kh5] A.Ya. Kheifets, *Hamburger moment problem: Parseval equality and Arov-singularity*, Journal of Functional Analysis, 141, 2 (1996) 374–420.

[BTr] J. Ball, T. Trent, *The abstract interpolation problem and commutant lifting: coordinate-free approach*, Operator Theory: Advances and Applications, 115 (2000) 51–83, Birkhäuser, Basel.

[NF] B.Sz.-Nagy, C. Foias, *Harmonic analysis of operators in Hilbert space.* North-Holland, Amsterdam, 1970.

[AG] D.Z. Arov, L.Z. Grossman, *Scattering matrices in the extension theory of isometric operators*, Soviet Math. Dokl., 27 (1983) 573–578.

[Kh6] A. Kheifets, *Parameterization of solutions of the Nehari problem and nonorthogonal dynamics*, Operator Theory: Advances and Applications, 115 (2000) 213–233, Birkhäuser, Basel.

A. Kheifets
The Research Institute
The College of Judea and Samaria
Ariel, 44837, Israel
e-mail: kheifets@yosh.ac.il

Operator Theory:
Advances and Applications, Vol. 134, 319–336

Chains of Space-Time Open Systems and DNA

M.S. Livšic

Dedicated to Harry Dym

Abstract. There is a striking resemblance between chains of space-time open systems and chains of nucleotides in molecular biology. It seems, hypothetically, that nucleotides can be treated as some kind of space-time systems. In particular, there exist attraction forces between corresponding links of two complementary chains of space-time systems going in the opposite directions. Chains of space-time systems can be replicated with the help of primers, elongations and templets. We show also that under some conditions the structure of a chain is a Bertrand curve — in particular, a double helix.

Introduction

"James Watson and Francis Crick showed that the structure of DNA is a double helix in which each helix is a chain of nucleotides held together by a phospodiester bond, and in which specific hydrogen bonds are formed by pairs of bases" [1, Chapter 11]. We show that some important properties of DNA can be given a natural explanation using the methods of system theory. In this way we can also obtain a new information about the properties of DNA.

Let us recall the basic results of the theory of operator vessels [2].

A collection

$$X = (A_1, A_2; H, \Phi, E; \sigma_1, \sigma_2)\,,$$

where A_1, A_2 are linear operators in a Hilbert space H ($\dim H = N \leq \infty$), E is a finite dimensional Hilbert space, $\Phi : H \longrightarrow E$ is a linear mapping, and σ_1, σ_2 are selfadjoint operators in E, is called a *colligation* if the following conditions hold:

$$\frac{1}{i}(A_j - A_j^*) = \Phi^* \sigma_j \Phi \quad (j = 1, 2). \tag{1}$$

The spaces H and E are called respectively the *inner space* and the *outer space* (or the *coupling space*) of the colligation X, and the mapping Φ is called the *window* of X. The colligation is said to be *commutative* if $A_1 A_2 = A_2 A_1$. The colligation is said to be *strict* if

1. $\Phi H = E$.
2. $\ker \sigma_1 \cap \ker \sigma_2 = 0$.

Systems

We consider also the following family of systems:

$$\mathcal{F}_\Gamma \ : \ \begin{cases} i\dfrac{df}{ds} + \left(A, \dfrac{dx}{ds}\right) f(s) = \Phi^* \left(\sigma, \dfrac{dx}{ds}\right) u(s), \ f(M_0) = f_0, \\ v(s) = u(s) - i\Phi f(s). \end{cases}$$

Here Γ: $x_k = x_k(s)$ ($k = 1, 2$, $s_0 \leq s \leq s_1$), is a (piecewise smooth) path in $\mathbb{R}^2$,

$$\left(A, \frac{dx}{ds}\right) = A_1 \frac{dx_1}{ds} + A_2 \frac{dx_2}{ds}, \quad \left(\sigma, \frac{dx}{ds}\right) = \sigma_1 \frac{dx_1}{ds} + \sigma_2 \frac{dx_2}{ds};$$

$u(s) \in E$ is the input, $f(s) \in H$ is the inner state, and $v(s) \in E$ is the output. For such systems the "energy" balance law holds:

$$\frac{d(f,f)}{ds} = \left(\left(\sigma, \frac{dx}{ds}\right) u(s), u(s)\right) - \left(\left(\sigma, \frac{dx}{ds}\right) v(s), v(s)\right).$$

Vortical systems

In the case of infinitely many revolutions along an infinitesimal circle, a so-called *vortical* system appears [3]. Besides the original window $\Phi : H \longrightarrow E$, the vortical system acquires its own "vortical" windows:

$$\operatorname{vort}^{\text{in}} \Phi = \sigma_1 \Phi A_2^* - \sigma_2 \Phi A_1^* = \begin{vmatrix} \sigma_1 \Phi & \sigma_2 \Phi \\ A_1^* & A_2^* \end{vmatrix},$$

and

$$\operatorname{vort}^{\text{out}} \Phi = \sigma_1 \Phi A_2 - \sigma_2 \Phi A_1 = \begin{vmatrix} \sigma_1 \Phi & \sigma_2 \Phi \\ A_1 & A_2 \end{vmatrix}.$$

Vessels

A collection

$$V = \left(X; \gamma^{\text{in}}, \gamma^{\text{out}}\right),$$

where X is a colligation and γ^{in}, γ^{out} are operators in E, is called a *vessel* if the following relations hold:

$$\operatorname{vort}^{\text{in}} \Phi = \gamma^{\text{in}} \Phi, \tag{2}$$

$$\operatorname{vort}^{\text{out}} \Phi = \gamma^{\text{out}} \Phi, \tag{2'}$$

$$\gamma^{\text{out}} - \gamma^{\text{in}} = \pi(X). \tag{3}$$

Here

$$\pi(X) = i(\sigma_1 \Phi \Phi^* \sigma_2 - \sigma_2 \Phi \Phi^* \sigma_1) = i \begin{vmatrix} \sigma_1 \Phi & \sigma_2 \Phi \\ \Phi^* \sigma_1 & \Phi^* \sigma_2 \end{vmatrix}.$$

It is known [2, Section 2.3] that conditions (1), (2), (3) are equivalent to the conditions (1), (2′), (3), i.e., in the figure below the left side of the quadrilateral is equivalent to the right side:

Theorem ([2, Section 2.3]). *Any commutative strict colligation can be embedded into a uniquely determined vessel. In this case* $\gamma^{in} = \gamma^{in*}$, $\gamma^{out} = \gamma^{out*}$.

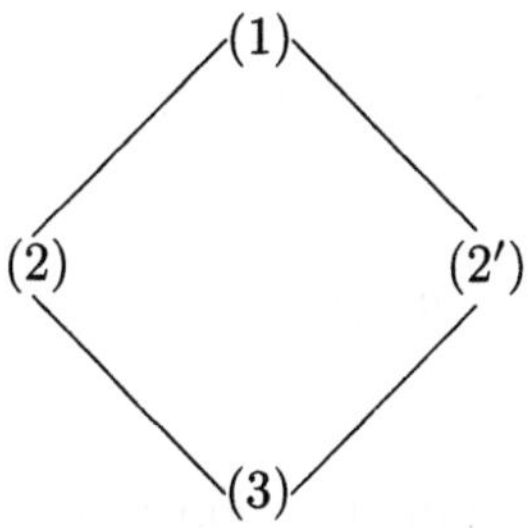

Conditions (2), (2′) can be viewed as the *adaptation* conditions of the vortical windows to the original window $\Phi : H \longrightarrow E$. It is interesting to notice that the adaptation factors γ^{in}, γ^{out} play also the role of *compatibility factors* for the overdetermined system of equations of a 2D system:

$$\mathcal{F}(X) \ : \ \begin{cases} i\dfrac{\partial f}{\partial x_1} + A_1 f(x_1, x_2) = \Phi^* \sigma_1 u(x_1, x_2), \\ i\dfrac{\partial f}{\partial x_2} + A_2 f(x_1, x_2) = \Phi^* \sigma_2 u(x_1, x_2), \\ v(x_1, x_2) = u(x_1, x_2) - i\Phi f(x_1, x_2). \end{cases} \tag{4}$$

Theorem ([2, Section 3.2]). *Let X be a commutative strict colligation. Then the equations of $\mathcal{F}(X)$ are compatible if and only if the input $u(x_1, x_2)$ satisfies the following PDE:*

$$\mathcal{D}^{in}(u) := \sigma_2 \frac{\partial u}{\partial x_1} - \sigma_1 \frac{\partial u}{\partial x_2} + i\gamma^{in} u = 0. \tag{5}$$

In this case the output $v(x_1, x_2)$ satisfies the PDE:

$$\mathcal{D}^{out}(v) := \sigma_2 \frac{\partial v}{\partial x_1} - \sigma_1 \frac{\partial v}{\partial x_2} + i\gamma^{out} v = 0. \tag{6}$$

Remark. If $\Phi H \neq E$, then the condition $\mathcal{D}^{\text{in}}(u) = 0$ is *sufficient* for the compatibility of the equations of $\mathcal{F}(X)$, and in this case the condition $\mathcal{D}^{\text{out}}(v) = 0$ holds as well.

1. Spectral analysis and synthesis of 2D systems

Spectral analysis

It is known [2, Section 2.4] that if $H^{(2)}$ is a common invariant subspace of the operators A_1 and A_2, then

$$A_j = \begin{pmatrix} A_j^{(1)} & 0 \\ C_j & A_j^{(2)} \end{pmatrix} \ (j = 1, 2), \quad H = \begin{pmatrix} H^{(1)} \\ H^{(2)} \end{pmatrix}, \tag{7}$$

where $C_j : H^{(1)} \longrightarrow H^{(2)}$, $\Phi = \begin{pmatrix} \Phi^{(1)} & \Phi^{(2)} \end{pmatrix}$, and $C_j = i\Phi^{(2)^*} \sigma_j \Phi^{(1)}$. The colligation X is the *coupling* $X = X^{(1)} \to X^{(2)}$, where

$$X^{(1)} = (A_1^{(1)}, A_2^{(1)}; H^{(1)}, \Phi^{(1)}, E; \sigma_1, \sigma_2), \ X^{(2)} = (A_1^{(2)}, A_2^{(2)}; H^{(2)}, \Phi^{(2)}, E; \sigma_1, \sigma_2),$$

and the corresponding 2D system is the coupling of $\mathcal{F}(X^{(1)})$ and $\mathcal{F}(X^{(2)})$, $\mathcal{F}(X) = \mathcal{F}(X^{(1)}) \to \mathcal{F}(X^{(2)})$:

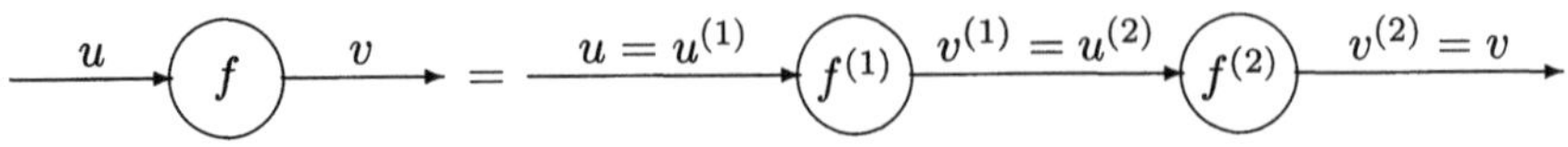

FIGURE 1. Cooperative States

If X is a commutative colligation with a finite dimensional inner space, then there exists a decreasing chain of common invariant subspaces

$$H = H^0 \supset H^1 \supset H^2 \supset \cdots \supset H^{N-1} \supset H^N = 0,$$

such that $\dim(H^{k-1} \ominus H^k) = 1$. In this case we obtain a spectral resolution of X into elementary colligations X^k with one dimensional inner spaces $\tilde{H}^k = H^{k-1} \ominus H^k$: $X = X^1 \to X^2 \to \cdots \to X^N$ and $\mathcal{F}(X) = \mathcal{F}(X^1) \to \mathcal{F}(X^2) \to \cdots \to \mathcal{F}(X^N)$.

Spectral synthesis

Unfortunately, the inverse problem of spectral synthesis is a difficult and even a "wild" problem, because the coupling of elementary colligations is generally speaking not commutative. To overcome this difficulty we will use the method of an *initial primer* and its *elongations.*

Let $X^0 = (A_1^0, A_2^0; H^0, \Phi^0, E; \sigma_1, \sigma_2)$ be a given commutative strict colligation, and let $\mathcal{M}' = (a_1', a_2')$ be a given point in $\mathbb{C}^2$. We will consider the following *elongation problem*: find an embedding of the point $\mathcal{M}' = (a_1', a_2')$ into an elementary colligation

$$X' = (A_1', A_2'; H', \Phi', E; \sigma_1, \sigma_2), \quad \dim H' = 1,$$

such that the coupling $X = X^0 \to X'$ is a commutative colligation. We remind that X^0 admits an embedding into a *uniquely determined* vessel $V^0 = (X^0; \gamma^{0\text{in}}, \gamma^{0\text{out}})$, where $\gamma^{0\text{in}}$, $\gamma^{0\text{out}}$ are the compatibility factors.

Elongation Theorem I. *If X^0 is a commutative strict colligation, then there exists a commutative elongation of X^0 with the help of a given point $\mathcal{M}' = (a_1', a_2')$ if and only if this point admits an embedding into a vessel*

$$V' = (a_1', a_2'; H', \Phi', E; \sigma_1, \sigma_2; \gamma'^{in}, \gamma'^{out}),$$

such that $\gamma'^{in} = \gamma^{0\,out}$.

Proof. By the definition of the coupling of colligations (see (7) and [2, Section 2.4]), $X^0 \to X' = X = (A_1, A_2; H, \Phi, E; \sigma_1, \sigma_2)$, where

$$A_j = \begin{pmatrix} A_j^0 & 0 \\ i\Phi'\sigma_j\Phi^0 & A_j' = a_j' \end{pmatrix} \quad (j = 1, 2),$$

$$\Phi H = \begin{pmatrix} \Phi^0 & \Phi' \end{pmatrix} \begin{pmatrix} H^0 \\ H' \end{pmatrix} = \Phi^0 H^0 + \Phi' H' = E.$$

The commutator

$$A_1A_2 - A_2A_1 = \begin{pmatrix} 0 & 0 \\ C_{21} & 0 \end{pmatrix},$$

where

$$C_{21} = i\Phi'^* \left[\sigma_1\Phi^0A_2^0 - \sigma_2\Phi^0A_1^0 - (a_2'\sigma_1 - a_1'\sigma_2)\Phi^0\right].$$

Using the vessel condition (2′) for $V^0 = (X^0; \gamma^{0\text{in}}, \gamma^{0\text{out}})$:

$$\sigma_1\Phi^0A_2^0 - \sigma_2\Phi^0A_1^0 = \gamma^{0\text{out}}\Phi^0,$$

we conclude that

$$C_{21} = i\Phi'^* \left[\gamma^{0\text{out}} - (a_2'\sigma_1 - a_1'\sigma_2)\right]\Phi^0.$$

Hence $C_{21} = 0$ if and only if $\Phi'^*(a_2'\sigma_1 - a_1'\sigma_2) = \Phi'^*\gamma^{0\text{out}}$ (Φ^0 can be cancelled because X^0 is strict). Taking the adjoints we obtain

$$(\bar{a}_2'\sigma_1 - \bar{a}_1'\sigma_2)\,\Phi' = \gamma^{0\text{out}}\Phi',$$

which is the vessel condition (2) for X' with $\gamma'^{\text{in}} = \gamma^{0\text{out}}$. Now we *define* γ'^{out} in accordance with the condition (3):

$$\gamma'^{\text{out}} = \gamma'^{\text{in}} + \pi(X'), \quad \pi(X') = i(\sigma_1\Phi'\Phi'^*\sigma_2 - \sigma_2\Phi'\Phi'^*\sigma_1).$$

Then the condition (2′) for X' holds as well, proving the Elongation Theorem. □

In terms of 2D systems, the Elongation Theorem means that the 2D field $\mathcal{F}(X^0)$ admits a consistent elongation $\mathcal{F}(X^0) \to \mathcal{F}(X')$ with the help of a given point $\mathcal{M}'(a_1', a_2')$ if and only if the elementary system $\mathcal{F}(X')$ *inherits* the compatibility factor $\gamma'^{\text{in}} = \gamma^{0\text{out}}$ from the primer $\mathcal{F}(X^0)$. Hence every 2D field that corresponds to a commutative strict colligation can be used as a primer.

It is known [2, Section 4.2] that for an arbitrary commutative vessel the following determinants are equal:

$$D(\lambda_1, \lambda_2) := \det(\lambda_1\sigma_2 - \lambda_2\sigma_1 + \gamma^{\text{in}}) = \det(\lambda_1\sigma_2 - \lambda_2\sigma_1 + \gamma^{\text{out}}).$$

It is known also (the Generalized Cayley-Hamilton Theorem) that the operators A_1, A_2 satisfy the equation $D(A_1, A_2) = 0$ (on the so-called principal subspace). The algebraic curve $D(\lambda_1, \lambda_2) = 0$ is called the *discriminant curve* of the given vessel V.

The Elongation Theorem implies that the primer X^0 defines uniquely the discriminant curve

$$D(\lambda_1, \lambda_2) = D^0(\lambda_1, \lambda_2) = \det(\lambda_1\sigma_2 - \lambda_2\sigma_1 + \gamma^{0\text{out}}).$$

If $\gamma'^{\text{in}} = \gamma^{0\text{out}}$, then the point $\mathcal{M}'(a_1', a_2')$ belongs to the same discriminant curve: $D^0(a_1', a_2') = 0$.

Elongation Theorem II. *Let X^0 be a primer such that $\sigma_2 > 0$, and let $\mathcal{M}'(a_1', a_2')$, $\operatorname{Im} a_2' > 0$, be a given point. Then there exists a commutative elongation of X^0 with the help of $\mathcal{M}'$ if and only if the point $\mathcal{M}'$ belongs to the discriminant curve of X^0.*

Proof. If $D^0(a_1', a_2') = 0$ then also $D^0(\bar{a}_1', \bar{a}_2') = 0$. Hence the equation

$$\left(\bar{a}_1'\sigma_2 - \bar{a}_2'\sigma_1 + \gamma^{0\text{out}}\right)\phi' = 0$$

has a nontrivial solution ϕ'. Multiplying from the left by ϕ'^* and taking imaginary parts we obtain

$$(\operatorname{Im} a_1')\phi'^*\sigma_2\phi' = (\operatorname{Im} a_2')\phi'^*\sigma_1\phi'.$$

There are two cases:

1. $\phi'^*\sigma_1\phi' \neq 0$; then $\operatorname{Im} a_1' \neq 0$ and
$$(\operatorname{Im} a_1') : \phi'^*\sigma_1\phi' = (\operatorname{Im} a_2') : \phi'^*\sigma_2\phi' > 0.$$
Multiplying ϕ' by an appropriate factor, we can normalize ϕ' in such a way that the two colligation conditions $\operatorname{Im} a_j' = \phi'^*\sigma_j\phi'$ are fulfilled, and we obtain a colligation which satisfies the conditions of Elongation Theorem I.
2. If $\phi'^*\sigma_1\phi' = 0$ then $\operatorname{Im} a_1' = 0$. The second colligation condition can be easily obtained by an appropriate normalization of ϕ'. □

Given a sequence $\mathcal{M}^k(a_1^k, a_2^k)$, $k = 1, \dots, N$, $\operatorname{Im} a_2^k > 0$, of points on the discriminant curve, one can synthesize step by step, using the Elongation Theorems, a commutative chain corresponding to the sequence $\mathcal{M}^k$, $k = 1, \dots, N$ (after the synthesis is completed the primer X^0 can be omitted):

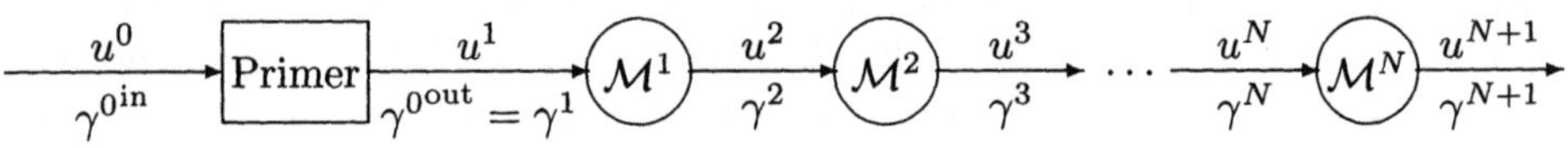

$$\begin{aligned}
&\gamma^1 = \gamma^{0\text{out}}, \\
&\left(\bar{a}_1^k\sigma_2 - \bar{a}_2^k\sigma_1 + \gamma^k\right)\phi^k = 0, \\
&\frac{1}{i}\left(a_j^k - \bar{a}_j^k\right) = \phi^{k*}\sigma_j\phi^k \ (j = 1, 2), \\
&\gamma^{k+1} = \gamma^k + i\left[\sigma_1\phi^k\phi^{k*}\sigma_2 - \sigma_2\phi^k\phi^{k*}\sigma_1\right]
\end{aligned} \tag{8}$$

$(k = 1, \dots, N)$.

In the case $\sigma_2 > 0$, the compatibility equations (5) are of the hyperbolic (parabolic) type and we can assume that $x_1 = t$, $x_2 = x$ are the *time coordinate* and the *space coordinate*, respectively. The equations of the links $\mathcal{F}(X^k)$ have the form:

$$\begin{cases}
i\dfrac{\partial f^k}{\partial t} + a_1^k f^k = \phi^k\sigma_1 u^k, \\
i\dfrac{\partial f^k}{\partial x} + a_2^k f^k = \phi^k\sigma_2 u^k, \\
u^{k+1} = u^k - i\phi^k f^k
\end{cases} \tag{9}$$

($k = 1, \ldots, N$), and the compatibility equations are

$$\sigma_2 \frac{\partial u^k}{\partial t} - \sigma_1 \frac{\partial u^k}{\partial x} + i\gamma^k u^k = 0. \tag{10}$$

We see that *in the presence of a primer, the spectral synthesis is a uniquely determined process.* But without a primer, the synthesized chain will, generally speaking, be in a chaotic (turbulent) state. Therefore, *2D synthesis can extend a chain but cannot start a chain.*

Damaged chains

Consider two strict consistent chains $\mathcal{F}_I$ and $\mathcal{F}_{II}$:

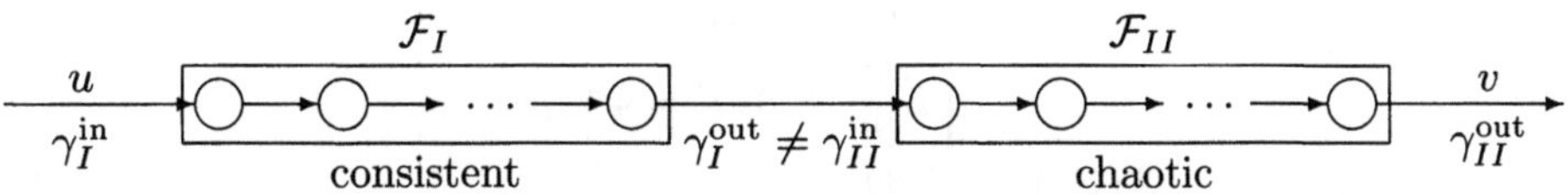

If the compatibility factors at the coupling location are distinct: $\gamma_I^{\text{out}} \neq \gamma_{II}^{\text{in}}$, then $\mathcal{F}_I$ *destroys* the consistency of $\mathcal{F}_{II}$. One can try *to repair* the chain $\mathcal{F}_I \to \mathcal{F}_{II}$ by *inserting* a third chain $\mathcal{F}_R$, such that $\gamma_R^{\text{in}} = \gamma_I^{\text{out}}$ and $\gamma_R^{\text{out}} = \gamma_{II}^{\text{in}}$.

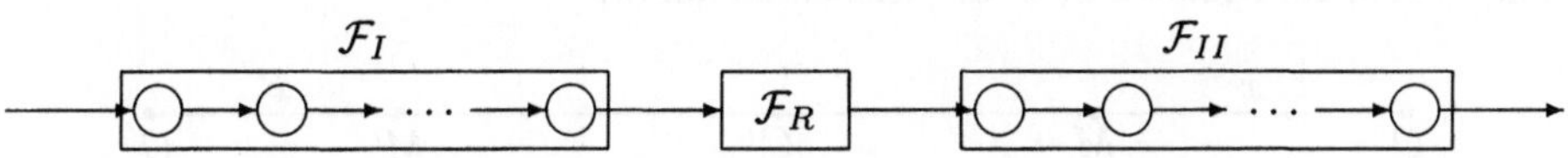

If the repair is possible then the discriminant curves $D_I = 0$, $D_{II} = 0$ coincide. In this case $D_R = D_I = D_{II}$.

2. Antiparallel reciprocity

Let us consider two mutually adjoint commutative strict vessels:

$$V = (A_1, A_2; H, \Phi, E; \sigma_1, \sigma_2; \gamma^{\text{in}}, \gamma^{\text{out}})$$

and

$$V^* = \left(A_1^*, A_2^*; H, \Phi', E; \sigma_1', \sigma_2'; \gamma'^{\text{in}}, \gamma'^{\text{out}}\right),$$

where

$$\Phi' = -\Phi, \quad \sigma_j' = -\sigma_j, \quad \gamma'^{\text{in}} = -\gamma^{\text{out}}, \quad \gamma'^{\text{out}} = -\gamma^{\text{in}}.$$

To these vessels there correspond two "complementary" space-time systems:

$$\sigma_2\frac{\partial u}{\partial t}-\sigma_1\frac{\partial u}{\partial x}+i\gamma^{\text{in}}u=0,$$

$$\mathcal{F}(V)\ :\ \begin{cases} i\dfrac{\partial f}{\partial t}+A_1 f=\Phi^*\sigma_1 u,\\ i\dfrac{\partial f}{\partial x}+A_2 f=\Phi^*\sigma_2 u,\\ v=u-i\Phi f,\end{cases}\tag{11}$$

$$\sigma_2\frac{\partial v}{\partial t}-\sigma_1\frac{\partial v}{\partial x}+i\gamma^{\text{out}}v=0,$$

and

$$\sigma_2\frac{\partial u'}{\partial t}-\sigma_1\frac{\partial u'}{\partial x'}+i\gamma^{\text{out}}u'=0,$$

$$\mathcal{F}(V^*)\ :\ \begin{cases} i\dfrac{\partial f'}{\partial t}+A_1^* f'=\Phi^*\sigma_1 u',\\ i\dfrac{\partial f'}{\partial x'}+A_2^* f'=\Phi^*\sigma_2 u',\\ v'=u'+i\Phi f',\end{cases}\tag{11$'$}$$

$$\sigma_2\frac{\partial v'}{\partial t}-\sigma_1\frac{\partial v'}{\partial x'}+i\gamma^{\text{in}}v'=0.$$

The spatial coordinates x, x' are chosen as follows

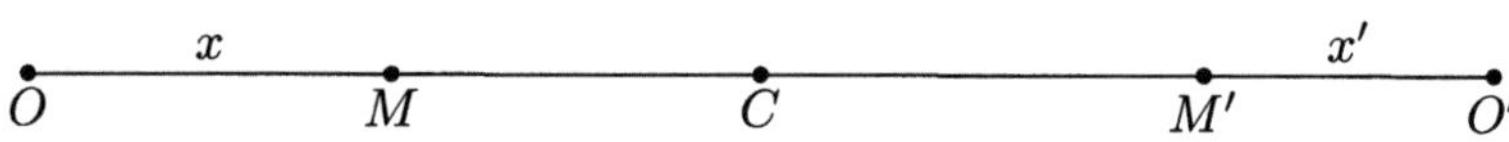

Here $x=\overline{OM}$, $x'=\overline{O'M'}$ are the distances between the corresponding points, the points O and O' are fixed, and the point C is located in the middle between O and O': $\overline{OC}=\overline{O'C}=a$. We will consider the case $x=x'=\rho$ $(0\le\rho\le a)$, when the points M and M' are symmetric with respect to the point C.

Reciprocity Theorem. *Assume that the output of $\mathcal{F}(V)$ is supplied as the input to $\mathcal{F}(V^*)$: $u'(t,\rho)=v(t,\rho)$, and that the initial values of the inner state are equal: $f(t_0,\rho_0)=f'(t_0,\rho_0)$. Then the inner state of $\mathcal{F}(V)$ and the inner state of $\mathcal{F}(V^*)$ coincide identically: $f(t,\rho)=f'(t,\rho)$ $(t_0\le t\le t_1,\ 0\le\rho\le a)$, and the output of $\mathcal{F}(V^*)$ coincides with the input of $\mathcal{F}(V)$: $v'(t,\rho)=u(t,\rho)$.*

Proof. Let $u'(t,\rho)=v(t,\rho)$ and $f'(t_0,\rho_0)=f(t_0,\rho_0)$. Consider the equations of $\mathcal{F}(V)$ and of $\mathcal{F}(V^*)$ at $t=t_0$:

$$\mathcal{F}(V)\ :\ \begin{cases} i\dfrac{\partial f}{\partial\rho}+A_2 f(t_0,\rho)=\Phi^*\sigma_2 u(t_0,\rho),\ \ f(t_0,\rho_0)=f_0,\\ v(t_0,\rho)=u(t_0,\rho)-i\Phi f(t_0,\rho),\end{cases}$$

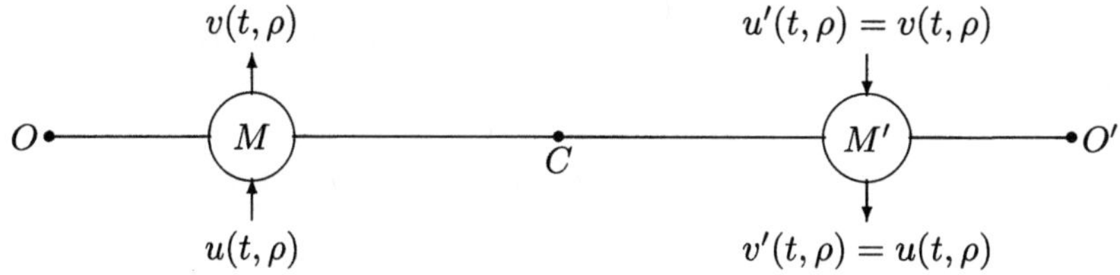

FIGURE 2. Reciprocity: $\overline{OM} = \overline{O'M'} = \rho$, $f(M) = f'(M')$

and

$$\mathcal{F}(V^*) \; : \; \begin{cases} i\dfrac{\partial f'}{\partial \rho} + A_2^* f' f(t_0, \rho) = \Phi^* \sigma_2 v(t_0, \rho), \;\; f'(t_0, \rho_0) = f_0, \\ v'(t_0, \rho) = v(t_0, \rho) + i\Phi f'(t_0, \rho). \end{cases}$$

Let $f(t_0, \rho)$ be the solution of the equations of $\mathcal{F}(V)$ (satisfying the initial condition $f(t_0, \rho_0) = f_0$). Substituting $f(t_0, \rho)$ on the left-hand side of the first equation of $\mathcal{F}(V^*)$ we obtain

$$i\frac{\partial f}{\partial \rho} + A_2^* f = i\frac{\partial f}{\partial \rho} + A_2 f + (A_2^* - A_2) f =$$
$$\Phi^* \sigma_2 u - i\Phi^* \sigma_2 \Phi f = \Phi^* \sigma_2 (u - i\Phi f) = \Phi^* \sigma_2 v. \quad (12)$$

Hence $f(t_0, \rho)$ satisfies the equations of $\mathcal{F}(V^*)$ (and the initial condition $f(t_0, \rho_0) = f_0$), and therefore $f(t_0, \rho) \equiv f'(t_0, \rho)$ $(0 \leq \rho \leq a)$. Analogously, using the time equations of $\mathcal{F}(V)$ and of $\mathcal{F}(V^*)$, we conclude that $f(t, \rho) \equiv f'(t, \rho)$. Then

$$v' = v + i\Phi f = u - i\Phi f + i\Phi f = u. \qquad \square$$

3. Attraction and repulsion forces

As we saw, to each commutative strict vessel V there corresponds an open 2D field $\mathcal{F}(V)$. It is easy to check that the compatibility equations $\mathcal{D}^{\text{in}}(u) = 0$ and $\mathcal{D}^{\text{out}}(v) = 0$ imply that

$$\frac{\partial}{\partial t}(\sigma_2 u, u) = \frac{\partial}{\partial x}(\sigma_1 u, u), \qquad \frac{\partial}{\partial t}(\sigma_2 v, v) = \frac{\partial}{\partial x}(\sigma_1 v, v).$$

From these relations it follows that there exist two potentials $W^{\text{in}}(t, x)$ and $W^{\text{out}}(t, x)$ such that

$$\begin{aligned} dW^{\text{in}} &= (\sigma_1 u, u)\, dt + (\sigma_2 u, u)\, dx, \\ dW^{\text{out}} &= (\sigma_1 v, v)\, dt + (\sigma_2 v, v)\, dx. \end{aligned}$$

From the relations

$$\begin{cases} \dfrac{\partial (f, f)}{\partial t} = (\sigma_1 u, u) - (\sigma_1 v, v), \\ \dfrac{\partial (f, f)}{\partial x} = (\sigma_2 u, u) - (\sigma_2 v, v), \end{cases} \quad (13)$$

it follows that $d(f,f) = dW^{\text{in}} - dW^{\text{out}}$, and we arrive at the *three potentials formula:*

$$(f,f) = W^{\text{in}} - W^{\text{out}} + C, \tag{14}$$

where (f,f) is the "inner" potential. We can consider now the "inner state force":

$$\mathcal{P}_x = -\kappa\,\text{grad}_x(f,f) = -\kappa\frac{\partial(f,f)}{\partial x} = \kappa\left[(\sigma_2 v, v) - (\sigma_2 u, u)\right], \tag{15}$$

where κ is a constant factor. If the conditions of the Reciprocity Theorem hold,

$$u(t,\rho) = v'(t,\rho), \quad f(t,\rho) = f'(t,\rho), \quad v(t,\rho) = u'(t,\rho),$$

in a domain $0 \le x = x' = \rho \le a$, $t_0 \le t \le t_1$, then the inner state force of the complementary system is

$$\mathcal{P}'_{x'}(t,\rho) = \kappa\left[(\sigma'_2 v', v') - (\sigma'_2 u', u')\right] = \kappa\left[(\sigma_2 v, v) - (\sigma_2 u, u)\right] = \mathcal{P}_x(t,\rho).$$

Assuming $\kappa > 0$, there are two possible cases:

1. $(\sigma_2 u(t,x), u(t,x)) < (\sigma_2 v(t,x), v(t,x))$ (in some domain, $(t,x) \in \Omega$). In this case $\mathcal{P}_x = \mathcal{P}'_{x'} > 0$. This is the case of *attraction* between the systems $\mathcal{F}(V)$ and $\mathcal{F}(V^*)$.
2. $(\sigma_2 u(t,x), u(t,x)) > (\sigma_2 v(t,x), v(t,x))$ $((t,x) \in \Omega)$. This is the case of *repulsion:* $\mathcal{P}_x = \mathcal{P}'_{x'} < 0$.

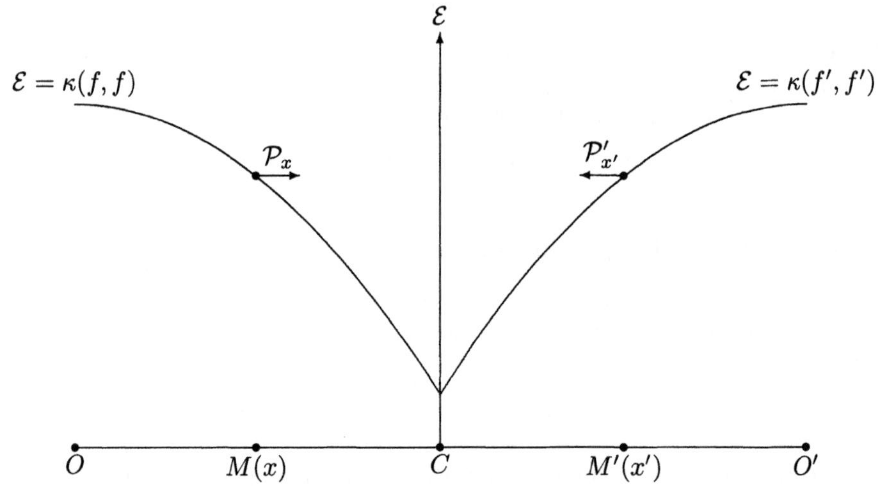

FIGURE 3. Attraction ($\kappa > 0$, $x = x'$, $\mathcal{P}_x = \mathcal{P}'_{x'} > 0$)

Remark. In the case $\sigma_2 > 0$, $\kappa > 0$, all the channels of the system $\mathcal{F}(V)$ are *direct* with respect to the space coordinate: energy enters the system through the input and leaves the system through the output, if the displacement $dx > 0$. All the channels of $\mathcal{F}(V^*)$ are *inverse* with respect to the displacement $dx' > 0$: energy enters through the output and leaves through the input. If the conditions of the Reciprocity Theorem hold, both systems emit or absorb energy simultaneously.

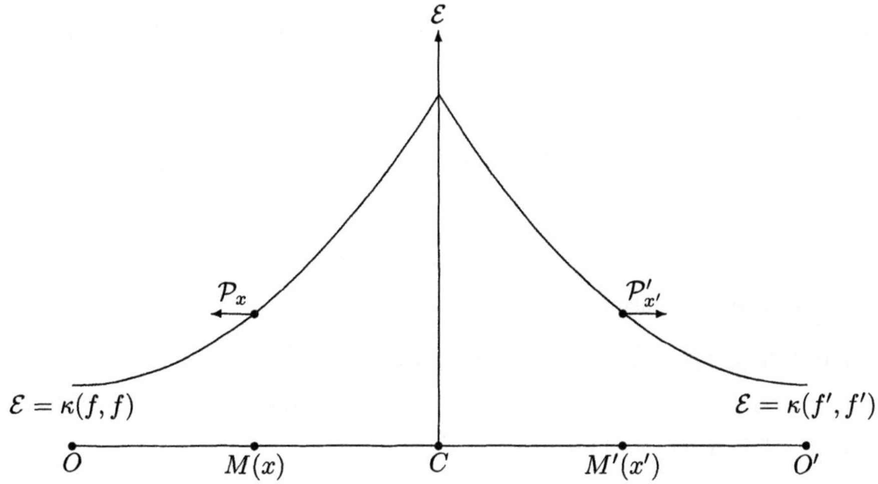

FIGURE 4. Repulsion ($\kappa > 0$, $x = x'$, $\mathcal{P}_x = \mathcal{P}'_{x'} < 0$)

Let us consider some special cases.

1. $u = v' = 0$, $\Phi f_0 \neq 0$. Then $v = u' = -i\Phi e^{i(tA_1+\rho A_2)} f_0 \neq 0$, and therefore, assuming $\sigma_2 > 0$, $\kappa > 0$, we have $0 = (\sigma_2 u, u) < (\sigma_2 v, v)$. This is the case of attraction. Analogously, if $u' = v = 0$, it will be the case of repulsion.

2. Let $u_\lambda = v'_\lambda = e^{i(t\lambda_1+\rho\lambda_2)} u^0_\lambda$, $(\lambda_1, \lambda_2) \in \mathbb{C}^2$. Assuming that the compatibility PDE $\mathcal{D}^{\text{in}}(u_\lambda) = 0$ is satisfied, we obtain the following algebraic equation

$$(\lambda_1\sigma_2 - \lambda_2\sigma_1 + \gamma^{\text{in}}) u^0_\lambda = 0. \tag{16}$$

If $u^0_\lambda \neq 0$, then the point $\mathcal{M}(\lambda_1, \lambda_2)$ belongs to the discriminant curve $D(z_1, z_2) = 0$, where

$$D(z_1, z_2) = \det(z_1\sigma_2 - z_2\sigma_1 + \gamma^{\text{in}}) = \det(z_1\sigma_2 - z_2\sigma_1 + \gamma^{\text{out}}).$$

The equations of $\mathcal{F}(V)$ have in this case a common solution $f_\lambda = e^{i(t\lambda_1+\rho\lambda_2)} f^0_\lambda$, where

$$f^0_\lambda = (A_1 - \lambda_1 I)^{-1}\Phi^*\sigma_1 u^0_\lambda = (A_2 - \lambda_2 I)^{-1}\Phi^*\sigma_2 u^0_\lambda. \tag{17}$$

The output is $v_\lambda = e^{i(t\lambda_1+\rho\lambda_2)} v^0_\lambda$, where

$$v^0_\lambda = S_1(\lambda_1) u^0_\lambda = S_2(\lambda_2) u^0_\lambda = \tilde{S}(\mathcal{M}) u^0_\lambda.$$

Here

$$S_j(\lambda_j) = I - i\Phi(A_j - \lambda_j I)^{-1}\Phi^*\sigma_j \quad (j = 1, 2) \tag{18}$$

are the characteristic functions of the (single-operator) colligations

$$X_j = (A_j; H, \Phi, E; \sigma_j),$$

and $\tilde{S}(\mathcal{M})$ is the joint characteristic (transfer) function; see [2, Sections 3.4 and 4.3]. The forces in this case are

$$\mathcal{P}_x = \kappa\left[(\sigma_2 v_\lambda, v_\lambda) - (\sigma_2 u_\lambda, u_\lambda)\right]$$
$$= \kappa e^{-2(t\,\mathrm{Im}\,\lambda_1 + \rho\,\mathrm{Im}\,\lambda_2)}\left((S_2(\lambda_2)^*\sigma_2 S_2(\lambda_2) - \sigma_2)u_\lambda^0, u_\lambda^0\right). \quad (19)$$

The equality

$$\frac{\partial(f,f)}{\partial x} = (\sigma_2 u_\lambda, u_\lambda) - (\sigma_2 v_\lambda, v_\lambda)$$

is equivalent in this case to

$$(2\,\mathrm{Im}\,\lambda_2)(f_\lambda^0, f_\lambda^0) = (\sigma_2 S_2(\lambda_2)u_\lambda^0, S_2(\lambda_2)u_\lambda^0) - (\sigma_2 u_\lambda^0, u_\lambda^0) = \left((S_2(\lambda_2)^*\sigma_2 S_2(\lambda_2) - \sigma_2)u_\lambda^0, u_\lambda^0\right).$$

Hence the following relations hold

$$\left((S_2(\lambda_2)^*\sigma_2 S_2(\lambda_2) - \sigma_2)u_\lambda^0, u_\lambda^0\right) \begin{cases} > 0, & \mathrm{Im}\,\lambda_2 > 0, \\ < 0, & \mathrm{Im}\,\lambda_2 < 0, \\ = 0, & \mathrm{Im}\,\lambda_2 = 0. \end{cases}$$

We obtain therefore, assuming $\kappa > 0$, the following classification:

2_+. $\mathrm{Im}\,\lambda_2 > 0$: attraction;
2_-. $\mathrm{Im}\,\lambda_2 < 0$: repulsion;
2_0. $\mathrm{Im}\,\lambda_2 = 0$: neutral.

If, in the case 2_0, we have also $\mathrm{Im}\,\lambda_1 = 0$, then $(f,f) \equiv \mathrm{const}$ for all (t,x).

Double chains of elementary systems

If the colligation X is the coupling of $X^{(1)}$ and $X^{(2)}$ as in (7),

$$X = (A_1, A_2; H, \Phi, E; \sigma_1, \sigma_2) = X^{(1)} \to X^{(2)},$$

then the adjoint colligation X^* is the coupling of $(X^{(2)})^*$ and $(X^{(1)})^*$,

$$X = (A_1^*, A_2^*; H, -\Phi, E; -\sigma_1, -\sigma_2) = (X^{(1)})^* \leftarrow (X^{(2)})^*.$$

Let us consider now two mutually adjoint chains of elementary vessels V^k and $(V^k)^*$ $(k = 1, 2, \ldots, N)$ as in (8), and the related chains of elementary space-time systems, $\mathcal{F}(V) = \mathcal{F}(V^1) \to \mathcal{F}(V^2) \to \cdots \to \mathcal{F}(V^N)$ and $\mathcal{F}(V^*) = \mathcal{F}((V^1)^*) \leftarrow \mathcal{F}((V^2)^*) \leftarrow \cdots \leftarrow \mathcal{F}((V^N)^*)$. If the conditions of the Reciprocity Theorem are fulfilled: $u = v'$, $f = f'$, $v = u'$, then these two chains of systems form a pair of *complementary* chains:

In the special case $u_\lambda^k(t,\rho) = e^{i(t\lambda_1 + \rho\lambda_2)}u_\lambda^k(0)$, it follows from the factorization theorem for characteristic functions [2, Theorem 3.4.6] that the sequence $u_\lambda^k(0)$ is given by $u_\lambda^k(0) = S_j^{k-1}(\lambda_j)\cdots S_j^1(\lambda_j)u_\lambda^1(0)$, where

$$S_j^k(\lambda_j) = I - i\frac{\phi^k\phi^{k*}}{a_j^k - \lambda_j}\sigma_j \quad (j = 1, 2;\ k = 1, 2, \ldots, N).$$

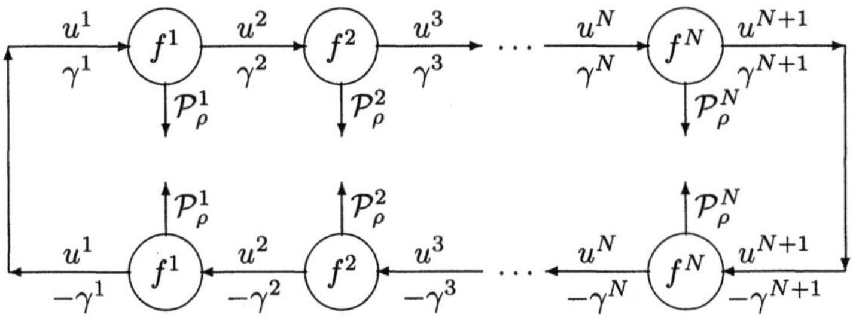

FIGURE 5. Two Snakes Bite Each Other

In the case of attraction $\operatorname{Im}\lambda_2 > 0$. The attraction forces between the links $\mathcal{F}(V^k)$ and $\mathcal{F}((V^k)^*)$ $(k = 1, 2, \ldots, N)$ of the two complementary chains are

$$\mathcal{P}^k_\rho = \kappa e^{-2(t \operatorname{Im}\lambda_1 + \rho \operatorname{Im}\lambda_2)} \left((S^k_2(\lambda_2)^* \sigma_2 S^k_2(\lambda_2) - \sigma_2) u^k_\lambda(0), u^k_\lambda(0)\right).$$

In the case of repulsion, $\operatorname{Im}\lambda_2 < 0$.

If only one of the two complementary chains is given, then this chain can be used as a *templet* for the synthesis of the other chain. If, for instance, $\mathcal{F} = \mathcal{F}^1 \to \mathcal{F}^2 \to \cdots \to \mathcal{F}^N$, then the joint transfer function is as above a product

$$\tilde{S}_V = \tilde{S}_{V^N} \cdots \tilde{S}_{V^2} \tilde{S}_{V^1}.$$

It is known [2, Proposition 3.4.5] that the transfer function of the adjoint vessel and the transfer function of a given vessel are related by $\tilde{S}_{V^*} = \left(\tilde{S}_V\right)^{-1}$. Hence

$$\tilde{S}_{V^*} = \left(\tilde{S}_{V^1}\right)^{-1} \left(\tilde{S}_{V^2}\right)^{-1} \cdots \left(\tilde{S}_{V^N}\right)^{-1}.$$

Therefore using the product expansion for $\tilde{S}_V$ step by step in the opposite direction, we can carry out the synthesis of the complementary chain.

The four elements case

Consider the case when all the points $\mathcal{M}^k$ $(k = 1, 2, \ldots, N)$ are chosen from some basic set of four points,

$$\mathcal{M}^k \in \{\mathcal{K}, \mathcal{L}, \mathcal{M}, \mathcal{N}\},$$

belonging to the given discriminant curve $D(\lambda_1, \lambda_2) = 0$. Since all the coefficients of $D(\lambda_1, \lambda_2)$ are real, $D(\overline{\lambda_1}, \overline{\lambda_2}) = \overline{D(\lambda_1, \lambda_2)}$. Assume that $D(\lambda_1, \lambda_2)$ is symmetrical also with respect to the transformation $[M(\lambda_1, \lambda_2)]' = M(\lambda_1, -\lambda_2)$: $D(\lambda_1, -\lambda_2) = D(\lambda_1, \lambda_2)$. We will choose four points $\mathcal{K}(b^1_1, b^1_2)$, $\mathcal{L}(b^2_1, b^2_2)$, $\mathcal{M}(c^1_1, c^1_2)$, $\mathcal{N}(c^2_1, c^2_2)$, such that

$$b^2_1 = \overline{b^1_1},\ b^2_2 = -\overline{b^1_2},\ c^2_1 = \overline{c^1_1},\ c^2_2 = -\overline{c^1_2},$$

and $\operatorname{Im} b^2_2 > 0$, $\operatorname{Im} c^2_2 > 0$. Evidently, $\mathcal{L}' = \mathcal{K}^*$, $\mathcal{M}' = \mathcal{N}^*$. The pairs $\{\mathcal{K}, \mathcal{L}'\}$ and $\{\mathcal{M}, \mathcal{N}'\}$ are pairs of corresponding points in the complementary chains $\mathcal{F}(V)$, $\mathcal{F}(V^*)$. As an example of the discriminant curve, we may consider a smooth cubic,

$$\lambda_2^2 \lambda_1 - (\lambda_1 - \alpha_1)(\lambda_1 - \alpha_2)(\lambda_1 - \alpha_3) = 0.$$

In this case [2, Section 1.4] $\dim E = 3$ and we may take

$$\sigma_1 = \begin{bmatrix} 1 & 0 & 0 \\ 0 & 0 & 0 \\ 0 & 0 & -1 \end{bmatrix}, \qquad \sigma_2 = I_3.$$

4. Self conjugate systems

If the conditions of the Reciprocity Theorem hold, then the equations of the two mutually complementary systems are

$$\mathcal{F}(V) \ : \ \begin{cases} i\dfrac{\partial h}{\partial t} + A_1 h = \Phi^* \sigma_1 u, \\ i\dfrac{\partial h}{\partial \rho} + A_2 h = \Phi^* \sigma_2 u, \\ v = u - i\Phi h, \end{cases}$$

and

$$\mathcal{F}(V^*) \ : \ \begin{cases} i\dfrac{\partial h}{\partial t} + A_1^* h = \Phi^* \sigma_1 v, \\ i\dfrac{\partial h}{\partial \rho} + A_2^* h = \Phi^* \sigma_2 v, \\ u = v + i\Phi h, \end{cases}$$

where $\rho = x = x'$, $h(t,\rho) = f(t,x)|_{x=\rho} = f'(t,x')|_{x'=\rho}$. Adding the respective equations for $\mathcal{F}(V)$ and for $\mathcal{F}(V^*)$, we obtain the following equations for $h(t,\rho)$:

$$\mathcal{F}(VV^*) \ : \ \begin{cases} i\dfrac{\partial h}{\partial t} + A_1' h = \Phi^* \sigma_1 p, \\ i\dfrac{\partial h}{\partial \rho} + A_2' h = \Phi^* \sigma_2 p, \end{cases} \qquad p = \frac{1}{2}(u+v),$$

where $A_j' = \operatorname{Re} A_j = \frac{1}{2}(A_j + A_j^*)$ $(j = 1, 2)$. It can be checked easily that the following relations hold:

$$\begin{cases} \dfrac{\partial (h,h)}{\partial t} = 4\operatorname{Re}(\sigma_1 p, q), \\ \dfrac{\partial (h,h)}{\partial \rho} = 4\operatorname{Re}(\sigma_2 p, q), \end{cases} \qquad q = \frac{1}{2}(u-v) = \frac{1}{2}\Phi h.$$

The system $\mathcal{F}(VV^*)$ is said to be *self conjugate*; $p(t,\rho)$ and $q(t,\rho)$ are the input and the output of $\mathcal{F}(VV^*)$, respectively.

In general, the real parts of commuting operators do not commute. It is known [2, Section 4.7] that the commutator of the real parts is given by

$$[A_1', A_2'] = \frac{i}{4}\Phi^*(\gamma^{\text{out}} - \gamma^{\text{in}})\Phi.$$

The coincidence of the compatibility factors at the input and at the output: $\gamma^{\text{in}} = \gamma^{\text{out}}$, is therefore sufficient (and for strict vessels, also necessary) for the commutativity of the real parts. For such "exclusive" systems, the compatibility PDEs at the input and at the output coincide,

$$\mathcal{D} \equiv \mathcal{D}^{\text{in}} \equiv \mathcal{D}^{\text{out}},$$

and the input $p(t,\rho)$ and the output $q(t,\rho)$ satisfy the same PDE: $\mathcal{D}(p) = 0$, $\mathcal{D}(q) = 0$. In the exclusive case we can take $p \equiv 0$; then the equations of $\mathcal{F}(VV^*)$ will be

$$\begin{cases} i\dfrac{\partial h}{\partial t} + A_1' h = 0, \\ i\dfrac{\partial h}{\partial \rho} + A_2' h = 0 \end{cases} \qquad (A_1'A_2' = A_2'A_1').$$

These equations have a common solution

$$h = e^{i(tA_1' + \rho A_2')} h_0,$$

such that $(h, h) = (h_0, h_0)$. Therefore, in the exclusive case $\gamma^{\text{in}} = \gamma^{\text{out}}$, when $u + v = 0$, $\mathcal{F}(VV^*)$ is a *closed conservative* system.

5. Tensor colligations and tube systems

Let us consider the following tensor of operators:

$$\begin{bmatrix} a_{11}A & a_{12}A & a_{13}A & 0 \\ a_{21}A & a_{22}A & a_{23}A & 0 \\ a_{31}A & a_{32}A & a_{33}A & 0 \\ 0 & 0 & 0 & B \end{bmatrix} \qquad (a_{\alpha\beta} = \overline{a}_{\alpha\beta}).$$

Assume that $(A, B; H, \Phi, E; \sigma(A), \sigma(B))$ is a strict commutative colligation. Let $\xi = (\xi_1, \xi_2, \xi_3)$, $\xi_1^2 + \xi_2^2 + \xi_3^2 = 1$, be a unit vector in $\mathbb{R}^3$. Then we can construct the following 4D colligation depending on ξ:

$$X_\xi = (A_\alpha = \eta_\alpha A, B; H, \Phi, E; \sigma_\alpha = \eta_\alpha \sigma(A), \sigma(B)),$$

where $\eta_\alpha = \sum_{\beta=1}^{3} a_{\alpha\beta}\xi_\beta$ (here and below the index α runs from 1 to 3; for details on the definitions and basic properties of 4D colligations, systems, and vessels, which are analogous to the 2D case, see [2, Chapters 2 and 3]). Assume that $\sigma(A) > 0$ and that $x_4 = t$ is the time coordinate. The corresponding 4D system is

$$\mathcal{F}_\xi \; : \; \begin{cases} i\dfrac{\partial f}{\partial x_\alpha} + \eta_\alpha A f(x_1, x_2, x_3, t) = \eta_\alpha \Phi^* \sigma(A) u(x_1, x_2, x_3, t), \\ i\dfrac{\partial f}{\partial t} + B f(x_1, x_2, x_3, t) = \Phi^* \sigma(B) u(x_1, x_2, x_3, t), \\ v(x_1, x_2, x_3, t) = u(x_1, x_2, x_3, t) - i\Phi f(x_1, x_2, x_3, t). \end{cases}$$

From the relations

$$\frac{1}{i}(A_\alpha A_\beta^* - A_\beta A_\alpha^*) = \eta_\alpha \eta_\beta \frac{1}{i}(AA^* - AA^*) = 0,$$

we conclude (as we assumed strictness) that $\gamma^{\text{in}}_{\alpha\beta} = 0$. The $x_\alpha x_\beta$ compatibility equations are

$$\sigma(A)\left(\eta_\beta \frac{\partial u}{\partial x_\alpha} - \eta_\alpha \frac{\partial u}{\partial x_\beta}\right) = 0.$$

Therefore $\dfrac{\partial u}{\partial x_\alpha} : \eta_\alpha = \dfrac{\partial u}{\partial x_\beta} : \eta_\beta = G$ and

$$du = (\eta_1\, dx_1 + \eta_2\, dx_2 + \eta_3\, dx_3)\, G,$$

and $u(x_1, x_2, x_3, t)$ is a function of t and $\rho = \sum(x_\alpha - x^0_\alpha)\eta_\alpha$: $u(x_1, x_2, x_3, t) = \tilde{u}(\rho, t)$. Then also $f(x_1, x_2, x_3, t) = \tilde{f}(\rho, t)$, $v(x_1, x_2, x_3, t) = \tilde{v}(\rho, t)$, and the equations of the 4D system $\mathcal{F}_\xi$ can be reduced to the equations of a 2D system:

$$\widetilde{\mathcal{F}}(\rho, t) \ : \ \begin{cases} i\dfrac{\partial \tilde{f}}{\partial \rho} + A\tilde{f}(\rho, t) = \Phi^*\sigma(A)\tilde{u}(\rho, t), \\ i\dfrac{\partial \tilde{f}}{\partial t} + B\tilde{f}(\rho, t) = \Phi^*\sigma(B)\tilde{u}(\rho, t), \\ \tilde{v}(\rho, t) = \tilde{u}(\rho, t) - i\Phi\tilde{f}(\rho, t). \end{cases}$$

We will consider the special case $a_{\alpha\beta} = \delta_{\alpha\beta}$. In this case $\eta_\alpha = \xi_\alpha$ and $\rho = ((x - x^0), \xi) = \sum(x_\alpha - x^0_\alpha)\xi_\alpha$. Assume that the system $\widetilde{\mathcal{F}}$ is defined in some cylindrical domain (tube) T in the direction of the unit vector ξ:

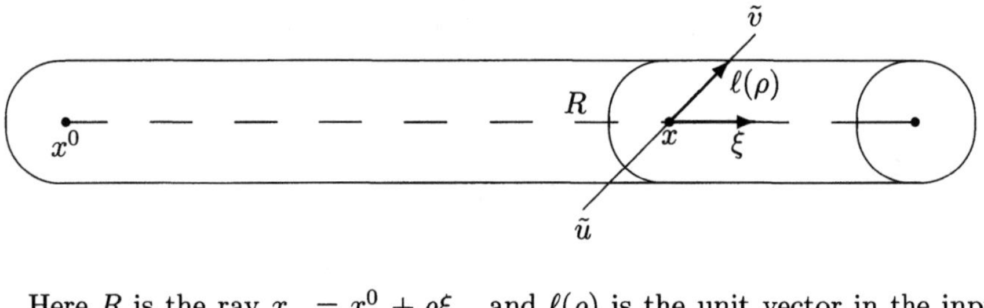

Here R is the ray $x_\alpha = x^0_\alpha + \rho\xi_\alpha$, and $\ell(\rho)$ is the unit vector in the input-output direction.

Let $\mathfrak{L}^\rho : x^\rho_\alpha = x^\rho_\alpha(s)$ $(s_0 \le s \le s_1,\ 0 \le \rho \le d)$ be a family of smooth curves, and let $\mathfrak{M}^s$ be the set of points of $\mathfrak{L}^\rho$ $(0 \le \rho \le d)$ corresponding to a fixed value of the parameter s.

Definition. We will say that a family of tubes T^s $(s_0 \le s \le s_1)$ is *attached* to the family of curves $\mathfrak{L}^\rho$ $(0 \le \rho \le d)$ if:

1. The sets $\mathfrak{M}^s$ $(s_0 \le s \le s_1)$ coincide with the corresponding rays R^s.
2. The directions of the principal normal vectors to the curves $\mathfrak{L}^\rho$ coincide with the directions of the corresponding rays.
3. The input-output directions of the tube systems T^s coincide with the directions of the tangent vectors to $\mathfrak{L}^\rho$ at the corresponding points.

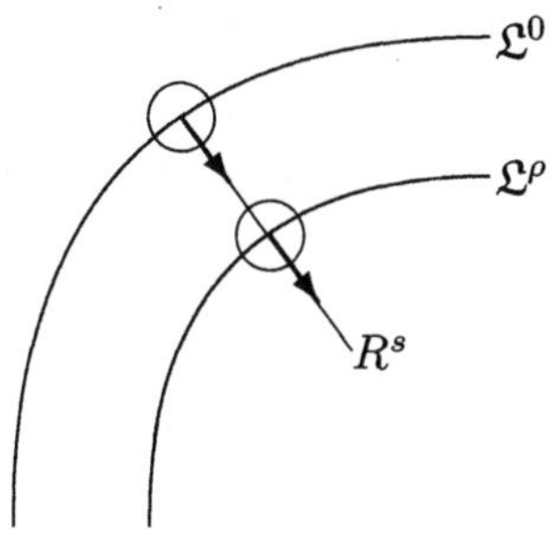

When the conditions 1–3 of the definition are fulfilled, the family of curves $\mathfrak{L}^\rho$ $(0 \leq \rho \leq d)$ is said to be a *tube-admissible* family.

From condition 2 we conclude that if $\mathfrak{L}^\rho$ are tube-admissible, then the principal normal vectors to the curves $\mathfrak{L}^\rho$ coincide at the points corresponding to the same value of the parameter s. Such curves are well known in classical differential geometry, and they are called Bertrand curves; see, e.g., [5], [4, p. 72]. The curvature $\kappa_1(s)$ and the torsion $\kappa_2(s)$ of the Bertrand curve $\mathfrak{L}^0$ are connected by the relation

$$\rho\kappa_1(s)\sin\alpha + \rho\kappa_2(s)\cos\alpha = \sin\alpha,$$

where α is the necessarily constant angle between the tangent vectors to the curves $\mathfrak{L}^0$ and $\mathfrak{L}^\rho$ at the corresponding points (which depends, generally speaking, on ρ).

Theorem. *Let $\mathfrak{L}^0$ be a tube-admissible curve. If all the principal normal vectors to $\mathfrak{L}^0$ lie in a fixed plane, then this curve is a simplex helix.*

Proof. It is well known that if all the principal normal vectors to a curve lie in a fixed plane and $\kappa_2 \neq 0$, then $\kappa_1/\kappa_2 = \text{const}$ along the curve. If in addition the curve is a Bertrand curve then from the relation

$$\rho\frac{\kappa_1}{\kappa_2}\sin\alpha + \rho\cos\alpha = \frac{1}{\kappa_2}\sin\alpha$$

it follows that $\kappa_1 = \text{const}$, $\kappa_2 = \text{const}$. In this case the equations of the curve can be brought to the form

$$x_1 = a\cos\theta,\ x_2 = a\sin\theta,\ x_3 = b\theta,$$

which are the equations of a simple helix. □

Remark. The case of a general tensor $a_{\alpha\beta}$ can be obtained from the case of $a_{\alpha\beta} = \delta_{\alpha\beta}$ that we have considered by an affine change of space variables.

6. Summary: Systems and genetics

We summarize in the following table the correspondence between the properties of 2D (space-time) systems that we described in this paper, and the standard properties of DNA.

I. Elementary 2D systems.	**I.** Nucleotides.
II. Chain synthesis of 2D systems goes in the direction from the *input* to the *output*.	**II.** Chain synthesis of nucleotides goes in the direction from $5'$ to $3'$.
III. Chain synthesis of 2D systems cannot start without a "primer" — an initial strict commutative colligation which defines the initial adaptation coefficient γ^0.	**III.** Chain synthesis of DNA cannot start without a "primer" — a short chain of nucleotides (oligonucleotides).
IV. Double chain of mutually adjoint elements going in the opposite directions; open systems forces.	**IV.** Double chain of mutually complementary nucleotides going in the opposite directions; hydrogen bonds.
V. Templets go in the direction from *out* to *in*; leading and lagging strands.	**V.** Templets go in the direction from $3'$ to $5'$; Okasaki fragments; leading and lagging strands.
VI. Tube systems and Bertrand curves.	**VI.** Double helix of nucleotides.

References

[1] A. Griffiths, J. Miller, D. Suzuki, H. Lewitin, and N. Gelbart, *An Introduction to Genetic Analysis*, Freeman and Co., New York, 1996 (sixth edition).

[2] M.S. Livšic, N. Kravitsky, A.S. Markus, and V. Vinnikov, *Theory of Commuting Nonselfadjoint Operators*, Mathematics and Its Applications, vol. 332, Kluwer, Dordrecht, 1995.

[3] M.S. Livšic, *Vortices of 2D systems*, Operator Theory, System Theory and Related Topics (D. Alpay and V. Vinnikov, eds.), Operator Theory: Adv. Appl., vol. 123, Birkhäuser Verlag, Basel, 2001, pp. 7–41.

[4] A.V. Pogorelov, *Differential Geometry*, Izdat. "Nauka", Moscow, 1974 (in Russian).

[5] J.K. Whittemore, *Bertrand curves and helices*, Duke Math. J. 6 (1940), 235–245.

M.S. Livšic
Department of Mathematics
Ben Gurion University of the Negev
POB 653
84105 Beer-Sheva, Israel

Operator Theory:
Advances and Applications, Vol. 134, 337–383

A Class of Robustness Problems in Matrix Analysis

André C.M. Ran and Leiba Rodman

Dedicated to Harry Dym on the occasion of his sixtieth birthday

Abstract. We present an overview of several results and a literature guide, prove some new results, and state open problems concerning description of all robust matrices in the following sense: Let be given a class of real or complex matrices $\mathcal{A}$, and for each $X \in \mathcal{A}$, a set $\mathcal{G}(X)$ is given. An element $Y_0 \in \mathcal{G}(X_0)$ will be called robust (relative to the sets $\mathcal{A}$ and $\mathcal{G}(X)$) if for every $X \in \mathcal{A}$ close enough to X_0 there is a $Y \in \mathcal{G}(X)$ that is as close to Y_0 as we wish. The following topics are covered, with respect to the robustness property: 1. Invariant subspaces of matrices; here the set $\mathcal{G}(X)$ is the set of all X-invariant subspaces. 2. Invariant subspaces of matrices with symmetries related to indefinite inner products. The invariant subspaces in question include semidefinite and neutral subspaces (with respect to an indefinite inner product). 3. Applications of invariant subspaces of matrices with or without symmetries. The applications include: general matrix quadratic equations, the continuous and discrete algebraic Riccati equations, minimal factorization of rational matrix functions with symmetries and the transport equation from mathematical physics. 4. Several matrix decompositions: polar decompositions with respect to an indefinite inner product, Cholesky factorizations, singular value decomposition.

Other related notions of robustness are studied as well, for example, a stronger notion of α-robustness, in which the magnitude of degree of closeness of Y and Y_0 (as measured in some appropriate metric) does not exceed the magnitude of $\|X - X_0\|^{1/\alpha}$.

1. The metaproblem

Many problems in applied mathematics, engineering, and physics require for their solutions that certain quantities associated with given matrices be computed. For example, these quantities may be solutions of a matrix equation with specified additional properties, an invariant subspace, a Cholesky factorization of a positive semidefinite matrix, or the set of singular values. From the point of view of

The work of the second author was partially supported by NSF grant DMS-9800704, and by a Faculty Research Assignment Grant from the College of William and Mary. The work of both authors was partially supported by NATO Grant 9600700.

computation or approximation of a solution of the problem at hand it is therefore important to know whether or not the required quantities can be in principle computed more or less accurately. This involves approximation of the required quantities of a matrix by the corresponding quantities of matrices that are perturbations (often restricted to a certain class of matrices) of the given matrix. To formalize this notion, we state a metaproblem:

Metaproblem. *Let be given a class of matrices $\mathcal{A}$, and for each $X \in \mathcal{A}$, a set of mathematical quantities $\mathcal{G}(X)$ is given. An element $Y_0 \in \mathcal{G}(X_0)$ will be called robust, or stable, with respect to $\mathcal{A}$ and the collection $\{\mathcal{G}(X)\}_{X\in\mathcal{A}}$, if for every $X \in \mathcal{A}$ which is sufficiently close to X_0 there exists an element $Y \in \mathcal{G}(X)$ that is as close to Y_0 as we wish. Give criteria for existence of a robust Y_0, and describe all of them.*

The statement of the metaproblem presumes a topology on $\mathcal{A}$ and on $\cup_{X\in\mathcal{A}}\mathcal{F}(X)$. In all cases, these will be natural, or standard, topologies. For example, if $\mathcal{A}$ is a subset of $n \times n$ matrices with complex entries, the topology induced by the operator norm will be assumed on $\mathcal{A}$. If $\cup_{X\in\mathcal{A}}\mathcal{F}(X)$ is a subset of the set of subspaces in $\mathbb{C}^n$ then the topology induced by the gap metric will be given on $\cup_{X\in\mathcal{A}}\mathcal{F}(X)$.

Assuming that Y_0 is robust, one might be interested in the degree of robustness, in other words, comparison of magnitudes between approximation of Y_0 by Y and approximation of X_0 by X. This approach leads to a more refined scale of robustness properties.

In the present paper we present an overview of several results and a literature guide concerning the metaproblem, including the degree of robustness, as well as prove some new results and state open problems. The main topics covered include: 1. Invariant subspaces of matrices; in this case we let $\mathcal{G}(X)$ be the set of invariant subspaces of a matrix X. 2. Invariant subspaces of matrices with symmetries related to indefinite inner products; for example, we let $\mathcal{G}(X)$ be the set of X-invariant subspaces that are simultaneously maximal semidefinite with respect to a given indefinite inner product. 3. Applications of invariant subspaces of matrices, with or without symmetries. 4. Matrix decompositions: polar decompositions with respect to an indefinite inner product, Cholesky factorizations, singular value decomposition.

Many of the results presented here are not new. For new results, at least an indication of a proof is given. To keep the paper within reasonable limits, we avoid detailed proofs (with few exceptions).

The report [62] contains a preliminary version of exposition in Sections 2 and 3.

Throughout the paper, we use the notation $\operatorname{Im} X$ for the range $\{Xx : x \in F^m\}$ of an $n \times m$ matrix X with entries in F, where $F = \mathbb{C}$, the field of complex numbers, or $F = \mathbb{R}$, the field of real numbers. Also,

$$\operatorname{Ker} X = \{x \in F^m : Xx = 0\}.$$

2. Stability of invariant subspaces

Invariant subspaces of a matrix or linear operator become increasingly important in applied problems. Recently, both theory and algorithms for computations of invariant subspaces have been developed, see, for example, [27], [40], [70]. The notion of a robust invariant subspace, as defined in the metaproblem, takes the following form (keeping in line with the terminology established here, we use "stability" rather than "robustness" in the sections concerning invariant subspaces): Let A be an $n \times n$ matrix, and let $\mathcal{M}$ be an invariant subspace of A, i. e., $Ax \in \mathcal{M}$ for every $x \in \mathcal{M}$. Then $\mathcal{M}$ is called *stable* if for every $\varepsilon > 0$ there is a $\delta > 0$ such that $\|A - B\| < \delta$ implies the existence of a B-invariant subspace $\mathcal{N}$ such that $\|P_{\mathcal{M}} - P_{\mathcal{N}}\| < \varepsilon$. Here $P_{\mathcal{M}}$, respectively $P_{\mathcal{N}}$, denote the orthogonal projection onto $\mathcal{M}$, respectively, $\mathcal{N}$.

Stable invariant subspaces were introduced and discussed in [14] and [3] (see also [4]) in the late seventies. A complete description of the stable invariant subspaces of a given matrix A was given there. Since then many refinements of this concept have been studied. See, for instance [22, 25, 49, 51, 63]. In [49, 51, 63] the rate of convergence to a stable invariant subspace was investigated. Although of less practical value from a numerical point of view, these results are interesting in their own right, and can also be applied in several situations as we shall show (see Sections 2.4 and 3.3).

The results on various notions of stability of invariant subspaces have applications (at least potentially) in the areas in which invariant subspaces play a key role. One such broad area is the state space methods in the theory of linear systems. More specifically, stability of solutions of matrix polynomial equations, factorizations of matrix polynomials, minimal and Wiener-Hopf factorizations of rational matrix functions, (see [4, 22]), cascade decompositions [26] have been studied in the literature using the invariant subspaces approach. In Section 2.4 we give some results in this direction concerning quadratic matrix equations.

It should be noted that the corresponding problems for invariant subspaces of linear bounded operators on infinite dimensional Banach spaces are also of interest, in particular, from the point of view of applications. However, not much is known here, and development of the theory of stable invariant subspaces of infinite dimensional operators is a challenge for future research. We note that Theorem 2.2 is valid in one direction in the infinite dimensional case, namely, every spectral invariant subspace is Lipschitz stable (there are examples of Lipschitz stable non-spectral invariant subspaces in infinite dimensions; these examples are constructed using one-sided resolvents, see [68]). Theorem 2.9 is valid in infinite dimensions as well. Some general results and additional references on stable invariant subspaces of infinite dimensional operators are found in [68].

2.1. General theory: The complex case

In this and the next subsection results concerning stability properties of matrices will be presented. We will study the concepts of stability, Lipschitz stability, α-stability and strong α-stability. Both the real and the complex case are considered. The results presented here have been proved in [3, 4, 14, 49, 63, 51].

We start with the complex case. For two subspaces $\mathcal{N}$ and $\mathcal{M}$ in the complex vector space $\mathbb{C}^n$ of n-dimensional column vectors, the *gap* $\theta(\mathcal{M},\mathcal{N})$ is defined as follows: $\theta(\mathcal{M},\mathcal{N}) = \|P_{\mathcal{M}} - P_{\mathcal{N}}\|$, where $P_{\mathcal{M}}$, respectively $P_{\mathcal{N}}$, denote the orthogonal projection onto $\mathcal{M}$, respectively, $\mathcal{N}$. Here and elsewhere in the paper we use the operator norm $\|X\|$, i.e., the maximal singular value of the matrix X (our results however are independent of the choice of the matrix norm). It is well known that $\theta(\mathcal{M},\mathcal{N})$ is a metric on the set of subspaces in $\mathbb{C}^n$ which makes this set a complete compact metric space (see, e.g., [22] for more details). Other measures of closeness between subspaces in $\mathbb{C}^n$ (spherical gap, minimal opening, canonical angles) have been extensively studied in the literature as well, also for the infinite dimensional Banach spaces; see the books [22, 4, 31, 70], for example.

Let A be a complex $n \times n$ matrix, and let $\mathcal{M} \subseteq \mathbb{C}^n$ be an invariant subspace of A. Then $\mathcal{M}$ is called *stable* if for every $\varepsilon > 0$ there is a $\delta > 0$ such that $\|A-B\| < \delta$ implies the existence of a B-invariant subspace $\mathcal{N}$ such that $\theta(\mathcal{M},\mathcal{N}) < \varepsilon$.

We introduce the following notation. Let A be an $n \times n$ matrix, and let λ be one of its eigenvalues. We denote by $\mathcal{R}_\lambda(A)$ the root subspace of A corresponding to λ, i.e., the subspace Ker $(A - \lambda I_n)^n$.

The following theorem describes stable invariant subspaces in the case of complex matrices.

Theorem 2.1. *Let A be an $n \times n$ complex matrix. An invariant subspace $\mathcal{M}$ is a stable A-invariant subspace if and only if for every eigenvalue λ of A for which the geometric multiplicity is larger than one, i.e., for which* dim Ker $(\lambda - A) > 1$, *we have $\mathcal{R}_\lambda(A) \cap \mathcal{M}$ is either $\{0\}$ or $\mathcal{R}_\lambda(A)$.*

For each eigenvalue λ of A for which the geometric multiplicity is equal to one $\mathcal{R}_\lambda(A) \cap \mathcal{M}$ is an arbitrary A-invariant subspace of $\mathcal{R}_\lambda(A)$.

Note that an A-invariant subspace is stable if and only if it is isolated (in the metric topology) in the set of all A-invariant subspaces (see [4]). The connection between stability and isolatedness is a recurrent theme in the study of stable invariant subspaces, although it is not always "if and only if": for example, in the real case (see the next subsection) every stable invariant subspace is isolated, but there are isolated invariant subspaces that are not stable.

We shall also use the following notion. An A-invariant subspace $\mathcal{M}$ is called *Lipschitz stable* if there are positive constants δ and K such that $\|A - B\| < \delta$ implies that B has an invariant subspace $\mathcal{N}$ with $\theta(\mathcal{M},\mathcal{N}) \leq K\|A - B\|$. The following theorem describes such subspaces (see [30]).

Theorem 2.2. *Let A be an $n \times n$ complex matrix. An invariant subspace $\mathcal{M}$ is Lipschitz stable if and only if it is a spectral subspace, i.e., for all eigenvalues λ of A we have $\mathcal{R}_\lambda(A) \cap \mathcal{M}$ is either $\{0\}$ or $\mathcal{R}_\lambda(A)$.*

An important issue in numerical analysis of Lipschitz stable invariant subspaces is finding good bounds for the constant K. Without going into details, we note that there exist bounds for K based on the concept of separation: Write A in the form $A = \begin{bmatrix} A_{11} & A_{12} \\ 0 & A_{22} \end{bmatrix}$ with respect to a suitable orthonormal basis in $\mathbb{C}^n$; it is assumed that the spectra of $A_{11} \in \mathbb{C}^{p\times p}$ and $A_{22} \in \mathbb{C}^{q\times q}$ do not intersect. The separation $\operatorname{sep}(A_{11}, A_{22})$ between A_{11} and A_{22} is defined as the norm of T^{-1}, where T is the invertible linear operator on $\mathbb{C}^{p\times q}$ defined by $T(X) = A_{11}X - XA_{22}$. The Lipschitz constant corresponding to the spectral A-invariant subspace spanned by the first p vectors of the orthonormal basis can be expressed in terms of $\operatorname{sep}(A_{11}, A_{22})$. We refer the reader to Chapter V of [70], for example, for complete details.

There are several possible notions that are weaker than Lipschitz stability and stronger than stability. One of them is connected to the following observation: if $\mathcal{M}$ is Lipschitz stable, and $A_n \to A$ is a sequence of matrices, then for *every* sequence of subspace $\mathcal{M}_n$, with $\mathcal{M}_n$ invariant under A_n, such that $\mathcal{M}_n$ is sufficiently close to $\mathcal{M}$, we have actually that $\theta(\mathcal{M}, \mathcal{M}_n) \le K\|A - A_n\|$.

We now introduce two closely related notions that are weaker than Lipschitz stability and stronger than stability. An A-invariant subspace $\mathcal{M} \subseteq \mathbb{C}^n$ is called *α-stable* if there exist constants $K > 0$, $\varepsilon > 0$, such that every matrix $B \in \mathbb{C}^{n\times n}$ with $\|A - B\| < \varepsilon$ has an invariant subspace $\mathcal{N} \subseteq F^n$ with the property that

$$\theta(\mathcal{M}, \mathcal{N}) \le K\|A - B\|^{\frac{1}{\alpha}}.$$

An A-invariant subspace $\mathcal{M} \subseteq \mathbb{C}^n$ is called *strongly α-stable* if for every sequence $A_n \to A$ and every sequence $\mathcal{M}_n \to \mathcal{M}$, where $\mathcal{M}_n$ is A_n-invariant, we have

$$\theta(\mathcal{M}, \mathcal{M}_n) \le K\|A - A_n\|^{\frac{1}{\alpha}}.$$

The following result that describes α-stable invariant subspaces was obtained in [63].

Before stating the results, let us introduce the following notation. For two natural numbers k and n, with $k < n$, we introduce a number $\gamma(k,\ n)$, as follows: $\gamma(k,\ n) = n$, whenever there is no set of k distinct n-th roots of unity that sum to zero, while $\gamma(k,\ n) = n - 1$ if such a set of k distinct n-th roots of unity does exist.

With these notations we first state the main result of [63].

Theorem 2.3. *Let A be an $n \times n$ matrix, and let $\mathcal{M} \subseteq \mathbb{C}^n$ be a nontrivial A-invariant subspace. The subspace $\mathcal{M}$ is α-stable for some positive α if and only if for every eigenvalue λ of A with geometric multiplicity larger than one either $\mathcal{M} \cap \mathcal{R}_\lambda(A) = (0)$, or $\mathcal{M} \cap \mathcal{R}_\lambda(A) = \mathcal{R}_\lambda(A)$.*

In that case $\mathcal{M}$ is α-stable if and only if

$$\gamma(\dim\ (\mathcal{M}\cap\mathcal{R}_\lambda(A)),\ \dim\ \mathcal{R}_\lambda(A))\leq\alpha,$$

for all eigenvalues λ of A such that

$$(0)\neq\mathcal{M}\cap\mathcal{R}_\lambda(A)\neq\mathcal{R}_\lambda(A).$$

If there are no such eigenvalues, then $\mathcal{M}$ is 1-stable.

In particular, $\mathcal{M}$ is α-stable for some α if and only if $\mathcal{M}$ is stable.

Next we describe the strongly α-stable subspaces (see [49]).

Theorem 2.4. *Let A be an $n\times n$ matrix, and let $\mathcal{M}\subseteq\mathbb{C}^n$ be a nontrivial A-invariant subspace. Let $\alpha>0$ be given. The subspace $\mathcal{M}$ is strongly α-stable if and only if either $\mathcal{M}\cap\mathcal{R}_\lambda(A)=(0)$, or $\mathcal{M}\cap\mathcal{R}_\lambda(A)=\mathcal{R}_\lambda(A)$, for every eigenvalue λ of A that satisfies one of the following two conditions:*

(i) *λ has geometric multiplicity larger than one,*
(ii) *λ has geometric multiplicity one and* $\dim\ \mathcal{R}_\lambda(A)>\alpha$.

For all other eigenvalues λ of A there is no restriction on $\mathcal{M}\cap\mathcal{R}_\lambda(A)$ other than that it is A-invariant.

Observe that, unlike the case of Lipschitz stability (where the set of strongly Lipschitz stable subspaces coincides with the set of Lipschitz stable subspaces), an α-stable subspace need not be strongly α-stable for $\alpha>1$. (See also [50] for an explicit example.)

The fact that the function γ, involving sums of n-th roots of unity, plays an essential role in α-stability may seem mysterious at first glance. Its role is perhaps best explained by considering the following example.

Example. Let $A=J_n(0)$, i.e., the $n\times n$ upper triangular Jordan block with zero eigenvalue. Consider the perturbation $A(\varepsilon)$ of A, which differs from A only in the $(n,1)$-entry where there is $\varepsilon>0$ instead of zero. The eigenvalues of $A(\varepsilon)$ are $\varepsilon^{\frac{1}{n}}\cdot\varepsilon_i$, where $\varepsilon_i,\ i=1,\cdots,n$ are the n-the roots of unity. The corresponding eigenvectors are

$$y_i(\varepsilon)=(1,\varepsilon_i,\varepsilon_i^2,\cdots,\varepsilon_i^{n-1})^T.$$

Consider the unique k-dimensional invariant subspace $\mathcal{M}$ of A, i.e., $\mathcal{M}=\text{span}\ \{e_1,\cdots,e_k\}$, where e_i denotes the i-th standard basis vector. Any k-dimensional invariant subspace of $A(\varepsilon)$ is given by

$$\mathcal{N}=\text{span}\ \{y_{i_1},\cdots,y_{i_k}\},$$

where $\{i_1,\cdots,i_k\}$ is a set of k different numbers from $\{1,\cdots,n\}$. So $\mathcal{N}$ is the range of the matrix $Y=[y_{i_1}\ \cdots\ y_{i_k}]$. Let $Y=\begin{bmatrix}Y_1\\Y_2\end{bmatrix}$, where Y_1 is a $k\times k$ matrix and Y_2 is an $(n-k)\times k$ matrix. Then $\mathcal{N}=\text{Im}\begin{bmatrix}I\\Y_2Y_1^{-1}\end{bmatrix}$. Computing the last column of $Y_2Y_1^{-1}$ we find in the $(1,k)$-entry the number $(\varepsilon_{i_1}+\cdots+\varepsilon_{i_k})\varepsilon^{\frac{1}{n}}$. Let u be the last column of $\begin{bmatrix} I \\ Y_2Y_1^{-1} \end{bmatrix}$. Then

$$\theta(\mathcal{M},\mathcal{N})^2 = \|P_{\mathcal{M}} - P_{\mathcal{N}}\|^2 \geq \|(P_{\mathcal{M}} - P_{\mathcal{N}})u\|^2 \cdot \frac{1}{\|u\|^2}.$$

One easily sees that for ε small enough $\|u\| < 2$, while $P_{\mathcal{M}}u = e_k$ and $P_{\mathcal{N}}u = u$. So $\theta(\mathcal{M},\mathcal{N}) \geq \frac{1}{2}\|u - e_k\| \geq \frac{1}{2}\varepsilon^{\frac{1}{n}}|\varepsilon_{i_1}+\cdots+\varepsilon_{i_k}|$. Therefore, if there is no set of k distinct n-th roots of unity summing to zero, we have for any k-dimensional subspace $\mathcal{N}$ of $A(\varepsilon)$ that $\theta(\mathcal{M},\mathcal{N}) \geq c\cdot\varepsilon^{\frac{1}{n}} = c\|A - A(\varepsilon)\|^{\frac{1}{n}}$.

2.2. General theory: The real case

In this subsection we study stability properties of real invariant subspaces of real matrices. Thus, every matrix is assumed to be real, and a subspace is in $\mathbb{R}^n$. Keeping these restrictions in mind, the definitions of stable, Lipschitz stable, α-stable and strongly α-stable do not change. For sake of brevity we therefore do not repeat these definitions.

For a real $n\times n$ matrix A, and real λ, we will interpret $\mathcal{R}_\lambda(A)$ in this subsection as a real subspace. Also, when A is real and $a \pm ib$ a complex conjugate pair of nonreal eigenvalues of A, we denote by $\mathcal{R}_{a\pm ib}(A)$ the real root subspace of A corresponding to $a \pm ib$, i.e.,

$$\mathcal{R}_{a\pm ib}(A) = \text{Ker}\ ((A^2 - 2aA + (a^2+b^2)I)^n) \subseteq \mathbb{R}^n.$$

We first state the analogue of Theorems 2.1 and 2.2.

Theorem 2.5. *Let A be a real $n \times n$ matrix, and let $\mathcal{M} \subseteq \mathbb{R}^n$ be a real invariant subspace of A. Then $\mathcal{M}$ is stable if and only if the following conditions hold.*

- (i) *$\mathcal{M} \cap \mathcal{R}_\lambda(A)$ is either $\{0\}$ or $\mathcal{R}_\lambda(A)$ for every real eigenvalue λ of A for which the geometric multiplicity is larger than one,*
- (ii) dim $(\mathcal{M} \cap \mathcal{R}_\lambda(A))$ *is even whenever* dim $\mathcal{R}_\lambda(A)$ *is even and λ is a real eigenvalue of A with geometric multiplicity one,*
- (iii) *$\mathcal{M} \cap \mathcal{R}_{a\pm ib}(A)$ is either $\{0\}$ or $\mathcal{R}_{a\pm ib}(A)$ for every pair of non-real eigenvalues $a \pm ib$ of A for which the geometric multiplicity is larger than one.*

For each real eigenvalue λ of A with geometric multiplicity one and odd algebraic multiplicity there are no restrictions on $\mathcal{M} \cap \mathcal{R}_\lambda(A)$ other than that it is an A-invariant subspace.

For each pair of non-real eigenvalues $a \pm ib$ of A with geometric multiplicity one there are no restrictions on $\mathcal{M}\cap\mathcal{R}_{a\pm ib}(A)$ other than that it is an A-invariant subspace.

Theorem 2.6. *Let A be a real $n \times n$ matrix, and let $\mathcal{M} \subseteq \mathbb{R}^n$ be an invariant subspace of A. Then $\mathcal{M}$ is Lipschitz stable if and only if $\mathcal{M}$ is a real spectral subspace, i.e., a sum of real root subspaces of A (each of them corresponding either to a real eigenvalue of A or to a pair of nonreal complex conjugate eigenvalues of A).*

To state the analogue of Theorem 2.3 in the real case (which can be found in [51]), we have to introduce some additional notation. A finite set of complex numbers $S = \{\zeta_1, \dots, \zeta_m\}$ will be called *zero sum self conjugate* if $\zeta_1 + \cdots + \zeta_m = 0$, and the non-real elements in S can be arranged in pairs of complex conjugate numbers. For two natural numbers k and n, with $k < n$, we define $\delta(k,\ n)$ as follows: $\delta(k,\ n) = n$ in the following three cases: (i) n is odd and there is no zero sum self conjugate set of k distinct n-th roots of 1, (ii) n is even and k is odd, (iii) n is even and divisible by 4, k is also even but not divisible by 4, and there is no zero sum self conjugate set of k distinct n-th roots of -1. In all other cases we define $\delta(k,\ n) = n - 1$.

Theorem 2.7. *Let A be an $n \times n$ real matrix, and let $\mathcal{M} \subseteq \mathbb{R}^n$ be a nontrivial A-invariant subspace. Let $\lambda_1, \dots, \lambda_r$ be the distinct real eigenvalues of A, and let $a_1 \pm ib_1, \dots, a_s \pm ib_s$ be the distinct pairs of complex conjugate non-real eigenvalues of A. Decompose $\mathcal{M}$ as follows:*

$$\mathcal{M} = \mathcal{N}_1 \dot{+} \dots \dot{+} \mathcal{N}_r \dot{+} \mathcal{L}_1 \dot{+}, \dots \dot{+} \mathcal{L}_s,$$

with $\mathcal{N}_j \subseteq \mathcal{R}_{\lambda_j}(A)$, $\mathcal{L}_j \subseteq \mathcal{R}_{a_j \pm ib_j}(A)$. The subspace $\mathcal{M}$ is α-stable if and only if all of the following conditions (i)–(v) are satisfied:

(i) $\mathcal{N}_j = (0)$ *or* $\mathcal{N}_j = \mathcal{R}_{\lambda_j}(A)$ *whenever* $\dim \operatorname{Ker} (\lambda_j I - A) > 1$.

(ii) $\mathcal{L}_j = (0)$ *or* $\mathcal{L}_j = \mathcal{R}_{a_j \pm ib_j}(A)$ *whenever the geometric multiplicity of* $a_j \pm ib_j$ *is larger than* 1,

(iii) *for every real eigenvalue λ_j of A such that* $\dim \operatorname{Ker} (\lambda_j I - A) = 1$ *and* $\dim \mathcal{R}_{\lambda_j}(A)$ *is odd, we have one of the three possibilities: (a)* $\mathcal{N}_j = 0$, *or (b)* $\mathcal{N}_j = \mathcal{R}_{\lambda_j}(A)$, *or (c)* $(0) \neq \mathcal{N}_j \neq \mathcal{R}_{\lambda_j}(A)$ *and* $\alpha \geq \delta(\dim \mathcal{N}_j, \dim \mathcal{R}_{\lambda_j}(A))$.

(iv) *for every real eigenvalue λ_j of A such that* $\dim \operatorname{Ker} (\lambda_j I - A) = 1$ *and* $\dim \mathcal{R}_{\lambda_j}(A)$ *is even we have one of the three possibilities: (a)* $\mathcal{N}_j = 0$, *or (b)* $\mathcal{N}_j = \mathcal{R}_{\lambda_j}(A)$, *or (c)* $(0) \neq \mathcal{N}_j \neq \mathcal{R}_{\lambda_j}(A)$, *and $\mathcal{N}_j$ is even dimensional and* $\alpha \geq \delta(\dim \mathcal{N}_j, \dim \mathcal{R}_{\lambda_j}(A))$.

(v) *for every pair of non-real eigenvalues $a_j \pm ib_j$ of A having geometric multiplicity one, we have one of the three possibilities: (a)* $\mathcal{L}_j = 0$, *or (b)* $\mathcal{L}_j = \mathcal{R}_{a_j \pm ib_j}(A)$, *or (c)* $(0) \neq \mathcal{L}_j \neq \mathcal{R}_{a_j \pm ib_j}(A)$ *and*

$$\alpha \geq \gamma(\frac{1}{2}\dim \mathcal{L}_j, \frac{1}{2}\dim \mathcal{R}_{a_j \pm ib_j}(A)).$$

Observe that in the situation described in (v) the dimensions of both $\mathcal{R}_{a_j \pm ib_j}(A)$ and $\mathcal{L}_j$ are even.

Finally, we give here the real analogue of Theorem 2.4.

Theorem 2.8. *Let A be a real $n \times n$ matrix and let $\mathcal{M}$ be a real A-invariant subspace. Denote by $\lambda_1, \dots, \lambda_r$ be the different real eigenvalues, and let $a_1 \pm ib_1, \dots, a_s \pm ib_s$ be the different non-real eigenvalues of A. Write*

$$\mathcal{M} = \mathcal{N}_1 \dot{+} \cdots \dot{+} \mathcal{N}_r \dot{+} \mathcal{L}_1 \dot{+} \cdots \dot{+} \mathcal{L}_s, \tag{2.1}$$

where $\mathcal{N}_j \subseteq \mathcal{R}_{\lambda_j}(A)$, $\mathcal{L}_j \subseteq \mathcal{R}_{a_j \pm ib_j}(A)$ *(the existence and uniqueness of representation* (2.1) *is well known). Then* $\mathcal{M}$ *is strongly* α*-stable if and only if the following conditions* (i)-(vii) *are all satisfied:*

(i) $\mathcal{N}_j = (0)$ *or* $\mathcal{N}_j = \mathcal{R}_{\lambda_j}(A)$ *whenever* $\dim \operatorname{Ker}(\lambda_j I - A) > 1$,

(ii) $\mathcal{L}_j = (0)$ *or* $\mathcal{L}_j = \mathcal{R}_{a_j \pm ib_j}(A)$ *whenever the geometric multiplicity of* $a_j \pm ib_j$ *is larger than 1,*

(iii) $\mathcal{N}_j = (0)$ *or* $\mathcal{N}_j = \mathcal{R}_{\lambda_j}(A)$ *whenever* $\alpha < \dim \mathcal{R}_{\lambda_j}(A)$ *and* $\dim \operatorname{Ker}(\lambda_j I - A) = 1$,

(iv) $\mathcal{L}_j = (0)$ *or* $\mathcal{L}_j = \mathcal{R}_{a_j \pm ib_j}(A)$ *whenever* $\alpha < \frac{1}{2} \dim \mathcal{R}_{a_j \pm ib_j}(A)$ *and* $a_j \pm ib_j$ *have geometric multiplicity one,*

(v) $\mathcal{N}_j$ *is an arbitrary A-invariant subspace contained in* $\mathcal{R}_{\lambda_j}(A)$ *whenever*

$$\dim \operatorname{Ker}(\lambda_j I - A) = 1,$$

$\alpha \geq \dim \mathcal{R}_{\lambda_j}(A)$ *and* $\dim \mathcal{R}_{\lambda_j}(A)$ *is odd,*

(vi) $\mathcal{N}_j$ *is an arbitrary even-dimensional A-invariant subspace contained in* $\mathcal{R}_{\lambda_j}(A)$ *whenever* $\dim \operatorname{Ker}(\lambda_j I - A) = 1$, $\alpha \geq \dim \mathcal{R}_{\lambda_j}(A)$ *and* $\dim \mathcal{R}_{\lambda_j}(A)$ *is even,*

(vii) $\mathcal{L}_j$ *is an arbitrary A-invariant subspace contained in* $\mathcal{R}_{a_j \pm ib_j}(A)$ *whenever* $a_j \pm ib_j$ *have geometric multiplicity one and* $\alpha \geq \frac{1}{2} \dim \mathcal{R}_{a_j \pm ib_j}(A)$.

2.3. Stability of subspaces that are invariant modulo a subspace

We consider here both the complex case and the real case, thus we let $F = \mathbb{C}$ or $F = \mathbb{R}$. Let $A \in F^{n \times n}$, and let $\mathcal{V} \subseteq F^n$ be a subspace. A subspace $\mathcal{M} \subseteq F^n$ is called *A-invariant modulo* $\mathcal{V}$ if $Ax \in \mathcal{M} + \mathcal{V}$ for every $x \in \mathcal{M}$. Thus, this is a generalization of the familiar notion of an A-invariant subspace (which is obtained as a particular case if $\mathcal{V} = (0)$).

Invariant subspaces modulo a subspace play a key role in solving many problems in linear systems theory, see, e.g., [77, 22], where they often appear as (A, B)-invariant subspaces. Given $A \in F^{n \times n}$, $B \in F^{n \times m}$, a subspace $\mathcal{M} \subseteq F^n$ is called (A, B)*-invariant* if $\mathcal{M}$ is $(A + BG)$-invariant for some $G \in F^{m \times n}$. It is well known that $\mathcal{M}$ is (A, B)-invariant if and only if $\mathcal{M}$ is A-invariant modulo Im B.

An (A, B)-invariant subspace $\mathcal{M}$ is called *stable* if for every $\varepsilon > 0$ there exists $\delta > 0$ such that for every pair of matrices (with entries in F) A' and B' satisfying $\|A - A'\| + \|B - B'\| < \delta$ there exists an (A', B')-invariant subspace $\mathcal{M}'$ such that $\theta(\mathcal{M}, \mathcal{M}') < \varepsilon$. The notions of α-stability and of Lipschitz stability are introduced analogously to the definitions given in Section 2.1.

In the controllable case, the Lipschitz stability is universal:

Theorem 2.9. *If the pair* (A, B) *is controllable, i.e.* $\sum_{j=0}^{\infty} \operatorname{Im}(A^j B) = F^n$, *then every* (A, B)*-invariant subspace is Lipschitz stable.*

This result is proved in [26] using the pole assignment theorem (in the complex case; the same proof works in the real case as well); see also [22].

Theorem 2.9 plays a key role in proving Lipschitz stability of linear fractional decompositions of rational matrix functions. These are decompositions of the form

$$U(\lambda) = W_{21}(\lambda) + W_{22}(\lambda)V(\lambda)(I - W_{12}(\lambda)V(\lambda))^{-1}W_{11}(\lambda),$$

where $U(\lambda, V(\lambda)$, and $W(\lambda) = \begin{bmatrix} W_{11}(\lambda) & W_{12}(\lambda) \\ W_{21}(\lambda) & W_{22}(\lambda) \end{bmatrix}$ are rational matrix functions of suitable sizes (with coefficients in F). Such decomposition represents the transfer function of the feedback (or cascade) connection of two systems given by their respective transfer functions $W(\lambda)$ and $V(\lambda)$ (see, e.g., Section 8.3 in [22]). The Lipschitz stability of linear fractional decompositions is proved in [26] (in the complex case) using the description of minimal linear fractional decompositions in terms of minimal realizations obtained in [2]. A complete proof is found also in [22]. We refer the reader to these sources for more details.

Without the controllability hypothesis in Theorem 2.9, the situation is more complicated, and so far there is no general description of stable, or Lipschitz stable, or α-stable (A, B)-invariant subspaces. We present one result in this direction taken from [67]. Denote by

$$\mathcal{J}(A, B) = \sum_{j=0}^{\infty} \text{Im } (A^j B)$$

the *controllable subspace* of the pair (A, B), and let $\mathcal{K}$ be a direct complement for $\mathcal{J}(A, B)$ in F^n. Let $\Pi_{\mathcal{K}} : F^n \to \mathcal{K}$ be the projection on $\mathcal{K}$ parallel to $\mathcal{J}(A, B)$. One easily verifies that the subspace $\Pi_{\mathcal{K}}(\mathcal{M})$ is $\Pi_{\mathcal{K}} A \mid_{\mathcal{K}}$-invariant for every (A, B)-invariant subspace $\mathcal{M}$.

Theorem 2.10. *Let $\mathcal{M}$ be an (A, B)-invariant subspace. If $\Pi_{\mathcal{K}}(\mathcal{M})$ is stable as a $\Pi_{\mathcal{K}} A \mid_{\mathcal{K}}$-invariant subspace, then $\mathcal{M}$ is a stable (A, B)-invariant subspace. Conversely, if $\mathcal{M}$ is a stable (A, B)-invariant subspace and $\mathcal{M} \cap \mathcal{J}(A, B) = (0)$, then $\Pi_{\mathcal{K}}(\mathcal{M})$ is a stable $\Pi_{\mathcal{K}} A \mid_{\mathcal{K}}$-invariant subspace.*

Note that the statement of Theorem 2.10 does not depend on the choice of the direct complement $\mathcal{K}$. We refer the reader to [67] for the proof of this theorem. In the same paper an example is given showing that the condition $\mathcal{M} \cap \mathcal{J}(A, B) = (0)$ is essential for the converse statement. Further results concerning stability of (A, B)-invariant subspaces are proved in the recent paper [73].

Returning to the invariance modulo a subspace, we say that a subspace $\mathcal{M}$ that is A-invariant modulo $\mathcal{V}$ is *stable* if for every $\varepsilon > 0$ there exists $\delta > 0$ such that there is an A'-invariant subspace $\mathcal{M}'$ modulo $\mathcal{V}'$ with $\theta(\mathcal{M}, \mathcal{M}') < \varepsilon$ as soon as the subspace $\mathcal{V}'$ and the matrix A' satisfy $\|A - A'\| + \theta(\mathcal{V}, \mathcal{V}') < \delta$. Another version of this notion is obtained if we keep $\mathcal{V}$ fixed, i.e., insist that $\mathcal{V} = \mathcal{V}'$; then we say that the stability is *with fixed* $\mathcal{V}$.

Problem 2.11. *Characterize Lipschitz stable and α-stable A-invariant subspaces modulo a subspace $\mathcal{V}$, also with fixed $\mathcal{V}$, not assuming controllability, i.e., not assuming that $\sum_{j=0}^{\infty} A^j(\mathcal{V}) = F^n$.*

Note that Theorem 2.10 remains valid for subspaces $\mathcal{M}$ that are stable A-invariant subspaces modulo $\mathcal{V}$ or stable A-invariant subspaces modulo $\mathcal{V}$ with fixed $\mathcal{V}$; in this case $\mathcal{J}(A,B)$ should be replaced by the smallest A-invariant subspace containing $\mathcal{V}$.

2.4. Quadratic matrix equation

Again, we consider two cases simultaneously: $F = \mathbb{C}$ or $F = \mathbb{R}$. Consider the equation

$$XBX + XA - DX - C = 0, \tag{2.2}$$

where A, B, C, D are matrices (over F) of sizes $n \times n$, $n \times m$, $m \times n$, and $m \times m$, respectively, and X is an $m \times n$ matrix (over F) to be found. A solution X of (2.2) is called *α-stable* if there exist $K > 0$ such that every equation

$$Y\tilde{B}Y + Y\tilde{A} - \tilde{D}Y - \tilde{C} = 0,$$

with coefficients $\tilde{A},\ \tilde{B},\ \tilde{C},\ \tilde{D}$ sufficiently close to A, B, C, D, respectively, has a solution Y satisfying

$$\|X - Y\| \le K(\|A-\tilde{A}\| + \|B-\tilde{B}\| + \|C-\tilde{C}\| + \|D-\tilde{D}\|)^{\frac{1}{\alpha}}.$$

Stability of solutions of (2.2) was studied in [14, 4].

There is a one-to-one correspondence between the solutions X and the n-dimensional subspaces $\mathcal{M} \subseteq F^{n+m}$ which are $\begin{bmatrix} A & B \\ C & D \end{bmatrix}$-invariant and are direct complements of the fixed m-dimensional subspace $\mathcal{M}_0 = \operatorname{Im} \begin{bmatrix} 0 \\ I_m \end{bmatrix}$. The correspondence is given by the formula

$$\mathcal{M}(X) = \operatorname{Im} \begin{bmatrix} I_n \\ X \end{bmatrix}.$$

Since the set of all direct complements of $\mathcal{M}_0$ is open, and for a fixed $m \times n$ matrix X_0 there is a constant $C > 0$ such that

$$C^{-1}\|X - X_0\| \le \theta(\mathcal{M}(X), \mathcal{M}(X_0)) \le C\|X - X_0\|,$$

for all X sufficiently close X_0 (see, e.g., formula (17.8.3) in [22]) it is easily seen that a solution X of (2.2) is α-stable if and only if the $\begin{bmatrix} A & B \\ C & D \end{bmatrix}$-invariant subspace $\mathcal{M}(X)$ is α-stable. Combining this observation with Theorem 2.3, the following result is obtained.

Theorem 2.12. *Case $F = \mathbb{C}$. A solution X of (2.2) is α-stable if and only if for every common eigenvalue λ of $A + BX$ and $D - XB$ the following two conditions are satisfied:*

(i) *λ has geometric multiplicity one as an eigenvalue of $\begin{bmatrix} A & B \\ C & D \end{bmatrix}$,*

(ii) *if k and l are the algebraic multiplicity of λ as an eigenvalue of $A + BX$ and $D - XB$, respectively, then $\gamma(k,\ k + l) \le \alpha$.*

The proof is done by applying similarity to the matrix $\begin{bmatrix} A & B \\ C & D \end{bmatrix}$ with the matrix $\begin{bmatrix} I & 0 \\ X & I \end{bmatrix}$ as the transformation matrix; one obtains the matrix

$$M = \begin{bmatrix} A+BX & B \\ 0 & D-XB \end{bmatrix}.$$

Clearly, $\mathcal{M}(X)$ is α-stable as a $\begin{bmatrix} A & B \\ C & D \end{bmatrix}$-invariant subspace if and only if the subspace $\mathcal{M} = \operatorname{Im} \begin{bmatrix} I_n \\ 0 \end{bmatrix}$ is α-stable as an M-invariant subspace.

Considering the set of all eigenvalues λ_0 of M for which

$$(0) \neq \mathcal{M} \cap \mathcal{R}_{\lambda_0}(M) \neq \mathcal{R}_{\lambda_0}(M). \tag{2.3}$$

and taking into account the special form of M and $\mathcal{M}$, one sees that

$$\mathcal{M} \cap \mathcal{R}_{\lambda_0}(M) = \left\{ \begin{bmatrix} x \\ 0 \end{bmatrix} \mid ((A+BX) - \lambda_0)^m x = 0 \right\}. \tag{2.4}$$

It is now easily seen that the set of eigenvalues for which (2.3) holds is precisely the set of common eigenvalues of $A + BX$ and $D - XB$. Clearly, condition (i) must hold for the subspace to be α-stable for some α. Next, one computes for such an eigenvalue $\gamma(\dim\,(\mathcal{M} \cap \mathcal{R}_{\lambda_0}(M)),\ \dim\, \mathcal{R}_{\lambda_0}(M))$. Since condition (i) holds, $\dim\, \mathcal{R}_{\lambda_0}(M))$ is the algebraic multiplicity of λ_0 as an eigenvalue of M. Because of (2.4) one has that $\dim\,(\mathcal{M} \cap \mathcal{R}_{\lambda_0}(M))$ is the algebraic multiplicity of λ_0 as an eigenvalue of $A+BX$ (taking into account that also $A+BX$ must have geometric multiplicity one at the eigenvalue λ_0). Considering $\det(M-\lambda) = \det((A+BX)-\lambda)\det((D-XB)-\lambda)$, it is seen that the algebraic multiplicity of λ_0 as an eigenvalue of M is precisely the sum of the algebraic multiplicities of λ_0 as an eigenvalue of $A+BX$ and $D-XB$. Thus $\gamma(\dim\,(\mathcal{M} \cap \mathcal{R}_{\lambda_0}(M)),\ \dim\, \mathcal{R}_{\lambda_0}(M)) = \gamma(k,\ k+l)$, and the theorem follows by applying Theorem 2.3.

The next theorem gives the analogue of Theorem 2.12 for the real case. The proof is analogous to the proof of Theorem 2.12.

Theorem 2.13. *Case $F = \mathbb{R}$. A solution X of* (3.1) *is α-stable if and only if for every common eigenvalue λ of $A+BX$ and $D-XB$ the following conditions are satisfied:*

(i) *λ has geometric multiplicity one as an eigenvalue of* $\begin{bmatrix} A & B \\ C & D \end{bmatrix}$,

(ii) *if λ is non-real and k and l are the algebraic multiplicity of λ as an eigenvalue of $A+BX$ and $D-XB$, respectively, then $\gamma(k,\ k+l) \leq \alpha$,*

(iii) *if λ is real, k and l are as above, and $k+l$ is odd, then $\delta(k,\ k+l) \leq \alpha$,*

(iv) *if λ is real, k and l are as above, and $k+l$ is even, then k is even and $\delta(k,\ k+l) \leq \alpha$.*

3. Stability of invariant subspaces with symmetries

Many applications making use of invariant subspaces of matrices involve symmetry in one or another way. Among these, we mention symmetric factorizations of selfadjoint matrix polynomials or selfadjoint rational matrix functions; the study of hermitian or positive semidefinite solutions to algebraic Riccati equations, and transport theory. In these applications an indefinite inner product space structure plays an important role. To give an example, let $H = H^*$ be an invertible selfadjoint $n \times n$ matrix. A matrix A is called H-selfadjoint if $HA = A^*H$. A subspace $\mathcal{M} \subseteq \mathbb{C}^n$ is called H-lagrangian if $H(\mathcal{M}) \stackrel{\text{def}}{=} \{Hx \ : \ x \in \mathcal{M}\} = \mathcal{M}^\perp$. Such subspaces appear naturally in the study of hermitian solutions of algebraic Riccati equations, as well as in the study of symmetric factorizations of selfadjoint matrix polynomials and selfadjoint rational matrix functions. It is therefore of interest to study the perturbation theory of such subspaces. This was started in [52], and in a long series of papers by the authors this study was carried through.

Let H be an invertible $n \times n$ matrix. Introduce on $\mathbb{C}^n$ the indefinite scalar product given by $[x, y] = \langle Hx, y\rangle$, where $\langle \cdot, \cdot \rangle$ stands for the standard scalar product in $\mathbb{C}^n$. An $n \times n$ matrix A will be called *H-selfadjoint* if it is selfadjoint in this indefinite scalar product, i.e., if $HA = A^*H$. It is well known that for such a matrix there is an A-invariant maximal H-nonnegative subspace, as well as an A-invariant maximal H-nonpositive subspace (see, e.g., [23], where such subspaces were explicitly constructed). Such subspaces play an important role in many applications, as we shall see in the following sections.

Some of the applications do lead in a natural way to the study of maximal semidefinite invariant subspaces for classes of matrices that are not selfadjoint in an indefinite scalar product, but have another special property, such as being H-unitary, i.e. $A^*HA = H$, or H-contractive ($A^*HA - H \leq 0$) or H-dissipative ($\frac{1}{2i}(HA - A^*H) \geq 0$). Explicit construction of invariant maximal semidefinite subspaces can be found in [23] for the case of H-unitary matrices and in [65] (see also [71]) for the case of H-dissipative matrices.

Several important applications in which stability, or robustness, of invariant subspaces with additional symmetries plays a key role, are not discussed in the present paper. These include factorizations of selfadjoint matrix polynomials [36], [21] (see also [23]) and their stability [53], factorizations of positive real rational matrix functions and of rational matrix functions of the form I + contraction on the real line (see [24]), and stability of those factorizations (see [64]). Stability properties of certain solutions of matrix polynomial equations, where the polynomial is weakly hyperbolic (see for the theory of such polynomials [37, 38, 42]), are studied in [57]; these results also will not be discussed here.

Problems of stability of invariant subspaces with symmetries in the context of infinite dimensional Hilbert spaces are also of interest. Again, a thorough investigation of these problems is a challenge for future research. A few results in this direction are already available, in connection with stability of stationary transport equations (see [58], [46]).

3.1. Selfadjoint matrices in indefinite inner products

We assume $F = \mathbb{C}$ in this and next subsections. We start with the canonical form of the pair (A, H), where A is H-selfadjoint, under congruent similarity, i.e., under transformation of the pair (A, H) to $(S^{-1}AS, S^*HS)$ for invertible matrices S.

Theorem 3.1. *Let H be an invertible hermitian $n \times n$ matrix, and let A be H-selfadjoint. Then there exists an invertible S such that $S^{-1}AS$ and S^*HS have the form*

$$S^{-1}AS = J_{k_1}(\lambda_1) \oplus \cdots \oplus J_{k_\alpha}(\lambda_\alpha) \\ \oplus J_{k_{\alpha+1}}(\lambda_{\alpha+1}) \oplus J_{k_{\alpha+1}}(\bar{\lambda}_{\alpha+1}) \oplus \cdots \oplus J_{k_\beta}(\lambda_\beta) \oplus J_{k_\beta}(\bar{\lambda}_\beta), \quad (3.1)$$

where $\lambda_1, \dots, \lambda_\alpha$ are real and $\lambda_{\alpha+1}, \dots, \lambda_\beta$ are non-real with positive imaginary parts, and

$$S^*HS = \varepsilon_1 Q_{k_1} \oplus \cdots \oplus \varepsilon_\alpha Q_{k_\alpha} \oplus Q_{2k_{\alpha+1}} \oplus \cdots \oplus Q_{2k_\beta}, \quad (3.2)$$

where $\varepsilon_1, \dots, \varepsilon_\alpha$ are ± 1. For a given pair (A, H), where A is H-selfadjoint, the canonical form (3.1), (3.2) *is unique up to permutation of orthogonal components in* (3.2) *and the same simultaneous permutation of the corresponding blocks in* (3.1).

Theorem 3.1 is well known and goes back to Weierstrass and Kronecker. A complete proof of this theorem can be found in many sources, see, e.g., [23, 72].

The signs ε_j in (3.2) form the *sign characteristic* of the pair (A, H). Thus, the sign characteristic consists of signs $+1$ or -1 attached to every partial multiplicity (= size of a Jordan block in the Jordan form) of A corresponding to a real eigenvalue.

From (3.1), (3.2) one can construct explicitly an A-invariant maximal H-nonnegative subspace as follows. Let $\{e_{ij}\}_{i=1}^{\alpha}{}_{j=1}^{k_i} \cup \{e_{ij}\}_{i=\alpha+1}^{\beta}{}_{j=1}^{2k_i}$ be the vectors from the Jordan basis with respect to which A and H have the form (3.1) and (3.2), respectively. In other words, these vectors are the columns of S. For $i = 1, \cdots, \alpha$ let

$$\mathcal{M}_i = \text{span}\,\{e_{i\,1}, \cdots, e_{i\,\frac{k_i}{2}}\} \text{ if } k_i \text{ is even},$$

$$\mathcal{M}_i = \text{span}\,\{e_{i\,1}, \cdots, e_{i\,k_i+\frac{1}{2}}\} \text{ if } k_i \text{ is odd and } \varepsilon_i = 1,$$

$$\mathcal{M}_i = \text{span}\,\{e_{i\,1}, \cdots, e_{i\,k_i-\frac{1}{2}}\} \text{ if } k_i \text{ is odd and } \varepsilon_i = -1.$$

For $i = \alpha + 1, \cdots, \beta$, let

$$\mathcal{M}_i = \text{span}\,\{e_{i\,1}, \cdots, e_{i\,k_i}\}.$$

Put

$$\mathcal{M} = \oplus_{i=1}^{\beta} \mathcal{M}_i. \quad (3.3)$$

Then $\mathcal{M}$ is A-invariant and maximal H-nonnegative (see, e.g., [23]).

Let A be H-selfadjoint. An A-invariant maximal H-nonnegative subspace $\mathcal{M}$ is called *stable* if for any H-selfadjoint matrix B with $\|A - B\|$ small enough there exists a B-invariant maximal H-nonnegative subspace $\mathcal{N}$ with $\theta(\mathcal{N}, \mathcal{M})$ as small as one wants.

An important property for the pair of matrices (A, H) connected to its canonical form is the following: the pair (A, H) is said to satisfy the *sign condition* if for any real eigenvalue λ_0 the signs in the sign characteristic of (A, H) corresponding to blocks of even order and eigenvalue λ_0 are all the same, and the same holds for blocks of odd order and eigenvalue λ_0. The sign condition was introduced and studied in [52] in connection with uniqueness of invariant maximal semidefinite subspaces (see Theorem 3.2 below).

Denote by $\mathcal{R}_+(A)$ (respectively $\mathcal{R}_-(A)$) the spectral invariant subspace of A corresponding to the open upper (respectively, lower) half plane, i.e., the span of the eigenvectors and generalized eigenvectors corresponding to eigenvalues of A in the open upper (respectively, lower) half plane.

The following theorem (which subsumes several results proved in [52]) gives necessary and sufficient conditions for the existence of stable invariant maximal semidefinite subspaces, and provides a full description of such subspaces.

Theorem 3.2. *Let A be an $n \times n$ H-selfadjoint matrix. Then the following statements are equivalent.*

(i) *There exists a stable A-invariant maximal H-nonnegative subspace.*
(ii) *There exists a stable A-invariant maximal H-nonpositive subspace.*
(iii) *There is a unique A-invariant maximal H-nonnegative subspace $\mathcal{M}$ such that the eigenvalues of $A|_{\mathcal{M}}$ are in the closed upper half plane.*
(iv) *There is a unique A-invariant maximal H-nonnegative subspace $\mathcal{M}$ such that the eigenvalues of $A|_{\mathcal{M}}$ are in the closed lower half plane.*
(v) *There is a unique A-invariant maximal H-nonpositive subspace $\mathcal{M}$ such that the eigenvalues of $A|_{\mathcal{M}}$ are in the closed upper half plane.*
(vi) *There is a unique A-invariant maximal H-nonpositive subspace $\mathcal{M}$ such that the eigenvalues of $A|_{\mathcal{M}}$ are in the closed lower half plane.*
(vii) *The pair (A, H) satisfies the sign condition.*

In that case the following are equivalent.

(i) *The A-invariant maximal H-nonnegative subspace $\mathcal{M}$ is stable.*
(ii) *$\mathcal{M} \cap \mathcal{R}_+(A)$ is stable as an A-invariant subspace.*
(iii) *$\mathcal{M} \cap \mathcal{R}_-(A)$ is stable as an A-invariant subspace.*
(iv) *For any eigenvalue λ_0 of A in the open upper half plane with geometric multiplicity larger than one either $\mathcal{R}_{\lambda_0}(A)$ is contained in $\mathcal{M}$ or it has zero intersection with $\mathcal{M}$.*

When (A, H) satisfies the sign condition the subspace defined by (3.3) is the unique A-invariant maximal H-nonnegative subspace $\mathcal{M}$ such that the eigenvalues of $A|_{\mathcal{M}}$ are in the closed upper half plane, and, according to the second part of the theorem, this subspace is stable.

An A-invariant maximal H-nonnegative subspace $\mathcal{M}$ is called *Lipschitz stable* if there are positive numbers K and δ such that for any H-selfadjoint matrix B

with $\|A - B\| < \delta$ there exists a B-invariant maximal H-nonnegative subspace $\mathcal{N}$ with $\theta(\mathcal{N}, \mathcal{M}) \leq K\|A - B\|$. The following theorem gives necessary and sufficient conditions for the existence of such subspaces and describes them.

Theorem 3.3. *Let A be an $n \times n$ H-selfadjoint matrix. Then the following conditions are equivalent.*

(i) *There exists a Lipschitz stable A-invariant maximal H-nonnegative subspace.*

(ii) *There exists a Lipschitz stable A-invariant maximal H-nonpositive subspace.*

(iii) *For every real eigenvalue λ_0 of A the subspace* Ker $(A - \lambda_0)$ *is H-definite (either positive definite or negative definite).*

In that case a maximal nonnegative invariant subspace is Lipschitz stable if and only if it is a spectral subspace.

Observe that condition (iii) in the theorem above also implies that the algebraic and geometric multiplicities of A at each of its real eigenvalues coincide. Moreover, the pair (A, H) satisfies the sign condition.

For several problems considered in the sequel A-invariant subspaces $\mathcal{M}$ that have the property that $H(\mathcal{M}) = \mathcal{M}^{\perp}$ play an important role. A subspace with this property is called an *H-lagrangian* subspace. Such subspaces are both maximal H-nonnegative and maximal H-nonpositive. We first describe the existence of A-invariant H-lagrangian subspaces in terms of the sign characteristic of the pair (A, H).

Theorem 3.4. *Let A be an $n \times n$ H-selfadjoint matrix. Then there exists an A-invariant H-lagrangian subspace if and only if for each real eigenvalue λ_0 of A the number of odd partial multiplicities of A at λ_0 is even, and the number of $+1$'s in the sign characteristic of the pair (A, H) corresponding to blocks in the canonical form with eigenvalue λ_0 and odd size is equal to the number of -1's in the sign characteristic of the pair (A, H) corresponding to those blocks.*

In particular, if there are only even partial multiplicities corresponding to real eigenvalues of A, then there exists an A-invariant H-lagrangian subspace.

An A-invariant H-lagrangian subspace $\mathcal{M}$ is called *unconditionally* stable if for every $\varepsilon > 0$ there is a $\delta > 0$ such that for every H-selfadjoint B with $\|A - B\| < \delta$ there exists a B-invariant H-lagrangian subspace $\mathcal{N}$ with $\theta(\mathcal{M}, \mathcal{N}) < \varepsilon$. The subspace $\mathcal{M}$ is called *conditionally* stable if for every $\varepsilon > 0$ there is a $\delta > 0$ such that for every H-selfadjoint B with $\|A - B\| < \delta$ and for which there exists a B-invariant H-lagrangian subspace, there is such a subspace $\mathcal{N}$ with $\theta(\mathcal{M}, \mathcal{N}) < \varepsilon$. Note that the difference is that in the latter case we restrict the class of matrices we allow as perturbations.

The following theorem is proved in [52].

Theorem 3.5. *Let A be an $n \times n$ H-selfadjoint matrix for which there exists an invariant H-lagrangian subspace.*

(a) There exists an unconditionally stable A-invariant H-lagrangian subspace if and only if A does not have real eigenvalues. In that case an A-invariant H-lagrangian subspace $\mathcal{M}$ is unconditionally stable if and only if $\mathcal{M} \cap \mathcal{R}_+(A)$ is stable as an A-invariant subspace.

(b) The following conditions are equivalent:

(i) *There exists a conditionally stable A-invariant H-lagrangian subspace,*

(ii) *A has only even partial multiplicities corresponding to real eigenvalues, and the pair (A, H) satisfies the sign condition,*

(iii) *for every A-invariant subspace $\mathcal{N} \subset \mathcal{R}_+(A)$ there is a unique A-invariant H-lagrangian subspace $\mathcal{M}$ with $\mathcal{M} \cap \mathcal{R}_+(A) = \mathcal{N}$.*

In that case an A-invariant H-lagrangian subspace $\mathcal{M}$ is conditionally stable if and only if $\mathcal{M} \cap \mathcal{R}_+(A)$ is stable as an A-invariant subspace.

When the conditions in part (b)(i) hold the A-invariant H-lagrangian subspaces $\mathcal{M}$ are in one-one correspondence with the set of A-invariant subspaces $\mathcal{N}$ contained in $\mathcal{R}_+(A)$, because of the equivalence with (iii). This parametrization is given by

$$\mathcal{M} = \mathcal{N} \dotplus \mathcal{M}_0 \dotplus (H\mathcal{N}^\perp \cap \mathcal{R}_-(A)),$$

where $\mathcal{M}_0$ is the A-invariant subspace spanned by first halves of Jordan chains of A corresponding to real eigenvalues.

Analogues of Theorems 3.1–3.5 for several real cases were studied in [54, 55, 56].

Problem 3.6. *Study the degree of stability (α-stability) of stable A-invariant maximal H-semidefinite subspaces and of (conditionally or unconditionally) stable A-invariant Lagrangian subspaces, both in the real and the complex cases.*

3.2. Dissipative matrices in indefinite inner products

We continue to assume $F = \mathbb{C}$ in this subsection. Recall that a matrix A is called H-dissipative, where $H = H^*$ if $\frac{1}{2i}(HA - A^*H)$ is positive semidefinite. For dissipative matrices in an indefinite inner product space there does not exist a canonical form; however, by applying similarity to A and congruence to H (with the same invertible matrix) we can transform the pair (A, H) into a simple form, from which it is easy to read off that invariant maximal semidefinite subspaces always exist. In fact, the simple form allows one to construct an A-invariant maximal H-nonnegative subspace in essentially the same way as in formula (4.3). The part of the construction pertaining to the real eigenvalues of A stays the same, yielding a subspace $\mathcal{M}_0$; an A-invariant maximal H-nonnegative subspace is then constructed by taking the direct sum of $\mathcal{M}_0$ with the spectral subspace of A corresponding to the open upper half plane. We refer the reader to [65], where the main results concerning the simple form are proved.

Let $H = H^*$ be invertible, and let A be H-dissipative. The A-invariant maximal H-nonnegative subspace $\mathcal{M}$ is called *stable* if for every $\varepsilon > 0$ there is a $\delta > 0$ such that every H-dissipative matrix B with $\|A - B\| < \delta$ has an invariant maximal H-nonnegative subspace $\mathcal{N}$ with $\theta(\mathcal{M}, \mathcal{N}) < \varepsilon$.

Before we can state the main result on stability of invariant maximal nonnegative subspaces for an H-dissipative matrix A we need to introduce the notion of the numerical range condition for the pair (A, H) (see [71], Chapter 3).

Let us denote the number of negative eigenvalues of H by κ. It is well known that the maximal length of a Jordan chain of A corresponding to a real eigenvalue is then $2\kappa + 1$.

Let λ be a real eigenvalue of A. Consider a Jordan basis for $R(A, \{\lambda\})$ and split it into the sets $J(\lambda, j)$, where $J(\lambda, j)$ consists of the basis vectors belonging to Jordan chains of length j. Denote by $n_{\lambda,j}$ the number of chains of length j. Now denote the basis vectors in $J(\lambda, j) \setminus \text{Ker } (A - \lambda)^{j-1}$ by $x_{j;1}, \ldots, x_{j;n_{\lambda,j}}$. Let m_j be $\frac{(j+1)}{2}$ in case j is odd, and $\frac{j}{2}$ in case j is even. Define $y_{j;k} = (A - \lambda)^{m_j - 1} x_{j;k}$.

From here on there is a difference in the situation for Jordan chains of odd length and Jordan chains of even length. We first deal with the case of Jordan chains of odd length.

For $j = 1, 3, \ldots, 2\kappa + 1$ let

$$CM_j = \begin{bmatrix} \langle Hy_{j;1}, y_{j;1} \rangle & \cdots & \langle Hy_{j;n_{\lambda,j}}, y_{j;1} \rangle \\ \vdots & & \vdots \\ \langle Hy_{j;1}, y_{j;n_{\lambda,j}} \rangle & \cdots & \langle Hy_{j;n_{\lambda,j}}, y_{j;n_{\lambda,j}} \rangle \end{bmatrix}$$

(here CM stands for *characteristic matrix*). Define

$$CM_{odd}(A, \lambda) = \text{diag } (CM_1, \ldots, CM_{2\kappa+1}),$$

and, finally, put

$$NR_{odd}(A, \lambda) = \{ \langle CM_{odd}(A, \lambda)x, x \rangle \mid x \neq 0 \}.$$

(here NR stands for numerical range).

It turns out that $CM_{odd}(A, \lambda)$ is hermitian and invertible and that $NR_{odd}(A, \lambda)$ is independent of the choice of the Jordan basis one starts with.

The pair (A, H) is said to satisfy the *odd numerical range condition* if $0 \notin NR_{odd}(A, \lambda)$ for all real eigenvalues of A.

For even length chains, let $j, k \in \{2, 4, \ldots, 2\kappa\}$ and put

$$CM_{j,k} = \begin{bmatrix} \langle H(A-\lambda)y_{j;1}, y_{k;1} \rangle & \cdots & \langle H(A-\lambda)y_{j;n_{\lambda,j}}, y_{k;1} \rangle \\ \vdots & & \vdots \\ \langle H(A-\lambda)y_{j;1}, y_{k;n_{\lambda,k}} \rangle & \cdots & \langle H(A-\lambda)y_{j;n_{\lambda,j}}, y_{k;n_{\lambda,k}} \rangle \end{bmatrix}.$$

One can show that for $j < k$ one has $CM_{j,k} = 0$, and that $CM_{j,j}$ is invertible. Put

$$CM_{even}(A, \lambda) = \begin{bmatrix} CM_{2,2} & \cdots & CM_{2\kappa,2} \\ \vdots & & \vdots \\ CM_{2,2\kappa} & \cdots & CM_{2\kappa,2\kappa} \end{bmatrix}.$$

Then $CM_{even}(A, \lambda)$ is an invertible block upper triangular matrix. Define

$$NR_{even}(A, \lambda) = \{ \langle CM_{even}(A, \lambda)x, x \rangle \mid x \neq 0 \}.$$

It turns out that $NR_{even}(A, \lambda)$ is independent of the choice of the Jordan basis one starts with.

The pair (A, H) is said to satisfy the *even numerical range condition* if $0 \notin NR_{even}(A, \lambda)$ for all real eigenvalues of A, and we say that (A, H) satisfies the *numerical range condition* if it satisfies both the odd numerical range condition and the even numerical range condition.

Even though the odd and even numerical ranges do not depend on the choice of the Jordan basis, and hence neither does the numerical range condition, a basis free definition does not exist so far.

It can be shown that if A is H-selfadjoint then the pair (A, H) satisfies the numerical range condition if and only if it satisfies the sign condition introduced in the previous subsection.

In [71], [64] the following result was proved:

Theorem 3.7. *The following are equivalent:*

(i) *there exists a stable A-invariant maximal H-nonnegative subspace,*
(ii) *there exists a stable A-invariant maximal H-nonpositive subspace,*
(iii) *the numerical range condition holds for the pair (A, H),*
(iv) *there is a unique A-invariant maximal H-nonnegative subspace $\mathcal{M}$ with $\sigma(A|_{\mathcal{M}})$ contained in the closed upper half plane,*
(v) *there is a unique A-invariant maximal H-nonpositive subspace $\mathcal{M}$ with $\sigma(A|_{\mathcal{M}})$ contained in the closed lower half plane.*

In that case, there is a unique stable A-invariant maximal H-nonnegative subspace, being the one with $\sigma(A|_{\mathcal{M}})$ contained in the closed upper half plane, and there is a unique stable A-invariant maximal H-nonpositive subspace, being the one with $\sigma(A|_{\mathcal{M}})$ contained in the closed lower half plane.

Problem 3.8. *Find the degree of stability of stable A-invariant maximal H-semidefinite subspaces described in Theorem* 3.7.

Problem 3.9. *Study invariant maximal semidefinite subspaces (existence, uniqueness, stability, degree of stability) of real $n \times n$ matrices that are dissipative with respect to an indefinite inner product in $\mathbb{R}^n$.*

The definition of dissipativity should be modified in the real case. We say that a real $n \times n$ matrix A is H-dissipative, where $H = H^*$ is an invertible real $n \times n$ matrix, if $HA + A^*H$ is positive semidefinite. One way to approach Problem 3.9 is by developing simple forms for real dissipative matrices analogous to those developed in [71], [65] for the complex case. Another notion of dissipativity in the real case is obtained by the condition that $HA - A^*H$ is positive semidefinite, where now the real matrix H is skew-symmetric (and invertible). Again, a study of simple forms, and of A-invariant subspaces H-neutral subspaces, for H-dissipative real matrices A, is an open problem.

3.3. Continuous algebraic Riccati equations

In this subsection we treat stability properties of solutions of the equation

$$XDX - XA - A^*X - C = 0, \tag{3.4}$$

where A, $D = D^*$, $C = C^*$ are $n \times n$ matrices (over $F = \mathbb{C}$ or $F = \mathbb{R}$). Equations of type (3.4), sometimes under additional assumptions that C is positive semidefinite, and D has the form $D = BR^{-1}B^*$ for an invertible positive definite $m \times m$ matrix R and an $n \times m$ matrix B, are known as continuous algebraic Riccati equations. Because of their central role in various control problems in continuous time, including linear quadratic optimal control, there is a voluminous literature on the equations (3.4), especially concerning numerical methods. We mention here only the books [7, 35, 47], where many additional references may be found. See, in particular, [35] for a background and general information concerning continuous algebraic Riccati equations.

Only hermitian solutions $X \in F^{n\times n}$ of (3.4) will be considered here. Such a solution is called *α-stable* if there exist $\varepsilon > 0$, $K > 0$ such that every equation with coefficients in $F^{n\times n}$

$$X\tilde{D}X - X\tilde{A} - \tilde{A}^*X - \tilde{C} = 0, \tag{3.5}$$

with $\tilde{D} = \tilde{D}^*$, $\tilde{C} = \tilde{C}^*$ and

$$\|D - \tilde{D}\| + \|A - \tilde{A}\| + \|C - \tilde{C}\| < \varepsilon$$

has a hermitian solution $Y \in F^{n\times n}$ such that

$$\|X - Y\| \le K(\|D - \tilde{D}\| + \|A - \tilde{A}\| + \|C - \tilde{C}\|)^{\frac{1}{\alpha}}.$$

In the rest of this subsection we consider the complex case: $F = \mathbb{C}$. The study of α-stability and other stability properties in this section will be given in terms of the Hamiltonian matrix

$$M = i\begin{bmatrix} A & -D \\ -C & -A^* \end{bmatrix}.$$

Observe that M is selfadjoint in the indefinite scalar product induced by $H = i\begin{bmatrix} 0 & I \\ -I & 0 \end{bmatrix}$.

It is well known (and easily seen) that $X \in \mathbb{C}^{n\times n}$ is a hermitian solution of (3.4) if and only if the graph subspace

$$\mathcal{M}(X) = \operatorname{Im}\begin{bmatrix} I \\ X \end{bmatrix}$$

is M-invariant and H-lagrangian. Thus, the investigation of stability properties of X is reduced to that of the M-invariant H-lagrangian subspace $\mathcal{M}(X)$. Recall that a H-lagrangian subspace $\mathcal{M}$ is certainly H-neutral, i.e., $\langle Hx, x\rangle = 0$ for all $x \in \mathcal{M}$.

There is an extensive literature on hermitian solutions of the equation (3.4), as well as the discrete Riccati equation ((3.9) below), in terms of invariant lagrangian graph subspaces, including numerical analysis and algorithms for computation of such subspaces. The numerical analysis concerns almost exclusively the graph subspaces that are actually spectral subspaces; see, for example, [12, 11, 32] for algorithms based on the sign function. A review of numerical methods for the algebraic Riccati equations is found in [40]. A Newton's method based algorithm that works also for some nonspectral subspaces was developed recently in [28]. In certain applications, for example, the well-known two Riccati equations approach for suboptimal H_∞ control problems [16], one encounters situations when the graph subspace in question is very close to being non-spectral. Thus, it is desirable to study stability of solutions of the Riccati equations (3.9) (below) and (3.4), also for nonspectral graph subspaces.

Before turning to α-stability let us consider stability. A solution X of (3.4) is called *conditionally stable* if for every $\varepsilon > 0$ there is a $\delta > 0$ such that every equation (3.5) such that

$$\|D-\tilde{D}\| + \|A-\tilde{A}\| + \|C-\tilde{C}\| < \delta$$

and which has a hermitian solution, also has a hermitian solution Y with $\|X-Y\| < \varepsilon$. The solution X of (3.4) is called *unconditionally stable* if in the above definition the restriction to equations (3.5) that have a hermitian solution is dropped, that is, every perturbed equation should have a solution.

These concepts where studied in [53] for $F = \mathbb{C}$ and in [59] for $F = \mathbb{R}$ (see also [60] for the analogue for the discrete algebraic Riccati equation).

Assume that $D \geq 0$ and (A, D) is *controllable*, that is

$$\operatorname{rank} \begin{bmatrix} D & AD & \dots & A^{n-1}D \end{bmatrix} = n.$$

Then one can show that there exists a solution to equation (3.4) if and only if there exists an M-invariant H-lagrangian subspace (see [35], also [69]). In that case it turns out that M has only even partial multiplicities at its real eigenvalues, and that the pair (M, H) satisifies the sign condition.

To state the main results in this direction we introduce the M-invariant subspace $\mathcal{M}_+$, which is the direct sum of the subspaces $\mathcal{R}_\lambda(M)$ with $\operatorname{Im}\lambda > 0$. We then have the following theorem.

Theorem 3.10. *Assume that (A, D) is controllable, $D \geq 0$ and that equation* (3.4) *has a hermitian solution X. Then*

(a) *There exists an unconditionally stable solution of equation* (3.4) *if and only if M does not have real eigenvalues. In that case the following statements are equivalent:*

(i) *The solution X is unconditionally stable.*

(iii) *The subspace* $\operatorname{Im}\begin{bmatrix} I \\ X \end{bmatrix}$ *is an unconditionally stable M-invariant H-lagrangian subspace.*

(iv) *The common eigenvalues of $i(A-DX)$ and its adjoint are eigenvalues of M of geometric multiplicity one.*

(b) *There exists a conditionally stable solution of* (3.4). *In that case the following statements are equivalent:*
   (i) *The solution* X *is conditionally stable.*
   (ii) *The subspace* $\operatorname{Im}\begin{bmatrix} I \\ X \end{bmatrix} \cap \mathcal{M}_+$ *is stable as an* M*-invariant subspace.*
   (iii) *The subspace* $\operatorname{Im}\begin{bmatrix} I \\ X \end{bmatrix}$ *is a conditionally stable* M*-invariant* H*-lagrangian subspace.*
   (iv) *The common non-real eigenvalues of* $i(A - DX)$ *and its adjoint are eigenvalues of* M *of geometric multiplicity one.*

Returning to α-stability, we need the following definition. An M-invariant, H-neutral subspace $\mathcal{M}$ is called *neutrally* α*-stable* if there is constant $K > 0$ such that every H-selfadjoint matrix M' which is sufficiently close to M has H-neutral invariant subspace $\mathcal{M}'$ such that

$$\theta(\mathcal{M}, \mathcal{M}') \leq K\|M - M'\|^{\frac{1}{\alpha}}.$$

In view of the above remark we obtain:

Proposition 3.11. *A hermitian solution* X *is* α*-stable if and only if the subspace* $\mathcal{M}(X)$ *is neutrally* α*-stable.*

Our main result on α-stability of hermitian solutions of (3.4) is the following.

Theorem 3.12. (a) *If there exists an* α*-stable hermitian solution* X *of* (3.4) *for some* α, *then* M *has no real eigenvalues.*
(b) *If* $D \geq 0$ *and the pair* (A, D) *is sign controllable i.e., for every eigenvalue* λ *of* A *at least one of the root subspaces* $\mathcal{R}_\lambda(A)$ *and* $\mathcal{R}_{-\bar{\lambda}}(A)$ *is contained in the controllable subspace* $\operatorname{Im}[D, AD, \cdots, A^{n-1}D]$ *of* (A, D), *then the necessary condition of part (a) is also sufficient; moreover, there exists a 1-stable hermitian solution* X *of* (3.4).
(c) *Assume that* M *has no real eigenvalues, and let* X *be a hermitian solution of* (3.4). *Let* σ_X *be the set of all eigenvalues of* M *in the upper half plane which have geometric multiplicity one and which are common eigenvalues of* $i(A+DX)$ *and* $-i(A+DX)^*$. *Then* X *is* α*-stable if and only if for all* $\lambda \in \sigma_X$ *we have*

$$\gamma(\dim\, (\mathcal{M}(X) \cap \mathcal{R}_\lambda(M)),\ \dim\, \mathcal{R}_\lambda(M)) \leq \alpha. \tag{3.6}$$

(d) *Assume that* M *has no real eigenvalues, and let* X *be a hermitian solution of* (3.4). *Then* X *is* α*-stable if and only if for each common eigenvalue* λ *of* $i(A + DX)$ *and its adjoint we have the following two conditions: (i)* λ *is an eigenvalue of* M *of geometric multiplicity one, (ii) if* k, *respectively* l, *denote the algebraic multiplicity of* λ *as an eigenvalue of* $i(A + DX)$, *respectively,* $-i(A + DX)^*$, *then* $\gamma(k,\ k + l) \leq \alpha$.

The α-stability of a hermitian solution X can fail because a nearby equation (3.5) has no hermitian solution at all. To eliminate this reason for absence of

α-stability, we introduce the following definition (which applies equally to the complex and the real case). A hermitian solution X of (3.4) is called *conditionally α-stable* if it satisfies the definition of α-stability under the additional assumption that the perturbed equation (3.5) has hermitian solutions to start with. Theorems 3.13 and 3.14 below were proved in [57].

Theorem 3.13. *A necessary condition for existence of a conditionally α-stable hermitian solution of* (3.4) *for some α is that all real eigenvalues of M (if any) have only even partial multiplicities, and for every real eigenvalue λ_0 of M there is a number $\varepsilon = \varepsilon(\lambda_0) = \pm 1$ such that for every Jordan chain $x_0, x_1, \ldots, x_m$ of eigenvector and generalized eigenvectors of M corresponding to λ_0 the inequality $\varepsilon\langle Hx_0, x_m\rangle \geq 0$ holds.*

In terms of the sign characteristic of the pair (M, H) the condition on the Jordan chains in Theorem 3.13 simply means that the signs in the sign characteristic of (M, H) corresponding to λ_0 are all equal.

We are able to obtain more detailed information in the particular (but generic) case when all real eigenvalues of M (if any) have geometric multiplicity one.

Theorem 3.14. *Case $F = \mathbf{C}$. Assume $D \geq 0$, the pair (A, D) is sign controllable, and the distinct real eigenvalues $\lambda_1, \ldots, \lambda_r$ of M (if any) have geometric multiplicity one and even algebraic multiplicities $m_1, \ldots, m_r$, respectively. Let*

$$\alpha_0 = \max(2, m_1 - 1, \ldots, m_r - 1), \tag{3.7}$$

or $\alpha_0 = 1$ if M has no real eigenvalues. Then

(i) *There exist conditionally α_0-stable hermitian solutions X of* (3.4). *They are characterized by the following properties:*

(a) *either* $\mathcal{R}_\lambda(M) \cap \left(\operatorname{Im}\begin{bmatrix} I \\ X \end{bmatrix}\right) = (0)$ *or* $\mathcal{R}_\lambda(M) \subseteq \operatorname{Im}\begin{bmatrix} I \\ X \end{bmatrix}$, *for every non-real eigenvalue λ of M having geometric multiplicity at least* 2, *or having geometric multiplicity* 1 *and algebraic multiplicity at least* $\alpha_0 + 2$,

(b) *for every non-real eigenvalue λ of M having geometric multiplicity* 1 *and algebraic multiplicity precisely $\alpha_0 + 1$ we have either*

$$\mathcal{R}_\lambda(M) \cap \left(\operatorname{Im}\begin{bmatrix} I \\ X \end{bmatrix}\right) = (0), \qquad \text{or} \qquad \mathcal{R}_\lambda(M) \subseteq \operatorname{Im}\begin{bmatrix} I \\ X \end{bmatrix}$$

or

$$\dim\left(\mathcal{R}_\lambda(M) \cap \left(\operatorname{Im}\begin{bmatrix} I \\ X \end{bmatrix}\right)\right) = k > 0,$$

where the positive integer k is such that there is a set of k distinct $(\alpha_0 + 1)$-th roots of unity that sum up to zero,

(c) *for every non-real eigenvalue λ of M having geometric multiplicity* 1 *and algebraic multiplicity precisely α_0 we have either*

$$\mathcal{R}_\lambda(M) \cap \left(\operatorname{Im}\begin{bmatrix} I \\ X \end{bmatrix}\right) = (0), \qquad \text{or} \qquad \mathcal{R}_\lambda(M) \subseteq \operatorname{Im}\begin{bmatrix} I \\ X \end{bmatrix} \tag{3.8}$$

or

$$\dim \left(\mathcal{R}_\lambda(M) \cap \left(\operatorname{Im} \begin{bmatrix} I \\ X \end{bmatrix} \right) \right) = k > 0,$$

where the positive integer k is such that there is no set of k distinct α_0-th roots of unity that sum up to zero.

(ii) *No hermitian solution X of* (3.4) *is conditionally α-stable with $\alpha < \alpha_0$.*

Conditions (a)–(c) in part (i) of Theorem 3.14 can be rephrased as follows. A hermitian solution X of (3.4) is α_0-stable if and only if for each common non-real eigenvalue λ of $i(A + DX)$ and its adjoint we have that λ is an eigenvalue of M of geometric multiplicity one, and if k, respectively l, denote the algebraic multiplicities of λ as an eigenvalue of $i(A+DX)$, respectively $-i(A+DX)^*$, then $\gamma(k, k+l) \leq \alpha_0$. (Compare part (d) of Theorem 3.12.)

A hermitian solution X of (3.4) is called *unmixed* if the spectrum of $i(A+DX)$ does not contain pairs of non-real complex conjugate eigenvalues; this concept was introduced and studied in [69]. It follows from Theorem 3.14 and the observation above that, under the conditions of Theorem 3.14, the unmixed hermitian solutions of (3.4) are α_0-stable.

To demonstrate typical techniques used in proofs of results concerning stability of invariant subspaces with symmetries, and of solutions of algebraic Riccati equations, we reproduce here a more or less detailed proof of Theorem 3.14. The proof is based on a series of lemmas.

Lemma 3.15. *For a fixed positive integer m, there is a real number $a > 1$ such that no sum of any k roots ($1 \leq k < m$) of the polynomial $p(x) = x^m - x + a$ is equal to zero.*

Proof. The existence of such an a close to zero was proved in Lemma 2.2 in [63]. Therefore, the product of all possible sums of less than m roots of $p(x)$, which is a polynomial function of a, is not identically zero, and hence there exists $a > 1$ with the required properties. □

We now fix an a as in Lemma 3.15.

Lemma 3.16. *Let J be an $m \times m$ nilpotent Jordan block, and for every real number u, $|u| \leq 1$, let $B(u)$ be an $m \times m$ matrix with the characteristic polynomial $x^m - u^{m-1}x + au^m$ and such that*

$$\|B(u) - J\| \leq C|u|^{m-1},$$

where the constant $C > 0$ is independent on u. Then there exists a positive constant K such that for every nontrivial (i.e., different from (0) *and the whole space) J-invariant subspace $\mathcal{N}$ and every $B(u)$-invariant subspace $\mathcal{M}$ the inequality*

$$\theta(\mathcal{M}, \mathcal{N}) \geq K\|B(u) - J\|^{\frac{1}{m-1}}$$

holds.

The proof of this lemma follows from the proof of Lemma 2.3 in [63].

Let H be an invertible hermitian $m \times m$ matrix, and let A be an H-selfadjoint $m \times m$ matrix. An A-invariant H-neutral subspace $\mathcal{M}$ will be called *conditionally neutrally α-stable* if it satisfies the definition of neutral α-stability (see the paragraph before Proposition 3.11) with the additional requirement that there is an M'-invariant H-neutral subspace of the same dimension as $\mathcal{M}$ to start with.

Lemma 3.17. *Assume that A has geometric multiplicity one for every real eigenvalue (if any). Let $\mathcal{M}$ be an $\frac{m}{2}$-dimensional, A-invariant, H-neutral subspace (it is implicitly assumed that such subspaces exist; in particular m is even). If $\mathcal{M}$ is not α-stable as an A-invariant subspace, then $\mathcal{M}$ is not conditionally neutrally α-stable as an $\frac{m}{2}$-dimensional, A-invariant, H-neutral subspace.*

Proof. Using the local principle (which is described in [53, 55], see especially Theorem 3.1 there), the proof is reduced to two separate cases: (1) $\sigma(A) = \{\lambda_0\}$, λ_0 is real; (2) $\sigma(A) = \{\lambda_0, \bar{\lambda}_0\}$, λ_0 non-real.

Let us first consider the case $\sigma(A) = \{\lambda_0\}$, λ_0 is real. We may assume that $\lambda_0 = 0$, and since A has geometric multiplicity one, we may further assume that A is the $m \times m$ Jordan block with eigenvalue zero, and H is the matrix with zeros everywhere except for the entries on the southwest-northeast diagonal, which are all ones. A justification of these assumptions is based on the canonical form of pairs of matrices (A, H), where A is H-selfadjoint (see, e.g., Theorem 1.3.3 in [23]). Then $\mathcal{M} = \text{span}\,\{e_1, \cdots, e_{\frac{m}{2}}\}$. As an A-invariant subspace, $\mathcal{M}$ is $(m-1)$-stable and is not α-stable for $\alpha < m-1$ (if $m > 2$); $\mathcal{M}$ is 2-stable and not α-stable for $\alpha < 2$ (if $m = 2$). In what follows we assume $m > 2$, leaving the relatively simple case $m = 2$ to the interested readers.

Fix $a > 1$ as in Lemma 3.15. For a real parameter u, $0 < |u| \leq 1$, let $C(u)$ be the companion $m \times m$ matrix having the characteristic polynomial $p(\lambda) = \lambda^m - u^{m-1}\lambda + au^m$, in other words, $C(u)$ is obtained from the nilpotent $m \times m$ Jordan block by adding $-au^m$ and u^{m-1} in the $(m, 1)$ and $(m, 2)$ positions, respectively. Let $S(u)$ be the $m \times m$ matrix obtained from the $m \times m$ identity matrix by adding $\frac{1}{2}u^{m-1}$ in the $(m, 1)$ position. A straightforward computation verifies that the matrix $B(u) = S(u)^{-1}C(u)S(u)$ is H-selfadjoint. Clearly,

$$\|B(u) - A\| \leq K|u|^{m-1},$$

where the positive constant K is independent of u. Elementary calculus shows that the polynomial $p(\lambda)$ has no real roots (indeed, since m is even and $u \neq 0$, there is only one minimum, call it λ_0, of $p(\lambda)$ on the real line, and because $a > 1$, the value $p(\lambda_0)$ is positive). Thus, there exist $\frac{m}{2}$-dimensional $B(u)$-invariant H-neutral subspaces. By Lemma 3.16, the subspace $\mathcal{M}$ is not conditionally neutrally α-stable for any $\alpha < m - 1$.

Next, consider the case $\sigma(A) = \{\lambda_0, \bar{\lambda}_0\}$, λ_0 non-real. In case dim Ker $(A - \lambda_0) > 1$ the result is known. Indeed, as an A-invariant subspace $\mathcal{M}$ is 1-stable or not stable at all, depending on whether or not it is a spectral subspace (recall that an A-invariant subspace $\mathcal{M}$ is called a *spectral subspace* if $\mathcal{M} = (0)$ or $\mathcal{M}$ is a direct sum of root subspaces $\mathcal{R}_\lambda(A)$). The same holds when we consider conditional neutral α stability, i.e., $\mathcal{M}$ is conditionally neutrally 1-stable or not conditionally neutrally stable at all, depending on whether or not $\mathcal{M}$ is a spectral subspace for A (see Theorem 1.11 in [50]). So, it remains to consider the case dim Ker $(A-\lambda_0) = 1$. Using the canonical form of (A, H), without loss of generality we assume that

$$A = \begin{bmatrix} J_m(\lambda_0) & 0 \\ 0 & J_m(\lambda_0)^* \end{bmatrix}, \qquad H = \begin{bmatrix} 0 & I \\ I & 0 \end{bmatrix},$$

where $J_m(\lambda_0)$ is the $m \times m$ Jordan block with eigenvalue λ_0. Let dim $\mathcal{M} \cap \mathcal{R}_{\lambda_0}(A) = k$. Then dim $\mathcal{M} \cap \mathcal{R}_{\bar{\lambda}_0}(A) = m - k$. If $k = 0$ or $k = m$ then again $\mathcal{M}$ is a spectral subspace, and the arguments of the preceding paragraph apply. So, let $1 \leq k \leq m-1$. By Theorem 2.1, for $\alpha < \gamma(k,\ m)$ $\mathcal{M}$ is not α-stable, while for $\alpha \geq \gamma(k,\ m)$ $\mathcal{M}$ is α-stable as an A-invariant subspace. So, we have to show that for $\alpha < \gamma(k,\ m)$ $\mathcal{M}$ is not conditionally neutrally α-stable as an H-neutral A-invariant subspace of dimension $\frac{m}{2}$. For this purpose let

$$A(u) = \begin{bmatrix} \lambda_0 I + B(u) & 0 \\ 0 & (\lambda_0 I + B(u))^* \end{bmatrix},$$

where $B(u)$ is the matrix constructed in the above proof of the case (1). As in that proof, we verify that there is a positive constant K such that for every $A(u)$-invariant subspace $\mathcal{N}$ the inequality

$$\theta(\mathcal{M}, \mathcal{N}) \geq K \|A(u) - A\|^{\frac{1}{m-1}}, \qquad -1 \leq u \leq 1$$

holds, and that there exist $A(u)$-invariant H-neutral subspaces of dimension $\frac{m}{2}$. □

Proof of Theorem 3.14. Let $\mathcal{M}$ be an n-dimensional, M-invariant, H-neutral subspace which is, in addition, a graph subspace, and is such that the spectral subspace of $M|\mathcal{M}$ corresponding to the non-real eigenvalues is a direct sum of root subspaces of M (the existence of such $\mathcal{M}$ is well known; see, for example, Lemma 7.2.6 and its proof in [35]). Clearly, $\lambda_1, \ldots, \lambda_r$ are the real eigenvalues of $M|\mathcal{M}$ with geometric multiplicity one and the algebraic multiplicities $\frac{1}{2}m_1, \ldots, \frac{1}{2}m_r$, respectively. Thus, by Theorem 2.1 the corresponding hermitian solution X of (5.3) given by $\mathcal{M} = \operatorname{Im} \begin{bmatrix} I \\ X \end{bmatrix}$ is conditionally α-stable. Here we use the fact that $\gamma(k,\ 2k) = 2k - 1$ for all $k > 1$, while $\gamma(1,\ 2) = 2$. The same Theorem 2.1 guarantees that every hermitian solution X satisfying the conditions (a) and (b) is conditionally α_0-stable.

That there are no other conditionally α_0-stable hermitian solutions, as well as the part (ii) of Theorem 3.14, follows from Lemma 3.17. □

Problem 3.18. *Characterize conditionally α-stable hermitian solutions of* (3.4) *in the general case.*

We refer to [57] for the proofs of other theorems in this subsection, as well as for analogous results concerning α-stability of real symmetric solutions of the equation (3.4) with real coefficients. The α-stability of hermitian solutions of the discrete algebraic Riccati equation

$$X = A^*XA - Q - A^*XB(R + B^*XB)^{-1}B^*XA, \tag{3.9}$$

in both the real and the complex cases is studied also in [57].

3.4. Positive semidefinite solutions of the algebraic Riccati equations

In this subsection we discuss stability properties of positive semidefinite solutions of the algebraic Riccati equation

$$XBB^*X - XA - A^*X - C^*C = 0. \tag{3.10}$$

All matrices in this subsection are assumed complex. It is proved in [17, 18] that (3.10) has a positive semidefinite solution if and only if the system (A, B, C) is output-stabilizable, i.e., there is a matrix G such that every solution $\xi(t)$ of a linear system of differential equations $\dot{x}(t) = (A + BG)x(t)$ satisfies $C\xi(t) \to 0$ as $t \to \infty$. However, following [39], the stronger hypothesis of controllability of the pair (A, B) will be imposed. This hypothesis guarantees the possibility to describe the set of positive semidefinite solutions in terms of invariant subspaces, and, moreover, to characterize the stable positive semidefinite solutions using the techniques of stable invariant subspaces; this analysis was carried out in [39]. Parametrization of positive semidefinite solutions in terms of certain invariant subspaces was given in [13, 33, 74, 75]. In [74] the parametrization is given under the assumption of output-stabilizability.

The controllability of (A, B) implies existence of a hermitian solution X_+ such that $A - BB^*X_+$ has all its eigenvalues in the closed left half plane. This solution is unique, it is positive semidefinite and it is the maximal hermitian solution, i.e. $X_+ \geq X$ for any other hermitian solution X of (3.10). See, e.g., [66], Theorem 2.1. If, in addition, (C, A) is observable, then there is just one positive semidefinite solution.

A positive semidefinite solution X of (3.10) is called *stably positive semidefinite* if for every $\varepsilon > 0$ there is a $\delta > 0$ such that $\|A - A_1\| + \|B - B_1\| + \|C - C_1\| < \delta$ implies that the algebraic Riccati equation

$$XB_1B_1^*X - XA_1 - A_1^*X - C_1^*C_1 = 0$$

has a positive semidefinite solution X_1 with $\|X - X_1\| < \varepsilon$.

The following result was proved in [39].

Theorem 3.19. *Assume (A, B) is controllable. Then there exists only one stably positive semidefinite solution of* (3.10), *being the maximal one.*

Let us turn now to the discrete algebraic Riccati equation

$$X = A^*XA + Q - A^*XB(R + B^*XB)^{-1}B^*XA, \tag{3.11}$$

where A is an $n \times n$ matrix, $Q \geq 0$ is also an $n \times n$ matrix, $R > 0$ is an $m \times m$ matrix and B is an $n \times m$ matrix. This equation plays a role in the study of LQ-optimal control for discrete time systems. We would like to have a parametrization of all positive semidefinite solutions X. We shall assume throughout that A is invertible and that $(A,\ B)$ is controllable. Introduce

$$T = \begin{bmatrix} A + BR^{-1}B^*A^{*-1}Q & -BR^{-1}B^*A^{*-1} \\ -A^{*-1}Q & A^{*-1} \end{bmatrix}. \tag{3.12}$$

Recall that

$$J_1 = \begin{bmatrix} 0 & I \\ -I & 0 \end{bmatrix} \quad \text{and} \quad J_2 = \begin{bmatrix} 0 & I \\ I & 0 \end{bmatrix}.$$

Straightforward computation yields that $T^*J_1T = J_1$, i.e., T is J_1-unitary. Again by straightforward computation it is checked that

$$\begin{aligned} J_2 - T^*J_2T &= 2\begin{bmatrix} Q + QA^{-1}BR^{-1}B^*A^{*-1}Q & -QA^{-1}BR^{-1}B^*A^{*-1} \\ -A^{-1}BR^{-1}B^*A^{*-1}Q & A^{-1}BR^{-1}B^*A^{*-1} \end{bmatrix} = \\ &= 2\left(\begin{bmatrix} Q & 0 \\ 0 & 0 \end{bmatrix} + \begin{bmatrix} Q \\ -I \end{bmatrix} A^{-1}BR^{-1}B^*A^{*-1} \begin{bmatrix} Q & -I \end{bmatrix}\right) \geq 0. \end{aligned} \tag{3.13}$$

Thus $T^*J_2T \leq J_2$, i.e., T is J_2-contractive.

We will use $\mathbb{C}_{in}$ for the set of complex numbers inside the open unit disc and $\mathbb{C}_{out}$ for the set of complex numbers outside the closed unit disc. The unit circle will be denoted by $\mathbb{T}$. In the rest of this section $\mathcal{V}$ denotes the maximal A invariant subspace contained in Ker Q. The subspace $\mathcal{V}$ contains the subspaces $\mathcal{V}_< = \mathcal{V} \cap \mathcal{R}(A, \mathbb{C}_{in})$, $\mathcal{V}_> = \mathcal{V} \cap \mathcal{R}(A, \mathbb{C}_{out})$, $\mathcal{V}_0 = \mathcal{V} \cap \mathcal{R}(A, \mathbb{T})$ and $\mathcal{V}_\leq = \mathcal{V} \cap \mathcal{R}(A, \mathbb{C}_{in} \cup \mathbb{T})$. The notation $\mathcal{V}$, etc., has been used for analogous subspaces but no confusion will arise.

It is known that the algebraic Riccati equation (3.11) has a positive semidefinite solution. In fact, there exists a positive semidefinite solution X_+ of (3.11), the maximal solution, such that $X_+ \geq X$ for any other hermitian solution X of (3.11) (see [66]).

Assume X is a hermitian solution of (3.11). Then it follows that X is positive semidefinite if and only if the corresponding subspace $\mathcal{M}$ is maximal J_2-nonnegative. It also follows that the matrix T has only even partial multiplicities for its eigenvalues on the unit circle (see [34]). Let $\mathcal{N}_0$ denote the T-invariant subspace spanned by the vectors that are in the first halves of Jordan chains of T corresponding to eigenvalues of T on the unit circle. Then $\mathcal{N}_0$ is J_1-neutral and J_2-neutral. Theorems 3.20 and 3.22 below were proved in [39].

Theorem 3.20. *Assume (A, B) is controllable, A is invertible, $R > 0$ and $Q \geq 0$. Then there is a one-one correspondence between the set of all A-invariant subspaces $\mathcal{N}$ contained in $\mathcal{V}_>$ and the set of all positive semidefinite solutions X of* (3.11). *More precisely, let $\mathcal{N}$ be such a subspace and let $\mathcal{M}$ be given by*

$$\mathcal{M} = P^*\mathcal{N} \dotplus \mathcal{N}_0 \dotplus ((J_1P^*\mathcal{N})^\perp \cap \mathcal{R}(T, \mathbb{C}_{in})). \tag{3.14}$$

Then

$$\mathcal{M} = \operatorname{Im} \begin{bmatrix} I \\ X \end{bmatrix} \tag{3.15}$$

for a positive semidefinite solution X of (3.11). *Conversely, assume X is a positive semidefinite solution of* (3.11) *and let $\mathcal{M}$ be as in* (3.15). *Then $\mathcal{M} \cap \mathcal{R}(T, \mathbb{C}_{out}) = P^*\mathcal{N}$ for some A-invariant subspace $\mathcal{N}$ contained in $\mathcal{V}_>$.*

Problem 3.21. *Obtain a description of positive semidefinite solutions of* (3.11) *analogous to that of Theorem* 3.20, *in a situation when A is not invertible.*

A positive semidefinite solution X_0 of (3.11) is called *stably positive semidefinite* if for every $\varepsilon > 0$ there is a $\delta > 0$ such that $\|A - A_1\| + \|B - B_1\| + \|Q - Q_1\| + \|R - R_1\| < \delta$ and $R_1^* = R_1$, $Q_1 \geq 0$ imply that the algebraic Riccati equation

$$X = A_1^* X A_1 + Q_1 - A_1^* X B_1 (R_1 + B_1^* X B_1)^{-1} B_1^* X A_1$$

has a positive semidefinite solution X_1 such that $\|X_0 - X_1\| < \varepsilon$.

The following theorem describes stably positive semidefinite solutions.

Theorem 3.22. *Assume (A, B) is controllable, A is invertible, $Q \geq 0$ and $R > 0$. Then the only stably positive semidefinite solution of* (3.11) *is the maximal solution.*

Problem 3.23. *Relax the hypothesis that A is invertible in Theorem* 3.22.

Problem 3.24. *Describe the degree of stability of the maximal solutions of* (3.10) *and of* (3.11).

As noted at the beginning of this subsection, the existence of positive semidefinite solutions of (3.10) is guaranteed if (A, B, C) is merely output-stabilizable. Thus:

Problem 3.25. *Characterize stable positive semidefinite solutions of* (3.10) *and determine their degrees of stability, assuming only output-stabilizability.*

Finally, development of stability results for real positive semidefinite symmetric solutions of equations (3.10) and (3.11) with real matrix coefficients is an open problem as well.

3.5. Application to minimal factorization of positive semidefinite matrix functions

Consider the proper rational $m \times m$ matrix function

$$W(\lambda) = D + C(\lambda - A)^{-1} B, \tag{3.16}$$

where A is an $n \times n$ matrix, B is an $n \times m$ matrix, C is an $m \times n$ matrix and finally, D is an $m \times m$ matrix. (All of these are considered as complex matrices here.) In this section we shall assume throughout that $W(\lambda)$ takes selfadjoint values on the real line. In particular, this implies that $D = W(\infty)$ is selfadjoint. When the realization (6.1) is minimal we can say more about the matrices in the realization by using the state space isomorphism theorem.

Proposition 3.26. *Let $W(\lambda) = D + C(\lambda - A)^{-1} B$ be a realization of a rational matrix function. If $D = D^*$ and there is an invertible hermitian matrix H such that*

$$HA = A^* H, \qquad HB = C^*, \tag{3.17}$$

then $W(\lambda)$ takes hermitian values on the real line.

Conversely, if the realization (3.16) *is minimal and* $W(\lambda)$ *takes hermitian values on the real line then* $D = D^*$ *and there is a unique invertible hermitian matrix* H *such that* (3.17) *holds.*

This proposition allows us to use the results and methods of indefinite inner product spaces to study selfadjoint rational matrix functions.

As a first result let us mention a factorization theorem that was first derived in [48]. To state the theorem, we introduce the matrix $A^\times = A - BD^{-1}C$. Recall that $W(\lambda)^{-1} = D - C(\lambda - A^\times)^{-1}B$.

Theorem 3.27. *Let* $W(\lambda) = D + C(\lambda I_n - A)^{-1}B$ *be a minimal realization of a rational matrix function having hermitian values on the real line, and assume that* D *is positive definite. Let* H *be the unique invertible hermitian matrix for which* (3.17) *holds. Then for any* A*-invariant maximal* H*-nonnegative subspace* $\mathcal{M}$ *and any* $A^\times$*-invariant maximal* H*-nonpositive subspace* $\mathcal{M}^\times$ *we have* $\mathcal{M} \dotplus \mathcal{M}^\times = \mathbb{C}^n$*. Thus to any such pair of subspaces corresponds a minimal factorization of* $W(\lambda)$*:*

$$W(\lambda) = W_1(\lambda)W_2(\lambda),$$

where

$$W_1(\lambda) = D^{\frac{1}{2}} + C(\lambda - A)^{-1}\Pi BD^{-\frac{1}{2}}, \quad W_2(\lambda) = D^{\frac{1}{2}} + D^{-\frac{1}{2}}C(I - \Pi)(\lambda - A)^{-1}B.$$

Here Π *is the projection on* $\mathcal{M}^\times$ *along* $\mathcal{M}$*.*

Theorem 3.27 illustrates the key role of invariant subspaces in study of minimal factorizations of rational matrix functions. This approach to minimal factorizations originates with [5], and has been extensive used since then; see, in particular, the books [4, 22].

The case where $W_2(\lambda) = W_1^*(\lambda)$ is of particular interest. Here, $W_1^*(\lambda)$ is the rational matrix function $W_1(\bar{\lambda})^*$. Clearly, if this holds we must have that $W(\lambda)$ takes positive semidefinite values on the real line. For this reason we first present a result concerning the poles and zeros of a rational matrix function having this property. The proof of this result can be found in [48], see also [23].

Proposition 3.28. *Let* $W(\lambda)$ *be a rational matrix function having positive semidefinite values on the real line. Let* $W(\lambda) = D + C(\lambda - A)^{-1}B$ *be a minimal realization of* $W(\lambda)$*. Then for each real pole* λ_0*, i.e. each real eigenvalue* λ_0 *of* A*, the partial multiplicities of* A *at* λ_0 *(which are the partial pole multiplicities of* $W(\lambda)$ *at* λ_0*) are even, and the sign characteristic of the pair* (H, A) *consists of* $+1$*'s only.*

For each real zero λ_0*, i.e., each real eigenvalue of* $A^\times$ *the partial multiplicities of* $A^\times$ *at* λ_0 *are even, and the sign characteristic of the pair* $(H, A^\times)$ *consists of* -1*'s only.*

This proposition, combined with Theorem 3.4 tells us that if (3.16) is a minimal realization of a positive semidefinite rational matrix function, then A has an invariant subspace $\mathcal{M}$ such that $H(\mathcal{M}) = \mathcal{M}^\perp$, and $A^\times$ has an invariant subspace $\mathcal{M}^\times$ such that $H(\mathcal{M}^\times) = \mathcal{M}^{\times\perp}$. Also Theorem 3.2 describes all such subspaces. We can employ this to obtain the following theorem.

Theorem 3.29. *Let $W(\lambda) = D + C(\lambda - A)^{-1}B$ be a minimal realization of a selfadjoint rational matrix function, and let H be the unique invertible hermitian matrix such that (3.17) holds. Assume that D is positive definite. Then the following are equivalent.*

(i) *$W(\lambda)$ has positive semidefinite values on the real line,*

((ii) *both A and $A^\times$ have only even partial multiplicities at their real eigenvalues,*

(iii) *there exists an A-invariant subspace $\mathcal{M}$ such that $H(\mathcal{M}) = \mathcal{M}^\perp$, and there exists an $A^\times$-invariant subspace $\mathcal{M}^\times$ such that $H(\mathcal{M}^\times) = \mathcal{M}^{\times\perp}$.*

Let W be a positive semidefinite rational matrix function with a minimal realization given by

$$W(\lambda) = I + C(\lambda - A)^{-1}B, \tag{3.18}$$

and let

$$W(\lambda) = L(\bar{\lambda})^* L(\lambda) \tag{3.19}$$

be a minimal symmetric factorization where L has a minimal realization given by $L(\lambda) = I + C_1(\lambda - A_1)^{-1}B_1$. This factorization is called an (*unconditionally*) *α-stable symmetric factorization* if there are positive constants ε and K such that whenever the triple of matrices (A', B', C') satisfies

$$\|A - A'\| + \|B - B'\| + \|C - C'\| < \varepsilon \tag{3.20}$$

and is such that

$$W'(\lambda) = I + C'(\lambda I_n - A')^{-1}B' \tag{3.21}$$

is hermitian on $\mathbb{R}$, then $W'(\lambda)$ has a minimal symmetric factorization

$$W'(\lambda) = L'(\bar{\lambda})^* L'(\lambda) \tag{3.22}$$

where $L'(\lambda) = I + C_1'(\lambda I_n - A_1')^{-1}B_1'$ with

$$\|A_1 - A_1'\| + \|B_1 - B_1'\| + \|C_1 - C_1'\| \le K \cdot (\|A - A'\| + \|B - B'\| + \|C - C'\|)^{\frac{1}{\alpha}}. \tag{3.23}$$

Note that it follows that $W'(\lambda)$ is a positive semidefinite rational matrix function.

The factorization (3.19) is called a (*conditionally*) *α-stable symmetric factorization* if for any *positive semidefinite* W' as in (3.21) for which (3.20) holds, we have (3.22) and (3.23). Observe that here we restrict attention to positive semidefinite perturbations only. It can be seen rather easily that both unconditional and conditional α-stability are independent of the particular minimal realization one starts with.

Theorem 3.30. *Let $W(\lambda)$ be positive semidefinite. Then the following holds:*

(a) There exists an unconditionally α-stable symmetric factorization for some α if and only if W has no real poles and zeros. In that case there exists a 1-stable symmetric factorization.

(b) If W has no real poles and zeros then the minimal symmetric factorization (3.19) is α-stable if and only if

(i) *for each common pole* λ_0 *of* L *and* L^* *the geometric multiplicity of* λ_0 *as a pole of* W *is one, and if* k*, respectively* l*, denote the algebraic multiplicities of* λ_0 *as a pole of* L*, respectively as a pole of* L^**, then* $\gamma(k,\ k+l) \leq \alpha$,

(ii) *for each common zero* λ_0 *of* L *and* L^* *the geometric multiplicity of* λ_0 *as a zero of* W *is one, and if* k*, respectively* l*, denote the algebraic multiplicities of* λ_0 *as a zero of* L*, respectively as a zero of* L^**, then* $\gamma(k,\ k+l) \leq \alpha$.

Concerning conditional α-stability we can only state a partial result in the (generic) case where the geometric pole and zero multiplicities of W at real poles and zeros, respectively, is one.

Theorem 3.31. *Let* $W(\lambda)$ *be positive semidefinite, and assume for every real pole and zero of* W *the geometric multiplicity is one. Let* $\lambda_1, \dots, \lambda_r$ *be the real poles of* W *and let* $\mu_1, \dots, \mu_s$ *be the real zeros of* W *with algebraic multiplicities* $m_1 \dots, m_r$ *and* $n_1, \dots, n_s$*, respectively. Let*

$$\alpha_0 = \max(2, m_1 - 1, \dots, m_r - 1, n_1 - 1, \dots, n_s - 1),$$

or $\alpha_0 = 1$ *if* W *does not have real poles or zeros. Then there exists a conditionally* α_0*-stable symmetric factorization of* W *and no symmetric minimal factorization is* α*-stable with* $\alpha < \alpha_0$.

The factorization (3.19) *is* α_0*-stable if and only if*

(i) *for each common non-real pole* λ_0 *of* L *and* L^* *the geometric multiplicity of* λ_0 *as a pole of* W *is one, and if* k*, respectively* l*, denote the algebraic multiplicities of* λ_0 *as a pole of* L*, respectively as a pole of* L^**, then* $\gamma(k,\ k+l) \leq \alpha_0$,

(ii) *for each common non-real zero* λ_0 *of* L *and* L^* *the geometric multiplicity of* λ_0 *as a zero of* W *is one, and if* k*, respectively* l*, denote the algebraic multiplicities of* λ_0 *as a zero of* L*, respectively as a zero of* L^**, then* $\gamma(k,\ k+l) \leq \alpha_0$.

Proofs of Theorems 3.30 and 3.31. As these results are new, we outline the proofs. First one observes that the factorization (3.19) is (conditionally, resp., unconditionally) α-stable if and only if the corresponding A-invariant subspace $\mathcal{M}$ and the corresponding $A^\times$-invariant subspace $\mathcal{M}^\times$ are (conditionally, resp., unconditionally) neutrally α-stable. Compare, e.g., the proof of Theorem 2.5 in [53], where this argument is exposed in detail. Using this observation Theorems 3.30 and 3.31 are proved in essentially the same way as Theorems 3.12 and 3.14 (see [57], Lemma 6.8 and the proofs of Theorems 3.27 and 3.30). □

Problem 3.32. *Characterize conditional* α*-stability of factorizations* (3.19) *in the non-generic case (i.e., not covered in Theorem* 3.31*).*

3.6. Application to transport theory

In this section we consider the following boundary value problem on $[0, \infty)$:

$$(T\psi)'(x) = -A\psi(x) \tag{3.24}$$

$$\lim_{x \downarrow 0} P_+\psi(x) = \phi_+; \quad \|\psi(x)\| \text{ bounded as } x \to \infty. \tag{3.25}$$

Here T is an $n \times n$ selfadjoint matrix with Ker $T = 0$, A is an $n \times n$ positive semidefinite hermitian matrix, and finally, P_+ is the spectral projection of T corresponding to its positive eigenvalues, and ϕ_+ is a given vector in $\operatorname{Im} P_+$.

Solutions of (3.24), (3.25) are important in astrophysics and in the theory of transport of neutrons, in these cases, however, it is the infinite dimensional analogue of (3.24), (3.25) which plays a role (see [43, 29]). The finite dimensional problem is relevant in the case where the scattering of particles is restricted to a finite number of directions (see, e.g., [76]). It was considered from several points of view in [61, 19].

Here we consider the question of what happens to bounded solutions if A and T are subject to perturbations. The infinite dimensional case was treated in [58], there only perturbations of A were allowed. Because of the technicalities involved in the infinite dimensional case we restrict our attention here to the finite dimensional situation (which, by the way, was also treated in [58]).

In treating this problem obviously the matrix $T^{-1}A$ plays an essential role. We shall denote by P_0 the spectral projection of $T^{-1}A$ with respect to $\{0\}$, by P_p its spectral projection with respect to $(0, \infty)$, and by P_m its spectral projection corresponding to $(-\infty, 0)$. The images of these projections will be denoted by H_0, H_p and H_m, respectively. We shall also denote by P_- the spectral projection of T corresponding to its negative eigenvalues and we denote $\operatorname{Im} P_- = H_-$, $\operatorname{Im} P_+ = H_+$.

As $T^{-1}A$ is nonnegative in the indefinite inner product given by T (in short, T-nonnegative), we first list some consequences of this fact; these follow easily by considering the canonical form (3.1), (3.2):

(i) all eigenvalues of $T^{-1}A$ are real,

(ii) the partial multiplicities of $T^{-1}A$ corresponding to non-zero eigenvalues are all one, and if $\lambda > 0$ (resp., $\lambda < 0$, is an eigenvalue of $T^{-1}A$, then $\langle Tx, x\rangle > 0$ (resp., < 0) for all $x \in \operatorname{Ker}\,(T^{-1}A - \lambda)$,

(iii) the partial multiplicities of $T^{-1}A$ with respect to the zero eigenvalue (if any) of $T^{-1}A$ are either 1 or 2, and the signs in the sign characteristic of the pair $(T^{-1}A, T)$ corresponding to Jordan blocks of order two (if any) are all $+1$.

Let us denote, as before, the spectral subspace of $T^{-1}A$ corresponding to its negative, zero, and positive eigenvalues, respectively, by H_m, H_0 and H_p, respectively. We shall say that the pair $(T^{-1}A, T)$ satisfies the *positive sign condition* (resp. *negative sign condition*) if Ker $T^{-1}A$ is T-nonnegative (resp., T-nonpositive). So, if $(T^{-1}A, T)$ satisfies either the positive or the negative sign condition then it certainly satisfies the sign condition. It can be shown that the pair $(T^{-1}A, T)$ satisfies the positive sign condition if and only if

$$\mathbb{C}^n = H_p \dot{+} \operatorname{Ker}\, T^{-1}A \dot{+} H_-. \tag{3.26}$$

In turn, this is equivalent to $H_p \dot{+} \operatorname{Ker}\, T^{-1}A$ being the unique $T^{-1}A$-invariant maximal T-nonnegative subspace. Likewise, $(T^{-1}A, T)$ satisfies the negative sign condition if and only if

$$\mathbb{C}^n = H_m \dot{+} \text{Ker } T^{-1}A \dot{+} H_+,$$

which, in turn, is equivalent to $\mathbb{C}^n = H_m \dot{+} \text{Ker } T^{-1}A$ being the unique $T^{-1}A$-invariant maximal T-nonpositive subspace.

Returning to equation (3.24) with boundary conditions (3.25), let us first describe the bounded solutions. These are given by

$$\psi(x) = e^{-xT^{-1}A}\phi_p + \phi_0, \qquad 0 < x < \infty, \tag{3.27}$$

where $\phi_p \in H_p$, $\phi_0 \in \text{Ker } A$ and $P_+(\phi_p + \phi_0) = \phi_+$. We have the following result (see [58], Corollary 3.2.2).

Proposition 3.33. *Given $\phi_+ \in H_+$ there exists a unique bounded solution of* (3.24), (3.25) *if and only if the pair $(T^{-1}A, T)$ satisfies the positive sign condition, or equivalently, if $\langle Tx, x\rangle \geq 0$ for all $x \in$ Ker A.*

Our main result concerns stability of the unique bounded solution under perturbations of T and A in case the positive sign condition holds; the statement is as follows:

Theorem 3.34. *Let $T = T^*$ be an invertible $n \times n$ matrix, and let $A \geq 0$ be an $n \times n$ matrix. Assume that $\langle Tx, x\rangle \geq 0$ for all $x \in$ Ker A, i.e., that the pair $(T^{-1}A, T)$ satisfies the positive sign condition. Let $T_k = T_k^*$ and $A_k \geq 0$ be such that T_k is invertible, $A_k \to A$, $T_k \to T$. Then for every bounded solution $\psi(x)$ of* (3.24), (3.25) *and for every sequence $\phi_{k,+} \in \text{Im}\, P_{k+}$, $k = 0, 1, 2, \ldots$, such that $\lim_{k\to\infty} \|\phi_{k,+} - \phi_+\| = 0$, there is a bounded solution $\psi_k(x)$ of*

$$(T_k\psi)'(x) = -A_k\psi(x), \quad \lim_{x\downarrow 0} P_{k+}\psi(x) = \phi_{k,+},$$

such that $\lim_{k\to\infty} \sup_{0<x\leq x_0} \|\psi_k(x) - \psi(x)\| = 0$ for every $x_0 > 0$.

Problem 3.35. *Is the converse true, i.e, if the positive sign condition is not satisfied, then no bounded solution is stable?*

Problem 3.36. *Find degrees of stability of bounded solutions $\psi(x)$ of* (3.24), (3.25), *under the hypotheses of Theorem* 3.34.

The proof of Theorem 3.34 is based on the observations preceding it, together with the fact that, according to Theorem 3.2, under assumption of the positive sign condition the unique $T^{-1}A$-invariant maximal T-nonnegative subspace is stable under perturbations of T and A.

The infinite dimensional version of this theorem justifies a numerical procedure for calculating the solution of (3.24), (3.25) that is in use in astrophysics. In that case $A = I - B$ with B a compact operator. The numerical procedure involves replacing B by a finite rank operator B_k, while keeping T fixed. The result in [58] (analogous to Theorem 3.34) then applies and tells us that the solution computed using this procedure will converge to the actual solution of the problem uniformly on compact intervals when $B_k \to B$.

In [46] robustness of solutions of the boundary value problem (3.24), (3.25) is studied for the physically interesting case when A is a matrix (or, in the infinite dimensional case, a compact perturbation of the identity operator) whose real part $\operatorname{Re} A$ is positive semidefinite and the condition $\operatorname{Ker} \operatorname{Re} A = \operatorname{Ker} A$ is satisfied.

4. Matrix decompositions

In this section, we study robustness properties of several matrix decompositions. First, we consider variants of the well-known polar decomposition. We do so in the case where the inner product space we work in is a finite dimensional Hilbert space, as well as for the case where the inner product is genuinly indefinite. Next, we discuss the Cholesky factorization of a positive semidefinite matrix. Finally, we consider the absolute value of a matrix and its singular value decomposition.

4.1. Polar decompositions

4.1.1. THE DEFINITE CASE It is well known that every $n \times n$ matrix X over F (where $F = \mathbb{C}$ or $F = \mathbb{R}$) admits a decomposition $X = UA$, where U is unitary (orthogonal if $F = \mathbb{R}$) and A is positive definite (and symmetric if $F = \mathbb{R}$).

We shall consider a more general situation. Let H be an invertible selfadjoint and positive definite $n \times n$ matrix over F. An *H-polar decomposition* of a matrix $X \in F^{n \times n}$ is, by definition, a factorization of the form

$$X = UA, \tag{4.1}$$

where U is H-unitary (i.e., $U^*HU = H$) and A is H-selfadjoint (i.e., $HA = A^*H$). This definition is more general than the standard definition in that we allow A to be H-selfadjoint (not just H-positive semidefinite) and the scalar product need not be the standard one.

It is a standard result that an H-polar decomposition always exists. Moreover, in this case one can take A having nonnegative spectrum; an H-polar decomposition of the form $X = UA$ with this property of A will be called a *nonnegative* H-polar decomposition. The factor A in a nonnegative H-polar decomposition is unique and coincides with $\sqrt[H]{X^H X}$. The factor U in a nonnegative H-polar decomposition is unique if and only if X is nonsingular.

We note the following perturbation bounds on the factors of the nonnegative polar decomposition for invertible matrices, assuming $H = I$. Let X and Y be invertible matrices with the nonnegative polar decompositions $X = UA$ and $Y = VB$. Then

$$\|U - V\| \leq \frac{2}{\|X^{-1}\|^{-1} + \|Y^{-1}\|^{-1}} \|X - Y\|; \tag{4.2}$$

$$\|A - B\| \leq \left(1 + \frac{2\|X\|}{\|X^{-1}\|^{-1} + \|Y^{-1}\|^{-1}}\right) \|X - Y\|. \tag{4.3}$$

Formula (4.2) is well known (see [41]; also [6], Theorem VII.5.1). As shown in [41], (4.2) is the best possible in the sense that the bound can be achieved. Formula (4.3) is an easy consequence of (4.2).

Next, we consider stability of polar decompositions $X = UA$, which are not necessarily nonnegative.

The polar decomposition (4.1) is called *stable* if for every $\varepsilon > 0$ there is $\delta > 0$ such that every pair of matrices (Y, G), where $Y \in F^{n\times n}$ and $G \in F^{n\times n}$ is Hermitian, admits a G-polar decomposition $Y = VB$ with $\|U-V\|+\|A-B\| < \varepsilon$, as soon as $\|Y-X\|+\|H-G\| < \delta$. Restricting this definition to perturbations of Y only, in other words, assuming $G = H$, we obtain the definition of *H-stability* of the polar decomposition. The polar decomposition is called *Lipschitz stable* (resp. *H-Lipschitz stable*) if there exist positive constants δ and K such that every $Y \in F^{n\times n}$ admits a G- (resp. H-) polar decomposition $Y = VB$ with $\|U-V\| + \|A-B\| \leq K(\|X-Y\| + \|G-H\|)$ (resp. $\|U-V\| + \|A-B\| \leq K\|X-Y\|$) as soon as $\|X-Y\| + \|G-H\| \leq \delta$ (resp. $\|X-Y\| \leq \delta$). Clearly, stability implies H-stability, Lipschitz stability implies stability and H-Lipschitz stability, and H-Lipschitz stability implies H-stability. The following result, proved in [44], shows in particular that these a priori distinct notions of stability are in fact equivalent.

Theorem 4.1. *(a) $F = \mathbb{C}$. There exist H-stable H-polar decompositions of X if and only if X is nonsingular. In this case, the following statements are equivalent for an H-polar decomposition $X = UA$:*

(i) *The decomposition is H-stable.*
(ii) *The decomposition is Lipschitz stable.*
(iii) *The H-selfadjoint matrices A and $-A$ have no common eigenvalues.*

(b) $F = \mathbb{R}$. There exist H-stable H-polar decompositions of X if and only if $\dim \operatorname{Ker} X \leq 1$. In this case, the following statements are equivalent for an H-polar decomposition $X = UA$:

(iv) *The decomposition is H-stable.*
(v) *The decomposition is Lipschitz stable.*
(vi) *The H-selfadjoint matrices A and $-A$ have no common nonzero eigenvalues.*

4.1.2. The indefinite case Now let H be invertible and selfadjoint. We shall assume that H is genuinely indefinite, i.e., H has at least one positive and one negative eigenvalue. As before, consider the (now indefinite) scalar product $[x, y] = \langle Hx, y\rangle$. Recall (see Section 3) that A is called H-selfadjoint if $HA = A^*H$ and U is H-unitary if $U^*HU = H$. As in the definite case, a decomposition $X = UA$, where A is H-selfadjoint and U is H-unitary, will be called an *H-polar decomposition* of X. In contrast to the definite case such decompositions need not exist for an arbitrary $n \times n$ matrix over F. Necessary and sufficient conditions for existence of an H-polar decomposition are given in [8].

Essentially, to prove the existence of and to actually construct an H-polar decomposition of a given $n\times n$ matrix X, one needs to find an H-selfadjoint matrix A such that

$$\begin{aligned} X^H X &= A^2 \\ \operatorname{Ker} X &= \operatorname{Ker} A, \end{aligned} \tag{4.4}$$

where $\operatorname{Ker} B$ stands for the null space of a matrix B [10, 8]. Once A is known, the map $Au \mapsto Xu$ is an H-isometry from the range $\operatorname{Im} A$ of A onto the range $\operatorname{Im} X$ of X, which can be extended to an H-unitary matrix U as a result of Witt's theorem [1, 9].

It turns out (see [45]) that, for the case $F = \mathbb{C}$, there exist stable H-polar decompositions of X if and only if $\sigma(X^H X) \cap (-\infty, 0] = \emptyset$. In the real case, this condition is only sufficient but not necessary (cf. Theorem 4.1). Nevertheless, for $F = \mathbb{R}$ we give necessary and sufficient conditions for the existence of stable H-polar decompositions of X. The results below are proved in [45].

The strategy in proving the first stability result of this section is to construct an H-selfadjoint matrix A satisfying (4.4) that depends continuously on X, if possible. The result is as follows

Theorem 4.2. *Let X be an $n \times n$ matrix over F such that $\sigma(X^H X) \cap (-\infty, 0] = \emptyset$. Then there exist an H-selfadjoint matrix A satisfying* (4.4), *an H-unitary matrix U satisfying $X = UA$ and constants $\delta, M > 0$, depending on X, A, H, and U only, such that for any pair of $n \times n$ matrices (Y, G) over F with G nonsingular selfadjoint and $\|X - Y\| + \|G - H\| < \delta$ there exists a G-polar decomposition $Y = VB$ of X satisfying*

$$\|A - B\| + \|U - V\| \leq M\,[\|X - Y\| + \|H - G\|]\,. \tag{4.5}$$

Moreover, such an A can be chosen with the additional property that $\sigma(A) \cap (-\infty, 0] = \emptyset$.

Conversely, let X be an $n \times n$ matrix over F having an H-polar decomposition and such that one of the following three conditions are satisfied:

(α) *$X^H X$ has negative eigenvalues;*

(β) $\sigma(X^H X) \cap (-\infty, 0] = \{0\}$ *and* $\operatorname{Ker} X^H X \neq \operatorname{Ker}(X^H X)^n$*;*

(γ) $\sigma(X^H X) \cap (-\infty, 0] = \{0\}$ *and* $\operatorname{Ker}(X^H X)^n = \operatorname{Ker} X^H X \neq \operatorname{Ker} X$.

Then in every neighborhood of X there is an $n \times n$ matrix Y over F such that Y does not have an H-polar decomposition. Moreover, Y can be chosen so that $Y^H Y$ does not have a G-selfadjoint square root for any invertible selfadjoint matrix G.

The case which is left out of the above theorem, namely, when $\sigma(X^H X) \cap (-\infty, 0] = \{0\}$ and the subspace $\operatorname{Ker} X^H X = \operatorname{Ker}(X^H X)^n = \operatorname{Ker} X$, will be considered next. In contrast with the above theorem, in this case perturbations of X that do not admit H-polar decompositions may not exist. Nevertheless, in many cases there are no stable H-polar decompositions.

An H-polar decomposition $X = UA$ is called *H-stable* if for every $\varepsilon > 0$ there is $\delta > 0$ such that every matrix $Y \in F^{n\times n}$ admits an H-polar decomposition $Y = VB$ with $\|U - V\| + \|A - B\| < \varepsilon$, as soon as $\|Y - X\| < \delta$. Note that this is a weaker condition than (4.5).

Theorem 4.3. *Assume that $X \in F^{n\times n}$ admits an H-polar decomposition, and assume furthermore that* $\dim \operatorname{Ker} X \geq 1$ *in the complex case or* $\dim \operatorname{Ker} X \geq 2$ *in the real case. Then no H-polar decomposition of X is H-stable.*

The results of the preceding two theorems show that no H-polar decomposition $X = UA$ is H-stable as soon as $X^H X$ has negative or zero eigenvalues, with the possible exception of the situation when $F = \mathbb{R}$, $\sigma(X^H X) \cap (-\infty, 0] = \{0\}$, the dimension of the kernel of X is one, and $\operatorname{Ker} X^H X = \operatorname{Ker} (X^H X)^n = \operatorname{Ker} X$. This exceptional situation requires special consideration. By analogy with the Hilbert space (see [44]) we expect here H-stability of H-polar decompositions, and this turns out to be the case indeed.

Theorem 4.4. *Let $F = \mathbb{R}$. Assume that X admits H-polar decomposition and $\operatorname{Ker} X^H X = \operatorname{Ker} (X^H X)^n = \operatorname{Ker} X$ is one-dimensional. Then there exists an H-stable H-polar decomposition $X = UA$. Moreover, every such H-polar decomposition is* Lipschitz H-stable, *i.e., every Y sufficiently close to X admits an H-polar decomposition $Y = VB$ with $\|U - V\| + \|A - B\| \leq C\|X - Y\|$, where the positive constant C is independent of Y.*

4.2. Cholesky decompositions

It is well known that every positive semidefinite $n \times n$ matrix A with entries in F ($F = \mathbb{C}$ or $F = \mathbb{R}$) admits a factorization

$$A = R^* R, \tag{4.6}$$

where R is an upper triangular $n \times n$ matrix with entries in F and such that the diagonal entries of R are all real nonnegative. Such factorizations will be called *Cholesky decompositions.* Note that a Cholesky decomposition of a given positive semidefinite matrix is unique if A is positive definite, but in general it is not unique.

A criterion for uniqueness of Cholesky decompositions will be given. For a positive semidefinite $n \times n$ matrix A let $\alpha_j(A)$ be the rank of the $j \times j$ upper left block of A; $j = 1, \cdots, n$. The inequalities

$$\alpha_j(A) \leq \alpha_{j+1}(A) \leq \alpha_j(A) + 1, \qquad j = 1, \cdots, n-1 \tag{4.7}$$

are evident.

Theorem 4.5. *A Cholesky decomposition* (4.6) *of A is unique if and only if either A is invertible or*

$$\alpha_{j_0}(A) = \alpha_{j_0+1}(A) = \cdots = \alpha_n(A),$$

where j_0 is the smallest index such that $\alpha_{j_0}(A) < j_0$.

Proof. We may assume that A is singular to start with. If $j_0 = 1$, then the top left entry of A is zero, and unless A is the zero matrix, a Cholesky decomposition of A is not unique:

$$\begin{bmatrix} 0 & 0 \\ 0 & B \end{bmatrix} = \begin{bmatrix} 0 & 0 \\ C_1^* & C_2^* \end{bmatrix} \begin{bmatrix} 0 & C_1 \\ 0 & C_2 \end{bmatrix}.$$

Here B is the lower right $(n-1) \times (n-1)$ block of A, and C_1, C_2 are any matrices of appropriate sizes such that C_2 is upper triangular and $B = C_1^* C_1 + C_2^* C_2$. Clearly, unless $B = 0$, the choice of C_1 and C_2 subject to the above conditions is not unique. This proves the theorem in the case $j_0 = 1$.

Assume now $j_0 > 1$. Write

$$A = \begin{bmatrix} A_1 & A_2 \\ A_2^* & A_3 \end{bmatrix}, \tag{4.8}$$

where A_1 is the (invertible by the definition of j_0) $(j_0 - 1) \times (j_0 - 1)$ upper left block of A. A consideration of the Schur complement of A_1, namely the formula

$$\begin{bmatrix} I & 0 \\ -A_2^*A_1^{-1} & I \end{bmatrix} \begin{bmatrix} A_1 & A_2 \\ A_2^* & A_3 \end{bmatrix} \begin{bmatrix} I & -A_1^{-1}A_2 \\ 0 & I \end{bmatrix} = \begin{bmatrix} A_1 & 0 \\ 0 & A_3 - A_2^*A_1^{-1}A_2 \end{bmatrix}, \tag{4.9}$$

shows that we can replace A by

$$\begin{bmatrix} A_1 & 0 \\ 0 & A_3 - A_2^*A_1^{-1}A_2 \end{bmatrix}$$

without loss of generality. Since A_1 is invertible, it is easy to see that every Cholesky factorization of

$$\begin{bmatrix} A_1 & 0 \\ 0 & A_3 - A_2^*A_1^{-1}A_2 \end{bmatrix}$$

must be of the form

$$\begin{bmatrix} A_1 & 0 \\ 0 & A_3 - A_2^*A_1^{-1}A_2 \end{bmatrix} = \begin{bmatrix} R_1^* & 0 \\ 0 & R_2^* \end{bmatrix} \begin{bmatrix} R_1 & 0 \\ 0 & R_2 \end{bmatrix}$$

for some upper triangular matrices R_1 and R_2, with R_1 having the size $(j_0 - 1) \times (j_0 - 1)$. As the $j_0 \times j_0$ upper left block of A is singular, the top left entry of $A_3 - A_2^*A_1^{-1}A_2$ is zero. We have reduced the proof to the already considered case when $j_0 = 1$ (as applied to the matrix $A_3 - A_2^*A_1^{-1}A_2$). □

A Cholesky decomposition (4.6) is called *robust* if for every $\varepsilon > 0$ there is a $\delta > 0$ such that $\|A - B\| < \delta$ and B positive semidefinite with entries in F implies the existence of a Cholesky decomposition $B = S^*S$ such that $\|S - R\| < \varepsilon$; (4.6) is called *α-robust* if there is a positive constant K such that every positive semidefinite B (with entries in F) sufficiently close to A admits a Cholesky decomposition $B = S^*S$ in which $\|S - R\| < K\|B - A\|^{\frac{1}{\alpha}}$. In the case of 1-robustness, we simply say that the Cholesky decomposition is *Lipschitz robust.*

Theorem 4.6.

(i) *A Cholesky decomposition of A is Lipschitz robust if and only if A is positive definite.*
(ii) *A Cholesky decomposition of A is 2-robust if and only if it is unique, i.e., the conditions of Theorem 4.5 are satisfied;*
(iii) *In all other cases, no Cholesky decomposition of A is robust.*

Proof. Assume first that A is positive definite. Using induction on the size $n \times n$ of A, we may assume that the Cholesky decomposition $A_1 = R_1^*R_1$ of the

$(n-1) \times (n-1)$ upper left block A_1 of A is Lipschitz robust. Now let B be a positive definite matrix sufficiently close to A. Partition

$$B = \begin{bmatrix} B_1 & B_2 \\ B_2^* & B_3 \end{bmatrix},$$

where B_1 is $(n-1) \times (n-1)$ and B_3 is a scalar. Then B has the Cholesky decomposition

$$\begin{bmatrix} B_1 & B_2 \\ B_2^* & B_3 \end{bmatrix} = \begin{bmatrix} Q_1^* & 0 \\ B_2^* Q_1 & \sqrt{B_3 - B_2^* B_1^{-1} B_2} \end{bmatrix} \begin{bmatrix} Q_1 & Q_1^{*-1} B_2 \\ 0 & \sqrt{B_3 - B_2^* B_1^{-1} B_2} \end{bmatrix}, \tag{4.10}$$

and using the Lipschitz robustness of the Cholesky decomposition of A_1, this property of the Cholesky decomposition of A easily follows from (4.10).

Assume now that A is not invertible but the conditions of Theorem 4.5 are satisfied. If $A = 0$, then $\|R\| = \|B\|^{\frac{1}{2}}$ for any Cholesky decomposition $B = R^*R$ of a positive semidefinite matrix B, hence the 2-robustness of the Cholesky decomposition of the zero matrix follows. If $A \neq 0$, then using the Schur complement, as in the proof of Theorem 4.5, we may assume without loss of generality that

$$A = \begin{bmatrix} A_1 & 0 \\ 0 & 0 \end{bmatrix},$$

where A_1 is invertible of size $p \times p$, $1 \leq p < n$. Now all Cholesky decompositions of any positive semidefinite matrix

$$B = \begin{bmatrix} B_1 & B_2 \\ B_2^* & B_3 \end{bmatrix}$$

(here B_1 is $p \times p$) sufficiently close to A are given by the formula

$$\begin{bmatrix} B_1 & B_2 \\ B_2^* & B_3 \end{bmatrix} = \begin{bmatrix} Q_1^* & 0 \\ B_2^* Q_1 & Q_3^* \end{bmatrix} \begin{bmatrix} Q_1 & Q_1^{*-1} B_2 \\ 0 & Q_3 \end{bmatrix}, \tag{4.11}$$

where $B_3 - B_2^* B_1^{-1} B_2 = Q_3^* Q_3$ is a Cholesky decomposition of the positive semidefinite matrix $B_3 - B_2^* B_1^{-1} B_2$. Since this matrix is close to the zero matrix, 2-robustness of the Cholesky decomposition of A follows from the already proved 2-robustness of the Cholesky decomposition of the zero matrix, and from the Lipschitz robustness of the Cholesky decomposition of A_1.

Finally, assume that $A = \tilde{R}^* \tilde{R} = R^* R$ are two Cholesky decompositions of A with $\tilde{R} \neq R$. For $\varepsilon > 0$ let $A(\varepsilon) = \left(\tilde{R} + \varepsilon I\right)^* \left(\tilde{R} + \varepsilon I\right)$. Then this is the unique Cholesky decomposition of $A(\varepsilon)$, but $\tilde{R} + \varepsilon I$ does not converge to R when $\varepsilon \to 0$. Therefore, $A = R^* R$ cannot be robust. □

It is easy to see from the proof of Theorem 4.6 that in case a Cholesky decomposition of A is unique but A is not invertible, the decomposition is not α-robust for any $\alpha < 2$.

4.3. Absolute values of matrices and singular value decompositions

For a (real or complex) $n \times n$ matrix A the absolute value $|A|$ is defined as the unique positive semidefinite square root of A^*A. It is well known that $|A|$ is a Lipschitz function of A; more precisely, the inequality

$$\||A| - |B|\|_2 \leq \sqrt{2}\|A - B\|_2 \tag{4.12}$$

holds for every two $n \times n$ matrices A and B, where $\| \cdot \|_2$ indicates the Frobenius norm (see, e.g., formula (VII.39) in [6]).

Lemma 4.7. *If $|A|$ has n distinct eigenvalues, then every orthonormal eigenbasis of $|A|$ is Lipschitz stable with respect to A; in other words, if $f_1, \dots, f_n$ is an orthonormal basis of F^n ($F = \mathbb{R}$ or $F = \mathbb{C}$) consisting of eigenvectors of $|A|$, then there exists a constant $K > 0$ such that for every B sufficiently close to A there exists an orthonormal basis $g_1, \dots, g_n$ of eigenvectors of $|B|$ such that $\|f_j - g_j\| \leq K\|A - B\|$.*

If $|A|$ has less than n distinct eigenvalues, then no orthonormal eigenbasis of $|A|$ is stable with respect to A.

Proof. The first statement of the lemma can be easily derived from well-known results on perturbations of eigenspaces of hermitian matrices (for example, the $\sin \Theta$ theorem [15], see also Section VII.3 in [6]), combined with (4.12). The constant K turns out to be proportional to

$$(\min\{\lambda_j - \lambda_{j-1} : 2 \leq j \leq n\})^{-1},$$

where $\lambda_n > \lambda_{n-1} > \cdots > \lambda_1$ are the eigenvalues of $|A|$.

For the second statement, we shall assume (without essential loss of generality) that $n = 2$, and we may also assume that $|A| = I_2$ (if $A = 0$, a slight modification of the subsequent reasoning is required). Take any orthonormal eigenbasis, then we may assume, changing the basis in $\mathbb{C}^2$ if necessary, that it is the standard basis. In other words, it suffices to prove that the standard orthonormal basis is not stable with respect to A as an eigenbasis of $|A|$. For this, consider perturbations $A(\varepsilon) = A \begin{bmatrix} 1+\varepsilon & 0 \\ 0 & 1 \end{bmatrix}$, where $\varepsilon > 0$; then $|A(\varepsilon)| = \begin{bmatrix} 1+\varepsilon & 0 \\ 0 & 1 \end{bmatrix}$. Letting $\varepsilon \to 0$ we see that the only orthonormal eigenbasis of $|A|$ that might be stable is the standard one (modulo, of course, multiplications of the vectors by complex numbers of modulus one). Now consider perturbation $B(\varepsilon)$ of A such that $|B(\varepsilon)| = \begin{bmatrix} 1 & \varepsilon \\ \varepsilon & 1 \end{bmatrix}$, which has orthonormal eigenbasis $\frac{\sqrt{2}}{2}\begin{bmatrix} 1 \\ 1 \end{bmatrix}$, $\frac{\sqrt{2}}{2}\begin{bmatrix} 1 \\ -1 \end{bmatrix}$ (again, modulo multiplications of the vectors by complex numbers of modulus one). Letting $\varepsilon \to 0$ we see that none of these orthonormal eigenbases will converge to the standard basis. So, no orthonormal eigenbasis of $|A|$ is stable with respect to A. □

Recall that a decomposition $A = UDV$ is called a *singular value decomposition* of an $n \times n$ matrix A if U and V are unitary (resp., real orthogonal in the real case) and D is a positive semidefinite diagonal matrix with the diagonal entries ordered in the non-increasing order. Obviously, such a decomposition is not unique: the columns of U and of V^* form orthonormal bases of eigenvectors of $|A^*|$ and $|A|$, respectively, which allows for considerable freedom. Note also that the singular values $s_1(A) \geq \cdots \geq s_n(A)$ of A, i.e., the eigenvalues of $|A|$, are always well behaved with respect to perturbations:

$$\max\{|s_j(A) - s_j(B)| : 1 \leq j \leq n\} \leq \|A - B\|. \tag{4.13}$$

(This fact is well known, see, e.g., [20] or p.78 in [6].) Lemma 4.7 leads therefore to the following result:

Theorem 4.8. *Let $A = UDV$ be a singular value decomposition of an $n \times n$ matrix A. If A has n distinct singular values, then there exist a constant $K > 0$ such that every matrix B admits a singular value decomposition $B = U'D'V'$ for which*

$$\|U - U'\| + \|D - D'\| + \|V - V'\| \leq K \cdot \|A - B\|. \tag{4.14}$$

If not all singular values of A are distinct, then there exist $\varepsilon > 0$ and a sequence of matrices $\{B_m\}_{m=1}^{\infty}$ converging to A such that

$$\|U - U'\| + \|D - D'\| + \|V - V'\| > \varepsilon$$

for every singular value decomposition $B_m = U'_m D'_m V'_m$ of B_m, and for every $m = 1, 2, \cdots$.

Proof. Assume that A has distinct singular values. It follows from Lemma 4.7 and the discussion preceding the theorem that there exists $\varepsilon_0 > 0$ such that (4.14) holds for all B satisfying $\|A - B\| < \varepsilon_0$. However, for B satisfying $\|A - B\| \geq \varepsilon_0$ the inequality (4.14) clearly holds, with K replaced by $\max\{K, 4\varepsilon_0^{-1}\} + 1$.

If not all singular values of A are distinct, then the statement of Theorem 4.8 follows from the proof of Lemma 4.7. □

References

[1] E. Artin. *Geometric Algebra.* Interscience Publishers, New York, 1957.

[2] J.A. Ball and J.W. Helton. The cascade decomposition of a given system vs the linear fractional decompositions of its transfer function. *Integral Equations and Operator Theory,* 5(1982), 341–385.

[3] H. Bart, I. Gohberg, and M.A. Kaashoek. Stable factorization of monic matrix polynomials and stable invariant subspaces. *Integral Equations and Operator Theory,* 1(1978), 496–517.

[4] H. Bart, I. Gohberg, and M.A. Kaashoek. *Minimal Factorization of Matrix and Operator Functions. Operator Theory: Advances and Applications* 1, Birkhäuser, Basel, 1979.

[5] H. Bart, I. Gohberg, M.A. Kaashoek, and P. Van Dooren. Factorizations of transfer functions. *SIAM J. Control and Optimization,* 18(1980), 675–696.

[6] R. Bhatia. *Matrix Analysis.* Springer Verlag, New York, etc., 1997.

[7] S. Bittanti, A.J. Laub, and J.C. Willems (eds.). *The Riccati Equation.* Springer Verlag, Berlin, 1991.

[8] Y. Bolshakov, C.V.M. van der Mee, A.C.M. Ran, B. Reichstein, and L. Rodman. Polar decompositions in finite dimensional indefinite scalar product spaces: General theory. *Linear Algebra and its Applications* **261**(1997), 91–141.

[9] Y. Bolshakov, C.V.M. van der Mee, A.C.M. Ran, B. Reichstein, and L. Rodman. Extension of isometries in finite dimensional indefinite scalar product spaces and polar decompositions. *SIAM Journal on Matrix Analysis and Applications* **18**(1997), 752–774.

[10] Y. Bolshakov and B. Reichstein. Unitary equivalence in an indefinite scalar product: An analogue of singular value decomposition. *Linear Algebra and its Applications,* **222**(1995), 155–226.

[11] R. Byers. Solving the algebraic Riccati equation with the matrix sign function. *Linear Algebra and its Applications,* 85(1987), 267–279.

[12] R. Byers, C. He, and V. Mehrmann. The matrix sign function method and the computation of invariant subspaces. *SIAM J. Matrix Anal. Appl.*, 18(1997), 615–632.

[13] F.M. Callier and J.L. Willems. Criterion for the convergence of the solution of the Riccati differential equation. *IEEE Trans. Automat. Control,* AC-26(1981), 1232–1242.

[14] S. Campbell and J. Daughtry. The stable solutions of quadratic matrix equations. *Proceedings Amer. Math. Soc.*, 74(1979), 13–23.

[15] C. Davis and W.M. Kahan. The rotation of eigenvectors by a perturbation III. *SIAM J. Numerical Anal.*, 7(1970), 1–46.

[16] J.C. Doyle, K. Glover, P.P. Khargonekar, and B. Francis. State-space solutions to standard H_2 and H_∞ control problems. *IEEE Trans. Autom. Control,* 16(1989), 831–847.

[17] T. Geerts. A necessary and sufficient condition for solvability of the linear-quadratic control problem without stability. *Systems & Control Letters,* 11(1988), 47–51.

[18] A.H.W. Geerts and M.L.J. Hautus. The output-stabilizable subspace and linear optimal control, in: *Robust Control of Linear Systems and Nonlinear Control,* MTNS-89 Proc., Vol. II, M.A. Kaashoek, J.H. van Schuppen and A.C.M. Ran, eds., Birkhäuser, Boston, 1990, 113–120.

[19] I. Gohberg and M.A. Kaashoek. The Wiener-Hopf method for the transport equation: a finite dimensional version. *Operator Theory: Advances and Applications* 51, Birkhäuser Verlag, Basel, 1991, 20–33.

[20] I. Gohberg and M.G. Krein. *Introduction to the Theory of Linear Non-selfadjoint Operators.* Amer. Math, Soc., 1969.

[21] I. Gohberg, P. Lancaster, and L. Rodman. Spectral analysis of selfadjoint matrix polynomials. *Annals of Mathematics,* 112(1980), 33–71.

[22] I. Gohberg, P. Lancaster, and L. Rodman. *Invariant Subspaces of Matrices with Applications.* J.Wiley & Sons, 1986.

[23] I. Gohberg, P. Lancaster, and L. Rodman. *Matrices and Indefinite Scalar Products. Operator Theory: Advances and Applications,* 8, Birkhäuser Verlag, Basel, 1983.

[24] I. Gohberg and A.C.M. Ran. On pseudo-canonical factorization of rational matrix functions. *Indagationes Mathematica N.S.* 4(1993), 51–63.

[25] I. Gohberg and L. Rodman. On distance between lattices of invariant subspaces of matrices. *Linear Algebra and its Applications*, 76(1986), 85–120.

[26] I. Gohberg and S. Rubinstein. Stability of minimal fractional decompositions of rational matrix functions. *Operator Theory: Advances and Applications* 18, I. Gohberg, ed., Birkhäuser Verlag, Basel, 1986, 249–270.

[27] G.H. Golub and C.F. Van Loan. *Matrix Computations* (2-nd edition). Johns Hopkins University Press, Baltimore, 1989.

[28] C.-H. Guo and P. Lancaster. Analysis and modification of Newton's method for algebraic Riccati equations. *Math. Comp.*, 67(1998), 1089–1105.

[29] W. Greenberg, C. van der Mee, and V. Protopopescu. *Boundary Value Problems in Abstract Kinetic Theory. Operator Theory: Advances and Applications* 23, Birkhäuser Verlag, Basel, 1987.

[30] M.A. Kaashoek, C.V.M. van der Mee, and L. Rodman. Analytic operator functions with compact spectrum, II. Spectral pairs and factorization. *Integral Equations and Operator Theory*, 5(1982), 791–827.

[31] T. Kato. *Perturbation Theory for linear Operators*, 2-nd edition. Springer Verlag, Berlin, 1976.

[32] C. Kenney and A.J. Laub. Matrix-sign algorithms for Riccati equations. *IMA J. Math. Control Inf.*, 9(1992), 331–344.

[33] V. Kucera. Algebraic Riccati equation: Hermitian and definite solutions. *The Riccati Equation*, S. Bittanti, A.J. Laub and J.C. Willems, eds., Springer, Berlin, 1991, 53–88.

[34] P. Lancaster, A.C.M. Ran, and L. Rodman. Hermitian solutions of the discrete algebraic Riccati equations. *Intern. J. Control*, 44(1986), 777–802.

[35] P. Lancaster and L. Rodman. *Algebraic Riccati Equations.* Oxford University Press, 1995.

[36] H. Langer. Factorization of operator pencils. *Acta Sci. Math.*, 38(1976), 83–96.

[37] H. Langer. Über eine Klasse polynomialer Scharen selbstadjungierter Operatoren im Hilbertraum I. *J. Funct. Anal.*, 12(1973), 13–29.

[38] H. Langer. Über eine Klasse polynomialer Scharen selbstadjungierter Operatoren im Hilbertraum II. *J. Funct. Anal.*, 16(1974), 221–234.

[39] H. Langer, A.C.M. Ran, and T.N.M. Temme. Nonnegative solutions of algebraic Riccati equations. *Linear Algebra and its Applications*, 261(1997), 317–352.

[40] A.J. Laub. Invariant subspace methods for the numerical solution of Riccati equations. *The Riccati Equation*, S. Bittanti, A.J. Laub, J.C. Willems, eds., Springer Verlag, Berlin, 1991, pp. 163–191.

[41] R.-C. Li. New perturbation bounds for the unitary polar factor. *SIAM J. Matrix Analysis and Appl.*, 16(1995), 327–332.

[42] A.S. Markus. *Introduction to the Spectral Theory of Polynomial Operator Pencils.* Amer. Math. Soc., Providence, Rhode Island, 1988, (translated from the Russian original, 1986).

[43] C.V.M. van der Mee. *Semigroup and Factorization Methods in Transport Theory.* Ph.D. Thesis, Free University Amsterdam, 1981.

[44] C.V.M. van der Mee, A.C.M. Ran, and L. Rodman. Stability of polar decompositions. *Glasnik Math.*, 35(2000), 137–148.

[45] C.V.M. van der Mee, A.C.M. Ran, and L. Rodman. Stability of selfadjoint square roots and polar decompositions in indefinite scalar product spaces. *Linear Algebra and its Applications*, 302–303(1999), 77–104.

[46] C.V.M. van der Mee, A.C.M. Ran, and L. Rodman. Stability of stationary transport equations with accretive collision operators; *Journal of Functional Analysis*, 174(2000), 478–512.

[47] V. Mehrmann. *The Autonomous Linear Quadratic Control Problem.* Lecture Notes in Control and Information Sciences, Vol. 163, Springer Verlag, Berlin, 1991.

[48] A.C.M. Ran. Minimal factorization of selfadjoint rational matrix functions. *Integral Equations and Operator Theory*, 5(1982), 850–869.

[49] A.C.M. Ran and L. Roozemond. On strong α-stability of invariant subspaces of matrices. *The Gohberg Anniversary Volume. Operator Theory: Advances and Applications* 40, H. Dym, S. Goldberg, M.A. Kaashoek, P. Lancaster, editors, Birkhäuser, 1989, 427–435.

[50] A.C.M. Ran and L. Rodman. On stability of invariant subspaces of matrices. *American Mathematical Monthly*, 97(1990), 809–823.

[51] A.C.M. Ran and L. Rodman. The rate of convergence of real invariant subspaces. *Linear and Algebra and its Applications*, 207(1994), 197–224.

[52] A.C.M. Ran and L. Rodman. Stability of invariant maximal semidefinite subspaces I. *Linear Algebra and its Applications*, 62(1984), 51–86.

[53] A.C.M. Ran and L. Rodman. Stability of invariant maximal semidefinite subspaces II. Applications: self-adjoint rational matrix functions, algebraic Riccati equations. *Linear Algebra and its Applications*, 63(1984), 133–173.

[54] A.C.M. Ran and L. Rodman. Stable real invariant semidefinite subspaces and stable factorizations of symmetric rational matrix functions. *Linear and Multilinear Algebra*, 22(1987), 25–55.

[55] A.C.M. Ran and L. Rodman. Stability of invariant Lagrangian subspaces I. *Topics in Operator Theory, Constantin Apostol Memorial Issue. Operator Theory: Advances and Applications* 32, Birkhäuser, Basel etc., (1988), 181–218.

[56] A.C.M. Ran, L. Rodman. Stability of invariant Lagrangian subspaces II. *The Gohberg anniversary collection. Operator Theory: Advances and Applications* 40, (H. Dym, S. Goldberg, M. A. Kaashoek, P. Lancaster, editors), Birkhäuser, Basel etc., 1989, 391–425.

[57] A.C.M. Ran and L. Rodman. Rate of stability of solutions of matrix polynomial and quadratic equations. *Integral Equations and Operator Theory*, 27(1997), 71–102.

[58] A.C.M. Ran and L. Rodman. Stability of solutions of the operator differential equation in transport theory. *Integral Equations and Operator Theory*, 8(1985), 75–118; Erratum, ibid, 894.

[59] A.C.M. Ran and L. Rodman. Stable solutions of real algebraic Riccati equations. *SIAM Journal of Control and Optimization*, 30(1992), 63–81.

[60] A.C.M. Ran and L. Rodman. Stable Hermitian solutions of discrete algebraic Riccati equations. *Mathematics of Control, Signals and Systems*, 5(1992), 165–193.

[61] A.C.M. Ran and L. Rodman. A matricial boundary value problem which appears in transport theory. *Journal of Mathematical Analysis and Applications*, 130(1988), 200–222.

[62] A.C.M. Ran and L. Rodman. Stability of invariant subspaces of matrices with applications. Report WS-488, Faculteit der Wiskunde en Informatica, Vrije Univertsiteit, Amsterdam, December 1997.

[63] A.C.M. Ran, L. Rodman, and A.L. Rubin. Stability index of invariant subspaces of matrices. *Linear and Multilinear Algebra*, 36(1993), 27–39.

[64] A.C.M. Ran, L. Rodman, and D. Temme. Stability of pseudospectral factorizations, *Operator Theory and Analysis, The M.A. Kaashoek Anniversary Volume. Operator Theory: Advances and Applications* 122 (H. Bart, I. Gohberg, A.C.M. Ran, editors), Birkhäuser Verlag, Basel, 2001, 359–383.

[65] A.C.M. Ran and D. Temme. Invariant semidefinite subspaces of dissipative matrices in an indefinite inner product space, existence, construction and uniqueness. *Linear Algebra and its Applications*, 212/213(1994), 169–214.

[66] A.C.M. Ran and R. Vreugdenhil. Existence and comparison theorems for algebraic Riccati equations for Continuous- and Discrete-Time Systems. *Linear Algebra Appl.*, 99(1988), 63–83.

[67] L. Rodman. Stable invariant subspaces modulo a subspace. *Operator Theory: Advances and Applications* 19, (H. Bart, I. Gohberg, M.A. Kaashoek, editors), Birkhäuser Verlag, Basel, 1986, pp. 399–413.

[68] L. Rodman. *An Introduction to Operator Polynomials. Operator Theory: Advances and Applications* 38, Birkhäuser Verlag, Basel, 1989.

[69] M.A. Shayman. Geometry of the algebraic Riccati equation. Parts I and II. *SIAM J. Control and Optim.*, 21(1983), 375–394 and 395–409.

[70] G.W. Stewart, J.-g. Sun. *Matrix Perturbation Theory.* Academic Press, 1990.

[71] T.N.M. Temme. *Dissipative Operators in Indefinite Scalar Product Spaces.* Ph.D. Thesis, Vrije Universiteit Amsterdam, 1996.

[72] R.C. Thompson. Pencils of complex and real symmetric and skew matrices. *Linear Algebra and its Applications*, 147(1991), 323–371.

[73] F.E. Velasco. Stable subspaces of matrix pairs. *Linear Algebra and its Applications*, 301(1999), 15–49.

[74] H.K. Wimmer. Decomposition and parametrization of semidefinite solutions of the continuous-time algebraic Riccati equation. *SIAM J. Control Optim.*, 32(1994), 995–1007.

[75] H.K. Wimmer. Isolated semidefinite solutions of the continuous-time algebraic Riccati equation. *Integral Equations Operator Theory*, 21(1995), 362–375.

[76] G.M. Wing. *An Introduction to Transport Theory.* Wiley, New York/London, 1962.

[77] W.M. Wonham. *Linear Multivariable Control: A Geometric Approach.* Springer Verlag, New York, 1974.

André C.M. Ran
Divisie Wiskunde en Informatica
Faculteit Exacte Wetenschappen
Vrije Universiteit Amsterdam
De Boelelaan 1081a
1081 HV Amsterdam, The Netherlands
e-mail: ran@cs.vu.nl

Leiba Rodman
Department of Mathematics
The College of William and Mary
Williamsburg, VA 23187-8795, USA
e-mail: lxrodm@math.wm.edu

Operator Theory:
Advances and Applications, Vol. 134, 385–401

Dual Discrete Canonical Systems and Dual Orthogonal Polynomials

L. Sakhnovich

To Harry Dym, with whom I began working on the notion of duality, with attachment and friendship

The string equation

$$-\frac{d^2\varphi(x,\lambda)}{dx^2} = \lambda\rho^2(x)\varphi(x,\lambda), \quad \rho(x) > 0, \quad 0 \le x \le l$$

can be written in the form

$$-\frac{d^2\varphi(x,\lambda)}{dx^2} = \lambda\frac{dM}{dx}\varphi(x,\lambda), \tag{0.1}$$

where

$$M(x) = \int_0^x \rho^2(t)\,dt.$$

The equation

$$-\frac{d^2\tilde{\varphi}(M,\lambda)}{dM^2} = \lambda\frac{dx}{dM}\tilde{\varphi}(M,\lambda) \tag{0.2}$$

is said to be dual to equation (0.1). The notion of a dual string was investigated by I.S. Kac and M.G. Krein [1]. Kac and Krein obtained the dual string equation from the original by interchanging the variables x and $M(x)$. Let us add conditions

$$\varphi(0,\lambda) = 1, \quad \varphi'(0,\lambda) = 0, \tag{0.3}$$

$$\tilde{\varphi}(0,\lambda) = 0, \quad \tilde{\varphi}'(0,\lambda) = 1 \tag{0.4}$$

to equations (0.1) and (0.2).

Then as it was shown in the work [1] there are spectral functions $\tau(\lambda)$ and $\tilde{\tau}(\lambda)$ of problems (0.1), (0.3) and (0.2), (0.4) such that

$$\tau(\lambda) = \tilde{\tau}(\lambda) = 0, \quad \lambda < 0; \qquad \tilde{\tau}(\lambda) = \int_0^\lambda \mu\,d\tau(\mu), \quad \lambda > 0 \tag{0.5}$$

We cannot transfer duality notion on the string matrix equations and on the canonical differential systems by changing variables. This problem was solved by H. Dym and the author in the joint work [2] due to the special form of the presentation of the original and dual systems. Let us note that relations (0.5) are preserved in the general situation as well. In this article we introduce the duality notion for the canonical discrete systems

$$W(k,z) - W(k-1,z) = izJ\gamma(k)W(k-1,z), \quad k \geq 1, \tag{0.6}$$

where $W(k,z)$, $\gamma(k)$, J are $(2m) \times (2m)$ matrices and

$$J = \begin{bmatrix} 0 & I_m \\ I_m & 0 \end{bmatrix}, \quad \gamma(k) \geq 0, \quad W(0,z) = I_{2m}. \tag{0.7}$$

The well-known recurrent relations

$$b_k\varphi(k+1,z) + a_k\varphi(k,z) + b_{k-1}^*\varphi(k-1,z) = z\varphi(k,z), \quad k \geq 0, \\ \varphi(-1,0) = 0, \tag{0.8}$$

in which b_k, a_k, $\varphi(k,z)$ are $m \times m$ matrices can be reduced to the canonical systems of the form (0.6), (0.7). The matrix polynomials $\varphi(k,z)$ are orthogonal with respect to the corresponding spectral matrix function $\tau(\lambda)$, i.e.

$$\int_0^\infty \varphi(k,\lambda)[d\tau(\lambda)]\varphi^*(l,\lambda) = \delta_{kl}I_m, \tag{0.9}$$

where δ_{kl} is the Kronecker symbol. In this article we present a method of constructing the system

$$\tilde{b}_k\tilde{\varphi}(k+1,z) + \tilde{a}_k\tilde{\varphi}(k,z) + \tilde{b}_{k-1}^*\tilde{\varphi}(k-1,z) = z\tilde{\varphi}(k,z) \tag{0.10}$$

which is dual to the original system (0.8). The dual system $\tilde{\varphi}(k,z)$ is orthogonal with respect to the corresponding spectral matrix function $\tilde{\tau}(\lambda)$, i.e.

$$\int_0^\infty \tilde{\varphi}(k,\lambda)[d\tilde{\tau}(\lambda)]\tilde{\varphi}^*(l,\lambda) = \delta_{kl}I_m. \tag{0.11}$$

The description of all the spectral matrix functions $\tau(\lambda)$ and $\tilde{\tau}(\lambda)$ satisfying relations (0.5) is given in this paper. The obtained results are new even for the scalar case ($m = 1$). In conclusion the results of the article are illustrated by a number of classic examples (Laguerre polynomials, Jacobi polynomials, Chebyshev polynomials).

1. Operator identities

The method of operator identities [3], [4] plays a significant role in this article. We shall write here the fundamental operator identities referring to the problem under consideration.

We denote by $l_m^2(N)$ the space of vector columns

$$\vec{f} = \mathrm{col}\,[f_0, f_1, \ldots, f_{N-1}]$$

with the norm

$$||\vec{f}||^2 = \sum_{k=0}^{N-1} f_k^* f_k,$$

where f_k are $m \times 1$ column vectors. In the space $l_m^2(N)$ we introduce the operators B and C:

$$\left(B\vec{f}\right)_k = q_k \sum_{j=0}^{k-1} p_j f_j, \qquad 1 \le k \le N-1, \tag{1.1}$$

$$\left(B\vec{f}\right)_0 = 0, \tag{1.2}$$

$$\left(C\vec{f}\right)_k = -p_k^* \sum_{j=0}^{k} q_j^* f_j, \qquad 0 \le k \le N-1. \tag{1.3}$$

Here p_k and q_k are $m \times m$ matrices. It follows from (1.1)–(1.3) that

$$\left[(B^* - C)\vec{f}\,\right]_k = p_k^* \sum_{j=0}^{N-1} q_j^* f_j, \qquad 0 \le k \le N-1. \tag{1.4}$$

Equality (1.4) can be written in the form

$$B^* - C = \Pi_2 \Pi_1^*, \tag{1.5}$$

where

$$(\Pi_1 g)_k = q_k g, \quad (\Pi_2 g)_k = p_k^* g, \qquad g \in G, \quad 0 \le k \le N-1 \tag{1.6}$$

(G is a space of the $m \times 1$ vectors). From identity (1.5) we deduce the relations

$$CB - B^*C^* = B^*\Pi_1\Pi_2^* - \Pi_2\Pi_1^* B, \tag{1.7}$$

$$BC - C^*B^* = \Pi_1\Pi_2^* B^* - B\Pi_2\Pi_1^*. \tag{1.8}$$

We introduce the operators

$$A = CB, \qquad \Phi_1 = B^*\Pi_1, \qquad \Phi_2 = i\Pi_2, \tag{1.9}$$

$$\tilde{A} = BC, \qquad \tilde{\Phi}_1 = -i\Pi_1, \qquad \tilde{\Phi}_2 = B\Pi_2. \tag{1.10}$$

Using notations (1.9) and (1.10) we can write relations (1.7) and (1.8) in the following form

$$A - A^* = i\,(\Phi_1\Phi_2^* + \Phi_2\Phi_1^*), \tag{1.11}$$

$$\tilde{A} - \tilde{A}^* = i\,(\tilde{\Phi}_1\tilde{\Phi}_2^* + \tilde{\Phi}_2\tilde{\Phi}_1^*). \tag{1.12}$$

From (1.1) and (1.2) we have

$$\left(A\vec{f}\right)_k = -p_k^* \sum_{j=1}^{k} q_j^* q_j \sum_{l=0}^{j-1} p_l f_l, \qquad k \ge 1, \tag{1.13}$$

$$\left(A\vec{f}\right)_0 = 0. \tag{1.14}$$

Formula (1.13) can be rewritten in the form

$$\left(A\vec{f}\right)_k = -p_k^* \sum_{l=0}^{k-1} \Big(\sum_{j=l+1}^{k} q_j^* q_j \Big) p_l f_l, \qquad k \geq 1. \tag{1.15}$$

Setting

$$L(k) = \sum_{j=1}^{k} q_j^* q_j, \quad k \geq 1, \quad L(0) = 0, \tag{1.16}$$

we represent (1.15) in the form

$$\left(A\vec{f}\right)_k = -p_k^* \sum_{j=0}^{k-1} \big[L(k) - L(j)\big] p_j f_j, \qquad k \geq 1. \tag{1.17}$$

Using (1.6) and (1.9) we obtain

$$\left(\Phi_2 g\right)_k = i p_k^* g, \qquad 0 \leq k \leq N-1, \tag{1.18}$$

$$\left(\Phi_1 g\right)_k = p_k^* \big[L(N-1) - L(k)\big] g, \qquad 0 \leq k \leq N-1. \tag{1.19}$$

According to (1.1) and (1.2) the equality

$$\left(\tilde{A}\vec{f}\right)_k = -q_k \sum_{j=0}^{k-1} \big[M(k-1) - M(j-1)\big] q_j^* f_j \tag{1.20}$$

is true. Here

$$M(k) = \sum_{j=0}^{k} p_j p_j^*, \quad k \geq 0, \quad M(-1) = 0.$$

From (1.6) and (1.10) we deduce that

$$\left(\tilde{\Phi}_1 g\right)_k = -i q_k g, \qquad 0 \leq k \leq N-1, \tag{1.21}$$

$$\left(\tilde{\Phi}_2 g\right)_k = q_k M(k-1) g, \qquad 0 \leq k \leq N-1. \tag{1.22}$$

Let orthogonal projectors $\mathcal{P}_k$ be defined by the equality

$$\mathcal{P}_k \vec{h} = \vec{h}_k, \quad 1 \leq k \leq N, \quad \mathcal{P}_0 \vec{h} = 0, \tag{1.23}$$

where

$$\vec{h} = \operatorname{col}\left[h_1, h_2, \ldots, h_{mN}\right],$$

$$\vec{h}_k = \operatorname{col}\left[h_1, h_2, \ldots, h_{mk}, 0, 0, \ldots, 0\right].$$

It is obvious that the following relations

$$A^* \mathcal{P}_k = \mathcal{P}_k A^* \mathcal{P}_k, \qquad (\mathcal{P}_k - \mathcal{P}_{k-1}) A (\mathcal{P}_k - \mathcal{P}_{k-1}) = 0, \tag{1.24}$$

$$\tilde{A}^* \mathcal{P}_k = \mathcal{P}_k \tilde{A}^* \mathcal{P}_k, \qquad (\mathcal{P}_k - \mathcal{P}_{k-1}) \tilde{A} (\mathcal{P}_k - \mathcal{P}_{k-1}) = 0 \tag{1.25}$$

are true.

2. Canonical systems (discrete case)

In this section the following systems of difference equations are considered.

$$W(k,z) - W(k-1,z) = izJ\gamma(k)W(k-1,z), \quad k \geq 1 \tag{2.1}$$

and

$$\tilde{W}(k,z) - \tilde{W}(k-1,z) = izJ\tilde{\gamma}(k)\tilde{W}(k-1,z), \quad k \geq 1, \tag{2.2}$$

where $W(k,z)$, $\tilde{W}(k,z)$, $\gamma(k)$ and $\tilde{\gamma}(k)$ are $(2m) \times (2m)$ matrices, $k = 0,1,2,\ldots,$

$$J = \begin{bmatrix} 0 & I_m \\ I_m & 0 \end{bmatrix}, \quad W(0,z) = \tilde{W}(0,z) = I_{2m}. \tag{2.3}$$

The matrices $\gamma(k)$ and $\tilde{\gamma}(k)$ are defined by the relations

$$\gamma(k) = \sigma(k) - \sigma(k-1), \qquad \tilde{\gamma}(k) = \tilde{\sigma}(k) - \tilde{\sigma}(k-1), \tag{2.4}$$

where

$$\sigma_k = \Pi^* \mathcal{P}_k \Pi, \qquad \tilde{\sigma}_k = \tilde{\Pi}^* \mathcal{P}_k \tilde{\Pi}, \qquad 1 \leq k \leq N. \tag{2.5}$$

Here we use the notations

$$\Pi = [\Phi_1, \Phi_2], \qquad \tilde{\Pi} = [\tilde{\Phi}_1, \tilde{\Phi}_2]. \tag{2.6}$$

In view of formulas (1.18), (1.19) and (1.21), (1.22) the following equalities

$$\gamma(k) = \begin{bmatrix} \{L(N-1) - L(k-1)\}p_{k-1} \\ \\ -ip_{k-1} \end{bmatrix} [p_{k-1}^* \{L(N-1) - L(k-1)\},\ ip_{k-1}^*], \tag{2.7}$$

$$\tilde{\gamma}(k) = \begin{bmatrix} iq_{k-1}^* \\ \\ M(k-2)q_{k-1}^* \end{bmatrix} [-iq_{k-1},\ q_{k-1}M(k-2)] \tag{2.8}$$

$(k \geq 1)$, are valid. It is obvious that

$$\gamma(k) \geq 0, \quad \tilde{\gamma}(k) \geq 0, \tag{2.9}$$

$$\gamma(k)J\gamma(k) = \tilde{\gamma}(k)J\tilde{\gamma}(k) = 0. \tag{2.10}$$

We shall call the system (2.2) to be dual to the system (2.1).

3. Spectral theory

Let us recall the main notions of the spectral theory [3] of systems (2.1). We suppose that

$$\operatorname{rank} q(k) = \operatorname{rank} p(k) = m, \qquad 0 \leq k \leq N-1. \tag{3.1}$$

With canonical systems (2.1) and (2.2) we associate the matrix functions

$$v(z) = i[a(z)\mathcal{R}(z) + b(z)Q(z)]\,[c(z)\mathcal{R}(z) + d(z)Q(z)]^{-1} \tag{3.2}$$

and

$$\tilde{v}(z) = i[\tilde{a}(z)\mathcal{R}(z) + \tilde{b}(z)Q(z)]\,[\tilde{c}(z)\mathcal{R}(z) + \tilde{d}(z)Q(z)]^{-1}. \tag{3.3}$$

The coefficient matrices of the linear-fractional transformations (3.2) and (3.3) have the forms

$$W^*(l,\overline{z}) = \begin{bmatrix} a(z) & b(z) \\ c(z) & d(z) \end{bmatrix} \tag{3.4}$$

and

$$\tilde{W}^*(l,\overline{z}) = \begin{bmatrix} \tilde{a}(z) & \tilde{b}(z) \\ \tilde{c}(z) & \tilde{d}(z) \end{bmatrix}. \tag{3.5}$$

Meromorphic $m \times m$ matrix functions $\mathcal{R}(z)$ and $Q(z)$ satisfy the relations

$$\det\left[\mathcal{R}^*(z)\mathcal{R}(z) + Q^*(z)Q(z)\right] \neq 0, \qquad \operatorname{Im} z > 0, \tag{3.6}$$

$$\mathcal{R}^*(z)Q(z) + Q^*(z)\mathcal{R}(z) \geq 0, \qquad \operatorname{Im} z > 0. \tag{3.7}$$

The matrix functions $v(z)$ and $\tilde{v}(z)$ belong to the Nevanlinna class and admit the representations

$$v(z) = \beta z + \alpha + \int_{-\infty}^{\infty} \left(\frac{1}{\lambda - z} - \frac{\lambda}{1+\lambda^2}\right) d\tau(\lambda) \tag{3.8}$$

and

$$\tilde{v}(z) = \tilde{\beta} z + \tilde{\alpha} + \int_{-\infty}^{\infty} \left(\frac{1}{\lambda - z} - \frac{\lambda}{1+\lambda^2}\right) d\tilde{\tau}(\lambda), \tag{3.9}$$

where $\alpha = \alpha^*$, $\tilde{\alpha} = \tilde{\alpha}^*$, $\beta \geq 0$, $\tilde{\beta} \geq 0$, $\tau(\lambda)$ and $\tilde{\tau}(\lambda)$ are monotonically increasing $m \times m$ matrix functions. We shall show that $\tau(\lambda)$ and $\tilde{\tau}(\lambda)$ are spectral matrix functions of canonical systems (2.1) and (2.2) respectively. Let us consider now the canonical system

$$Y(k,z) - Y(k-1,z) = izJ\gamma(k)Y(k-1,z), \quad 1 \leq k \leq N, \tag{3.10}$$

where

$$Y(k,z) = \operatorname{col}\left[Y_1(k,z),\, Y_2(k,z)\right],$$

$Y_1(k,z)$, $Y_2(k,z)$ are vector functions of the $m \times 1$ order. We add the following boundary condition

$$D_2 Y_1(0,z) + D_1 Y_2(0,z) = 0. \tag{3.11}$$

Here D_1 and D_2 in (3.11) are matrices of the $m \times m$ order. We shall suppose that

$$D_1 D_2^* + D_2 D_1^* = 0, \quad D_1 D_1^* + D_2 D_2^* = I_m. \tag{3.12}$$

We denote by $l_m^2(\gamma, N)$ the space of the vectors

$$\vec{g} = \operatorname{col}\left[g(0), g(1), \ldots, g(N-1)\right],$$

where $g(k)$ are vector columns of the order $2m$. The norm in $l_m^2(\gamma, N)$ is defined by the equality

$$||\vec{g}\,||_\gamma^2 = \sum_{k=0}^{N-1} g^*(k)\gamma(k+1)g(k).$$

We associate with system (3.10) and conditions (3.11) the operator

$$V_N\vec{g} = \sum_{k=0}^{N-1} [D_1, D_2] W^*(k,u)\gamma(k+1)g(k), \tag{3.13}$$

that maps vectors from $l^2_m(\gamma, N)$ into vectors $f(u)$ $(-\infty < u < \infty)$ of the order m.

Definition 3.1. *A monotonically increasing matrix function $\tau(u)$ $(-\infty < u < \infty)$ of the $m \times m$ order is called a spectral matrix function of system* (3.10), (3.11) *if the corresponding operator V_N maps $l^2_m(\gamma, N)$ isometrically into $l^2_m(\tau)$.*

The inner product in $l^2_m(\tau)$ is defined by formula

$$\big(f_1(u), f_2(u)\big) = \int_{-\infty}^{\infty} f_2^*(u)[d\tau(u)]f_1(u).$$

Without loss of generality (see [3]) we can suppose that

$$D_1 = 0, \qquad D_2 = I_m,$$

i.e. the boundary condition has the form

$$Y_1(0,z) = 0. \tag{3.14}$$

Let us consider the system

$$\tilde{Y}(k,z) - \tilde{Y}(k-1,z) = i\, z\, J\, \tilde{\gamma}(k)\, \tilde{Y}(k-1,z), \quad 1 \le k \le N \tag{3.15}$$

and the boundary condition

$$\tilde{Y}_1(0,z) = 0. \tag{3.16}$$

We denote by $\tilde{\tau}(u)$ the spectral matrix function of system (3.15), (3.16). The following theorem follows directly from results of the book ([3], Ch. 8).

Theorem 3.1. *Let operators A and $\tilde{A}$ be defined by formulas* (1.17) *and* (1.20), *respectively, and let the following conditions be fulfilled:*

$$\operatorname{rank} p_k = \operatorname{rank} q_k = m. \tag{3.17}$$

Then the following assertions are true.

1. *Let $v(z)$ and $\tilde{v}(z)$ admit representations* (3.2) *and* (3.3), *respectively. Then parameters β and $\tilde{\beta}$ from* (3.8) *and* (3.9) *are equal to zero. The matrix functions $\tau(u)$ and $\tilde{\tau}(u)$ from* (3.8) *and* (3.9) *are spectral matrix functions of systems* (2.1) *and* (2.2), *respectively.*
2. *Let $\tau(u)$ and $\tilde{\tau}(u)$ be spectral $m \times m$ matrix functions of systems* (2.1) *and* (2.2). *Then there exist α and $\tilde{\alpha}$ such that corresponding*

$$v(z) = \alpha + \int_{-\infty}^{\infty} \left(\frac{1}{u-z} - \frac{u}{1+u^2}\right) d\tau(u)$$

and

$$\tilde{v}(z) = \tilde{\alpha} + \int_{-\infty}^{\infty} \left(\frac{1}{u-z} - \frac{u}{1+u^2} \right) d\tilde{\tau}(u)$$

can be represented in forms (3.2) *and* (3.3), *respectively.*

4. Classic systems

In this section we shall show how systems (2.1) and (2.2) can be reduced to the classic form

$$b_k\varphi(k+1,z) + a_k\varphi(k,z) + b_{k-1}^*\varphi(k-1,z) = z\varphi(k,z), \quad 0 \le k \le N-1, \tag{4.1}$$
$$\varphi(-1,z) = 0,$$

where a_k, b_k are $m \times m$ matrices and $a_k = a_k^*$, $\det b_k \neq 0$. Let us represent the solution $W(k,z)$ of system (2.1) in the block form

$$W(k,z) = \{w_{ij}(k,z)\}_{i,j=1}^{2},$$

where all the blocks $w_{ij}(k,z)$ are of $m \times m$ order.

We consider the $m \times m$ matrix functions

$$\varphi_1(k,z) = r_1^*(k)w_{11}(k,z) + r_2^*(k)w_{21}(k,z), \qquad 0 \le k \le N-1, \tag{4.2}$$

$$\varphi_2(k,z) = r_1^*(k)w_{12}(k,z) + r_2^*(k)w_{22}(k,z), \qquad 0 \le k \le N-1, \tag{4.3}$$

where

$$r_1(k) = [L(N-1) - L(k)]p_k, \qquad r_2(k) = -ip_k. \tag{4.4}$$

It follows from (2.1) that

$$w_{1,s}(k,z) - w_{1,s}(k-1,z) = izr_2(k-1)\varphi_s(k-1,z), \tag{4.5}$$

$$w_{2,s}(k,z) - w_{2,s}(k-1,z) = izr_1(k-1)\varphi_s(k-1,z) \tag{4.6}$$

$$(s = 1,2).$$

From (4.3) we obtain

$$r_2^{*-1}(k)\varphi_2(k,z) = r_2^{*-1}(k)r_1^*(k)w_{12}(k,z) + w_{22}(k,z), \qquad 0 \le k \le N-1. \tag{4.7}$$

Using the notation

$$\Delta\varphi(k) = \varphi(k) - \varphi(k-1)$$

we deduce from (4.7) the relation

$$\Delta[r_2^{*-1}(k)\varphi_2(k,z)] = \Delta[r_2^{*-1}(k)r_1^*(k)w_{12}(k,z)] + \Delta w_{22}(k,z), \ 1 \le k \le N-1, \tag{4.8}$$

$$\varphi_2(0,z) = r_2^*(0) = ip_0^*. \tag{4.9}$$

In view of (4.4), (4.5) and (4.8) the following relation

$$\Delta[r_2^{*-1}(k)\varphi_2(k,z)] = \left(\Delta[r_2^{*-1}(k)r_1^*(k)]\right)w_{12}(k,z) \tag{4.10}$$

holds. We have taken into account that

$$r_1^*(k-1)r_2(k-1)+r_2^*(k-1)r_1(k-1)=0. \tag{4.11}$$

It follows from (1.16) and (4.4) that

$$\Delta[r_2^{*-1}(k)r_1^*(k)]=iq_k^*q_k, \qquad 1\le k\le N-1. \tag{4.12}$$

According to (4.9), (4.10) and (4.12) we have

$$-\Delta\{q_k^{-1}q_k^{*-1}\Delta[p_k^{*-1}\varphi(k,z)]\}=zp_{k-1}\varphi(k-1,z), \quad 2\le k\le N-1, \tag{4.13}$$

$$\varphi(0,z)=p_0^*, \qquad \varphi(1,z)=p_1^*(1-zq_1^*q_1p_0p_0^*), \tag{4.14}$$

where

$$\varphi(k,z)=-i\varphi_2(k,z). \tag{4.15}$$

The equation (4.13) is an analogue of the matrix string equation. The equation (4.13) can be rewritten in the classic form (4.1), where

$$a_s=p_s^{-1}(q_{s+1}^{-1}q_{s+1}^{*-1}+q_s^{-1}q_s^{*-1})p_s^{*-1}, \qquad s\ge 1, \tag{4.16}$$

$$a_0=p_0^{-1}q_1^{-1}q_1^{*-1}p_0^{*-1}, \tag{4.17}$$

$$b_s=-p_s^{-1}q_{s+1}^{-1}q_{s+1}^{*-1}p_{s+1}^{*-1}, \qquad s\ge 0. \tag{4.18}$$

We note that the second boundary condition in (4.14) can be omitted, it follows from formulas (4.1) and (4.17). In terms of equation (4.1) formula (3.13) takes the form

$$F(u)=V\vec{f}=\sum_{k=0}^{N-1}\varphi^*(k,u)f_k, \tag{4.19}$$

where

$$f_k=r_1^*(k)g_1(k)+r_2^*(k)g_2(k).$$

According to Definition 3.1 the spectral matrix function $\tau(u)$ of system (4.1) can be characterized by the relation

$$\int_{-\infty}^{\infty}F^*(u)[d\tau(u)]F(u)=\sum_{k=0}^{N-1}f_k^*f_k,$$

as the following relation

$$\sum_{k=0}^{N-1}g^*(k)\gamma(k+1)g(k)=\sum_{k=0}^{N-1}f_k^*f_k$$

holds. Now we shall consider the dual system (2.2) and introduce the matrix function

$$\tilde{\varphi}_2(k,z)=\tilde{r}_1^*(k)\tilde{w}_{12}(k,z)+\tilde{r}_2^*(k)\tilde{w}_{22}(k,z), \tag{4.20}$$

where

$$\tilde{r}_1(k)=q_k^*, \quad \tilde{r}_2(k)=-iM(k-1)q_k^*. \tag{4.21}$$

As in the case of system (2.1) we obtain the relations

$$-\Delta\left\{p_{k-1}^{*-1}p_{k-1}^{-1}\Delta\left[q_k^{-1}\tilde{\varphi}(k,z)\right]\right\} = zq_{k-1}^{*}\varphi(k-1,z), \quad 2\le k\le N-1, \tag{4.22}$$

$$\tilde{\varphi}(0,z)=0, \quad \tilde{\varphi}(1,z)=p_0p_0^{*}, \tag{4.23}$$

where

$$\tilde{\varphi}(k,z)=i\tilde{\varphi}_2(k,z). \tag{4.24}$$

It follows from (4.22) that

$$z\psi(k,z)=\tilde{b}_k\psi(k+1,z)+\tilde{a}_{k+1}\psi(k,z)+b_{k-1}^{*}\psi(k-1,z), \quad 0\le k\le N-2, \tag{4.25}$$

$$\psi(0,z)=p_0p_0^{*}\,. \tag{4.26}$$

Here we use the following notations

$$\psi(k,z)=\tilde{\varphi}(k-1,z), \quad 0\le k\le N-2, \tag{4.27}$$

$$\tilde{b}_k=-q_{k+1}^{*-1}\,p_{k+1}^{*-1}\,p_{k+1}^{-1}\,q_{k+2}^{-1}, \quad k\ge 0 \tag{4.28}$$

$$\tilde{a}_k=q_{k+1}^{*-1}\,\left(p_{k+1}^{*-1}\,p_{k+1}^{-1}+p_k^{*-1}\,p_k^{-1}\right)\,q_{k+1}^{-1}, \quad k\ge 0. \tag{4.29}$$

According to Definition 3.1 the spectral matrix $\tilde{\tau}(\lambda)$ of system (4.25), (4.26) can be defined by the relation

$$\int_{-\infty}^{\infty}\tilde{F}^{*}(u)[d\tilde{\tau}(u)]\tilde{F}(u)=\sum_{k=0}^{N-2}f_k^{*}f_k\,, \qquad \text{where} \qquad \tilde{F}(u)=\tilde{V}\vec{f}=\sum_{k=0}^{N-2}\psi^{*}(k,u)f_k\,.$$

5. On the connection between $\tau(u)$ and $\tilde{\tau}(u)$

Now we shall consider the following interpolation problem.

Problem 5.1. Let the operator identities (1.11), (1.12) be fulfilled. It is necessary to find monotonically increasing $m\times m$ matrix functions $\tau(u)$ and $\tilde{\tau}(u)$ such that representations

$$I_H=\int_{-\infty}^{\infty}(I_H-Au)^{-1}\Phi_2[d\tau(u)]\Phi_2^{*}(E-A^{*}u)^{-1} \tag{5.1}$$

$$I_H=\int_{-\infty}^{\infty}(I_H-\tilde{A}u)^{-1}\tilde{\Phi}_2[d\tilde{\tau}(u)]\tilde{\Phi}_2^{*}(I_H-\tilde{A}^{*}u)^{-1} \tag{5.2}$$

are true and

$$\tau(u)=\tilde{\tau}(u)=0, \quad u<0; \quad \tilde{\tau}(u)=\int_0^u s\,d\tau(s), \quad u>0. \tag{5.3}$$

(Here $H=l_m^2(N)$, I_H is identity operator in the space H.)

Definition 5.1. *We shall say that the pair $\{\mathcal{P}(z),\mathcal{Q}(z)\}$ of $m\times m$ matrix functions are Stieltjes if $\mathcal{P}(z)$ and $\mathcal{Q}(z)$ are meromorphic in $\mathbb{C}\setminus\mathbb{R}_+$ and if the following* *three inequalities are fulfilled at every point* $z \in \mathbb{C} \setminus \mathbb{R}_+$ *at which* $\mathcal{P}(z)$ *and* $Q(z)$ *are holomorphic:*

1. $$\mathcal{P}^*(z)\mathcal{P}(z) + Q^*(z)Q(z) > 0,$$

2. $$\frac{\mathcal{P}^*(z)Q(z) + Q^*(z)\mathcal{P}(z)}{i(\overline{z} - z)} \geq 0,$$

3. $$\frac{\overline{z}\mathcal{P}^*(z)Q(z) + Q^*(z)z\mathcal{P}(z)}{i(\overline{z} - z)} \geq 0.$$

Let us associate the $(2m) \times (2m)$ matrix function

$$Q(z) = \left\{ I_{2m} + z \begin{bmatrix} \Pi_1^* C^* \\ -\Pi_2^* \end{bmatrix} (I - B^* C^* z)^{-1} [\Pi_2, C\Pi_1] \right\} \Gamma \tag{5.4}$$

with operator identity (1.5). Here

$$\Gamma = \begin{bmatrix} I_m & \Pi_1^* \Pi_1 \\ 0 & I_m \end{bmatrix}. \tag{5.5}$$

We represent $\Theta(z)$ in the block form

$$\Theta(z) = \{\Theta_{ij}(z)\}_{i,j=1}^2 ,$$

where all the blocks $\Theta_{ij}(z)$ are of the $m \times m$ order.

From the results of article [4] we directly deduce the following assertions.

Theorem 5.1. *Let operators* B, C *and* Π_1, Π_2 *be defined by formulas* (1.1)–(1.3) *and* (1.6), *the following condition being fulfilled*

$$\det p_k \neq 0, \quad \det q_k \neq 0, \quad k \geq 0 \tag{5.6}$$

The matrix functions $\tau(\lambda)$ *and* $\tilde{\tau}(\lambda)$ *are solutions of interpolation Problem* 5.1 *if and only if the matrix function*

$$s(z) = \int_0^\infty \frac{d\tau(\lambda)}{\lambda - z} \tag{5.7}$$

can be represented in the form

$$s(z) = [\Theta_{11}(z)\mathcal{R}(z) + \Theta_{12}(z)Q(z)] [\Theta_{21}(z)\mathcal{R}(z) + \Theta_{22}(z)Q(z)]^{-1} \tag{5.8}$$

where $\mathcal{R}(z), Q(z)$ *are a Stieltjes pair.*

Theorem 5.2. *Let the conditions of Theorem* 5.1 *be fulfilled. Matrix functions* $\tau(\lambda)$ *and* $\tilde{\tau}(\lambda)$ *satisfying relations* (5.3) *are spectral matrix functions of corresponding systems* (4.14), (4.15) *and* (4.25), (4.26) *if and only if the matrix function* $s(z)$ *defined by formula* (5.7) *can be represented in form* (5.8).

It follows from Theorems 5.1 and 5.2 that the set of the solutions of interpolation Problem 5.1 coincides with the set of the solutions of the spectral problem for the corresponding systems.

Remark 5.1. *Let the conditions of Theorem* 5.1 *be fulfilled. Then the following assertions are valid.*

1. *If* $\tau(\lambda)$ *is a spectral* $m \times m$ *matrix function of system* (4.14), (4.15) *such that* $\tau(\lambda) = 0$ *when* $\lambda < 0$, *then*

$$\tilde{\tau}(\lambda) = \int_0^\lambda s\, d\tau(s) \tag{5.9}$$

is a spectral matrix function of system (4.25), (4.26).

2. *If* $\tilde{\tau}(\lambda)$ *is a spectral* $m \times m$ *matrix function of system* (4.25), (4.26) *such that* $\tilde{\tau}(\lambda) = 0$ *when* $\lambda < 0$, *then there exists a spectral matrix function of system* (4.14), (4.15) *connected with* $\tilde{\tau}(\lambda)$ *by relation* (5.9).

6. On roots of matrix orthogonal polynomials

As it is known [5] the spectral $m \times m$ matrix function $\tau(\lambda)$ and the sequence of the matrix polynomials $\varphi_n(z)$ $(n = 0, 1, 2, \dots)$ correspond to difference system (4.1). The matrix polynomials $\varphi_n(z)$ are such that $\varphi_0(z) = I_m$ and

$$b_k \varphi_{k+1}(z) + a_k \varphi_k(z) + b_{k-1}^* \varphi_{k-1}(z) = z\varphi_k(z), \tag{6.1}$$

where $a_k = a_k^*$, $\det b_k \neq 0$. The polynomials form an orthogonal system, i.e.

$$\int_\alpha^\beta \varphi_j(\lambda)[d\tau(\lambda)]\varphi_k^*(\lambda) = \delta_{jk} I_m, \tag{6.2}$$

where $-\infty \leq \alpha$, $\beta \leq \infty$, $0 \leq j$, $k < \infty$.

Theorem 6.1. *The roots of the polynomials* $\det \varphi_n(z)$ *are real and are located in the interval* (α, β).

Proof. Let z_0 be a root of $\det \varphi_n(z)$. Then for some constant $m \times 1$ vector h $(h \neq 0)$ the equality

$$h^* \varphi_n(z_0) = 0 \tag{6.3}$$

is fulfilled.

Let us also note that in view of (6.2) the relation

$$\int_\alpha^\beta \varphi_n(\lambda)[d\tau(\lambda)]\psi_l^*(\lambda) = 0, \tag{6.4}$$

where $\psi_l(z)$ is a $m \times m$ matrix of degree l and $l < n$, is valid. It follows from relations (6.3) and (6.4) that

$$h^* \int_\alpha^\beta \varphi_n(\lambda)[d\tau(\lambda)] \frac{\varphi_n^*(\lambda)}{\lambda - z_0} h = 0. \tag{6.5}$$

We shall write relations (6.5) in the form

$$\overline{z}_0 h^* \int_\alpha^\beta \frac{\varphi_n(\lambda)}{\lambda - \overline{z}_0} [d\tau(\lambda)] \frac{\varphi_n^*(\lambda)}{\lambda - z_0} h = h^* \int_\alpha^\beta \lambda \frac{\varphi_n(\lambda)}{\lambda - \overline{z}_0} [d\tau(\lambda)] \frac{\varphi_n^*(\lambda)}{\lambda - z_0} h = 0. \tag{6.6}$$

From formula (6.2) we deduce the following representation

$$h^* \varphi_n(\lambda)/(\lambda - \overline{z}_0) = \sum_{k=0}^{n-1} c_k^* \varphi_k(\lambda), \tag{6.7}$$

where c_k are $m \times 1$ vectors. Hence the inequality

$$h^* \int_\alpha^\beta \frac{\varphi_n(\lambda)}{\lambda - \overline{z}_0} [d\tau(\lambda)] \frac{\varphi_n^*(\lambda)}{\lambda - z_0} h = \sum_{k=0}^{n-1} c_k^* c_k > 0 \tag{6.8}$$

is true. Thus formula (6.6) signifies that $\overline{z}_0$ is the centre of gravity of the mass distribution on the segment $[\alpha, \beta]$. Thus the estimation $\alpha \leq z_0 \leq \beta$ holds. Let us show that $z_0 \neq \alpha$. We shall suppose that $z_0 = \alpha$. Then we have

$$h^* \int_\alpha^\beta (\lambda - \alpha) \frac{\varphi_n(\lambda)}{\lambda - \alpha} [d\tau(\lambda)] \frac{\varphi_n^*(\lambda)}{\lambda - \alpha} h = 0. \tag{6.9}$$

If $\beta < \infty$ then the inequality

$$\int_\alpha^\beta (\lambda - \alpha) \frac{\varphi_n(\lambda)}{\lambda - \alpha} [d\tau(u)] \frac{\varphi_n^*(\lambda)}{\lambda - \alpha} h \geq \int_\alpha^\beta \varphi_n(\lambda) [d\tau(\lambda)] \varphi_n^*(\lambda) > 0 \tag{6.10}$$

is valid. As relations (6.9) and (6.10) contradict one another, then $z_0 \neq \alpha$. It is proved in the same way that $z_0 \neq \beta$. If $\beta = \infty$ then for some finite $\tilde{\beta}$ relation (6.10) is fulfilled, i.e. $z_0 \neq \alpha$ in this case too. The theorem is proved. □

Further we shall consider the case when the spectrum of system (6.1) is nonnegative, i.e. $\alpha \geq 0$. From Theorem 6.1 we deduce the following assertions.

Corollary 6.1. *If the spectrum of system* (6.1) *is nonnegative then all the roots of the polynomial* det $\varphi_n(z)$ *are positive.*

Corollary 6.2. *If the spectrum of system* (6.1) *is nonnegative then all the* $m \times m$ *matrices* $\varphi_n(0)$ *are invertible.*

7. Recurrent formula

In the scalar case ($m = 1$) the recurrent formula for the orthogonal polynomials is written in the following form (see [6])

$$\Phi_{n+1}(z) = (A_n z + B_n)\Phi_n(z) - C_n\Phi_{n-1}(z), \tag{7.1}$$

where

$$A_n = \overline{A}_n \neq 0, \quad B_n = \overline{B}_n, \quad C_n = \overline{C}_n \neq 0. \tag{7.2}$$

Setting

$$h_n = \int_{-\infty}^{\infty} |\Phi_n(\lambda)|^2 d\tau(\lambda),$$

we shall introduce the normalized polynomials

$$\varphi_n(z) = \Phi_n(z)/\sqrt{h_n}. \tag{7.3}$$

If follows from formula (7.1) that

$$z\varphi_n(z) = b_n\varphi_{n+1}(z) + a_n\varphi_n(z) + b_{n-1}\varphi_{n-1}(z), \tag{7.4}$$

where

$$b_n = \sqrt{\frac{h_{n+1}}{h_n}} \Big/ A_n, \quad a_n = -\frac{B_n}{A_n}. \tag{7.5}$$

Example 7.1. Laguerre polynomials. In the case of Laguerre polynomials $L_n^\gamma(z)$ we have ([7], Ch. 10):

$$\alpha = 0, \quad \beta = \infty, \quad \tau'(\lambda) = e^{-\lambda}\lambda^\gamma \ (\lambda > 0), \quad \gamma > -1, \tag{7.6}$$

$$A_n = -1/(n+1),\ B_n = -(2n+\gamma+1)/(n+1),\ C_n = (n+\gamma)/(n+1), \tag{7.7}$$

$$h_n = \Gamma(\gamma+n+1)/n!, \quad L_n^\gamma(0) = \Gamma(\gamma+n+1)/[n!\,\Gamma(\gamma+1)]. \tag{7.8}$$

Using formulas (7.5) and (7.7), (7.8) we obtain

$$a_n = 2n+\gamma+1, \quad b_n = -\sqrt{(n+1)(n+\gamma+1)}. \tag{7.9}$$

8. Method of calculating parameters p_k and q_k of system

In Section 4 we have shown how system (2.1) can be reduced to the classical system (4.1). Here the coefficients a_k and b_k are expressed by the parameters p_k and q_k (see (4.17)–(4.19)). In this section we find a simple connection between the parameters p_k, q_k of system (4.1) and the values of the matrix polynomials $\varphi_n(z)$ in the point $z = 0$. We shall need the following assertion.

Lemma 8.1. *If the spectrum of system* (4.1) *is nonnegative, then*

$$T_k = \varphi_k^*(0)b_k\varphi_{k+1}(0) < 0, \quad k \geq 0. \tag{8.1}$$

Proof. It follows from relation (6.1) that

$$b_k\varphi_{k+1}(0) + a_k\varphi_k(0) + b_{k-1}^*\varphi_{k-1}(0) = 0, \tag{8.2}$$

$$\text{i.e.} \qquad \varphi_k^*(0)a_k\varphi_k(0) = -(T_k + T_{k-1}^*). \tag{8.3}$$

We shall use the relation

$$a_k = \int_0^\infty \lambda\varphi_k(\lambda)\,[d\tau(\lambda)]\,\varphi_k^*(\lambda), \tag{8.4}$$

$$b_k = \int_0^\infty \lambda\varphi_k(\lambda)\,[d\tau(\lambda)]\,\varphi_{k+1}(\lambda), \tag{8.5}$$

which follow directly from (6.1) and from the fact that the system of the matrix polynomials $\varphi_n(z)$ is orthogonal and normalized. Similarly to the deduction (6.10) we deduce

$$a_k > 0, \quad k \geq 0 \tag{8.6}$$

from formula (8.4). As

$$T_0 = -\varphi_k^*(0)\,a_k\,\varphi_k(0) = T_0^* < 0,$$

it follows from (8.3) that

$$T_k = T_k^*, \quad k \geq 0. \tag{8.7}$$

Now we shall consider the auxiliary matrix function

$$\tau_\nu(\lambda) = (1-\nu)\tau_0(\lambda) + \nu\tau(\lambda), \quad 0 \leq \nu \leq 1, \tag{8.8}$$

where

$$\tau_0(\lambda) = \begin{cases} 0, & \lambda < 0 \\ -e^{-\lambda}I_m, & \lambda \geq 0 \end{cases}.$$

Laguerre polynomials $L_n^0(\lambda)I_m$ correspond to the matrix $\tau_0(\lambda)$. The matrices $T_k(\nu)$ correspond to the spectral matrix $\tau_\nu(\lambda)$. In view of (7.8) and (7.9) we obtain

$$T_k(0) < 0. \tag{8.9}$$

It follows from relation (8.8) that

$$\int_0^\infty \psi_n(\lambda)[d\tau_\nu(\lambda)]\psi_n^*(\lambda) > 0, \quad 0 \leq \nu \leq 1, \tag{8.10}$$

where $\psi_n(z)$ is an arbitrary matrix polynomial of the degree n $(n = 0, 1, \ldots)$ with the leading coefficient equal to I_m. As it is known (see [5]) this fact implies the existence of the orthogonal and normalized system of the polynomials $\varphi_n(\lambda,\nu)$ continuously dependent on the parameter ν. It means that the matrices $T_k(\nu)$ are also continuous. From Corollary 6.2 and inequality (8.6) we obtain that

$$\det T_k(\nu) \neq 0, \tag{8.11}$$

Relations (8.7), (8.9), (8.11) and continuity of $T_k(\nu)$ imply that $T_k(\nu) < 0$. The lemma is proved. □

In view of (8.1) and (8.2) the following assertion holds.

Theorem 8.1. *If the spectrum of system* (4.1) *is nonnegative then the coefficients* a_k *and* b_k *can be represented in form* (4.17)–(4.19), *where*

$$p_k = \varphi_k^*(0), \quad q_{k+1} = U_k(-T_k)^{-1/2} \tag{8.12}$$

and U_k *are arbitrary unitary* $m \times m$ *matrices.*

9. Laguerre polynomials

We deduce from formulas (7.8), (7.9) and (8.12) that the following equalities

$$p_n = L_n^\gamma(0)/\sqrt{h_n} = \sqrt{\frac{\Gamma(\gamma+n+1)}{n!}}\,\frac{1}{\Gamma(\gamma+1)}, \tag{9.1}$$

$$q_{n+1} = \sqrt{\frac{n!}{\Gamma(\gamma+n+2)}}\,\Gamma(\gamma+1) \tag{9.2}$$

are true for Laguerre polynomials $L_n^\gamma(z)$. Let us consider the dual problem corresponding to the case of Laguerre polynomials. In view of (4.28), (4.29) and (9.1), (9.2) the equalities

$$\tilde{a}_n = 2n+\gamma+2, \quad \tilde{b}_n = -\sqrt{(\gamma+n+2)(n+1)} \tag{9.3}$$

are valid. Comparing formulas (7.9) and (9.3) we deduce the following assertion.

Proposition 9.1. *The dual system of Laguerre polynomials* $L_n^{\gamma+1}(z)$ *corresponds to the original system of Laguerre polynomials* $L_n^\gamma(z)$.

10. Jacobi polynomials

In the case of Jacobi polynomials $\Phi_n^{(\alpha,\beta)}(z)$ we have (see [7], Ch. 10)

$$a = -1, \quad b = 1, \quad \tau'(\lambda) = (1-\lambda)^\alpha(1+\lambda)^\beta, \tag{10.1}$$

where $\alpha > -1$, $\beta > -1$. In order to have a system with a nonnegative spectrum we shall shift z, i.e. we shall consider the polynomial system $\Phi_n^{(\alpha,\beta)}(z-1)$. For this new system formulas (10.1) have the form

$$a = 0, \quad b = 2, \quad \tau'(\lambda) = (2-\lambda)^\alpha\lambda^\beta. \tag{10.2}$$

Similarly to Proposition 9.1 the following assertion can be proved.

Proposition 10.1. *The dual system of the polynomials* $\Phi_n^{\alpha,\beta+1}(z-1)$ *corresponds to the original system of Jacobi polynomials* $\Phi_n^{(\alpha,\beta)}(z-1)$.

In conclusion we shall write parameters of some special cases of Jacobi polynomials.

I: Let $\alpha = \beta = -\frac{1}{2}$, i.e. we shall consider Chebyshev polynomials. In this case we shall obtain

$$b_n = \frac{1}{2}, \quad n \geq 1; \quad b_0 = \frac{1}{\sqrt{2}}, \quad a_n = 1, \quad n \geq 0$$

$$p_n = (-1)^n \sqrt{\frac{2}{\pi}}, \quad n \geq 1; \quad p_0 = \frac{1}{\sqrt{\pi}}, \quad q_n = \sqrt{\pi}(-1)^n, \quad n \geq 0.$$

II: Let $\alpha = \beta = 0$, i.e. we shall consider Legendre polynomials. We shall obtain

$$b_n = \frac{n+1}{\sqrt{(2n+1)(2n+3)}}, \quad a_n = 1, \quad n \geq 0;$$

$$p_n = (-1)^n \sqrt{n + \frac{1}{2}}, \quad q_n = (-1)^n \sqrt{\frac{2}{n+1}}, \quad n \geq 0.$$

References

[1] Kac I.S., Krein M.G., *On the spectral function of the string*, Amer. Math. Soc. Translation **103** (1974), 19–102.

[2] Dym H., Sakhnovich L.A., *On dual canonical systems and dual matrix string equation*, to appear.

[3] Sakhnovich L.A., Interpolation Theory and its applications, Kluwer, Dordrecht, 1997.

[4] Bolotnikov V., Sakhnovich L.A., *On an operator approach to interpolation problems for Stieltjes functions, Integral Equations and Operator Theory*, v. **35**, No. **4** (1999), 423–470.

[5] Berezanskii Yu.M. Expansion in Eigenfunctions of Self-adjoint Operators, Amer. Math. Soc., Providence, 1968.

[6] Szegő G., Orthogonal polynomials, NY, 1959.

[7] Bateman H., Erdelyi A., Higher transcendental functions, Vol. 2, 1953.

L. Sakhnovich
735 Crawford Ave.
Brooklyn, NY 11223
USA
e-mail: Lev.Sakhnovich@verizon.net

Operator Theory:
Advances and Applications, Vol. 134, 403–414

Non-Selfadjoint Sturm-Liouville Operators with Multiple Spectra

Vadim Tkachenko

We consider Sturm-Liouville operators

$$H = -\frac{d^2}{dx^2} + q(x), \qquad q(x) \in \mathcal{L}^2[0,\pi], \tag{1}$$

with one of the following boundary conditions

$$\begin{array}{llll} D)\, y(0)=0, & N)\, y(0)=0, & P)\ y(0)=y(\pi), & AP)\ y(0)=-y(\pi), \\ \quad y(\pi)=0; & \quad y'(\pi)=0; & \quad y'(0)=y'(\pi); & \quad y'(0)=-y'(\pi). \end{array}$$

Spectrum of each problem is discrete and behaves asymptotically as the spectrum of the corresponding operator with $q(x) \equiv 0$. Namely,

$$\lambda_n(D) = \left(n + \frac{Q}{n} + \frac{f_n(D)}{n}\right)^2, \qquad Q \in \mathbf{C};$$

$$\lambda_n(N) = \left(n - \frac{1}{2} + \frac{Q}{n} + \frac{f_n(N)}{n}\right)^2, \qquad \{f_n(\cdot)\}_{n=1}^{\infty} \in \ell^2;$$

$$\lambda_n^{\pm}(P) = \left(2n + \frac{Q}{2n} \pm \frac{f_n(P)}{n} + \frac{g_n(P)}{n^2}\right)^2, \qquad \{g_n(\cdot)\}_{n=1}^{\infty} \in \ell^2; \tag{2}$$

$$\lambda_n^{\pm}(AP) = \left(2n+1 + \frac{Q}{2n+1} \pm \frac{f_n(AP)}{n} + \frac{g_n(AP)}{n^2}\right)^2.$$

Our aim is to find additional conditions, if any, which guarantee that a given sequence satisfying one of equations (2) is the spectrum of the corresponding boundary problem. In particular, we would like to know whether some points of the spectra may be multiple and whether there are some restrictions on their multiplicities, i.e., on dimensions of corresponding root subspaces.

Given a complex-valued function $f(\lambda, x)$ of $\lambda \in \mathbf{C}$ and $x \in [0,\pi]$, we denote by $f'(\lambda, x)$ its partial derivative with respect to the spacial variable x. Let $U(\lambda, x)$ be the fundamental matrix of operator H :

$$\begin{pmatrix} y(\lambda,x) \\ y'(\lambda,x) \end{pmatrix} = U(\lambda,x)\begin{pmatrix} y_0 \\ y_1 \end{pmatrix}, \qquad U(\lambda,x) = \begin{pmatrix} c(\lambda,x) & s(\lambda,x) \\ c'(\lambda,x) & s'(\lambda,x) \end{pmatrix},$$

where $y(\lambda, x)$ is the solution of the Cauchy problem

$$-y''(x) + q(x)y(x) = \lambda^2 y(x), \qquad 0 < x < \pi, \tag{3}$$

$$y(\lambda,0) = y_0, \qquad y'(\lambda,0) = y_1,$$

$c(\lambda, x)$ and $s(\lambda, x)$ are solutions of (3) satisfying $c(\lambda, 0) = s'(\lambda, 0) = 1, c'(\lambda, 0) = s(\lambda, 0) = 0$, and let $U(\lambda) = U(\lambda, \pi)$ be the monodromy matrix of H. Using entries of the monodromy matrix, the above spectra are

$$\begin{aligned} \lambda_n(D) &= (\nu_n(D))^2; & s(\nu_n(D), \pi) &= 0, \\ \lambda_n(N) &= (\nu_n(N))^2; & s'(\nu_n(N), \pi) &= 0, \\ \lambda_n^{\pm}(P) &= (\nu_n^{\pm}(P))^2; & u_+(\nu_n^{\pm}(P)) &= 1, \\ \lambda_n^{\pm}(AP) &= (\nu_n^{\pm}(AP))^2; & u_+(\nu_n^{\pm}(AP)) &= -1, \end{aligned}$$

where $u_+(\lambda) = (c(\lambda, \pi) + s'(\lambda, \pi))/2$ is the Hill determinant of H. It is known [1] that the multiplicity of μ as a point of some of above spectra coincides with its multiplicity as a zero of $s(\sqrt{\mu}, \pi)$, $s'(\sqrt{\mu}, \pi)$, $u_+(\sqrt{\mu})$ or $u_-(\sqrt{\mu})$, respectively.

In the self-adjoint case, i.e., $\Im q(x) = 0$, all above spectra are real and their complete description is well known (cf., [2]–[7]). In particular, for such potentials $q(x)$ the spectra of problems D and N are simple, i.e., all eigenspaces are of dimension 1, there are not associated root functions, and the interlacing conditions

$$\lambda_n(N) < \lambda_n(D) < \lambda_{n+1}(N)$$

are satisfied. The spectra of problems P and AP are at most of multiplicity 2, i.e., eigenspaces may be of dimensions either 1 or 2, there are no root functions, and the interlacing conditions

$$\ldots \lambda_{2n-2}(D) \leq \lambda_{n-1}^+(P) < \lambda_n^-(AP) \leq \lambda_{2n-1}(D) \leq \lambda_n^+(AP)$$

$$< \lambda_n^-(P) \leq \lambda_{2n}(D) \leq \lambda_n^+(P) < \ldots$$

are fulfilled.

If a spectral gap collapses, then the multiplicity of either $\lambda_n^{\pm}(P)$ or $\lambda_n^{\pm}(AP)$ is equal to 2.

Representations (2) show that, for large n, the multiplicities of spectral points of non-selfadjoint operator (1) are just the same as in the self-adjoint case. It is commonly accepted (cf., [3]) that an operator with complex-valued potential may have a finite number of spectral points of arbitrary multiplicities. Such points create difficulties in investigating dynamical flows generated by the operator (cf., [4]), but their possible presence must be taken into account in some important characterization problems, (cf., [5]). In [6], Ch.2, a class of Schrödinger operators in $L^2[0, \infty)$ was constructed with continuous spectrum coinciding with $[0, \infty)$ and eigenvalue $\lambda = 0$ of arbitrary order $n > 0$. Nevertheless, to the best of our knowledge, there is no single explicit example of operator (1) on a finite interval with the multiple Dirichlet or Neumann spectra.

The main aim of the present paper is to show that there is no restrictions either on a (finite) number of multiple points or their multiplicities in non-selfadjoint case, i.e., for potentials $q(x)$ taking on non-real values. Namely, the following statements are true:

Proposition 1. *For every finite set $\mathcal{D} = \{\lambda_1, \dots, \lambda_n\}$ of pairwise distinct points in* $\mathbf{C}$ *and every set $M(\mathcal{D}) = \{m_1, \dots, m_n\}$ of positive integers there exists a Sturm-Liouville operator* (1) *for which $\{\lambda_1, \dots, \lambda_n\}$ are points of the Dirichlet spectrum with multiplicities equal to respective numbers from $M(\mathcal{D})$.*

Proposition 2. *For every finite set $\mathcal{P} = \{\lambda_1^+, \dots, \lambda_n^+\}$ of pairwise distinct points in* $\mathbf{C}$ *and every set $M(\mathcal{P}) = \{m_1, \dots, m_n\}$ of positive integers there exists a Sturm-Liouville operator* (1) *for which $\{\lambda_1^+, \dots, \lambda_n^+\}$ are points of the periodic spectrum with multiplicities equal to respective numbers from $M(\mathcal{P})$.*

The similar statement is valid for anti-periodic spectrum.

Before stating the next proposition let us clarify possible relations between multiplicities of a point in the Dirichlet and periodic spectra of the same Sturm-Liouville operator. Namely, let λ_k be a point of the Dirichlet spectra of multiplicity $m > 2$ and a point of the periodic spectra of multiplicity $p < m$. If $\nu_{\pm k} = \pm\sqrt{\lambda_k}$, then $u_+(\lambda) - 1 = a_k(\lambda - \nu_k)^p(1 + o(1)), a_k \neq 0$. On the other hand,

$$\begin{aligned}
u_+(\lambda) - 1 &= \frac{c(\lambda, \pi) + s'(\lambda, \pi)}{2} - 1 \\
&= \frac{1}{2}\left(c(\lambda, \pi) + \frac{1}{c(\lambda, \pi)}\right) - 1 + \frac{c'(\lambda, \pi)s(\lambda, \pi)}{c(\lambda, \pi)} \\
&= \frac{1}{2}\left(1 + c_1(\lambda - \nu_k) + \cdots + \frac{1}{1 + c_1(\lambda - \nu_k)\cdots}\right) - 1 + \frac{c'(\lambda, \pi)s(\lambda, \pi)}{c(\lambda, \pi)} \\
&= \frac{1}{2}\left((c_1(\lambda - \nu_k) + \cdots)^2 - (c_1(\lambda - \nu_k) + \cdots)^3 + \cdots\right) \\
&\quad + s_m(\lambda - \nu_k)^m(1 + o(1)).
\end{aligned}$$

Hence

$$a_k(\lambda - \nu_k)^p(1 + o(1)) = c_q^2(\lambda - \nu_k)^{2q}(1 + o(1)) + s_m(\lambda - \nu_k)^m(1 + o(1)),$$

with some $c_q \neq 0$, and we arrive at the necessary condition stating that p is an even number.

Proposition 3. *Let $\mathcal{S} = \{\lambda_1, \dots, \lambda_n\}$ be a set of pairwise distinct points in* $\mathbf{C}$ *and let $M(\mathcal{D}) = \{m_1, \dots, m_n\}$ and $M(\mathcal{P}) = \{p_1, \dots, p_n\}$ be two sets of positive integers such that either p_k is an even number and $p_k < m_k$ for all $k = 1, \dots, n$, or $p_k > m_k$ for all $k = 1, \dots, n$. Assume that if $\lambda_k = \overline{\lambda}_j$ then $m_k = m_j, p_k = m_j$. Then there exists a Sturm-Liouville operator (1) such that every number $\lambda_j \in \mathcal{S}$ for $j = 1, \dots, n$ is a point of its Dirichlet spectrum of multiplicity $m_j \in M(\mathcal{D})$ and a point of its periodic spectrum of multiplicity $p_j \in M(\mathcal{P})$.*

To construct Sturm-Liouville operators with above properties we will prove the following statement which solves an inverse problem using Dirichlet and Neumann spectra.

Theorem 1. *Two sequences $\{\lambda_n\}_{n=1}^{\infty}$ and $\{\mu_n\}_{n=0}^{\infty}$ of complex numbers are the Dirichlet and Neumann spectra of some operator* (1) *if and only if the following conditions are fulfilled:*

i) $\lambda_n \neq \mu_k$ *for all* n *and* k, *and representations*

$$\lambda_n = \left(n + \frac{Q}{n} + \frac{f_n}{n}\right)^2; \qquad \mu_n = \left(n - \frac{1}{2} + \frac{Q}{n} + \frac{g_n}{n}\right)^2; \qquad n \geq 1, \tag{4}$$

are valid with $Q \in \mathbf{C}$, $\{f_n\}_{n=1}^{\infty} \in \ell^2, \{g_n\}_{n=1}^{\infty} \in \ell^2$;

ii) *if*

$$s(\lambda) = \pi \prod_{n=1}^{\infty} \frac{\lambda_n - \lambda^2}{n^2}, \qquad c(\lambda) = \prod_{n=0}^{\infty} \frac{\mu_n - \lambda^2}{(n - 1/2)^2}, \tag{5}$$

and

$$F(x,t) = \sum_n \left\{ \operatorname{res}_{s(\nu_n)=0} \left(\frac{c(z)}{zs(z)} \sin zx \sin zt \right) - \frac{1}{\pi} \sin nx \sin nt \right\}$$

with residii over all distinct points ν_n, *then the homogeneous Gelfand-Levitan equation*

$$h(t) + \int_0^x F(t,s)h(s)ds = 0, \qquad 0 \leq t \leq x, \tag{6}$$

has only the trivial solution $h(s) \in L^2(0,x)$ *for every* $x \in [0,\pi]$.

For real-valued potential $q(x)$ all numbers $\lambda_n, \mu_n, Q, f_n, g_n$ are real, the sequences $\{\lambda_n\}_{n=1}^{\infty}$ and $\{\mu_n\}_{n=0}^{\infty}$ are interlacing, and (6) is reduced to the form

$$\sum_{k=1}^{\infty} \frac{c(\sqrt{\lambda_k})}{\sqrt{\lambda_k}\dot{s}(\sqrt{\lambda_k})} \left| \int_0^x h(s) \sin \sqrt{\lambda_k} s \, ds \right|^2 = 0.$$

Here and in what follows $\dot{s}(\lambda)$ means the derivative of $s(\lambda)$. Interlacing conditions imply

$$\frac{c(\sqrt{\lambda_k})}{\sqrt{\lambda_k}\dot{s}(\sqrt{\lambda_k})} > 0, \qquad k = 1, 2, \ldots.$$

Hence

$$\int_0^x h(s) \sin \sqrt{\lambda_k} \, sds = 0, \qquad k = 1, 2, \ldots$$

and $h(s) = 0$. In other words, in the self-adjoint case, i.e., for real $q(x)$, interlacing conditions and representations (4) imply *ii*) and Theorem 1 coincides with classical results of B. Levitan and V. Marchenko [1960–1964].

Proof of Theorem 1. For complex-valued potentials all above parameters may be nonreal and properties *i*) and *ii*) are independent one from another. The necessity of condition *i*) for such potentials is well known [6], and necessity of *ii*) was proved in [8].

To prove that conditions *i*) and *ii*) are sufficient for the existence of potential $q(x)$ we will use a parametrization theorem from [8].

In what follows we denote by $\mathcal{P}W_\pi$ the Paley-Wiener class of all entire functions of exponential type π which are square integrable on the real line.

Given two sequences $\{\lambda_n\}_{n=1}^{\infty}$ and $\{\mu_n\}_{n=0}^{\infty}$ satisfying (4) we first conclude, similar to [9], that the functions $s(\lambda)$ and $c(\lambda)$ defined by (5) may be represented in the form

$$s(\lambda) = \frac{\sin \pi\lambda}{\lambda} - \frac{\pi Q \cos \pi\lambda}{\lambda^2} + \frac{h(\lambda)}{\lambda^2}, \tag{7}$$

and

$$c(\lambda) = \cos \pi\lambda + \frac{\pi Q \sin \pi\lambda}{\lambda} + \frac{g(\lambda)}{\lambda}. \tag{8}$$

Here Q is the same constant as in (4) and $h(\lambda), g(\lambda) \in \mathcal{P}W_\pi$. For the sake of simplicity we assume that $\lambda_n \neq 0$ for all n and set $\nu_{\pm n} = \pm\sqrt{\lambda_{|n|}}$, $\tau_{\pm n} = \pm\sqrt{\mu_{|n|}}$. According to $i)$ we have $\tau_n \neq \nu_k$ for all n, k, and $c(\nu_n) \neq 0$ for all n.

Suppose that an operator (1) corresponding to $s(\lambda)$ and $c(\lambda)$ does exist and ν_n is a zero of $s(\lambda)$ of multiplicity m_n. According to the Liouville identity we have $c(\lambda,\pi)s'(\lambda,\pi) - c'(\lambda,\pi)s(\lambda,\pi) \equiv 1$ and whence

$$u_+(\lambda) = \frac{c(\lambda,\pi) + s'(\lambda,\pi)}{2} = \frac{1}{2}\left(c(\lambda,\pi) + \frac{1}{c(\lambda,\pi)}\right) + O((\lambda - \nu_n)^{m_n}).$$

Having in mind the latter relation we define the even meromorphic function

$$\varphi(\lambda) = \frac{1}{2}\lambda^2\left(c(\lambda) + \frac{1}{c(\lambda)}\right) - \lambda^2 \cos \pi\lambda - \lambda\pi Q \sin \pi\lambda + \frac{1}{2}\pi^2 Q^2 \cos \pi\lambda.$$

Using (7) and (4) we find

$$c(\nu_n) = \cos \pi\left(\nu_n - \frac{Q}{\nu_n}\right) + \left(1 - \cos\frac{\pi Q}{\nu_n}\right)\cos \pi\nu_n - \left(\sin\frac{\pi Q}{\nu_n} - \frac{\pi Q}{\nu_n}\right)\sin \pi\nu_n$$
$$+ \frac{g(\nu_n)}{\nu_n} = (-1)^n + \frac{g(\nu_n)}{\nu_n} + \frac{\sigma_n}{n^2}$$

for all sufficiently big $|n|$ with some bounded sequence $\{\sigma_n\}$. Hence

$$(c(\nu_n))^{-1} = (-1)^n - \frac{g(\nu_n)}{\nu_n} - \frac{\sigma_n}{n^2} + \frac{\kappa_n}{n^2}, \qquad \sum_n |\kappa_n| < \infty.$$

Similarly

$$\nu_n^2 \cos \pi\nu_n + \nu_n \pi Q \sin \pi\nu_n - \frac{1}{2}\pi^2 Q^2 \cos \pi\nu_n = (-1)^n + \frac{\chi_n}{n^2}, \qquad \sum_n |\chi_n| < \infty,$$

and hence $\{\varphi(\nu_n)\}_{n=-\infty}^{\infty} \in \ell^1$.

Representation (7) shows that $\lambda s(\lambda)$ is a sine-type entire function [10] and therefore there exists the unique even entire function $f(\lambda) \in \mathcal{P}W_\pi$ satisfying

$$f(0) = \varphi(0) = \frac{1}{2}\pi^2 Q^2;\ f^{(k)}(\nu_n) = \varphi^{(k)}(\nu_n),\ k = 0, 1, \ldots, m_n - 1,\ n = \pm 1, \pm 2, \ldots$$

Using the subharmonic properties of $|f(\lambda)|$ we obtain

$$\{f(n)\}_{n=-\infty}^{\infty} \in \ell^1.$$

Let us set

$$u_+(\lambda) = \cos \pi\lambda + \frac{\pi Q}{\lambda} \sin \pi\lambda - \frac{\pi^2 Q^2}{2\lambda^2} \cos \pi\lambda + \frac{f(\lambda)}{\lambda^2}. \quad (9)$$

It is evident that $u_+(\lambda)$ is an even entire function of exponential type π. Since $f(\lambda) - \varphi(\lambda) = O((\lambda - \nu_n)^{m_n})$ for every $n = \pm 1, \pm 2, \dots$, we have

$$\begin{aligned} u_+(\lambda) &= \frac{1}{\lambda^2}(\lambda^2 \cos \pi\lambda + \lambda\pi Q \sin \pi\lambda - \frac{1}{2} Q^2 \pi^2 \cos \pi\lambda + f(\lambda)) \\ &= \frac{1}{2}\left(c(\lambda) + \frac{1}{c(\lambda)}\right) + O((\lambda - \nu_n)^{m_n}) \quad (10) \end{aligned}$$

If now

$$\psi(\lambda) = \frac{\lambda}{2}\left(c(\lambda) - \frac{1}{c(\lambda)}\right),$$

then (8) implies $\{\psi(\nu_n)\}_{n=-\infty}^{\infty} \in \ell^2$, and there exists the odd entire function $v(\lambda) \in \mathcal{P}W_\pi$ such that

$$v(0) = 0;\ v^{(k)}(\nu_n) = \psi^{(k)}(\nu_n),\ k = 0, 1, \dots, m_n - 1,\ n = \pm 1, \pm 2, \dots$$

The function $u_-(\lambda) = \lambda^{-1} v(\lambda)$ is an even entire function from $\mathcal{P}W_\pi$ and for every $n = \pm 1, \pm 2, \dots$, we have

$$u_-(\lambda) = \frac{1}{\lambda}(\psi(\lambda) + O((\lambda - \nu_n)^{m_n})) = \frac{1}{2}\left(c(\lambda) - \frac{1}{c(\lambda)}\right) + O((\lambda - \nu_n)^{m_n}). \quad (11)$$

Therefore $u_+^2(\lambda) - 1 - u_-^2(\lambda) = O((\lambda - \nu_n)^{m_n})$ which means that $(u_+^2(\lambda) - 1 - u_-^2(\lambda))s(\lambda)^{-1}$ is an entire function. Representations (10) and (11) show that the entire function $u_+(\lambda) + u_-(\lambda) - c(\lambda)$ vanishes at every point ν_n, $n = \pm 1, \pm 2, \dots$, with all its derivatives up to the order $m_n - 1$, and hence $\omega(\lambda) \equiv (u_+(\lambda) + u_-(\lambda) - c(\lambda))s(\lambda)^{-1}$ is also an entire function. It follows from (7)–(9) and the definition of $u_-(\lambda)$ that $\omega(\lambda)$ vanishes as $|\lambda| \to \infty$ outside of some strip $\{\lambda : |\Im\lambda| \leq C\}$, which implies that $c(\lambda) = u_+(\lambda) + u_-(\lambda)$.

We conclude that the triple $\{s(\lambda), u_+(\lambda), u_-(\lambda)\}$ satisfies all conditions of Theorem 1 from [8] and therefore there exists the unique operator (1) whose Dirichlet and Neumann spectra are $\{\lambda_n\}_{n=1}^{\infty}$ and $\{\mu_n\}_{n=0}^{\infty}$, respectively.

The following proposition describes some class of sequences which may be Dirichlet spectra of Sturm-Liouville operators with complex-valued potentials.

Theorem 2. *Let $\{\lambda_n\}_{n=1}^{\infty}$ be a sequence of complex numbers symmetric with respect to the real axis* [1] *and representable in the form*

$$\lambda_n = \left(n + \frac{Q}{n} + \frac{f_n}{n}\right)^2; \qquad n \geq 1, \quad (12)$$

with some $Q \in \mathbf{R}$ and $\{f_n\}_{n=1}^{\infty} \in \ell^2$. Then there exists a Sturm-Liouville operator $H = -d^2/dx^2 + q(x)$ with $q(x) \in \mathcal{L}^2[0, \pi]$ for which $\{\lambda_n\}_{n=1}^{\infty}$ coincides with the Dirichlet spectrum.

---

[1] which means, in particular, that nonreal numbers λ_n and $\overline{\lambda}_n$ have the same multiplicity

Proof. Given a sequence $\{\lambda_n\}_{n=1}^{\infty}$, we again define the function $s(\lambda)$ by (5) and find that it is real on the real line and representation (7) is valid with $f(\lambda) \in \mathcal{PW}_\pi$. We will construct a function $c(\lambda)$ which together with $s(\lambda)$ satisfies all conditions of Theorem 1.

Without loss of generality we assume that $\lambda_n \neq 0$ for all n and that all real numbers from the sequence $\{\lambda_n\}_{n=1}^{\infty}$ are positive and set $\nu_{\pm n} = \pm\sqrt{\lambda_n}$. Since the sequence $\{\lambda_n\}_{n=1}^{\infty}$ is symmetric with respect to the real axis and satisfies (12), we can fix N sufficiently big such that all ν_n with $|n| \geq N$ are real simple zeros of $s(\lambda)$. Let us set

$$c_{n,0} = \pi^{-1}\nu_n(1 + i\sigma_n)\dot{s}(\nu_n), \qquad |n| \geq N, \tag{13}$$

where all numbers $\sigma_n = \sigma_{-n}$ are real, positive and such that

$$\sum_{n \geq N} n^2\sigma_n^2 < \infty.$$

For every entire n is such that $|n| < N$ and for the zero ν_n of $s(\lambda)$ of multiplicity $m_n \geq 1$, we fix real numbers

$$c_{n,0} = 1, \quad c_{n,1}, \quad c_{n,2}, \ldots, \quad c_{n,m_n-1}, \tag{14}$$

the same both for ν_n and $\overline{\nu_n}$ if the latter numbers are not real. Using (13) and (14) we introduce the interpolation data

$$g_{n,k} = \nu_n c_{n,k} + k c_{n,k-1} - \varphi^{(k)}(\nu_n), \qquad k = 0, 1, \ldots, m_n - 1, \tag{15}$$

where $\varphi(\lambda) = \lambda \cos \pi\lambda + \pi Q \sin \pi\lambda$. From (7) and (13) we obtain, for sufficiently big $|n|$,

$$c_{n,0} = (-1)^n + \kappa_n, \; \varphi(\nu_n) = (-1)^n\nu_n + \chi_n, \; \sum_{|n| \geq N} \left(|n\kappa_n|^2 + |\chi_n|^2\right) < \infty.$$

Therefore

$$\sum_{|n| \geq N} |g_{n,0}|^2 < \infty.$$

Denote by $g(\lambda)$ the entire function from class $\mathcal{PW}_\pi$ interpolating (15), i.e., satisfying conditions

$$g^{(k)}(\nu_n) = g_{n,k}, \qquad k = 0, 1, \ldots, m_n - 1,$$

and such that $g(0) = 0$, and define the function $c(\lambda)$ by (8). It is evident that $c(\lambda)$ is an entire function of exponential type π. According to interpolation conditions (15) we have

$$\nu_n c^{(k)}(\nu_n) + k c^{(k-1)}(\nu_n) = \nu_n c_{n,k} + k c_{n,k-1}, \; k = 0, 1, \ldots, m_n - 1,$$

and hence $c^{(k)}(\nu_n) = c_{n,k}$, $k = 0, 1, \ldots, m_n - 1$. Since $\Re c(\nu_n) \neq 0$ for all $n = \pm 1, \pm 2, \ldots$, the zero sets of $s(\lambda)$ and $c(\lambda)$ do not intersect, and since $g(\lambda) \in \mathcal{PW}_\pi$, it follows from (8) that zeros τ_n of $c(\lambda)$ form a sequence such that

$$\tau_n = -\tau_{-n} = n - \frac{1}{2} + \frac{Q}{n} + \frac{g_n}{n}, \qquad \{g_n\}_{n=1}^{\infty} \in \ell^2.$$

Finitely, the sequence $\mu_n = \tau_n^2, n = 0, \pm 1, \pm 2, \ldots$ is of the form (4).

If $h(s) \in \mathcal{L}^2[0,x]$ is a solution of the homogeneous Gelfand-Levitan equation (6), then

$$\sum_n \mathrm{res}_{s(\nu_n)=0} \left\{ \frac{c(z)}{zs(z)} \int_0^x h(s) \sin zs \, ds \sin zt \right\} = 0, \qquad 0 \le t \le x,$$

and

$$\sum_n \mathrm{res}_{s(\nu_n)=0} \left\{ \frac{c(z)}{zs(z)} \int_0^x h(s) \sin zs \, ds \int_0^x \overline{h(s)} \sin zs \, ds \right\} = 0. \tag{16}$$

According to assumptions of Theorem 2 and relations (14) the partial sum

$$\sum_{|n|<N} \mathrm{res}_{s(\nu_n)=0} \left\{ \frac{c(z)}{zs(z)} \int_0^x h(s) \sin zs \, ds \int_0^x \overline{h(s)} \sin zs \, ds \right\} \tag{17}$$

is real. Separating the imaginary parts in equation (16) we obtain

$$\sum_{|n|\ge N} \sigma_n \left| \int_0^x h(s) \sin \nu_n s \, ds \right|^2 = 0, \tag{18}$$

and since σ_n are positive for all $n, |n| \ge N$, it follows

$$\int_0^x h(s) \sin \nu_n s \, ds = 0, \qquad |n| \ge N.$$

Therefore the homogeneous Gelfand-Levitan equation is reduced to the form

$$\sum_{|n|<N} \mathrm{res}_{s(\nu_n)=0} \left\{ \frac{c(z)}{zs(z)} \int_0^x h(s) \sin zs \, ds \sin zt \right\} = 0, \qquad 0 \le t \le x,$$

and we obtain

$$\int_0^x s^k h(s) \sin \nu_n s \, ds = 0, \qquad k = 0, \ldots, m_n - 1, \qquad |n| < N.$$

The system

$$\{\sin \nu_n s, s \sin \nu_n s, \ldots, s^{m_n - 1} \sin \nu_n s\}_{n=1}^{\infty}$$

is complete in $\mathcal{L}^2[0,x]$ for every $x \in [0,\pi]$ and therefore $h(s) = 0$ for $s \in [0,x]$. As a result the sequences $\{\lambda_n\}_{n=1}^{\infty}$ and $\{\mu_n\}_{n=0}^{\infty}$ with $\mu_n = \tau_n^2$ satisfy all conditions of Theorem 2 and according to its statement they are the Dirichlet and Neumann spectra, respectively, of some Sturm-Liouville operator.

Proof of Proposition 1. Let a finite set $\mathcal{D} = \{\lambda_1, \dots, \lambda_n\}$ of pairwise distinct points in $\mathbf{C}$ and a set $\mathcal{M}(\mathcal{D}) = \{m_1, \dots, m_n\}$ of positive integers be given. If for $\lambda_j \in \mathcal{D}$ there exists no $\lambda_k \in \mathcal{D}$ such that $\overline{\lambda}_k = \lambda_j$, then we join $\overline{\lambda}_j$ to the set $\mathcal{D}$ with multiplicity m_j; otherwise instead of m_j and m_k we include in $M(\mathcal{D})$ the number $M = \max\{m_j, m_k\}$. The resulting set $\mathcal{D}^*$ is symmetric with respect to the real axis and we extend it to a sequence $\{\lambda_n\}_{n=1}^{\infty}$ satisfying the conditions of Theorem 2. According to this theorem there exists an operator (1) with Dirichlet spectrum $\{\lambda_n\}_{n=1}^{\infty}$. Some points $\lambda_j \in \mathcal{D}^*$ may have an excessive multiplicity $M > m_j$. If it occurs, we replace $s(\lambda)$ by a function $s^*(\lambda)$ for which every such λ_j is a zero of multiplicity m_j and another number λ_j^* is an additional zero of multiplicity $M - m_j$. If all such λ_j^*'s are sufficiently close to λ_j, then the norm $\|s(\lambda) - s^*(\lambda)\|_{\mathcal{P}W_\pi}$ is sufficiently small, and by Theorem 2 from [8] the functions $s^*(\lambda)$ and $c(\lambda)$ generate operator (1) with the required properties.

Proof of Proposition 2. is based on the following description [11] of Hill determinants of operators (1).

Theorem 3. *For a function $u_+(\lambda)$ to be Hill's determinant of some Sturm-Liouville operator* (1) *it is necessary and sufficient that it be an even entire function of exponential type π which may be represented in the form* (9) *where Q is a complex number, and $f \in \mathcal{P}W_\pi$ is an entire function of exponential type not exceeding π and satisfying condition*

$$\sum_{n=-\infty}^{+\infty} |f(n)| < \infty.$$

Given a set $\mathcal{P} = \{\lambda_1^+, \dots, \lambda_n^+\}$ of pairwise distinct points in $\mathbf{C}$ and a set $M(\mathcal{P}) = \{m_1, \dots, m_n\}$ of positive integers, we define polynomials

$$Q(\lambda) = (\lambda^2 - a^2) \prod_{k=1}^{n} (\lambda^2 - \lambda_k^+)^{m_k},$$

$$R(\lambda) = \prod_{k=1}^{N+1} (\lambda^2 - 4k^2), \quad N = \sum_{k=1}^{n} m_k\,,$$

and the function

$$u_+(\lambda) = \frac{Q(\lambda)}{R(\lambda)} (\cos \lambda\pi - 1) + 1.$$

With a proper choice of a complex number $a \in \mathbf{C}$, $u_+(\lambda)$ is an entire function of the form $u_+(\lambda) = \cos \lambda\pi + f(\lambda)/\lambda^4$ where $f(\lambda)$ is bounded on the real line. The set of points where $u_+(\lambda) = 1$ contains $\mathcal{P}$ with multiplicities from $M(\mathcal{P})$. According to [11] there exists some Sturm-Liouville operator for which $u_+(\lambda)$ is the Hill determinant and hence $\mathcal{P}$ is a part of the periodic spectrum with multiplicities from $M(\mathcal{P})$.

Proof of Proposition 3. For the sake of simplicity we assume that $n = 1$, $\mathcal{S} = \{\lambda_1\}$, $M(\mathcal{D}) = \{m\}$ and $M(\mathcal{P}) = \{p\}$. Suppose first that $m > 2$ and $p = 2q < m$. Denote by $\{\lambda_n\}_{n=1}^{\infty}$ an arbitrary sequence of complex numbers symmetric with respect to the real axis, satisfying (12) and containing the point λ_1 of multiplicity m. As before we set $\nu_{\pm n} = \pm\sqrt{\lambda_n}$, define a function $s(\lambda)$ by (5) and construct $c(\lambda)$ using interpolation data (13)–(14) with

$$c_{-1,j} = c_{1,j} = \bar{c}_{-1,j} = \bar{c}_{1,j} = \begin{cases} 1 & j = 0 \\ q!c_q \neq 0 & j = q \\ 0 & j = 0, \ldots, m-1; j \neq 0, q. \end{cases}$$

As shown above in the proof of Theorem 2, there exists operator (1) for which

$$c(\lambda, \pi) = c(\lambda) = 1 + c_q(\lambda - \nu_1)^q + O((\lambda - \nu_1)^m).$$

If we substitute the expansion $s'(\lambda, \pi) = 1 + s_1(\lambda - \nu_1) + \ldots + s_{m-1}(\lambda - \nu_1)^{m-1} + O((\lambda - \nu_1)^m)$ into identity $c(\lambda, \pi)s'(\lambda, \pi) = 1 + O((\lambda - \nu_1)^m)$, we obtain

$$s'(\lambda, \pi) = 1 - c_q(\lambda - \nu_1)^q + c_q^2(\lambda - \nu_1)^{2q} + O((\lambda - \nu_1)^{2q+1}).$$

Finally,

$$u_+(\lambda) - 1 = \frac{c(\lambda, \pi) + s'(\lambda, \pi)}{2} - 1 = \frac{1}{2}\, c_q^2(\lambda - \nu_1)^{2q}(1 + o(1)),$$

which shows that $\lambda_1 = \nu_1^2$ is a point of the periodic spectrum of multiplicity $2q$.

Let now p and m be two integers, $p \geq m \geq 1$, and let a complex number $\lambda_1 \neq \bar{\lambda}_1$ be given. Following the proof of Proposition 2 we find

$$u_+(\lambda) = \cos\lambda\pi + (\cos\lambda\pi - 1)\frac{(\lambda^2 - \lambda_1)^p(\lambda^2 - \bar{\lambda}_1)^p(\lambda^2 - \alpha) - R(\lambda)}{R(\lambda)}$$

with

$$R(\lambda) = \prod_{k=1}^{2p+1} (\lambda^2 - 4k^2)$$

is an even entire function of exponential type π, taking on real values on the real line. It is evident that $u_+(\lambda) = 1 + c_p(\lambda - \sqrt{\lambda_1})^p + o((\lambda - \sqrt{\lambda_1})^p)$ with $c_p \neq 0$. If

$$\alpha = 4\sum_{k=1}^{2p+1} k^2 - 2p\Re\lambda_1,$$

then $u_+(\lambda)$ may be represented in the form

$$u_+(\lambda) = \cos\lambda\pi + \frac{w(\lambda)}{\lambda^4} \tag{19}$$

where $w(\lambda)$ is an entire function bounded on the real line. In particular it means that (9) is valid with $Q = 0$.

We set $\nu_0 = 0$, $\nu_k = -\nu_{-k} = \sqrt{\lambda_1}$, $\nu_{k+m} = -\nu_{-k-m} = \sqrt{\bar{\lambda}_1}$, $k = 1, \ldots, m$. It follows from the definition of the function $u_+(\lambda)$ that there exists an integer $N > |\sqrt{\lambda_1}|$ such that all its critical points $\{\mu_n\}$ outside of the disk $\{\lambda : |\lambda| \leq N + 1/2\}$ are real and simple and are located in a small neighborhood of integers. If a critical point $\mu_n, |\mu_n| > N + 1/2$ is such that $u_+^2(\mu_n) < 1$, we set $\nu_n = \mu_n$; otherwise

$u_+^2(\mu_n) \geq 1$ and we choose ν_n to be the solution to the equation $u_+^2(\lambda) - 1 = -n^{-4}$, the most close to μ_n. It follows from (19) that

$$\nu_n = n + O\left(\frac{1}{n^2}\right), \quad u_+^2(\nu_n) - 1 = O\left(\frac{1}{n^4}\right).$$

If necessary, we define pairwise distinct real numbers $\nu_n = -\nu_{-n}$ for $2m < n < N + 1/2$ in such a way that the inequality $u_+^2(\nu_n) - 1 < 0$ is valid for all n, $|n| > 2m$.

With the sequence $\{\nu_n\}_{n=-\infty}^{\infty}$ being fixed we set

$$s_0(\lambda) = \pi\lambda \prod_{n=1}^{\infty} \frac{\lambda_n - \lambda^2}{n^2}.$$

The function $s_0(\lambda)$ is an odd entire function of exponential type π, bounded on the real line. The S. Bernstein theorem on the derivatives of such functions [10] shows that $s_0(n) = s_0(n) - s_0(\nu_n) = O(\nu_n - n) = O(n^{-2})$ and hence

$$s_0(\lambda) = \sin\lambda\pi + \frac{1}{\pi\lambda} \sum_{n \neq 0} (-1)^n \frac{n s_0(n) \sin\lambda\pi}{\lambda - n}.$$

We conclude that $s_0(\lambda) = \sin\lambda\pi + \lambda^{-1}g(\lambda)$, $g(0) = 0$, $g(\lambda) \in \mathcal{PW}_\pi$ and the function $s(\lambda) = \lambda^{-1}s_0(\lambda)$ has the form (8) with $Q = 0$.

The function $s_0(\lambda)$ is a sine-type function and we can define the function $v(\lambda) \in \mathcal{PW}_\pi$ by the interpolation data

$$v(0) = 0,\ v(\pm\nu_1) = v(\pm\overline{\nu}_1) = \ldots = v^{(m-1)}(\pm\nu_1) = v^{(m-1)}(\pm\overline{\nu}_1) = 0,$$

$$v(\nu_n) = \mathrm{i} \cdot \mathrm{sign}\,\{\dot{s}(\nu_n)\}\nu_n\sqrt{1 - u_+^2(\nu_n)}, \qquad |n| > 2m,$$

and set $u_-(\lambda) = \lambda^{-1}v(\lambda)$. According to such definition we have

$$u_+^2(\lambda) - 1 - u_-^2(\lambda) = \begin{cases} O((\lambda - (\pm\nu_1)^m) & \lambda \to \pm\nu_1 \\ O((\lambda - (\pm\overline{\nu}_1)^m) & \lambda \to \pm\overline{\nu}_1 \\ O(\lambda - \nu_n) & \lambda \to \nu_n,\ |n| > 2m, \end{cases}$$

which means that $(u_+^2(\lambda) - 1 - u_-^2(\lambda))s^{-1}(\lambda)$ is an entire function.

Let us consider the Gelfand-Levitan equation (6) with $c(\lambda) = u_+(\lambda) + u_-(\lambda)$. Since $u_+(\lambda) = 1 + O((\lambda - (\pm\nu_1))^p)$, $u_-(\lambda) = O((\lambda - (\pm\nu_1))^m)$, and $u_+(\lambda) = 1 + O((\lambda - (\pm\overline{\nu}_1))^p)$, $u_-(\lambda) = O((\lambda - (\pm\overline{\nu}_1))^m)$ as $\lambda \to \pm\nu_1$ and $\lambda \to \pm\overline{\nu}_1$, respectively, we have $c(\lambda) = 1 + O((\lambda - (\pm\nu_1))^m)$ and $c(\lambda) = 1 + O((\lambda - (\pm\overline{\nu}_1))^m)$. It implies that the sum

$$\sum_{|n| \leq m} \mathrm{res}_{s(\nu_n)=0} \left\{ \frac{c(z)}{z s(z)} \int_0^x h(s) \sin zs\, ds \int_0^x \overline{h(s)} \sin zs\, ds \right\}$$

is real. On the other hand, $\Im c(\nu_n) = \Im u_-(\nu_n) = \operatorname{sign}\{\dot{s}(\nu_n)\}\sqrt{1-u_+^2(\nu_n)}$ for $|n| > m$. Similar to (18) we obtain

$$\sum_{|n|>m} \frac{\operatorname{sign}\{\dot{s}(\nu_n)\}}{\dot{s}(\nu_n)}\sqrt{1-u_+^2(\nu_n)}\left|\int_0^x h(s)\sin\nu_n s\,ds\right|^2 = 0,$$

which yields $h(s) = 0$, $0 \le s \le x \le \pi$. Therefore the triple $\{s(\lambda),\ u_+(\lambda),\ u_-(\lambda)\}$ satisfies all conditions of Theorem 1 from [8] and there exists an operator (1) for which λ_1 is a point of the Dirichlet and periodic spectra of prescribed multiplicities which completes the proof.

In particular, we proved that *a complex number may be simultaneously a point of the Dirichlet spectrum of multiplicity m and a point of the periodic spectrum of multiplicity p of some Sturm-Liouville operator if and only if either $m \le p$ or $m > p$ and p is an even number.*

References

[1] Naimark, M.A. [1967] Linear Differential Operators, Part I. Frederick Ungar Publishing Co., New York.

[2] Titchmarsh, E.C. [1958] Eigenfunction expansions associated with second-order differential equations, v. II, Clarendon Press, Oxford.

[3] Birnir, B. [1986] Complex Hill's equation and the complex periodic Korteweg-de Vries equations, Commun. Pure and Appl. Math., **39**, 1–49.

[4] Birnir, B. [1986] Singularities of the complex Korteweg-de Vries flows, Commun. Pure and Appl. Math., **39**, 283–305.

[5] Gesztezy, F., and Weikard, R. [1996] Picard potentials and Hill's equation on a torus, Acta Math., **176**, 73–107.

[6] Marchenko, V.A. [1986] Sturm-Liouville operators and applications, Birkhäuser, Basel.

[7] Levitan, B.M. [1977] Inverse Sturm-Liouville problems, Nauka, Moscow, 1–240.

[8] Sansuc, J.-J., and Tkachenko, V. [1996] Spectral parametrization of non-selfadjoint Hill's operators, Journ. Diff. Equat., **125**, 2, 366–384.

[9] Levin, B., and Ostrovskii, I.V. [1980] On small perturbations of the set of zeros of functions of sine type, Math. USSR Izvestia, **14**, 1.

[10] Levin, B. [1996] A.M.S., **150**, 248 pp.

[11] Tkachenko, V. [1992] Spectral analysis of non-selfadjoint Hill operator, Doklady AN SSSR, **322**, 2, 248–252.

Vadim Tkachenko
Department of Mathematics
Ben-Gurion University of the Negev
Beer–Sheva 84105, Israel

MIX
Papier aus verantwortungsvollen Quellen
Paper from responsible sources
FSC® C105338

If you have any concerns about our products,
you can contact us on
ProductSafety@springernature.com

In case Publisher is established outside the EU,
the EU authorized representative is:
Springer Nature Customer Service Center GmbH
Europaplatz 3, 69115 Heidelberg, Germany

Printed by Libri Plureos GmbH
in Hamburg, Germany